Handbook of Food Processing

Handbook of Food Processing

Editor: Lisa Jordan

CALLISTO REFERENCE

www.callistoreference.com

Callisto Reference,
118-35 Queens Blvd., Suite 400,
Forest Hills, NY 11375, USA

Visit us on the World Wide Web at:
www.callistoreference.com

ISBN: 978-1-64116-056-8 (Hardback)

Cataloging-in-Publication Data

Handbook of food processing / edited by Lisa Jordan.
    p. cm.
Includes bibliographical references and index.
ISBN 978-1-64116-056-8
1. Food industry and trade. 2. Food--Preservation . 3. Food processing machinery.
4. Food processing plants. I. Jordan, Lisa.
TP373 .H36 2019
664.02--dc23

Table of Contents

Preface

Food processing is the transformation of raw or cooked food into marketable food products that can be stored and consumed. It combines commercial practices such as thermal processing, freezing, etc. and household practices such as pickling, mincing or macerating among many others. Research in this field includes analyses of processing stages and their applicability with respect to industrial standards of quality, environmental impact and sustainability. This book outlines the processes and methodologies of food processing. It also discusses the fundamentals as well as modern approaches to food preservation. This book is meant for students, academicians and professionals who are looking for an elaborate reference text on food processing.

This book is a result of research of several months to collate the most relevant data in the field.

When I was approached with the idea of this book and the proposal to edit it, I was overwhelmed. It gave me an opportunity to reach out to all those who share a common interest with me in this field. I had 3 main parameters for editing this text:

1. Accuracy – The data and information provided in this book should be up-to-date and valuable to the readers.

2. Structure – The data must be presented in a structured format for easy understanding and better grasping of the readers.

3. Universal Approach – This book not only targets students but also experts and innovators in the field, thus my aim was to present topics which are of use to all.

Thus, it took me a couple of months to finish the editing of this book.

I would like to make a special mention of my publisher who considered me worthy of this opportunity and also supported me throughout the editing process. I would also like to thank the editing team at the back-end who extended their help whenever required.

Editor

A Comparative Study of the Physico-Chemical Characteristics of the Ready-to Eat Coconut Based Snack

Sivasakthi M* and Sangeetha N

Department of Food Science and Technology, Pondicherry University, Puducherry-605014, Tamil Nadu, India

Abstract

Coconut slices (thickness 0.8 ± 0.1 mm) were osmotically dehydrated (0 min to 720 min) using filtrates of functional ingredients such as beet root, carrot, ginger and mint as impregnating solutions. During the osmotic process, the diffusion of minerals, organic acids, phenolics and vitamins between coconut slices and osmotic medium might have contributed various physico-chemical changes in acidity, TSS (Total soluble solids) of the osmotic medium and rehydration characteristics of the snack. The kinetics of osmotic dehydration process revealed that significant changes in mass transfer parameters WR (Weight reduction), SG (Solid gain) and WL (Water loss) observed during 450th minutes of the process, irrespective of the functional filtrates and dehydration methods used (hot-air oven at temperature 45-60°C for about 6-7 hours and freeze drying at temperature (-40 to 30°C) for a duration of 14-16 hours). Texture analysis revealed that osmotic dehydration induced modifications in cell structure of the coconut based snack with respect to the drying methods and the use of sugar as osmotic infusions.

Keywords: Osmotic dehydration; Coconut slices; Functional ingredients; Physico-chemical; Drying methods

Introduction

The roots carrots (*Daucus carota*) and beet roots (*Beta vulgaris*), the spices like mint (*Mentha piperita*) and ginger (*Zingiber officinale*) always contributed an imperative role in the social, economic lives and food habits all over the nation. In India, the production of beet roots was estimated about 1.16% to 33.42% in the year 2011-12 [1]. Carrot production in India was estimated about 1.15 million tonnes in the year 2012-13 [1]. Carrots and beet roots generally contains considerable amount of natural sugars, phenolic compounds such as carotenoids, anthocyanins, flavonoids, betacyanin and betaxanthin, the natural antioxidants essential for potential health benefits. The spices mint and ginger are consumed worldwide. In India the production of ginger was 33% of the global production and 12,000 tonnes of mint oil was produced from 75% of menthol mint estimated in the year 2001 [2]. Mint and ginger are well known for their pungency principles menthol, menthone, terpenoids, curcumin, shogaols, gingerols and antioxidant vitamins and minerals which play vital task in therapeutic applications.

Coconuts are grown widely in subtropical and tropical regions and considered as a major ingredient in daily cuisines. Coconuts are rich in inorganic ions like potassium, sodium, phosphorus and calcium etc, which helps in blood coagulation. He also reported that coconuts composed of fatty acids like caprylic acid (8%), capric acid (7%), lauric acid (49%), myristic acid (18%), palmitic acid (8%), stearic acid (2%), oleic acid (6%), linoleic acid (2%) and profuse nature of presence of medium chain saturated fatty acids (MCFAs) which enables them to be absorbed directly from the intestines and straightly sent to the liver, which then metabolized and energy is produced [3].

Coconuts are used as ready to eat food products in confectionary food items and ice-creams in the form of pie filling, fruit ice cream with the assistance of dehydration techniques like osmotic dehydration (OD), Kabara [4] and further drying proceeded such as air drying, hot air oven drying, freeze drying and microwave drying etc., to get dried fruits, snacks and other products with extended keeping quality. The consumers prefer dried food products in terms of convenience, good health and pleasure.

Osmotic dehydration is a process in which partial removal of water from tissues of plant by dipping in hypotonic solutions. During the osmotic dehydration process, two major counter current passes due to water and the activity of the solute: the diffusion of water from food to the osmotic medium and solute from solution to the food. The mass transfer mechanism is mainly attributed to the activity of water and solute across the membrane of cell enables equilibrium between the solute and water [5]. Conventionally OD process has been carried out in atmospheric pressure. Now-a-days OD plans a significant role in industrial sector for pre-drying purposes to achieve better textural characteristics with associated changes in color, Total soluble solids (TSS). The color changes in dried food products could be in reliant with the presence of poly phenols and other sugars. The mass kinetics are generally proceeds in the DIS (Dehydration Impregnation Soaking) phenomenon which influences variables like temperature, concentration of osmotic medium, sample shape, size and agitation [6].

This paper documents the development of ready-to eat coconut based snack and the changes occurred during the process of OD of the coconut slices like TSS, sugar-acid ratio, acidity and mass transfer kinetics WR, SG and WL and rehydration characteristics using filtrates of functional ingredients such as *Beta vulgaris*, *Daucus carota*, *Zingiber officinale* and *Mentha piperita* as osmotic medium as a pre-treatment followed by assisted drying methods namely hot air oven drying and freeze drying to obtain healthy snack enriched with functional properties.

Materials and Methods

Solvents and reagents

Gallic acid, phosphotungsto molybdic acid, folin- Denis reagent (FD) and Na_2CO_3, folin-ciocalteu's Reagent, monobasic potassium

*Corresonding author: Sivasakthi M, Department of Food Science and Technology, Pondicherry University, Puducherry-605014, Tamil Nadu, India
E-mail: sivasakthiphd2015@gmail.com

phosphate, magnesium carbonate, sodium sulphate, acetone, hydrochloric acid, acetate buffer, 2,4,6-tripyridyl-s-triazine), ferric chloride, were purchased from Scientific suppliers (Pondicherry, India). Chromatographic grade methanol, ethanol, acetonitrile, glacial acetic acid, phosphoric acid, formic acid, 5% tetrahydrofuran, were purchased from Merck (Mumbai, India). Standards: All standards were of analytical grade. Quercetin RS, kaempferol, myricetin, apigenin, luteolin, hesperidin and isorhamnetin were obtained as described by Crozier et al. [7].

Selection and pre-processing of raw materials

Matured, fresh and good quality coconuts were procured from the shop Pazhamudir Nilayam, Pondicherry to carry out research on osmotic dehydration. The steps included in preliminary processing of coconuts such as selection of coconuts (10-12 months old), husk removal, breaking coconuts into two halves, endosperm separation from shell and testa removal with sharp knife. The processed coconuts were further standardized to get optimum thickness using the slicers to conduct experiments.

Using screw gauge the pre-processed coconuts were sliced ranged from thickness 0.6 mm to 1.0 mm to get optimum thickness. To avoid microbial contamination, the coconut slices were kept immersed in water in a vessel before starting the experimental analysis. Steam blanching of the coconut slices was done at the temperature 80 to 90°C for 5 minutes to prevent enzymatic actions. Osmotic dehydration of the pre-processed slices was performed using sugar as the osmotic agent. Since it is available everywhere, easy to procure, low cost, possessing good diffusion and mass transfer properties according to Lerici et al. [8].

Sugar was used as one of the main ingredient in osmotic medium along with functional filtrates namely *Daucus carota*, *Beta vulgaris*, *Mentha piperita* and *Zingiber officinale* were procured from Pazhamudir Nilayam, Puducherry and the process of osmotic dehydration was carried out. The functional ingredients such as beet root, carrot, ginger and mint were pre-processed using simple steps like washing, peeling, grating, grinding into fine paste and filtered to get filtrates using 0.08 mm sieve. The functional filtrates were taken in the range from every 5% addition up to 100% and infused with sugar as the main ingredient in osmotic medium (100%). For standardization purpose the coconut slices, sugar and filtrates of functional ingredients were taken in three ratios like 1:1:1, 1:2:2 and 1:3:3.

Process involved in the formulation of ready-to-eat coconut based snack

The process of osmotic dehydration was performed with or without the infusion of functional filtrates for duration of 0 min to 720 min. The finalization of duration of the osmotic process was based on the physico-chemical characteristics of the coconut samples like maximum decline in brix, weight of the osmo dehydrated coconut slices and the extent of impregnation of functional properties.

Hot air drying and freeze-drying were adopted in the present study. The coconut samples were dehydrated in hot air oven dryer by allowing hot air which flushed inside the chamber and the food gets exposed to hot air, the desired moisture content achieved due to removal of moisture up to 2-3%. The desired moisture content was attained with standardized temperature 45-50°C and time 6-7 hours which was dependent on the nature of the product, whereas most of the fruits and vegetables are dehydrated in the temperature range 50-60°C to get dried foods without much degradation reactions [9]. The coconut samples were dehydrated using freeze dryer (Model: Del Vac) by spreading

the samples uniformly in the trays and kept inside the chamber where all the valves are closed and checked whether the instrument working under vacuum condition. The low temperature -40°C to 30°C was applied over a period of time 14-16 hours, which found to preserve the color and texture of the samples through thermostat adjustment. The instrument worked under an automatic mode with respect to vacuum and temperature. As soon as the drying process is completed, the instrument is switched off and a slow release of vacuum was ensured, finally the samples were collected. The dehydrated coconut based snacks developed using hot air oven and freeze drying methods were packed in aluminium foil laminated LDPE pouches with 100% infusion of nitrogen gas and kept for storage at temperature 35°C prior to analytical determinations.

Finalized process parameters

The process parameters were represented in the ratio 1:2:2 after several permutations and combinations from the above methodology. The ratio 1:2:2 represents thickness of coconut slices, sweetness of sugar and concentration of various filtrates obtained from the functional ingredients, which were essential further processing. The infusion of sugar and impregnation functional filtrates at 100% concentration contributed the coconut based snack with most acceptable texture, taste, sweetness and crispiness; hence finally this proportion was selected. However the control snack represents the process of osmotic dehydration of the coconut slices in which the filtrates of functional ingredients were not impregnated.

Quality analysis of the ready-to-eat coconut based snack

The physico-chemical characteristics of the coconut based snack were analyzed after the finalization of various process parameters as per standard methods and protocols. All analytical determinations were carried out in triplicate Values were represented as mean ± standard deviation.

Effect of drying methods and impregnation of filtrate of functional ingredients on the physico-chemical characteristics of the ready-to-eat coconut based snack and the osmotic solution

Osmotic process is influenced by the pH of the osmotic solution. Whereas gradual removal of water from the food materials subjected to osmosis depends mainly on acidity of the osmotic solution which has impact on textural properties of fruits and vegetables.

Physico-chemical properties of the osmotic solution:

Acidity (%): The physico-chemical property acidity of the osmotic solution was analyzed by titrimetric method [10], where 0.1 N of sodium hydroxide is titrated against 0.1 N oxalic acid using phenolphthalein as an indicator.

Total soluble solids (TSS) (°Bx): About 5 g dried samples were extracted with 250 ml of acetone for 24 hr. For sterol analysis, acetone extracts were used frequently. Gas chromatography was used to separate individual sterols by 1.80-m column, 6 mm id., packed with 5 z OV-101 on Anakrom ABS 80- to 90-mesh. The temperature of the column was 250°C and the temperature of flash heater was kept 50°C above that of the column. The temperature of flame ionization detector was 275°C. The carrier gas was Helium at a flow rate of 100 ml/min [11].

Sugar-acid ratio: Sugar acid ratio of the osmotic solutions were determined using the formula,

TSS/Acid ratio = TSS/% acidity [12].

Physico-chemical properties of the ready-to-eat coconut based snack: The mass fluxes occurred through osmotic dehydration not only cause changes in the structure of tissues due to cellular alterations by immersion of foods in concentrated solutions, promotes water loss, deformation and breakage of cellular components associated with gas-liquid exchanges, but also result in conformity with the product texture and appearance [13].

Texture analysis-crispiness (N): Texture analysis of the ready-to-eat coconut based snack was performed with a 5 mm HDP-CFS cylindrical ball probe by Texture Analyzer (Model No. 5197, stable Micro Systems HD Plus, Gold alming, Surrey, GU71YL, UK). Each slice of the coconut snack was placed on the heavy duty platform and the test speed was set to 1 mm/sec and the probe compressed 50% of the snack to measure the hardness. Recording of maximum force is calculated as the hardness of the slices. Maximum breaking force (N) and deformation were measured from the force-deformation curve.

Analysis of color using Hunter lab color flex: Color of the ready-to-eat coconut based snack including 5 or 6 slices from each treatment were selected to determine using the Hunter's Lab Colorimeter (Model: CX2748, Easy Match QC, Software Version 4.0, Hunter Lab, USA) with spectral reflectance. The Hunter Labs color space is a 3-dimensional rectangular color space based on the opponent-colors theory. The color determinations were reported as L^*, a^*, b^*, whereas (lightness) axis - 0 is black, 100 is white, a^*(redness-greenness) axis -positive values are red; negative values are green and 0 is neutral, b^* (yellowness-blueness) - positive values are yellow; negative values are blue and 0 is neutral values and ΔE indicates the overall average color according to Olajide [14].

The hue angles were calculated as the arctangent of b^*/a^* expressed as degrees and the chroma values were also calculated as the square root of the sum of the squared values of both CIE a^* and CIE b^*. The chroma and Hue angle were calculated by the formulas given below.

$$Chroma = \sqrt{a^2 + b^2}$$

$$Hue\ angle = \tan^{-1}(b/a)$$

Mass transfer kinetics

The process of osmotic dehydration generally results in exchange of water from food to the solution and from the solution to the food and also migration of natural solutes like minerals, vitamins, organic acids, sugars, reducing sugars, some flavor compounds, volatiles, etc., from the food materials to the solution which affects nutritional and sensory quality of the foods. Osmotic dehydration reduces the water activity in foods by effective moisture removal process due to driving force exerted by the osmotic agent in the solution; thereby microbial entry is prevented [15]. Thus the mass transfers help in overall yield and quality of the dehydrated food products. Due to the semi permeable nature of the plant tissue, the flux of water removal is greater owing to smaller molecular size of water molecules. This paved a way for the decrement in water content in foods with respect to time till equilibrium gradient is attained. Hence weight of the food materials decreased corresponds to water activity. Rastogi and Raghavararo [16] examined that about 50% of reduction in fresh weight of vegetables and fruits obtained through osmotic dehydration with strengthened structure and enable easy transportation, occupy less storage space.

The coconut slices of 0.8 ± 0.1 mm were removed from the immersion solution at selected time intervals of 5, 10, 15, 20, 25 and 30 min and quickly rinsed (with distilled water) during preliminary trial processes, later on the time increased up to 720 minutes and readings taken at every 90 minutes interval. The excess of solution at the surface of the slices was removed with absorbent paper.

Weight reduction (%)

The coconut slices were weighed before and after osmotic dehydration to calculate the percentage of weight reduction (WR). The final weight of food samples is affected by the overall exchange mechanism between the solid and liquid of the sample. The moisture content was determined to calculate water loss (WL) and solid gain (SG) based on the following equations [17].

$$WR\ (\%) = (W_i - W_f)/W_i * 100$$

Where, W_i is the initial mass of the sample (g), W_f is the final mass of the sample (g).

Solids gain (%)

Solids gain represents the amount of solids gained from the osmotic medium that the amount of sugar uptake by the slices from the osmotic solution. During osmotic dehydration, water flows from the food materials into the concentrated osmotic medium, and solute transferred from the osmotic medium into the food material [18].

$$SG\ (\%) = (W_{sf} - W_{si}/W_i) * 100$$

Where, W_{si} is the initial total solids content in the sample (g), W_{sf} is the final total solids content in the sample (g).

Water loss (%)

Water loss represents the total amount of weight reduction as well as the solids gain taking place when the osmotic dehydration is taking place. WL is the net loss of weight of the foods on initial weight basis occurred during the process of osmotic dehydration [15].

$$WL = SG + WR$$

Rehydration characteristics

Rehydration is defined as a complex procedure affected by a number of factors such as chemical composition of the product, drying techniques and conditions, immersion medium composition, temperature etc., which measures the quality characteristics of the product. Rehydration depends on the nature and degree of pre-treatment of the foods [19-21].

Best quality dehydrated snack prepared from the above treatments samples were used for studying the rehydration characteristics with four replications. Rehydration of the coconut snack slices is ensured by soaking 10 g of the samples in a beaker containing boiled distilled water (50°C) of five times the weight of the dehydrated snack according to the procedure of Lewicki [22] with slight modifications. At time intervals of 5,10,15,20,25 and 30 minutes, the rehydrated slices were taken out from the beaker. After removing the adhering water by using the blotting paper, weight of the slices were noted and then placed in the water for further rehydration. The rehydration character of the snack is better than that of the desiccated coconut powder. The rehydration parameters were determined according to Lewicki [22] as rehydration ratio and coefficient of rehydration.

Rehydration ratio (RR) = WR/WD, Coefficient of rehydration (CR) = WR/Wo

Where WR = drained weight of rehydrated sample (g), WD = weight of dehydrated sample (g), Wo = weight of sample before dehydration (g).

Statistical interpretation of the data

The analyses on sensory, physical, chemical, functional,

phytochemical and shelf-life characteristics were done using triplicate samples. The data on experimental results were subjected to Analysis of Variance (ANOVA) and differences between means were assessed by LSD and independent sample 't' test using the statistical package SPSS (18 version) to compare the means to determine the most acceptable treatment ($p \leq 0.05$).

Results and Discussion

Physico-chemical characteristics of the osmotic solution

Acidity (%) value of the osmotic solutions: Figure 1 explains the percent acidity value of the osmotic solutions. The percentage acidity values of the control solution were 0.03 during 90 min and 0.2 in the 450 min beyond which it remained stable. The acidity level of the filtrate of functional ingredients ranged from 0.03 to 0.06 (90th min) to 0.22 to 0.30 (450th min) beyond which there was no change in the acidity levels. The percent acidity range observed at 90th min reveals the presence of minimum to maximum percentage acidity in the impregnated solutions D (*Beta vulgaris*) and A (*Mentha piperita*). Whereas the percent acidity value of filtrate B ranged from 0.054 to 0.27 at 90th and 450th min respectively, whilst filtrate C ranged from 0.059 to 0.25 at 90th min and 450th min which was observed as an intermediate acidity percent value between solutions D and A.

The acidity levels of the control solution were lesser than the functional filtrate solutions which could be due to the diffusion of organic acids from the sample into the solution. The outcome of the present study was in par with the finding of Dahiya and Dhawan [23] who reported that osmotic dehydrated aonlas found to contain lower acidity values, whereas osmotic solution contains higher acidity values. The analysis of variance of the results of the present study indicated that the impregnation of filtrates of functional ingredients through osmotic dehydration had a significant ($p \leq 0.05$) effect on acidity towards better product development and moreover acidity factor acts as an important attribute for enrichment of flavor as well [24].

Total Soluble Solids (°Bx) of the osmotic solutions: Figure 2 reveals the TSS of the osmotic solutions. The TSS (°Bx) values were found to decrease as the time increases during osmotic dehydration. The descending tendency of the TSS in the osmotic solutions observed in the present study was in par with the findings of Spiess et al. [25] who noted significant changes such as transfer of water from grape fruit to the solution and solute uptake by the grape fruit.

The TSS values of the control osmotic solution were comparatively lesser than all functional filtrates due to the natural presence of sugars in the filtrate. The TSS value of control solution ranged from 52.11 (90th min) to 45.41 (450th min), whereas among functional filtrates, the range was slightly higher especially in infusion of *Daucus carota* (55.00-90th min to 49.2- 450th min) and *Beta vulgaris* infusion (53.40-90th min to 49.00-450th min). The changes in TSS of *Mentha piperita* and *Zingiber officinale* filtrate infusions ranged from 51.50 to 45.41 and 52.68 to 46.40 which was observed for duration from every 90 min to 450 min and showed significant difference at $p \leq 0.05$, beyond which no change was observed. Thus analysis of variance applied between the sugar infused osmotic solutions either with or without the impregnation of filtrates of functional ingredients had significant effect on TSS of ready-to-eat coconut based snack. During osmotic dehydration, removal of water enhanced in the coconut slices and remarkable increment in TSS was noted. The outcome was consistent with the findings of Raoult-Wack [12] who has determined that utilization of hypertonic solutions result in 70% of removal of water without any heat applications.

Azuara [26] reported that better weight reduction, solid gain and water loss was obtained by the usage of concentrated sugar solutions in the apple disks. The change in TSS might be due to the migration of osmotic solution from higher sugar concentration to lower sugar concentration and uptake of solutes by the samples from the solution. The finding of the present study was similar to the result of Raoult-Wack et al. Delgado and Rubiolo [27] and Fathi et al. [28], where significant changes in balance of concentration between solid foods and solution subjected to osmotic dehydration, especially in case of pomegranate seeds was observed.

Sugar acid ratio: The sugar acid ratio of the osmotic solution is discussed in Figure 3. The sugar acid ratio of all the osmotic solutions was found to decrease with increase in time corresponding to the increased acidity and decreased TSS levels. However the SAR of the control solution was significantly greater ($p \leq 0.05$) than all samples impregnated with filtrate of functional ingredients. The sugar acid ratio of the control solution ranged from 2605.50 to 225.55 and that of sample D impregnated with *Beta vulgaris* was highest ranging from 2341.00 to 215.86, followed by *Beta vulgaris* impregnated osmotic solution D. The succeeding order of sugar acid ratio was found in impregnated osmotic solutions C (*Daucus carota*), B (*Zingiber oficinale*) and A (*Mentha piperita*) possessing initial and final values in the range of 2052.00 to 193.70, 1173.00 to 173.78 and 914.00 to 158.22 respectively, which represents the prolific presence of sugars and organic acids in the functional ingredients. Thus the results of variance of analysis showed that the osmotic solutions either with or without impregnation of functional ingredients had a significant ($p \leq 0.05$) effect on sugar acid ratio of ready-to-eat coconut based snack. This parameter plays a vital role in assessing the overall quality of dehydrated products with respect to flavor enrichment.

Physico-chemical properties of the ready-to-eat coconut based snack

Texture analysis-crispiness (N) of the ready-to-eat coconut based snack: Figure 4 depicts the textural characteristics of the ready-to-eat

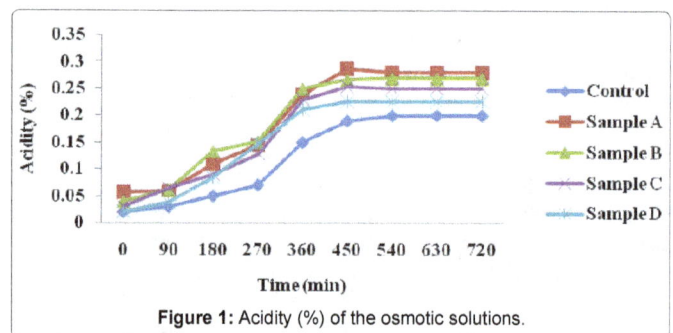

Figure 1: Acidity (%) of the osmotic solutions.

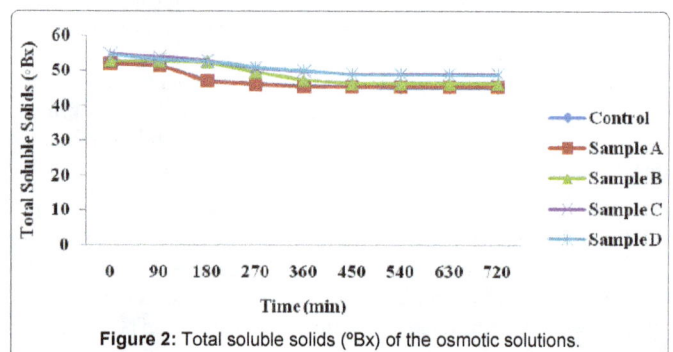

Figure 2: Total soluble solids (°Bx) of the osmotic solutions.

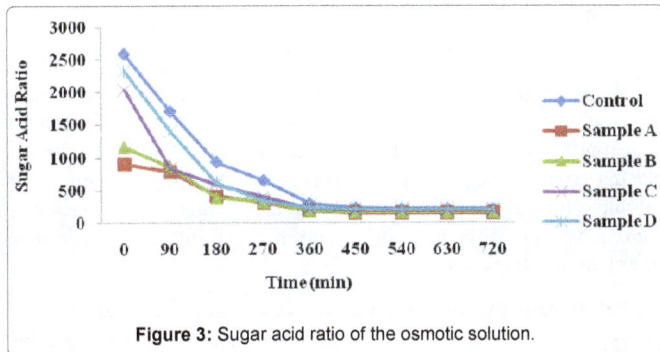

Figure 3: Sugar acid ratio of the osmotic solution.

coconut based snack. Texture analysis of the samples was performed with a HDP-CFS cylindrical ball probe using Texture Analyzer. Maximum breaking force (N) and deformation were measured from the force-deformation curve. Almost the breaking force for all the samples required around 5.7 to 5.9 N, hence there was no significant difference observed ($p \geq 0.05$). The optimum force required to break the snack was dependent on the hardness of the tissue and cell wall matrices. The concept was made in agreement with Lewicki [22] who stated that textural behavior is mainly related to the structural and biophysical properties of foods. Also the increase in time and concentration of solutes affect the texture of the osmotic dehydrated foods. Sucrose present in fruits and vegetables acts as an important factor influencing the texture of foods.

Color analysis of the ready-to-eat coconut based snack: Table 1 reveals the color characteristics of ready-to-eat coconut based snack. On analyzing the color parameters, the control samples possessed significantly lesser ($p \leq 0.05$) chroma values (T_1Control-4.64, T_2Control-4.94) than samples impregnated with functional filtrates with respect to the changes in the intensity of hue values. Irrespective of the drying method adopted, higher chroma values was observed in the impregnated samples D (T_1D-17.40, T_2D-17.14) and C (T_1C-13.9, T_2C-13.94) which could be due to the presence of colored pigments with grater intensity when compared with sample A (T_1A-10.50, T_2A-9.70) and sample B (T_1B-6.65, T_2B-5.87).

The lightness of the samples exposed to freeze drying was found to contain higher values due to the retention of nutrients without much degradation of pigments as drying was carried out at low temperature. Moreover the increased lightness values represents the preservation of pigments and nutrients without enzymatic deterioration in freeze dried samples than hot air oven dried which is related to the outcome of Krokida and Maroulis [29] who stated that the preservation of color of foods might be observed by its constant lightness L parameter thereby in chroma values. The finding of the present study was similar to the result of Larrauri et al. [30] who reported that L* values of the hot-air dried samples were significantly lower than the freeze-dried samples of the red grape pomace peels.

However the hot air oven dried samples A and B subjected to temperature 45°C ± 10°C resulted in decrement of degradation of pigments revealed more or less similar a* values when compared with freeze dried samples with slight significant difference at $p \leq 0.05$. The a* values of the hot air oven dried samples C and D found to possess slightly higher values than freeze dried samples due to enzymatic degradation of pigments which was in agreement with the finding of Krokida et al. [31] who reported that the a* chroma value of air dried samples increase significantly during air drying. The increase of the a-value denotes a more red color, which is indicative of the browning

reaction. A small increment in b* values (yellowness) was observed in all functionally infused coconut samples of both drying methods than control samples corresponding to the increase in a* values which was consistent with the result of Krokida et al. [29] who indicated that there was an increment in the b* values (yellowness) which was observed in untreated samples of banana whereas slower increase in b* value was observed in osmotic treated samples.

The chroma value of the hot air dried and freeze dried control samples were found to hold slightly decreased values than functionally infused samples, which represents the nature of pigments present in the functional ingredients. Nevertheless the chroma value of the coconut samples of both the drying methods retained almost similar values with slight significant difference at $p \leq 0.05$. However the chroma and total color difference of the freeze dried infused samples found to possess slightly lesser values than hot air dried samples indicating meager degradation of pigments. The increment in the total color difference was measured with reference to the 3 co-ordinates L*, a* and b* where lower lightness and higher a* and b* values were observed according to the concept of Cortes et al. [32].

The hue angle value of sample D of both the drying methods was found lowest than standard 90° hue angle which represents the presence of orange-red color, followed by hot air dried sample C with hue angle of 83.34, freeze dried sample C of hue angle -85.39, followed by hue angles of sample T_1B and T_2B (-82.83 and -80.52), finally control snack and sample T_1A and T_2A possessing hue angle values nearer to 90° which specifies the yellowness and greenness of the samples. The increase in chroma and total color difference values represents the extent of change in color during osmotic dehydration and also reduction of pigments occurred during drying process which could be related to the concept of Larrauri et al. [30].

Mass transfer kinetics: Bekele and Ramaswamy [33] reported that osmotic dehydration (OD) process involves soaking of foods in hypertonic solution to bring down the moisture level prior to drying process. During the process of osmotic dehydration, two counter current passes, where water flow from food into the solution and water soluble solute transfer from solution into the food takes place thereby infusion of vitamins, minerals, sugars, organic acids, volatiles and flavor compounds from food into the solution was observed. These changes are responsible for the overall quality of the dehydrated product. The percent weight reduction (WR), solid gain (SG) and water loss (WL) of all the T_1 and T_2 samples resulted in significant increments with the increase in time from 90th to 540th min and did not increase beyond 540th min up to 720th min of osmotic dehydration, hence the values remained plateau ($p \leq 0.05$). The WR, SG and WL were found to be slightly increased in T_1 and T_2 samples impregnated with functional

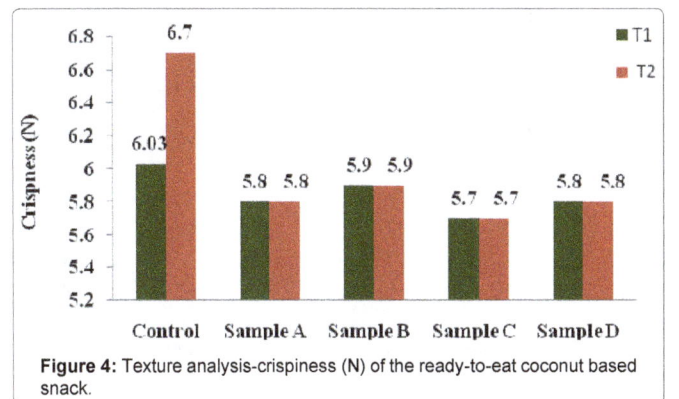

Figure 4: Texture analysis-crispiness (N) of the ready-to-eat coconut based snack.

filtrate of C and D. This is due to the uptake of more sugars from the carrot and beetroot infusion representing the plentiful sources of sugars.

Weight reduction (%) of the ready-to-eat coconut based snack: Table 2 describes the weight reduction property of the ready-to-eat coconut based snack. It was depicted that gradual decrease in weight reduction was shown from every 90th min to 540th min. The most significant changes was observed in weight reduction in control coconut snack from 90th min (7.11) to 540th min (42.70) when subjected to T1, while in case of T_2 samples, the range was 7.52-42.68.

Similarly for all the functionally impregnated ready-to-eat coconut based snacks, the maximum weight reduction was observed with significant difference at $p \leq 0.05$ from 90 min to 540 min, viz., T_1A- (7.16 to 43.11), T_2A- (9.24 to 44.13), T_1B- (7.16 to 43.21), T_2B- (9.33 to 43.42), T_1C- (7.22 to 43.36), T_2C- (9.64 to 44.14) and T_1D- (6.84 to 43.36), T_2D (9.64 to 44.14). However the weight reduction of all the coconut samples did not increase beyond 540 minutes of osmotic dehydration. The weight reduction in the functional filtrate impregnated with samples C and D was found to be slightly increased. This might be due to the uptake of more sugars from the carrot and beetroot infusions, representing the presence of sugars in these roots which is also supported by Rastogi et al. [34]. The finding was similar to the outcome of Torreggiani et al. [1], who stated that due to extra solute incorporation there is an increase in the weight reduction with increase in time. The result of the present study was in par with the outcome of Bchiret al. [35] was stated that increment in WR of pomegranate seeds was noted with increase in immersion time by using various concentration of date juices at the initial state of osmotic dehydration whilst the maximum changes in the above parameters took place at 20

min of the osmotic dehydration process. The greatest weight reduction was observed up to 40% until 120 minutes of osmotic dehydration.

Welti et al. [5] stated that the solute concentration of fruits increased by the most constructive technology i.e. osmotic dehydration or osmotic concentration. This technique includes the mechanism of placing fruits or vegetables in hypertonic salt or sugar solutions of soluble solids, ensuing water removal and increment in soluble solids content by solute absorption, which ultimately results in net reduction in weight of solid foods.

Solids gain (%) of the ready-to-eat coconut based snack: The solids gain of the ready-to-eat coconut based snack is discussed in Table 3. On analyzing the solid gain of the samples using analysis of variance reveals that irrespective of the osmotic impregnations and drying methods adopted all the impregnated samples and control samples found to contain of increased values corresponding to the progress in osmotic dehydration process which was observed for a duration from every 90th min to 540th min, beyond which no change was noted significantly. The mean solid gain content of impregnated samples C and D subjected to drying methods T_1 and T_2 (T_1C-0.41 to 2.44, T_2C-0.49 to 2.97, T_1D-0.41 to 2.44, T_2D-0.49 to 2.97) were observed with significant increments during osmotic dehydration process (90 to 540 min) which were greater ($p \leq 0.05$) than solid gain content possessed by impregnated samples A (T_1A-0.23 to 1.39, T_2A-0.29 to 2.23) B (T_1B-0.31 to 1.78, T_2 B-0.29 to 2.23) and Control (T_1Control-0.225 to 1.35, T_2Control-0.263 to 1.88).

The slight increment of solid gain in impregnated samples C and D could be due to the uptake of more sugars from filtrates of carrot and beetroot infusions which is described as chief sources of

Filtrate of functional ingredients	L* Drying method		p-value	a* Drying method		p-value	b* Drying method		p-value	Chroma Drying method		ΔE Drying method		Hue angle Drying method	
	T1	T2		T1	T2		T1	T2		T1	T2	T1	T2	T1	T2
Control	66.71 ± 0.02	69.83 ± 0.07	0.000*	-.643 ± 0.02	-0.78 ± 0.01	0.000*	4.60 ± 0.003	4.88 ± 0.03	0.000*	4.64	4.94	-	-	-82.04	-80.92
A	57.45 ± 0.13	61.45 ± 0.06	0.000*	-1.90 ± 0.11	-1.19 ± 0.04	0.000*	8.60 ± 0.040	9.63 ± 0.17	0.001*	10.50	9.70	10.17	9.64	-77.55	-82.95
B	61.80 ± 0.35	61.80 ± 0.10	1.00[NS]	-0.83 ± 0.005	-0.967 ± 0.01	0.000*	6.60 ± 0.30	5.79 ± 0.07	0.010*	6.65	5.87	5.3	8.08	-82.83	-80.52
C	60.01 ± 0.01	69.94 ± 0.06	0.000*	1.621 ± 0.0	1.12 ± 0.01	0.000*	13.9 ± 0.003	13.9 ± 0.003	1.00[NS]	13.90	13.94	11.68	9.21	83.34	-85.39
D	53.89 ± 0.06	57.90 ± 0.004	0.000*	11.05 ± 0.04	10.09 ± 0.01	0.000*	13.45 ± 0.16	13.86 ± 0.08	0.017*	17.40	17.14	14.92	18.5	50.59	53.95

All values are means of triplicate determinations± standard deviation (SD), T_1- Hot Air Oven Drying, T_2- Freeze Drying Sample A-*Mentha piperita*, Sample B-*Zingiber officinale*, Sample C-*Daucus carota* and Sample D-*Beta vulgaris* filtrate impregnated coconut based snack, NS- Not Significant, * Significant at $p \leq 0.05$

Table 1: Color values of the ready-to-eat coconut based snack.

Samples	Drying method	Time (min)									p-value
		0	90	180	270	360	450	540	630	720	
Control	T1	1.42 ± 0.02	7.11 ± 0.09	14.23 ± 0.06	21.35 ± 0.12	28.46 ± 0.1	35.58 ± 0.17	42.7 ± 0.6	42.7 ± 0.6	42.70 ± 0.6	0.001*
	T2	1.624 ± 0.03	7.52 ± 0.02	14.45 ± 0.013	21.53 ± 0.103	28.56 ± 0.15	35.59 ± 0.31	42.68 ± 0.2	42.68 ± 0.2	42.68 ± 0.2	0.002*
A	T1	1.440 ± 0.08	7.16 ± 0.02	14.11 ± 0.02	21.49 ± 0.19	28.86 ± 0.13	35.89 ± 0.14	43.11 ± 0.1	43.11 ± 0.1	43.11 ± 0.1	0.003*
	T2	2.098 ± 0.01	9.24 ± 0.02	18.12 ± 0.01	27.13 ± 0.120	32.01 ± 0.09	38.42 ± 0.06	44.13 ± 0.1	44.13 ± 0.1	44.13 ± 0.1	0.001*
B	T1	1.44 ± 0.04	7.16 ± 0.01	14.11 ± 0.08	22.01 ± 0.05	28.98 ± 0.21	36.07 ± 0.05	43.21 ± 0.1	43.21 ± 0.10	43.21 ± 0.1	0.005*
	T2	2.15 ± 0.02	9.33 ± 0.02	18.21 ± 0.01	27.22 ± 0.15	32.14 ± 0.08	37.82 ± 0.10	43.42 ± 0.15	43.42 ± 0.15	43.42 ± 0.15	0.001*
C	T1	1.44 ± 0.01	7.22 ± 0.01	14.44 ± 0.12	22.08 ± 0.30	29.09 ± 0.20	36.08 ± 0.17	43.36 ± 0.10	43.36 ± 0.10	43.36 ± 0.1	0.003*
	T2	2.38 ± 0.09	9.64 ± 0.01	18.41 ± 0.01	27.52 ± 0.110	32.34 ± 0.11	38.11 ± 0.09	44.14 ± 0.1	44.14 ± 0.1	44.14 ± 0.1	0.001
D	T1	1.44 ± 0.020	6.84 ± 0.02	14.24 ± 0.02	22.08 ± 0.150	29.09 ± 0.23	36.08 ± 0.20	43.36 ± 0.2	43.36 ± 0.24	43.36 ± 0.24	0.002
	T2	2.38 ± 0.09	9.64 ± 0.01	18.41 ± 0.01	27.52 ± 0.110	32.34 ± 0.11	38.11 ± 0.09	44.14 ± 0.1	44.14 ± 0.1	44.14 ± 0.1	0.002*

All values are means of triplicate determinations± standard deviation (SD), T_1-Hot Air Oven Drying, T_2-Freeze Drying Sample A-*Mentha piperita*, Sample B-*Zingiber officinale*, Sample C-*Daucus carota* and Sample D-*Beta vulgaris* filtrate impregnated coconut based snack, Rows followed by different alphabets are *Significantly different ($p \leq 0.05$), NS- Not Significant by LSD

Table 2: Weight reduction (%) of the ready-to-eat coconut based snack.

sugars according to Pavlov et al. [36]. The finding was similar to the outcome of Torreggiani et al. [1]who explained that due to extra solute incorporation the solid gain increases with increase in time. Time has considerable effect on solid gain (sugar uptake) and moisture loss. The finding of the present study was consistent with the outcome of Bchiret al. [35] who stated that osmotic dehydration resulted in increment of WR, SG and WL of pomegranate seeds with increase in immersion time by using various concentrations of date juices. The maximum changes in the above parameters took place at 20 min of the osmotic dehydration process. The increment in WR was observed in the pomegranate seeds whereas SG increased at significant level up to 6% during OD process and remained stable at the end of the process.

The significant increase ($p \leq 0.05$) in SG in the present study was coinciding with the fact suggested by Delgado and Rubiolo [27] and Fathi et al. [28] who observed that SG during osmotic dehydration might be obtained by the movement of sucrose to the pomegranate seeds through cell wall and accumulation between cell wall and its membranes due to existence of sugar between the food and the osmotic medium. The osmotic medium found to contain more sucrose than the seeds. Bchiret al. [35] stated that the sucrose present in date juice influenced the texture of the seeds and was contributing about 39% of the total sucrose content added additionally in the osmotic syrup which is the contributing factor for increase in solid gain. Also Lombard et al. [37] studied the effects of osmotic dehydration on mass transfer kinetics of pine apple cylinders. The cylinders found to contain increased solid content (Solid gain) after 240th min of osmotic dehydration process [38].

Water loss (%) of the ready-to-eat coconut based snack: Table 4 discusses the water loss property of the ready-to-eat coconut based snack. The water loss property of the samples was computed based on the summation of weight reduction and solid gain values. It was inferred from the above results, samples C and D infused with Daucus carota and Beta vulgaris extracts possessed greater SG and WR values which ultimately resulted in maximum water loss property (T_1C-7.6 to 45.79, T_2C-10.2 to 47.11, T_1D-7.25 to 45.8, T_2D-10.2 to 47.11) which was observed with significant difference at $p \leq 0.05$ at 90th min and 540th min on account of their prolific natural presence of sugars, whereas a slight decline in values were noted in samples A (T_1A-7.39 to 44.51, T_2A-9.52 to 46.36), B (T_1B -7.47 to 44.99, T_2B -9.42 to 45.65) and Control (T_1Control-7.34 to 44.05, T_2Control-7.79 to 44.57). The finding was similar to the outcome of Torreggiani et al. [1] who stated that due to extra solute incorporation there was an increase in the water loss (%) with increase in time. Thus the results of analysis of variance showed that the sugar infused osmotic solutions either with or without impregnation of functional ingredients had a significant ($p \leq 0.05$) effect on water loss of the ready-to-eat coconut based snack.

The increment in water loss of the samples in the present research was in reliant with the findings of Kowalska et al. [12] who reported that sucrose, glucose and starch syrup upon osmotic dehydration of pumpkin for 180th min showed an increment of maximum water loss and principally showed increment in solid gains when compared with samples with no pre-treatments carried out like blanching and freezing. Delgado and Rubiolo [27] and Fathi et al. [28] reported that osmotic dehydration generally results in water loss in with significant reduction in weight and solid gain accompanied by structural changes in cell wall tissue matrix of the solid foods with compaction of the surface layers with enhanced mass transfer resistance for solids and water.

In the present study, the 100% sugar infusion showed maximum water loss property during osmotic dehydration, which was related to

the concept of Fathi et al. [28] where osmotic dehydration is favored over solid uptake that lead to mass loss of pear fruit. It was found that increase in syrup concentration from 35 to 45° Brix and syrup fruit ratio from 1 to 2, there was substantial alteration in water loss from 18.09% to 23.18%, mass loss from 9.26% to 20.06% and solid gain from 13.59 to 16.38%. Bchir et al.[35] reported that, the various temperatures 30, 40 and 50°C had different impact on mass transfer kinetics of pomegranate seeds, which was immersed in 50°Brix glucose, sucrose and the mixture of the two (50:50). Effective water loss was noted at 37% in glucose, 46% in sucrose and 41% in mixture of solution (50:50).

The findings of the present study is in line with Santacruz-Vázquezet al.[39] who reported that there exists strong relationship between higher concentrations of sugar and increment in water loss ratios observed in some fruits of tropical origin such as melon, papaya and guava and also agreed that efficient osmotic dehydration achieved depends on concentration of sugar used. Lombard et al. [3] examined the sensory and stability of the osmotic dehydrated products which depend extremely on efficient water loss and solute gain. The net yield of the osmotic dehydration process will result in the summation of two counter current mass fluxes. Optimization process plays a vital role in obtaining maximum yield with optimum sensory traits of the food products.

Rehydration characteristics of the ready-to-eat coconut based snack: The drying process is evaluated by two mandatory physical factors such as appearance of the vegetables and the time required for reconstituting the vegetables. Also the reconstitution of dried fruits or vegetables depends mainly on the internal structure of dried food pieces and the degree of damage of water-holding components on drying. The reconstitution of dried vegetables become faster, when the food has an open and porous structure. The intrinsic factors which influence rehydration of dried food products involves product chemical composition, pre-drying treatment, formulation of products, drying conditions and techniques etc. The extrinsic factors constitute composition of osmotic medium, temperature and hydrodynamic conditions. These factors affect the tissue composition of plants, thereby causing impairment to the rehydration characteristics.

Rehydration ratio of the ready-to-eat coconut based snack: Table 5 explains the rehydration ratio of the ready-to-eat coconut based snack rehydration consists of two processes such as water absorption and leaching of solutes. A perfect rehydration yields a product with similar proximate principles of the raw foods.

The significant increase in Rehydration Ratio (RR) was observed from every 5th min to 20th min of water absorption process at $p \leq 0.05$ level. The superficial changes in RR was observed during 5 min (1.21) to 20 min (1.92) in case of T_1 samples and among the T_2 samples, the range was 1.59 during 5th min -2.06 during 20th min was obtained within 15 minutes itself, beyond which no water absorption was noted. The quick absorption of water by freeze dried samples was due to its porous structure.

In the present study, the coconut slices were pre-treated using sugar solution which increased the rehydration property. This outcome was similar to the finding of Leszczyk et al. [40] who indicated that rehydration of osmotic treated (salt and sucrose) celery prevented shrinkage due to the mechanical and structural strength provided by the osmotic agents on drying, there by increased water absorption. The pre-treatment of coconut slices using osmotic dehydration followed by drying methods increased the water absorption phenomenon. Leszczyk et al. [40] reported that rehydration increased water absorption of dried

Samples	Drying method	Time (min)									p-value
		0	90	180	270	360	450	540	630	720	
Control	T1	0.045 ± 0.008	0.225 ± 0.009	0.451 ± 0.005	0.675 ± 0.006	0.901 ± 0.008	1.13 ± 0.017	1.35 ± 0.007	1.35 ± 0.007	1.35 ± 0.007	0.001*
	T2	0.066 ± 0.009	0.263 ± 0.005	0.487 ± 0.009	0.683 ± 0.004	1.467 ± 0.012	1.67 ± 0.009	1.88 ± 0.026	1.88 ± 0.026	1.88 ± 0.026	0.002*
A	T1	0.048 ± 0.011	0.23 ± 0.01	0.468 ± 0.012	0.699 ± 0.02	0.930 ± 0.06	1.161 ± 0.01	1.39 ± 0.01	1.39 ± 0.01	1.39 ± 0.01	0.001*
	T2	0.087 ± 0.006	0.29 ± 0.01	0.524 ± 0.003	0.777 ± 0.008	1.529 ± 0.016	1.98 ± 0.009	2.23 ± 0.02	2.23 ± 0.02	2.23 ± 0.02	0.003*
B	T1	0.063 ± 0.005	0.31 ± 0.01	0.599 ± 0.001	0.907 ± 0.006	0.996 ± 0.009	1.489 ± 0.012	1.78 ± 0.01	1.78 ± 0.01	1.78 ± 0.01	0.004*
	T2	0.088 ± 0.01	0.29 ± 0.015	0.527 ± 0.01	0.778 ± 0.013	1.528 ± 0.016	1.975 ± 0.009	2.23 ± 0.01	2.23 ± 0.01	2.23 ± 0.0	0.002*
C	T1	0.085 ± 0.01	0.41 ± 0.02	0.816 ± 0.006	1.219 ± 0.013	1.629 ± 0.018	2.028 ± 0.01	2.44 ± 0.01	2.44 ± 0.01	2.44 ± 0.01	0.001*
	T2	0.283 ± 0.00	0.49 ± 0.013	0.875 ± 0.012	0.977 ± 0.011	1.966 ± 0.009	2.544 ± 0.008	2.97 ± .001	2.97 ± 0.01	2.97 ± 0.01	0.003*
D	T1	0.088 ± 0.009	0.41 ± 0.006	0.82 ± 0.06	1.22 ± 0.011	1.635 ± 0.023	2.032 ± 0.007	2.44 ± 0.005	2.44 ± 0.005	2.44 ± 0.005	0.005*
	T2	0.283 ± 0.005	0.49 ± 0.013	0.875 ± 0.012	0.977 ± 0.011	1.966 ± 0.009	2.544 ± 0.008	2.97 ± 0.01	2.97 ± 0.01	2.97 ± 0.01	0.003*

All values are means of triplicate determinations± standard deviation (SD), T_1-Hot Air Oven Drying, T_2-Freeze Drying Sample A-*Mentha piperita*, Sample B-*Zingiber officinale*, Sample *C-Daucus carota* and Sample *D-Beta vulgaris* filtrate impregnated coconut based snack, Rows followed by different alphabets are *Significantly different (p ≤ 0.05), NS- Not Significant by LSD

Table 3: Solids gain (%) of the ready-to-eat coconut based snack.

Samples	Drying method	Time (min)									p-value
		0	90	180	270	360	450	540	630	720	
Control	T1	1.47 ± 0.09	7.34 ± 0.06	14.68 ± 0.14	22.02 ± 0.2	29.4 ± 0.00	36.71 ± 0.02	44.05 ± 0.02	44.05 ± 0.02	44.05 ± 0.02	0.001*
	T2	1.69 ± 0.01	7.79 ± 0.02	14.94 ± 0.02	22.22 ± 0.1	30.03 ± 0.01	37.26 ± 0.01	44.57 ± 0.01	44.57 ± 0.01	44.57 ± 0.01	0.005*
A	T1	1.49 ± 0.01	7.39 ± 0.02	14.58 ± 0.01	22.19 ± 0.02d	29.79 ± 0.08	37.05 ± 0.04	44.51 ± 0.02	44.51 ± 0.02	44.51 ± 0.02	0.001*
	T2	2.18 ± 0.01	9.52 ± 0.01	18.65 ± 0.2	27.91 ± 0.3	33.5 ± 0.01	40.39 ± 0.01	46.36 ± 0.04	46.36 ± 0.04	46.36 ± 0.01	0.003*
B	T1	1.50 ± 0.01	7.47 ± 0.10	14.71 ± 0.1	22.9 ± 0.05	29.98 ± 0.01	37.56 ± 0.01	44.99 ± 0.01	44.99 ± 0.01	44.99 ± 0.01	0.005*
	T2	2.24 ± 0.00	9.42 ± 0.05	18.74 ± 0.06	27.9 ± 0.07	33.67 ± 0.01	39.79 ± 0.02	45.65 ± 0.00	45.65 ± 0.00	45.65 ± 0.00	0.001*
C	T1	1.53 ± 0.00	7.6 ± 0.01	15.3 ± 0.009	23.29 ± 0.01	30.7 ± 0.01	38.11 ± 0.07	45.79 ± 0.02	45.79 ± 0.02	45.79 ± 0.02	0.001*
	T2	2.66 ± 0.01	10.2 ± 0.00	19.3 ± 0.01	28.5 ± 0.01	34.3 ± 0.01	40.65 ± 0.02	47.11 ± 0.01	47.11 ± 0.01	47.11 ± 0.01	0.003*
D	T1	1.53 ± 0.00	7.25 ± 0.04	15.06 ± 0.00	23.30 ± 0.01	30.72 ± 0.01	38.11 ± 0.07	45.8 ± 0.01	45.8 ± 0.01	45.8 ± 0.01	0.003*
	T2	2.66 ± 0.01	10.2 ± 0.00	19.3 ± 0.01	28.5 ± 0.01	34.3 ± 0.01	40.65 ± 0.01	47.11 ± 0.01	47.11 ± 0.01	47.11 ± 0.01	0.004*

All values are means of triplicate determinations± standard deviation (SD), T_1- Hot Air Oven Drying, T_2-Freeze Drying Sample A-*Mentha piperita*, Sample B-*Zingiber officinale*, Sample *C-Daucus carota* and Sample *D-Beta vulgaris* filtrate impregnated coconut based snack, Rows followed by different alphabets are *Significantly different (p ≤ 0.05), NS-Not Significant by LSD

Table 4: Water loss (%) of the ready-to-eat coconut based snack.

Samples	Drying method	Time (min)							p-value
		0	5	10	15	20	25	30	
Control	T1	1.11 ± 0.006	1.21 ± 0.009	1.232 ± 0.006	1.430 ± 0.07	1.92 ± 0.011	1.92 ± 0.011	1.92 ± 0.011	0.001*
	T2	1.319 ± 0.010	1.595 ± 0.004	1.614 ± 0.003	1.944 ± 0.006	1.98 ± 0.013	1.98 ± 0.013	1.98 ± 0.013	0.003*
A	T1	1.24 ± 0.005	1.417 ± 0.006	1.497 ± 0.012	1.503 ± 0.01	1.901 ± 0.005	1.901 ± 0.005	1.901 ± 0.005	0.002*
	T2	1.312 ± 0.003	1.582 ± 0.006	1.608 ± 0.001	1.95 ± 0.007	1.99 ± 0.007	1.99 ± 0.007	1.99 ± 0.007	0.004*
B	T1	1.435 ± 0.006	1.508 ± 0.184	1.79 ± 0.01	1.89 ± 0.011	1.92 ± 0.012	1.916 ± 0.012	1.916 ± 0.012	0.000*
	T2	1.346 ± 0.005	1.59 ± 0.008	1.63 ± 0.007	1.97 ± 0.003	2.06 ± 0.00	2.06 ± 0.005	2.06 ± 0.005	0.002*
C	T1	1.3 ± 0.014	1.467 ± 0.013	1.595 ± 0.013	1.878 ± 0.012	1.914 ± 0.008	1.914 ± 0.008	1.914 ± 0.008	0.001*
	T2	1.32 ± 0.009	1.59 ± 0.017	1.618 ± 0.01	1.960 ± 0.003	2.06 ± 0.006	2.06 ± 0.00	2.06 ± 0.006	0.002*
D	T1	1.12 ± 0.00	1.502 ± 0.00	1.55 ± 0.00	1.71 ± 0.00	1.892 ± 0.00	1.892 ± 0.00	1.89 ± 0.00	0.003*
	T2	1.30 ± 0.014	1.580 ± 0.004	1.59 ± 0.006	1.94 ± 0.006	1.984 ± 0.003	1.984 ± 0.003	1.98 ± 0.003	0.003*

All values are means of triplicate determinations± standard deviation (SD), Sample A-*Mentha piperita*, T_1- Hot Air Oven Drying, T_2- Freeze Drying Sample B-*Zingiber officinale*, Sample *C-Daucus carota* and Sample *D-Beta vulgaris* filtrate impregnated coconut based snack, Rows followed by different alphabets are *Significantly different (p≤0.05), NS- Not Significant by LSD

Table 5: Rehydration ratio of the ready-to-eat coconut based snack.

foods up to 42.5% to 60.5%. He studied that rehydration increased the water absorption property of celery after the pre-treatment process of osmotic dehydration. In addition, Nindo et al. [21] reported that soaking of dried onions for a period of 20 minutes resulted in an increase in reconstitution property.

Coefficient of Rehydration of the ready-to-eat coconut based snack: The Coefficient of Rehydration (CR) of the samples showed increment from every 5 to 20 min. The significant change in CR (p ≤ 0.05) was noted in T_1 samples during 5 min (0.559) to 20 min (0.9), whilst the T_2 samples attained the maximum values within 5 min (0.64) to 15 min (0.95) itself. The quicker water absorption characteristics RR and CR were noted in freeze dried samples due to its porous and hygroscopic nature of tissue matrix (Table 6). Tunde-Akintunde [41] reported that rehydration ratio and coefficient of rehydration characteristics could be increased with increase in temperature and time. In the present study, the samples were dried using hot-air oven

Samples	Drying method	Time (min)							p-value
		0	5	10	15	20	25	30	
Control	T1	0.515 ± 0.01	0.559 ± 0.014	0.57 ± 0.012	0.66 ± 0.09	0.9 ± 0.00	0.9 ± 0.00	0.9 ± 0.00	0.003*
	T2	0.618 ± 0.01	0.75 ± 0.040	0.756 ± 0.01	0.921 ± 0.01	0.94 ± 0.03	0.94 ± 0.03	0.94 ± 0.03	0.002*
A	T1	0.579 ± 0.01	0.661 ± 0.007	0.696 ± 0.00	0.70 ± 0.01	0.886 ± 0.01	0.886 ± 0.01	0.886 ± 0.009	0.003*
	T2	0.613 ± 0.01	0.74 ± 0.03	0.752 ± 0.01	0.912 ± 0.01	0.93 ± 0.04	0.93 ± 0.04	0.93 ± 0.04	0.004*
B	T1	0.51 ± 0.005	0.56 ± 0.01	0.57 ± 0.01	0.66 ± 0.05	0.88 ± 0.009	0.88 ± 0.009	0.88 ± 0.009	0.001*
	T2	0.61 ± 0.000	0.64 ± 0.0	0.77 ± 0.01	0.93 ± 0.003	0.95 ± 0.005	0.95 ± 0.005	0.95 ± 0.005	0.003*
C	T1	0.61 ± 0.006	0.68 ± 0.004	0.74 ± 0.004	0.87 ± 0.004	0.887 ± 0.01	0.887 ± 0.01	0.887 ± 0.01	0.002*
	T2	0.63 ± 0.007	0.76 ± 0.007	0.77 ± 0.01	0.93 ± 0.005	0.95 ± 0.04	0.95 ± 0.04	0.95 ± 0.04	0.004*
D	T1	0.54 ± 0.011	0.71 ± 0.001	0.74 ± 0.009	0.81 ± 0.003	0.90 ± 0.001	0.90 ± 0.001	0.90 ± 0.001	0.005*
	T2	0.62 ± 0.007	0.75 ± 0.004	0.76 ± 0.02	0.93 ± 0.009	0.94 ± 0.006	0.94 ± 0.006	0.94 ± 0.006	0.001*

All values are means of triplicate determinations± standard deviation (SD), Sample A-*Mentha piperita*, T$_1$- Hot Air Oven Drying, T$_2$- Freeze Drying Sample B-*Zingiber officinale*, Sample C-*Daucus carota* and Sample D-*Beta vulgaris* filtrate impregnated coconut based snack, Rows followed by different alphabets are *Significantly different (p≤0.05), NS- Not Significant by LSD

Table 6: Coefficient of rehydration of the ready-to-eat coconut based snack.

and freeze drier which resulted in quick reaction and moreover within a minimum duration, where maximum water absorption took place at 20th min of rehydration process according to the concept of Maskan [42].

Conclusion

The filtrate of functional ingredients with sugar infusion was considered as a good impregnating solution used to carry out OD of coconut slices. In fact, the use of functional ingredients contributed a high nutritive value product with antioxidant properties. On concerning this, the functional ingredients could be endorsed for the preservation of many fruits and vegetables using OD technology. The rate of various OD parameters such as mass transfer kinetics, TSS, SAR, acidity and rehydration properties was in reliant with time and temperature and also diffusion mechanism. The diffusion lead to the process of leaching of natural solutes occurred from the coconut slices in to the solution which is not negligible quantitatively, but contributed towards the nutritional and organoleptic traits of the coconut slices and functional filtrates. OD process not only leads to change in composition of tissues but resulted in modification of cell structure, which was analyzed using texture analyzer. OD was noted as a good method to valorize coconut slices, suggesting utilization of filtrates of functional ingredients in formulations of nutritious healthy snack food items.

Acknowledgement

The authors express our gratitude towards University Grants Commission, New Delhi for the award of 'Junior Research Fellowship' for meritorious students.

References

1. Torreggiani D (1993) Osmotic dehydration in fruit and vegetable processing. Food Research International 26: 59-68.

2. Pathak RK, Ram RA (2002) Approaches for green good production in horticulture. In Souvenir, National Seminar-cum-workshop on Hi-Tech Horticulture and Precision Farming 33-35.

3. Lombard GE, Oliveira JC, Fito P, Andrés A (2008) Osmotic dehydration of pineapple as a pre-treatment for further drying. Journal of Food Engineering 85: 277-284.

4. Kabara JJ (1984) Antimicrobial agents derived from fatty acids. Journal of the American Oil Chemists Society 61: 397-403.

5. Welti J, Palou E, Lopez-Malo A, Balseira A (1995) Osmotic concentration - Drying of mango slices. Drying Technology 13: 405-416.

6. Taiwo KA, Angersbach A, Knorr D (2002) Influence of high intensity electric field pulses and osmotic dehydration on the rehydration characteristics of apple slices at different temperatures. Journal of Food Engineering 52: 185-192.

7. Crozier A, Lean MEJ, McDonald MS, Black C (1997) Quantitative analysis of flavonoid content of commercial tomatoes, onions, lettuce, and celery. Journal of Agricultural and Food Chemistry 45: 590-595.

8. Lerici CR, Pinnavaia G, Rosa MD, Bartolucci L (1985) Osmotic dehydration of fruit: Influence of osmotic agents on drying behavior and product quality. Journal of Food science 50: 1217-1219.

9. Demirel D, Turhan M (2003) Air-drying behavior of dwarf cavendish and gros michel banana slices. Journal of Food Engineering 59: 1-11.

10. AOAC (1990) Official Method of Analysis of the Official Analytical Chemist: 25th Ed. Virgin.

11. Evans L (2007) Senior scientific associate expert committee (DSN) Dietary Supplements: Non-botanicals, USP27–NF22, Pharmacopeia Forum.

12. Raoult-Wack AL (1994) Recent advances in the osmotic dehydration of foods. Trends in Food Science & Technology 5: 255-260.

13. Chiralt A, Fito P (2003) Transport mechanisms in osmotic dehydration. The role of the structure. Food Science and Technology International 9: 179-186.

14. Ozen BF, Dock LL, Ozdemir M, Floros JD (2002) Processing factors affecting the osmotic dehydration of diced green peppers. International Journal of Food Science and Technology 37: 497-502.

15. Ranganna S (1986) Vitamins: In Hand book of Analysis and quality control for fruit and vegetable products. (2ndedn), Tata McGraw Hill publishing Co.ltd, New Delhi.

16. Rebecca OPS, Boyce AN, Chandran S (2010) Pigment identification and antioxidant properties of red dragon fruit (Hylocereus polyrhizus). African Journal of Biotechnology 9: 1450-1454.

17. Mazza G (1983) Dehydration of carrots. International Journal of Food Science and Technology 18: 113-123.

18. Lima AWO, Cal-Vidal J (1983) Hygroscopic behaviour of freeze dried bananas. International Journal of Food Science & Technology 18: 687-696.

19. Nieto A, Salvatori D, Castro MA, Alzamora SM (1998) Air drying behaviour of apples as affected by blanching and glucose impregnation. Journal of Food Engineering 36: 63-79.

20. Kaymak-Ertekin F (2002) Drying and rehydrating kinetics of green and red peppers. Journal of Food Science 67: 168-175.

21. Nindo CI, Sun T, Wang SW, Tang J, Powers JR (2003) Evaluation of drying technologies for retention of physical quality and antioxidants in asparagus (Asparagus officinalis L.). LWT-Food Science and Technology 36: 507-516.

22. Lewicki PP (1998) Some remarks on rehydration of dried foods. Journal of Food Engineering 36: 81-87.

23. Dahiya S, Dhawan SS (2003) Effect of drying methods on nutritional composition of dehydrated aonla fruit (Emblica officinalis, Garten) during storage. Plant Foods for Human Nutrition 58: 1-9.

24. Imran M, Khan H, Shah M, Khan R, Khan F (2010) Chemical composition and antioxidant activity of certain Morus species. Journal of Zhejiang University Science B 11: 973-980.

25. Spiess, WEL, Behsnilian D (1998) Osmotic treatments in food processing. Current State and Future Needs 98: 47056.

26. Azuara E, Gutiérrez CI, Gutierrez GF (2002) Osmotic dehydration of apples by immersion in concentrated sucrose/maltodextrin solutions. Journal of Food Processing and Preservation 26: 295-306.

27. Delgado AE, Rubiolo AC (2005) Microstructural changes in strawberry after freezing and thawing processes. LWT- Food Science and Technology 38: 135-142.

28. Fathi M, Mohebbi M, Razavi SMA (2011) Effect of osmotic dehydration and air drying on physicochemical properties of dried kiwifruit and modeling of dehydration process using neural network and genetic algorithm. Journal of Food and Bioprocess Technology 4: 1519-1526.

29. Krokida M, Maroulis ZB (2000) Quality changes during drying of food materials. In drying technology in agricultural and food sciences. AS Mujumdar (Edn), Enfield Science Publishers.

30. Larrauri JA, Sánchez-Moren C, Saura-Calixto F (1998) Effect of temperature on the free radical scavenging capacity of extracts from red and white grape pomace peels. Journal of agricultural and food chemistry 46: 2694-2697.

31. Krokida MK, Oreopoulou V, Maroulis, ZB (2000) Water loss and oil uptake as a function of frying time. Journal of Food Engineering 44: 39-46.

32. Waliszewski K, Cortes HD, Pardio VT, Garcia MA (1999) Color parameter changes in banana slices during osmotic dehydration. Drying Technology 17: 955-960.

33. Bekele Y, Ramaswamy H (2010) Going beyond conventional osmotic dehydration for quality advantage and energy savings. Engineering Journal and Science Technology (EJAST) 1: 1-15.

34. Rastogi NK, Raghavarao KSMS (2004) Mass transfer during osmotic dehydration of pineapple: considering Fickian diffusion in cubical configuration. LWT-Food Science and Technology 37: 43-47.

35. Bchir B, Besbes S, Karoui R, Paquot M, Attia H, et al. (2012) Osmotic dehydration kinetics of pomegranate seeds using date juice as an immersion solution base. Food and Bioprocess Technology 5: 999-1009.

36. Pavlov A, Kovatcheva P, Georgiev V, Koleva I, Ilieva M (2002) Bio synthesis and radical scavenging activity of betalains during the cultivation of red beet (Beta vulgaris) hairy root cultures. Journal of Biotechnology 57: 640-644.

37. Olajide PS, Samuel OA, Sanni LO, Bamiro FO (2010) Optimization of pre-fry drying of yam slices using response surface methodology. Journal of Food Process Engineering 33: 626-648.

38. Lakshmi B, Vimala V (2000) Nutritive value of dehydrated green leafy vegetable powders. Journal of Food Science and Technology 37: 465-471.

39. Santacruz-Vazquez V, Santacruz V, Azquez C, Welti-Chanes J, Farrera-Rebollo, et al. (2008) Effects of air drying on the shrinkage, surface temperatures and structural features of apples slabs by means of fractal analysis. Revista Mexicana de Ingenier´ıaQu´ımica 7: 55-63.

40. Kowalska H, Lenart A, Leszczyk D (2008) The effect of blanching and freezing on osmotic dehydration of pumpkin. Journal of Food Engineering 86: 30-38.

41. Tunde-Akintunde TY (2011) Mathematical modeling of sun and solar drying of chilli pepper. Renewable Energy 36: 2139-2145.

42. Maskan M (2001) Drying, shrinkage and rehydration characteristics of kiwifruits during hot air and microwave drying. Journal of Food Engineering 48: 177-182.

Effect of Refined Milling on the Nutritional Value and Antioxidant Capacity of Wheat Types Common in Ethiopia and a Recovery Attempt with Bran Supplementation in Bread

Heshe GG[1], Haki GD[2] and Woldegiorgis AZ[1]*

[1]*Center for Food Science and Nutrition, College of Natural Sciences, Addis Ababa University-1176, Ethiopia*
[2]*Department of Food Science and Technology, Botswana College of Agriculture, Private Bag 0027, Gaborone, Botswana*

Abstract

The effect of wheat flour refined milling on nutritional and antioxidant quality of of two types of wheat (hard and soft) grown in Ethiopia was first evaluated. Then a recovery was attempted on bread prepared with the supplementation of the white wheat flour with different levels (0%, 10%, 20% and 25%) of wheat bran. Whole wheat flour (100% extraction) and white wheat flour (68% extraction) were subjected to proximate, mineral and antioxidant analysis. Results indicated that at low extraction rate (68%) the value of protein, fat, fiber, ash, Iron, zinc and phosphorous and antioxidant content of the samples were significantly affected (decreased) ($P<0.05$) by milling. The Total Phenolic Content (TPC) of white wheat flours, which ranged from 3.34 to 3.49 mg GAE/g were significantly lower ($p<0.005$) than those of whole wheat flours (ranging 7.66 to 8.20 GAE/g). At the concentration of 50 mg/mL, the DPPH scavenging effect of wheat extracts decreased in the order of soft whole, hard whole, soft white and hard white wheat flour, which was 90.39, 89.89, 75.80, and 57.57%, respectively. Moreover, the protein, fat, ash, fiber, iron and zinc contents of the bran supplemented breads increased significantly ($P<0.05$) with the progressive increase in the bran level of the bread. The highest value for protein, fat, fiber, ash, iron and zinc was 12.04, 2.61, 2.48, 3.27 g/100 g and 4.84 and 2.33 mg/100 g respectively were found in 25% bran supplemented bread. The sensory evaluation of bread showed that all level of supplementation had mean score above 4 on a 7 point hedonic scale on over all acceptances. The results indicated that refined milling at 68% extraction significantly reduces the nutritional and antioxidant activity of wheat flour. Bread of good nutritional and sensory qualities could also be produced from 10% and 20% bran supplementation.

Keywords: Wheat; Whole flour; White flour; Bran; Refined milling; Nutritional; Antioxidant

Introduction

Wheat has accompanied humans since remote times (as far back as 3000 to 4000 BC) in their evolution and development, evolving itself (in part by nature and in part by manipulation) from its primitive form (emmer wheat) into the presently cultivated species [1]. Wheat crop is widely adapted to a variety of environments and is cultivated in tropical, subtropical and temperate areas [2]. It is widely consumed by humans, in the countries of primary production (which number over 100 in the FAO production statistics for 2004) and in other countries where wheat cannot be grown [3]. It also occupies 27 percent of the total cereal production worldwide [1]. It is thus, an important agricultural commodity which is consumed in large amount all over the world among all grains.

Ethiopia is the largest wheat producer in sub-Saharan Africa [4]. Nationally, wheat ranks fourth in total area coverage (1,389, 215.00 ha). It is also third in productivity (after maize and sorghum) among cereals [5]. It is one of the most important crops grown and consumed in Ethiopia both in terms of total production (2.85 million MT in 2010/11) [6] and the proportion of total calories consumed in the country (19.6% of calories consumed) [7].

Wheat possesses several health benefits, especially when utilized as a wholegrain product. According to Kumar [8], wheat provides protection against diseases such as constipation, ischaemic, heart disease, diverticulum, appendicitis, diabetes and obesity. These benefits are attributed in part to the presence of different compounds such as dietary fibers, phytochemicals, proteins, vitamins and minerals [9].

Whole wheat grain consists of bran, germ and endosperm. When refined, only carbohydrate rich endosperm is retained. This results in a big loss of many nutritionally valuable biochemical compounds such

as dietary fiber, vitamins, minerals and antioxidant compounds which play an important role in reducing cardiovascular disease (CVD) [10]. When white flour is produced, many important nutrients and fibre are removed, because these components are mainly located in bran and germ [11]. Wheat bran is rich in protein (~14%), carbohydrates (~27%), minerals (~5%) and fat (~6%) [12]. In addition, bran is the main by-product produced by milling. Wheat bran is a most important fiber source which is inexpensive and available. It is a good source of not only dietary fiber but also for other major nutrients. The loss of vitamins and minerals in the refined wheat flour has led to widespread prevalence of constipation and other digestive disturbances and nutritional disorders [8].

Milling is the critical process affecting the concentrations of nutrients in wheat-derived food products. The outer parts of the kernel, especially the aleurone layer and the germ are richer in minerals. Conventional milling reduces nutritional content in flour and concentrates them in the milling residues [13]. White flour with the extraction rate 68% meaning up to 32% of the original grain is not in the flour. Whole grain flour includes all parts of the seed and is 100%

---

*Corresonding author: Woldegiorgis AZ, Center for Food Science and Nutrition, College of Natural Sciences, Addis Ababa University-1176, Ethiopia
E-mail: ashuyz1@yahoo.com

extraction. Milling of wheat into highly refined flours not only preclude considerable amounts of nutrients from human consumption, but the remaining flours have a much poorer nutritive value than flour made from whole wheat.

Over the past twenty years, wheat production and consumption have both increased in Ethiopia [14]. On top of this wheat flour covers a substantial proportion of the population. According to the survey conducted in Ethiopia, 28% of consumers purchase flour and flour products. It is estimated that wheat flour reaches about 22 million people [15]. However, no published information is available regarding the effect of refining on nutritional and antioxidant capacity of wheat that are commonly grown in Ethiopia though it is widely distributed and consumed. The nutritional value and antioxidant properties of wheat grain are significantly influenced by soil type and richness, growing temperatures, moisture levels, other climatic differences, and genotype [16].

It is therefore, very important to understand the nutritional value of wheat grown in Ethiopia and evaluate after the effect of refined milling. In addition, it is necessary to find a way to improve the nutrient quality of the refined wheat flour products without compromising the palatability of the product. Hence, in this study an attempt was also made to recover the nutrients through supplementation of bran in wheat bread making.

Materials and Methods

Samples

Hard wheat (*Kubsa*) and soft wheat (*ET-13*) samples were obtained from Kebron food complex (Oromia region) and Wedera farmers cooperative (Debrebrhan), respectively in Ethiopia. Bran sample was obtained from Universal Food Complex (Addis Ababa, Ethiopia) (Figure 1).

Milling of wheat

The amount of water required for tempering was calculated according to AACC [17]. One kilogram of each sample from both soft and hard wheat were cleaned and tempered separately to 14% moisture level and kept for 6 and 24 h, respectively at ambient temperature in a closed plastic jar. After tempering, wheat samples were milled at the extraction rate of 68% based on the capacity of the milling machine and 100% by using Buhler Automatic mill (Deutschland). The milling of flour was conducted at Kokeb Flour and Pasta Factory and extraction rate was calculated according to Slavin [18].

Formulation of bread

Flour blends were prepared by mixing wheat flour with wheat bran in the proportions of 100: 0, 90:10, 80:20 and 75:25 (wheat flour to bran) using homogenizer and 100% white wheat flour was used as control. The formulation was made based on the preliminary test. The four flour samples were packaged in black low density polyethylene bags and stored in plastic containers at room temperature from where samples were taken bread production.

Processing of bread

Bread was prepared by partially replacing the wheat flour with 0%, 10%, 20%, and 25% of bran sample. The dough was prepared based on the method described by Hertzberg [19] with some modification by weighing 400 g of wheat flour, 8 g sugar, 4 g salt, 8 g of oil and 1.6 g yeast. A mixer (Linkrich-B15) was used to mix the ingredients in order to homogenize the mixture for 30 min and then water was added to the dough until the desired consistency was achieved. The dough was weighed and divided into 3 equal portions for replications. These were placed in baking pans and left for 1 h. Then they were transferred into an oven pre-heated to about 180 - 250°C and allowed to bake for 20 min. The baked products were left to cool.

Nutritional analysis of wheat flour, bran and bread

All samples were analyzed for moisture, crude protein, crude fat, crude fiber and total ash according to their respective standard methods as described below.

Determination of moisture content: Dishes used for the moisture determination were first dried at 105°C for 1 hour in drying oven. It was then transferred to the desiccators, cooled for 30 minutes, and weighed. The prepared samples were mixed thoroughly and about 5 g of the flour samples were transferred to the dried and weighed dishes. The dishes and their contents were placed in the drying oven and dried for 3 hrs at 105°C, and then the dishes and their contents were cooled in desiccators to room temperature and reweighed [20].

Determination of total ash content: Total ash was determined according to AOAC [20]. The porcelain dishes used for the analysis were placed into a muffle furnace for 30 min at 550°C. The dishes were removed and cooled in desiccators for about 30 minutes to room temperature; each dish was weighed. About 2.5 g of the flour and bread samples were added into each dish. The dishes were placed on a hotplate under a fume hood and the temperature was slowly increased until smoking ceases and the samples become thoroughly charred. The dishes were placed inside the muffle furnace at 550°C for 5 hours and removed from the muffle and cooled in the desiccator and weighed. The amount of total ash was calculated by using the following formula. Weight of total ash was calculated by difference and expressed as percentage of sample.

Determination of crude protein: Protein (N × 6.25) was determined by the Kjeldahl method [20]. All nitrogen is converted to ammonia by digestion with a mixture of concentrated sulfuric acid containing copper sulfate and potassium sulfate as a catalyst. The ammonia released after alkalinization with sodium hydroxide is steam distilled into boric acid and titrated with hydrochloric acid.

Determination of crude fat content: Crude fat was determined by exhaustively extracting a known weight of sample in diethyl ether (boiling point, 55°C) in a soxhlet extractor [20]. The ether was evaporated from the extraction flask. The amount of fat was quantified gravimetrically and calculated from the difference in weight of the extraction flask before and after extraction as percentage.

Determination of crude fiber content: 1 g of sample (W_1) and 1 g of celite (sand) was weighed into a crucible and placed the crucible in Fibertec hot extraction and added 150 ml of hot 0.64 N of sulfuric acid [20]. Add 2 - 4 drops of n-octanol to prevent foaming and boiled for

Figure 1: Hard wheat, bran and soft wheat.

10 min. In step 2 of hot extraction add 150 ml of 0.556 N NaOH and put 2 - 4 droplets of n-octanol and boil it for 10 min. In both process washing with hot deionized water was applied. Evaporate solvent, dry the crucibles at 130°C for 2 hours and cooled in desiccator and weigh. Ash the sample in the crucible at 550°C for 3 hours and cooled down and weigh.

Determination of mineral contents

Determination of Fe, and Zn: Iron and zinc were determined according to the standard method of AOAC [20] using flame Atomic Absorption Spectrophotometer. Ash was obtained from dry ashing of the samples. The ash was wetted completely with 5 ml of 6 N HCl, and dried on a low temperature on hot plate. A 7 ml of 3 N HCl was added to the dried ash and heated on the hot plate until the solution just boiled. The ash solution was cooled to room temperature in a hood and was filtered using filter paper (Whatman 45). A 5 ml of 3 N HCl was added into each crucible dishes and was heated until a solution boiled then cooled and filtered into the flask. The crucible dishes are again washed three times with deionized water, the washing was filtered into a flask. Then the solution was cooled and diluted to 50 ml with de-ionized water. A blank was prepared by taking the same procedure as the sample.

Determination of phosphorus: Phosphorus was determined using the molybdovanadate method [21]. Briefly, 5 ml of aliquot was pipetted from the sample digest into a 100 ml volumetric flask. Ten ml of the molybdate and vanadate solutions was added to the samples and the standards and made up to volume with distilled water. After 10-30 minutes color developed and measured on the absorbance of the blank, sample and standards by spectrophotometer at a wavelength of 460 nm.

Antioxidant capacity determination

Sample extraction: Samples were extracted based on the procedures as outlined by Woldegiorgis [22]. The powdered wheat samples were homogenized and weighed in to ten gram was then extracted by stirring with 100 ml of methanol at 25°C at 150 rpm for 24 h using an incubator shaker (ZHWY-103B) and then filtered through Whatman No. 1 filter paper. The residue was then extracted with two additional 100 ml portions of methanol as described above. The combined methanolic extracts were evaporated at 40°C to dryness using a rotary evaporator and re-dissolved in methanol at a concentration of 50 mg/ml and stored at 4°C for further use.

Determination of free radical scavenging activity: The hydrogen atoms or electrons donation ability of the corresponding extracts and some pure compounds were measured from the bleaching of purple colored methanol solution of DPPH [23]. Antioxidant activity of the methanol extracts was determined by DPPH radical scavenging method as described by Woldegiorgis [22]. A 0.004% solution of DPPH radical solution in methanol was prepared and then 2 ml of DPPH solution was mixed with 1 ml of various concentrations (0.1-50 mg/ml) of the extracts in methanol. Finally, the samples were incubated for 30 min in the dark at room temperature. Scavenging capacity was read spectrophotometrically by monitoring the decrease in absorbance at 517 nm. Ascorbic acid was used as a standard and mixture without extract was used as the control. Inhibition of free radical DPPH in percent (I%) was then calculated.

Total phenolics determination: Phenolic compounds concentration in the wheat was estimated with Folin-Ciocalteu reagent according to the Singleton and Rossi method [24] as described by Woldegiorgis [22]. One milliliter of sample (5000 μg) was mixed with 1 ml of Folin and Ciocalteu's phenol reagent. After 3 min, 1 ml of saturated sodium carbonate (20%) solution was added to the mixture and adjusted to 10 ml with distilled water. The reaction was kept in the dark for 90 min, after which the absorbance was read at 725 nm. Gallic acid was used to construct the standard curve (5-80 μg/ml). The results were mean values ± standard error of mean and expressed as mg of gallic acid equivalents/g of extract (GAEs).

Sensory evaluation of bread

Sensory evaluation was conducted for the freshly baked breads by 30 semi-trained panelists consisting of students male and female aged from 23-43 years old from Food Science and Nutrition Center of Addis Ababa University. The samples were presented randomly in identical containers, coded with three digit numbers. The sensory was conducted using a seven point hedonic scale. Where 1=dislike very much, 2 = dislike moderately, 3 = dislike slightly, 4 = neither like nor dislike, 5 = like slightly, 6 = like moderately and 7= like very much. The sensory attributes which were evaluated were taste, odour, colour, texture and overall acceptability. Those samples were considered as acceptable which their average score for the overall acceptability were greater than 4 which mean neither like nor dislike [25].

Statistical analysis

The data were subjected to analysis of variance (ANOVA) and Duncan's multiple range tests were used for mean separation at p < 0.05. Linear regression analysis was used to calculate IC50 value. Pearson correlation between DPPH scavenging (%) and total phenolic content was considered at p < 0.05.

Results and Discussion

Proximate composition of wheat, wheat bran and bread

The mean value for moisture, crude protein, crude fat, total ash and crude fiber of wheat bran, wheat flour (hard and soft), white wheat flour (hard and soft) and bran supplemented bread are presented in Tables 1-3. The mean values for moisture contents of different whole wheat and white wheat flours are presented in Table 1. It ranged from 10.48 to 12.30%. The highest moisture level, 12.30%, was found in hard white wheat flour. The moisture content varies significantly between whole and white flour and between hard and soft white flour (P<0.05). The increment on moisture content of both soft and hard white wheat as compared to the whole wheat flour could be due to the addition of water during the tampering process to facilitate milling of wheat which resulted in retaining more water in refined wheat flour than whole wheat flour.

There is also a significant difference (P<0.05) on total ash contents of all flour samples (Table 1). The result indicated that highest ash content ranged from 1.62 to 1.41%. The highest ash content (1.62%) was found in hard whole wheat flour where as the soft white flour showed the lowest (0.38%) ash content. The results were comparable to Azizi [26] values obtained from different extraction rate of wheat flour which ranged 1.51% to 0.54% ash content at 93% and 70% extraction rate respectively.

The result for crude fat content is shown in Table 1 and the values showed significant difference (P<0.05) between whole wheat flours and white wheat flours. The fat content decreased in white wheat flour. The highest fat content, 1.83%, was found in whole wheat flour (100% extraction rate); whereas, the lowest, 1.32%, was found in white wheat flour (low extraction rate). The high percentage of fat in whole wheat

Parameters	Wheat samples			
	HWF	HWWF(refined)	SWF	SWWF(refined)
Moisture (%)	10.75 ± 0.38[a]	12.30 ± 0.09[c]	10.48 ± 0.10[a]	11.60 ± 0.23[b]
Ash (%)	1.62 ± 0.03[d]	0.65 ± 0.01[b]	1.41 ± 0.07[c]	0.38 ± 0.07a
Fat (%)	1.82 ± 0.04[b]	1.43 ± 0.18[a]	1.78 ± 0.10[b]	1.32 ± 0.11[a]
Protein (%)	14.40 ± 0.30[d]	11.91 ± 0.087[c]	9.11 ± 0.12[b]	7.13 ± 0.06a
Fiber (%)	2.6 ± 0.08[b]	0.42 ± 0.06[a]	2.5 ± 0.08[b]	0.36 ± 0.07[a]

HWF- Hard whole wheat flour, HWWF- Hard white wheat flour (refined), SWF -Soft whole wheat flour, SWWF- Soft white wheat flour (refined).

Data are average of triplicate ± SE .Mean value with different superscript in the same rows are significantly different (P<0.05)

Table 1: Proximate composition of wheat.

Parameter	Composition (g/100g)
Protein	15.26 ± 0.35
Fat	3.12 ± 0.7
Fiber	9.97 ± 0.27
Ash	4.5 ± 0.16

Table 2: Proximate composition of wheat bran.

flour is because wheat germ is ground along with endosperm during milling [27].

The results of this study also indicated that the protein contents for all flours varied significantly (P<0.05). The protein contents decreased with in both hard and soft white wheat flour; at the same time there was significant difference between hard and soft whole wheat flour. The highest protein content found on hard whole wheat flour was 14.40%; whereas the protein content of soft whole wheat flour was 9.11%. The result of this study was in line with the value of hard and soft whole wheat reported by Blakeney [28].

The mean value for fiber content of hard whole wheat, hard white wheat, soft whole wheat and soft white wheat flour are 2.6, 0.42, 2.5 and 0.36 g/100 g, respectively (Table 1). The result of this study showed that there is significant difference (P<0.05) between whole wheat flour and white wheat flour. Both hard and soft whole wheat flour exhibited high crude fiber content (2.6 and 2.5%). The white wheat flour showed less fiber content because wheat bran was removed during milling process which decreased the amount of fiber in flour. The result of this study is in agreement with Azizi [26], who reported crude fiber in the range of 0.30 to 2.24% white wheat and whole flour respectively.

The proximate composition of wheat bran samples are given in Table 2. Wheat bran was found to contain highest amounts of crude protein, fat, fiber and ash with mean values of 15.26%, 3.12%, 9.97%, and 4.45% respectively (Table 2). The objective of milling is to separate the bran and germ from the starchy endosperm so that the endosperm can be ground into flour. The aleurone layer, which is rich in protein, minerals and vitamins, usually breaks away with the outer layer of the bran in the milling process, thus, contributing significantly to the nutritional quality of the bran fraction [29].

Proximate composition of different bran supplemented bread and control were also analyzed for proximate composition. The mean values for moisture contents of the bread samples are presented in Table 3 which ranged from 30.92 to 32.83%. The highest moisture level, 32.83%, was found in 25% bran supplemented bread. The moisture content of the control bread decreased from the three samples bread significantly (P<0.05).

The statistical analysis for crude protein is presented in Table 3. The mean value for protein content of all the study samples of bread ranged

from 9.42 for control to 12.04 for WBB (75:25). The protein contents for three of the breads (0%, 10%, 20% bran supplemented bread) varied significantly (P<0.05). However, there was no significant difference between 20% and 25 % bran supplemented bread. The result of protein contents are in agreement with the findings of Butt [30] who reported an increase contents of protein with an increase in bran proportion.

The result of this study indicate that crude fat showed significant difference (P<0.05) among all breads. The fat content increased with an increase in bran level. The highest fat content, 2.61%, is found in 25% bran supplemented bread, whereas, the lowest, 1.56%, was found in the control bread. The increase in fat content is because of the germ which is grounded along with bran and endosperm during milling, results in bread with higher fat content than the control bread [27].

The mean values of crude fiber contents of different bread samples are given in Table 3. The statistical analysis showed significant (P<0.05) effect on the quantity of crude fiber. The crude fiber contents ranged from 0.38% to 3.27%. The 25% bran supplemented bread exhibited the highest crude fiber (3.27%), whereas, the control bread contain the lowest crude fiber (0.38%). The crude fiber increased with an increase in bran supplementation rate. The control showed less fiber contents because the bread was made from refined bread with no addition of bran.

Ash is the mineral residue remaining after a sample has been completely oxidized in a manner such that all organic volatile material is driven off, while preventing any mineral from being lost [29,30]. Ash varied significantly among all the bread (Table 3). The statistical analysis showed significant (P<0.05) effect on total ash contents. The results indicate that ash content ranged from 1.38 to 2.48%. The highest ash content (2.48%) was found in 25% bran supplemented bread, whereas the control showed the lowest (1.38%) ash content. The addition of 10 to 25% wheat bran to the bread increased the ash content.

Mineral content of wheat flour and bread

The mineral content of the whole and white flour samples are shown in Table 4. According to the results of this study, the iron content level in hard and soft whole wheat flour is significantly different (p<0.05) from hard and soft white wheat flour. The iron content of whole wheat flour ranged from 2.95 to 4.15 mg/100 g whereas the iron content of the white wheat flour ranged from 2.51 to 3.35 mg/100 g.

Dewettinck [31] reported that the iron content of the whole wheat was (1-5 mg/100 g) which is in agreement with the result of this study. However, the level of iron in the white flour decreases significantly. The milling process removes many important nutrients when white flour is produced. The bran and the germ are relatively rich in minerals and the milled products contain less of these than the original grain. As a result of milling the palatability is increased but the nutritional value of the products is decreased [32].

Zinc content varied significantly among all whole wheat and white wheat flour (Table 4). The zinc content of the whole wheat ranged from 3.59 to 2.47 mg/100 g. However, the white flour contained in the range of 0.58 to 1.39 mg/100 g. The highest Zn content 3.59 mg/100 g was found in hard whole wheat flour. This study showed that low rate of extraction of wheat reduce the zinc content of the wheat significantly (P<0.05). The hard whole wheat flour zinc content was 3.59 mg/100 g where as milling reduced the zinc content to 1.39 mg/100 g. Whereas in case of soft wheat the decrease was from whole wheat to white wheat flour 2.47 to 0.58 mg/100 g of zinc respectively. According to Lopez [33] 80% of the total amounts of minerals are concentrated in the

Samples	Moisture %	Protein %	Fat %	Ash %	Fiber %
Control(WFB)	30.92 ± 0.38[a]	9.42 ± .22[a]	1.56 ± .06[a]	1.38 ± 0.10[a]	0.38 ± 0.02[a]
WF:BR (90:10)	33.14 ± 0.26[b]	10.70 ± 0.55[b]	2.14 ± 0.032[b]	2.04 ± 0.32[b]	2.13 ± 0.05[b]
WF:BR (80:20)	33.24 ± 0.69[b]	11.66 ± 0.89[c]	2.36 ± 0.04[c]	2.29 ± 0.04[c]	3.11 ± 0.06[c]
WF:BR (75:25)	32.83 ± 0.58[b]	12.04 ± 0.84[c]	2.61 ± 0.06[d]	2.48 ± 0.02[d]	3.27 ± 0.008[d]

WFB-(white wheat flour bread)- control, WF:BR 90:10, 10% bran supplemented bread ,WF:BR 80:20- 20% bran supplemented bread, WF:BR 75:25 - 25% bran supplemented bread

Data are average of triplicate ± SE. Mean value with different superscript in the same column are significantly different (P<0.05).

Table 3: Proximate composition of Bran supplemented bread and control (Dry weight basis).

Wheat samples	Mineral content mg/100 g		
	Iron	Zinc	Phosphorous
HWF	4.15 ± 0.12[c]	3.59 ± 0.063[d]	337.99 ± 0.56[d]
HWWF	2.51 ± 0.16[a]	1.39 ± 0.036[b]	144.69 ± 0.61[b]
SWF	3.35 ± 0.17[b]	2.47 ± 0.04[c]	313.98 ± 1[c]
SWWF	2.95 ± 0.26[a]	0.58 ± 0.01[a]	77.03 ± 0.51[a]

HWF- hard whole wheat flour, HWWF- Hard white wheat flour, SWF -soft whole wheat flour. SWWF-Soft white wheat flour

Mean value with different superscript in the same column are significantly different (P<0.05).

Table 4: Mineral composition of whole and refined wheat flour.

aleurone layer of pericerap (bran) which was removed during milling process while only 20% minerals are present in endosperm.

The mean total content of phosphorous on hard whole and soft whole wheat flour was 337.99 and 313.98 mg/100 g respectively however, there was significant decrease of phosphorous content (P<0.05) on the refined milled product of hard and soft white wheat flour which was 144.69 and 77.03 mg/100 g respectively. Wheat is one of the cereals which are classified as rich sources of phosphorous. The result of the whole wheat flour was in line with the finding of Dewettinick [31] which reported phosphorous from 200-1200 mg/100 g.

The replacement effect of different levels of wheat bran on the mineral content of bread is shown in Table 5. All the mean value for iron content varied among the bread samples. Results showed that mineral progressively increased when levels of bran were increased. Control contained 1.98 mg/100 g iron and at the level of 25% bran replacement, the value increased to 4.84 mg/100 g of iron. The result of this study was in agreement with the study of Butt [30].

The replacement effect of different levels of wheat bran on the zinc content of bread is shown in Table 5. Results showed that zinc progressively increased when levels of bran were increased. The result ranged from 0.93 to 2.33 mg /100 g of zinc. The value of zinc at the bran supplementation of between 20 and 25% was not significantly different (P>0.05).

Micronutrient malnutrition greatly increases mortality and morbidity rates, diminishes cognitive abilities of children, lowers labor productivity and reduces the quality of life for all those affected. Deficiency of micronutrients, such as iron and zinc, is critical and major problem. It could be concluded that the addition of bran improves the nutritional quality of bread and could be a means of providing adult their daily requirements of iron and zinc. Although supplementation of bran improves the mineral content of the bread due to increase of phytic acid content, the availability of these minerals may be reduced. Therefore, such issues would need further evaluation.

Antioxidant capacity of wheat

The percentage yields of extracts were 7.58% w/w (Hard refined wheat), 8.2% w/w (Hard whole flour wheat) and 7.7% w/w (Soft refined wheat flour) and 7.3% w/w (soft whole flour wheat).

It has been recognized that the total phenolic content of plant extracts is associated with their antioxidant activities due to their redox properties, which allow them to act as reducing agents, hydrogen donors and singlet oxygen quenchers. Total phenolic content (TPC) was expressed as milligrams of Gallic acid equivalent (GAE) per gram (mg/g) of dry flour samples. As shown in Table 6, the total phenolic content (TPC) in whole wheat was highest. The TPC of white wheat flours (refined), which ranged from 3.34 to 3.49 mg GAE/g which were significantly lower (p<0.005) than those of whole wheat flours (range 7.66 to 8.20 mg GAE/g). However, the mean content did not vary much between whole hard and soft wheat type. Also there was no significant variation between soft and hard white wheat flour. The difference in the total phenolic content between whole and white wheat flour could be due to the process of milling. Research found antioxidants in wheat concentrated mostly in the aleurone layer of bran with some in the pericarp, nuclear envelope and germ [34,35].

The ability of wheat extracts to quench reactive species by hydrogen donation was measured through the DPPH radical scavenging activity test. The antioxidants can react with DPPH, a violet colored stable free radical, converting it into a yellow colored α,α-diphenil-β-picrylhydrazine. The discoloration of the reaction mixture can be quantified by measuring the absorbance at 517 nm, which indicates the radical-scavenging ability of the antioxidant. The antioxidant capacity whole and refined wheat was measured as the DPPH scavenging activity.

The DPPH radical scavenging effects of wheat methanol extracts was shown in Figure 2. As the concentration of sample increased, the percent inhibition of DPPH radical also increased [36]. At the concentration of 50 mg/mL, the scavenging effect of ascorbic acid, and wheat extracts, on the DPPH radical scavenging decreased in the order of L- ascorbic acid > soft whole > hard whole > soft white > hard white wheat flour, which were 92.53, 90.39, 89.89, 75.80, 57.57 % respectively. Therefore, the percentage of DPPH radical scavenging capacity of soft whole and hard whole wheat extracts are comparable with commercial antioxidants, L-ascorbic acid at concentration of 50 mg/mL. This suggested that whole wheat contain compounds that can donate electron/hydrogen easily and stabilizes free radicals.

The IC_{50} values of all the extracts were calculated from plotted graph of percentage scavenging activity against concentration of the extracts (Figure 2). The lower the IC_{50} value, the higher is the scavenging potential. The IC_{50} values ranged from 10.56 mg/mL for whole wheat extracts to 41.25 mg/mL for hard white wheat extracts. Strongest scavenging activity (lower IC_{50} values) was recorded for whole hard and soft wheat extracts which appeared more than four times stronger than that of hard white flour and two times stronger than that of soft white wheat extracts. IC_{50} value of ascorbic acid well known antioxidant was relatively more pronounced than that of the extracts (Figure 2). The results of this study demonstrate that the antioxidant content of wheat has been affected by the refined extraction/milling process. According

Samples	Iron mg/100g	Zinc mg/100g
Control (WFB)	1.98 ± 0.056 [a]	0.93 ± 0.064 [a]
WF:BR(90:10)	2.34 ± 0.159[b]	1.697 ± 0.108[b]
WF:BR (80:20)	2.95 ± 0.0753[c]	2.28 ± 0.131[c]
WF:BR (75:25)	4.83 ± 0.074[d]	2.33 ± 0.066[c]

WFB-(white wheat flour bread)- control, WF:BR 90:10, 10 % bran supplemented bread ,WF:BR 80:20- 20% bran supplemented bread, WF:BR 75:25 - 25% bran supplemented bread.

Data are average of triplicate ± SE Mean value with different superscript in the same column are significantly different (P<0.05)

Table 5: Mineral analysis of bread (Dry weight basis).

Sample	Yield % (g/100g)	Total Phenolics (mg GAE/g)
HWF (whole)	8.2	7.66 ± 0.70[b]
HWWF(refined)	7.58	3.49 ± 0.86[a]
SWWF (refined)	7.7	3.34 ± 0.14[a]
SWF(whole)	7.3	8.20 ± 0.35[b]

HWF- hard whole wheat flour, HWWF- Hard white wheat flour (refined), SWF -Soft whole wheat flour. SWWF-Soft white wheat flour (refined)

Data are average of triplicate ± SE. Values in the same column with different superscript are statistically significant (p< 0.05).

Table 6: Percent yield and Total Phenolics Content.

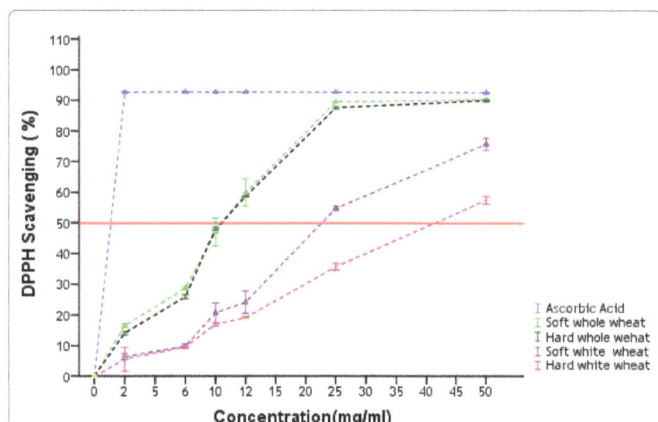

Figure 2: DPPH radical scavenging activity (%) of wheat extracts and standard (Values are average of triplicate measurements (mean ± SEM)).

to Fikreyesus [37] the DPPH (IC$_{50}$) for whole wheat flour is 15.56 which is different form the result obtained in this research. It is known that the antioxidant properties of wheat grain are significantly influenced by the genotype and environmental conditions [16]. To the best of our knowledge, there are no or few studies conducted on antioxidant content of Ethiopian wheat and particularly on comparison of antioxidant content on the whole and white wheat flour.

A relationship between phenolic content and antioxidant activity was extensively investigated, and both positive and negative correlations were reported. Bakchiche [38], Petra [39] and many other research groups stated that there was a positive correlation. However, a few evidences of no significant correlation were confronted [40]. In this study, the dependence of DPPH scavenging activity (%) in relation to the total phenolic content was also evaluated. The total phenolic content correlated significantly with DPPH scavenging activity (R^2=0.637, p<0.05). Thus the phenolics from the wheat extracts showed a good hydrogen-donating capacity, as well as high reactivity to free radicals, leading to the stabilization and termination of the radical chain reactions.

Sensory analysis of the bread

The sensory attributes of bread made from bran (Figure 3) using different ratio were evaluated using 7-point hedonic scale at Addis Ababa University; Center for Food Science and Nutrition by semi-trained panelists of first year M.Sc program students of Food Science and Nutrition stream and the mean scores of evaluated sensory attributes were presented in Table 7.

Taste is an important parameter when evaluating sensory attribute of food. The product without acceptable good test is likely to be unacceptable by consumers. The observed mean score of taste in experiential bran supplemented bread ranged from 4.43 to 5.93 (Table 7). Control (100% white wheat flour bread) had the highest mean sore in taste (5.93) followed by 10% bran supplemented bread (5.93). The 100% white wheat flour (control) bread had significant difference (p<0.05) with 10% bran (5.26), 20% bran bread (4.80) and 25% bran bread (4.43). The control bread had significant difference (p<5) from three of the bran supplemented bread (10%, 20%, 25%). The 10% bran supplemented bread scores >5 indicating that it is moderately likable by panelists and 20% and 25% bran supplemented bread were rated as neither like nor dislike by the panelists (scored as 4.80 and 4.43 respectively). This is due to the addition of the bran to the bread.

The mean score of odor of bread ranged from 4.76 to 5.53 (Table 7). Most of the samples were similar in odor while only 25% supplemented bread significantly (P<0.05) decreased. Three of the samples were liked moderately whilst the 25% bran supplemented bread was rated as neither like nor dislike.

As is shown in Table 7 colour of bread had low score as a result of increasing the level of wheat bran. The colour of control and 10% supplemented bread were similar in appearance while 20% and 25% bran supplemented bread were decreased (4.76 and 4.13) significantly (p<0.05). The results indicated that no significant difference (p>0.05) were observed by panelists between the control and 10% supplemented bread. The texture of the control and 10% bran supplemented bread were relatively most preferred (liked moderately) by the panelists. While the bread prepared from increasing level of bran supplement from 25% were scored as 4.5.

Generally, among the bread products, the control was highly acceptable by the panelists, with a score of 5.93. Next to this the 10% and 20% bran supplemented bread scored, 5.43 and 4.96 respectively. These two are liked moderately. The result obtained for 25% bran

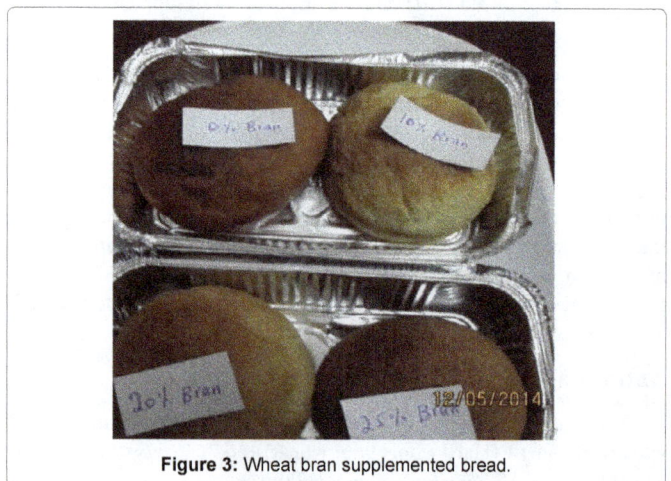

Figure 3: Wheat bran supplemented bread.

Sample	Test	Odour	Colour	Texture	Overall acceptability
Control (WFB)	5.93 ± 0.14[c]	5.53 ± 0.21[b]	5.93 ± 0.18[c]	5.70 ± 0.19[b]	5.93 ± 0.11[c]
WF:BR (90:10)	5.26 ± 0.16[b]	5.43 ± 0.18[b]	5.46 ± 0.14[c]	5.30 ± 0.13[b]	5.43 ± 0.15[b]
WF:BR (80:20)	4.80 ± 0.24[b]	5.33 ± 0.23[b]	4.76 ± 0.22[b]	4.60 ± 0.20[a]	4.96 ± 0.16[b]
WF:BR (75:25)	4.43 ± 0.30[a]	4.76 ± 0.22[a]	4.13 ± 0.24[a]	4.36 ± 0.23[a]	4.50 ± 0.22[a]

WFB-(white wheat flour bread) - control, WF:BR 90:10, 10% bran supplemented bread ,WF:BR 80:20- 20% bran supplemented bread, WF:BR 75:25 - 25% bran supplemented bread.

Data are average of triplicate ± SE Mean value with different superscript in the same column are significantly different (P<0.05).

Table 7: Sensory characteristics of wheat bran supplemented Bread.

supplemented bread was significantly different from the previous two. The latter was neither like nor disliked by the panelists and it scored 4.5. In relation to this Lazardiou [25] reported that those samples were considered as acceptable which their average score for the overall acceptability were greater than 4 which mean neither like nor dislike. Thus, whole wheat flour with high extraction rate (100%) needed to be given high emphasis by consumers because of its nutritional and antioxidant capacity of the product. Commercialization of wheat bran as a value added food ingredient to benefit from its high nutrition content.

Conclusion

Wheat and wheat products are important staple foods that are commonly consumed in Ethiopia. Consumption of whole grains as part of the diet is recommended for health reasons because they are good source of minerals, fibers, protein and antioxidants. There are no studies on the effect of refining on the nutritional content and antioxidant capacity of wheat grown in Ethiopia. This study showed that wheat extraction/refining at the lower rate have significantly reduces the proximate composition as well as the antioxidant content of wheat in both hard and soft wheat samples.

Addition of wheat bran to white wheat flour improves the nutritional value of the bread. Based on obtained results, the incorporation of wheat bran in the ratio of 10 to 20% showed better sensory acceptability though the proximate composition and mineral content increased at 25% bran supplemented bread. This indicates that bread of good nutritional and sensory qualities could be produced from 10% and 20% bran supplementation. The result of this study also indicated that wheat bran as a good source of minerals and fibers and can be used to supplement bread.

Acknowledgments

Authors would like to acknowledge Addis Ababa University for covering the research costs.

References

1. Curtis B, Rajaram S, Macpherson H (2002) Bread Wheat Improvement and Production. FAO, Rome.

2. Hussain A, Larsson H, Kuktaite R, Johansson E (2010) Mineral composition of organically grown wheat genotypes: contribution to daily minerals intake. International J Environ Res Public Health 7: 3442-3456.

3. Shewry P (2009) Wheat. Journal of Experimental Botany 60: 1537-1553.

4. MOA (2011) Animal and Plant Health Regulatory Directorate Crop Variety Register. No.4. Addis Ababa, Ethiopia.

5. CSA (2005) Report on area and production of major crops. Statistical Bulletin. Addis Ababa, Ethiopia.

6. CSA (2011) Agricultural Sample Survey Report-Production. Federal Democratic Republic of Ethiopia, Central Statistical Agency Addis Ababa: Ethiopia.

7. Rashid S, Getnet K, Lemma S (2010) Maize value chain potential in Ethiopia. IFPRI.

8. Kumar P, Yadava R, Gollen B, Kumar S, Verma R, et al. (2011) Nutritional contents and medicinal properties of wheat: A Review. Life Sciences and Medicine Research 22: 1-10.

9. Ragaee S, Guzar I, Abdel-Aal E-SM, Seetharaman K (2012) Bioactive components and antioxidant capacity of Ontario hard and soft wheat varieties. Canadian Journal of Plant Science 92: 19-30.

10. Mellen P, Walsh T, Herrington D (2008) Whole grain intake and cardiovascular disease: A meta-analysis. Nutr Metab Cardiovasc Dis 18: 283-290

11. Banu I, Stoenescu G, Ionescu VS, Aprodu I (2012) Efect of the Addition of wheat bran stream on dough rheology and bread quality. The Annals of the University Dunarea de Jos of Galati Fascicle VI - Food Technology 36: 39-52.

12. Haque A, Ud-Din S, Anwarul H (2002) The Effect of Aqueous Extracted Wheat Bran on the Baking Quality of Biscuit. International Journal of Food Science and Technology 37: 453-462.

13. Cubadda F, Aureli F, Raggi A, Carcea M (2009) Effect of milling, pasta making and cooking on minerals in durum wheat. Journal of Cereal Science 49: 92-97.

14. Bergh K, Chew A, Gugerty MK, Anderson CL (2012) Wheat Value Chain: Ethiopia. University of Washington: Evans School Policy Analysis and Research (EPAR), No 204.

15. FDRE (2011) Ministry of Health Assessment of Feasibility and Potential Benefits of Food Fortification in Ethiopia. Addis Ababa.

16. Adom K, Sorrells M, Liu R (2003) Phytochemical profiles and antioxidant activity of wheat varieties. Journal of Agricultural and Food Chemistry 51: 7825-7834.

17. AACC (2000) Approved Methods of American Association of Cereal Chemists. Am Assoc Cereal Chem (15thedn). Arlington, USA.

18. Slavin JL, Jacobs D, Marquart L (2000) Grain processing and nutrition. Crit Rev Food Sci Nutr 40: 309-326.

19. Hertzberg J, Francois Z (2007) Artisan Bread in Five Minutes a Day: The Discovery that Revolutionizes Home Baking, St. Martin's Press, New York.

20. AOAC (2000) Official Methods of Analysis. The Association of Official Analytical Chemists. (17thedn), Arlington, USA.

21. AOAC (1990) Official Methods of analysis. Association of Official Analytical Chemists. (15thedn), Washington DC.

22. Woldegiorgis A (2014) Antioxidant property of edible mushrooms collected from Ethiopia. Food Chemistry 157: 30-36.

23. Gursoy N, Sarikurkeu C, Tepe B, Solak M (2010) Evaluation of Antioxidant Activities of 3 Edible Mushrooms: Ramaria flava (Schaef: Fr.) Quel, Rhizopogon roseolus (Corda) TM Fries, and Russula delica Fr. Food Science and Biotechnology 19: 691-696.

24. Singleton VL, Rossi JA (1965) Colorimetry of total phenolics with phospho molybdic phosphotungstic acid reagents. American Journal of Enology and Viticulture 16: 144-158.

25. Lazaridou A, Duta D, Papageorgiou M, Belc N, Biliaderis C, et al. (2007) Effects of hydrocolloids on dough rheology and bread quality parameters in gluten-free formulations. Journal of Food Engineering 79: 1033-1047.

26. Azizi MH, Sayeddin S, Payghambardoost S (2006) Effect of flour extraction rate on flour composition, dough rheology characteristics and quality of flat breads. Journal of Agricultural Science and Technology 8: 323-330.

27. Farooq Z, Rehman S, Bilal MQ, Butt MS (2001) Suitability of wheat varieties/lines for the production of leavened flat bread (Naan). J Res Sci 12: 171-179.

28. Blakeney A, Cracknell R, Crosbie G, Jefferies S, Miskelly D, et al. (2009) Understanding Wheat Quality: A basic introduction to Australian wheat quality. Wheat Quality Objectives Group. Kingston, Australia.

29. Posner ES (2000) Wheat. In: Kulp K, Ponte Jr, Handbook of Cereal Science and Technology, (2ndedn) Marcel Dekker Inc, New York.

30. Butt B, Ihsanullah M, Anjum FM, Aziz A, Randhawa AM, et al. (2004) Development of minerals enriched brown flour by utilizing wheat milling by-products. Internet Journal of Food Safety 3: 15-20.

31. Dewettinck K, Bockstaele FV, Kuhne B, Walle DV, Courtens T, et al. (2008) Nutritional value of bread: Influence of processing, food interaction and consumer perception. Journal of Cereal Science 48: 243-257.

32. Hoseney RC (1992) Principles of Cereal Science and Technology. AACC, Minnesota.

33. Lopez HW, Krespine V, Lemairet A, Coudray C (2003) Wheat variety has a major influence on mineral bioavailability: studies in rats. Journal of Cereal Science 37: 257-266.

34. Fulcher R, Duke T (2002) Whole-grain structure and organization: Implications for nutritionists and processors. American Association of Cereal Chemists, St. Paul, MN, 9-45.

35. Zilic S, Serpen A, Akillioglu G, Jankovic M, Gokmen V, et al. (2012) Distributions of phenolic compounds, yellow pigments and oxidative enzymes in wheat grains and their relation to antioxidant capacity of bran and debranned flour. Journal of Cereal Sciences 56: 652-658.

36. Haung D, Ou B, Prior RL (2005) The chemistry behind antioxidant capacity assays. Journal of Agricultural and Food Chemistry 53: 1841-1856.

37. Fikreyesus S, Vasantha H, Astatke T (2013) Antioxidant capacity, total phenolics and nutritional content in selected Ethiopian staple food ingredients. Int J Food Sci Nutr 64: 915-920.

38. Bakchiche B, Gherib A, Smail A, Custodia G, Grac M, et al. (2013) Antioxidant activities of eight Algerian plant extracts and two essential Oils. Industrial Crops and Products 46: 85- 96.

39. Petra T, Barbara C, Natasa P, Helena A (2012) Studies of the correlation between antioxidant properties and the total phenolic content of different oil cake extracts. Industrial Crops and Products 39: 210- 217.

40. Ebrahimzadeh MA, Pourmorad F, Bekhradnia AR (2008) Iron chelating activity, phenol and flavonoid content of some medicinal plants from Iran. African Journal of Biotechnology 7: 3188-3192.

3

Detection of *invA* Gene in Isolated Salmonella from Marketed Poultry Meat by PCR Assay

Indu Sharma* and Kashmiri Das

Microbiology Department, Assam University, Silchar-788011, India

Abstract

Background: The aims of the current study were to detect the *invA* gene from chicken samples meant for human consumption from N.E India.

Materials and method: After *Salmonella* sp. identification with culture method, PCR assay were developed for detection of pathogenic genes and antibiotic resistance genes of *Salmonella* sp.

Results: *Salmonella* was detected in 80 samples of poultry carcasses from main poultry markets in Silchar, Assam, N.E India. A total of 40 *Salmonella* isolates were found in chicken samples (43%) and the isolates had a growth on brilliant green agar and De-oxycholate citrate agar medium, were oxidase negative and catalase positive and exhibited no changes in the colour of the medium with 100% motility. All the strains were subjected to Salmonella-specific gene (*invA*) and were confirmed as *Salmonella* positive by the predicted product of 284-bp DNA fragment. Salmonella isolates recovered from poultry samples were tested for antibiotic susceptibility against 5 selected antibiotics of which ciprofloxacin was observed to be highly susceptible (77.5%).

Conclusion: Our results recommended the use of PCR for detection of pathogenic genes of bacteria as a safe, rapid, and accurate method in laboratories. High levels of Salmonellosis infections in poultry farms has raised an eye amongst the poultry management personnel's to consider various effective control programs to prevent the economic loss resulting from mortality and spreading of infection.

Keywords: Poultry meat; Salmonella; PCR; *InvA* gene

Introduction

Poultry provides an immense supply of food for the entire world's population where poultry meat and eggs in majority are all time preferred to other kinds of animal food products for a variety of reasons, but most of these have been implicated as a major source of *Salmonella* infection in human [1]. Today, reports on mishandling of poultry products and raw poultry carcasses, uncooked poultry meat are one of the frequent causes of human infection caused by *Salmonella* species [2]. Also, a number of *Salmonella* serotypes can be transferred to poultry from sources, such as feedstuffs, breeding flocks, rodents, and wild birds, etc. of which may act as an important hindering factor for the economy and growth of a country. Thus in the present time, several new techniques in this regard have been developed for the rapid detection of *Salmonella* serovars such as selective culture medium and enzyme-linked immunosorbent assay [3]. However, today the Polymerase Chain Reaction (PCR) method has emerged as a powerful, rapid and a reliable tool for detection and identification of food- borne pathogens such as *Salmonella*. Malorny [4] where several chromosomal genes including *invA* are target genes for PCR amplification of *Salmonella* species [5]. The *invA*, gene of *Salmonella* contains those sequences that are unique to this genus and has been proved as a suitable PCR target with potential diagnostic applications [6]. Thus, keeping the above in view, the present investigation highlights the screening of poultry meat samples for *Salmonella* sp. and their confirmation by *invA* specific PCR methods.

Materials and Methods

Clinical isolates and phenotypic identification

A total of 80 poultry samples were collected from various parts of Silchar, Cachar District, Assam, comprising of meat, liver, intestine scrapings and faecal materials. The samples mostly belonged to highly crowded areas, slaughter houses, swabs from butchers knife used for cutting chicken. Samples were aseptically enriched in selenite F broth (Himedia) and incubated at 37°C for 24 hours. A loopful from the broth was then streaked on brilliant green agar (Himedia) and De-oxycholate citrate agar (Himedia) for further incubation at 37°C for 24 hours. The distinct isolated colonies were identified on the basis of their morphological and biochemical characters. The shape and colours of the colonies were examined under the microscope after Gram staining. Isolates were biochemically confirmed for the activities of oxidase, catalase, methyl red, voges-prauskauer test, urease, motility, indole production, citrate utilization, TSI and sugar fermentation tests through a series of conventional biochemical tests. Colonies that depicted biochemical reactions like *Salmonella* were transferred to nutrient agar slant (Himedia) and incubated at 37°C for 24 hrs.

Antibiotic susceptibility test

Antibiotic susceptibility of the isolates was assayed according to Kirby-Bauer disc diffusion method on Mueller-Hinton agar (MHA) [7]. The isolates were freshly inoculated on saline water for detection of turbidity via comparison with 0.5 Mc. Farland solution (1.5 x 108 CFU/ml). The antibiotic disks used were of HI-MEDIA Laboratories, Mumbai, India, consisting of the following: Kanamycin (30 μg), Ciprofloxacin (5 μg), Ampicillin (25 μg), Norfloxacin (10 μg) and Nalidixic acid (30 μg). Positive tests were indicated by zones of

---

*Corresponding author: Indu Sharma, Microbiology department, Assam University, Silchar-788011, India, E-mail: drsharma7652@gmail.com

inhibition which were measured by using the zone size interpretative tables provided by the manufacturer of the discs.

Genotypic screening of *invA* gene from isolated strain by PCR assay

DNA of *Salmonella* isolates was extracted and purified using Genomic DNA Mini Kit (cultured cell).

Primers set and PCR amplification program: The primers (Forward and Reverse) used in PCR were specific for *invA* gene which was chosen according to Rahn [6]. The sequence of primer used in this study is presented in Table 1.

Electrophoresis of PCR products

PCR for amplification of *invA* gene was performed using 12.5 µl of Tag Green Master Mix 2x DNA polymerase [8], specific primers for *invA* gene (1 µl forward and 1 µl reverse), 2 µl of DNA extract as a template, 8.5 µl of nuclease free water to make the final reaction volume up to 25 µl. The PCR conditions started with thermocycler (Eppendorf) program that is initial denaturation at 95°C for 2 minutes, 34 cycles of denaturation at 95°C for 15 seconds, 45°C for 1 minute, and 72°C for 45 seconds and final extension at 72°C for 7 minutes.

Agarose gel electrophoresis

Five microliters of the reaction mixture and 1 µl marker DNA (100 bp DNA ladder mix; MBI Fermentas, USA) mixed separately with 1 µl of 6X gel loading dye (MBI Fermentas, USA) and were analyzed by submarine gel electrophoresis in 1.2% agarose (Hi-Media, Mumbai, India) at 60 V for 1 hour and 20 min. or until the second dye marker had run 3/4th of the gel. The reaction products were visualized with UV light after staining with ethidium bromide. The identities of the amplicons were confirmed by comparison of the amplicon sizes with the predicted sizes and photographed [8].

Results

A total of 80 chicken samples were procured for this study, of which 40 isolates of *Salmonella* sp. were recovered. Table 2 represents the total number of isolates and their percentage.

All the 40 isolates had a growth on brilliant green agar and De-oxycholate citrate agar medium, were oxidase negative and catalase positive and exhibited no changes in the colour of the medium. 100% motility was observed in all the isolates; indole and methyl-red tests were observed as positive and voges-Proskauer test as negative; all of them represented citrate utilization, urease test as negative, were non-lactose fermenters on MacConkey agar medium and fermented glucose, fructose and maltose. TSI was observed as K/A^{G+H2S}+. All the 40 isolates were identified as *Salmonella* organisms.

Forty *Salmonella* isolates recovered from poultry samples were tested for antibiotic susceptibility against 5 selected antibiotics. Percentages of *Salmonella* isolates susceptible to antibiotics are represented in Table 3.

Forty *Salmonella* strains were isolated from poultry specimens, by culturing in selenite F and then transferring to Salmonella-Shigella agar, when subjected to Salmonella specific-PCR using primers *invA* F and *invA* R where all isolates including positive control and a single 284 bp amplified DNA fragment, on 1.2% agarose gel (Figure 1).

Discussion

The present study evaluated the microbial quality of poultry meat sold at various retail markets of Silchar, were heavily contaminated with *Salmonella* sp. (43%) [9-11]. Presence of bacteria in meat is of great public health concern [12] indicating a potential breakdown of hygiene at various stages of the food processing and distribution chain and/or a lack of refrigeration of meat. As observed during the course of this study, the methods of slaughtering of animal are responsible for this microbial contamination. Traditional method of butchering using knives and cutting lines appears more capable of minimizing faecal contamination then modern mechanized system which are manned by a team of operators. This was inferred because poultry samples collected from the local abattoir were less contaminated, than those collected from the areas where modern equipment is used [13].

Biochemical reactions are very important for speciating the isolates, while isolates could get a correct identification only when based on genetic methods. In the present study, all the 40 (43%)

Name of primer	Target gene	Primer sequence (5'-3')	Amplified product size	Reference
InvA F	invA	GTGAAATTATCGCCACGTTCGGGCAA	284 bp	Rahn et al. [15]
InvA R		TCATCGCACCGTCAAAGGAACC		

Table 1: Sequences of primers for *invA* gene.

SL No.	Samples from different parts of poultry	Samples procured	Total no. of isolates isolated	Percentage of *Salmonella sp*
	Meat	29	15	52%
	Liver	23	12	52%
	Scrapping	11	5	45%
	Fecal	17	8	47%
	Total	80	40	43%

Table 2: Percentage of *Salmonella* sp. contamination in poultry chicken samples from the different areas of Silchar.

Antibiotics	Total susceptible isolates	% of susceptible isolates
Ampicillin	25	62.5%
Ciprofloxacin	31	77.5%
Kanamycin	30	75%
Norfloxacin	23	57.5%
Nalidixic acid	26	65%

Table 3: Percentages of *Salmonella* isolates susceptible to antibiotics from poultry samples.

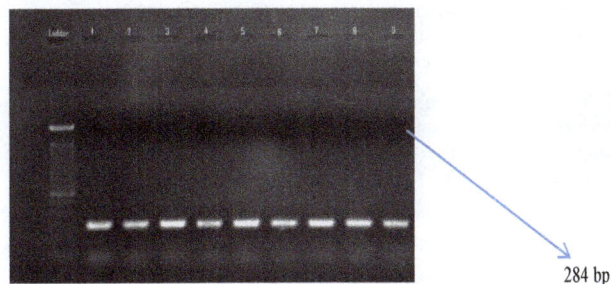

Figure 1: Gel image showing amplification of *inv*A gene. Ladder used is 100 bp ladder. Lanes 1,2,3,4,5,6,7,8 and 9 exhibits the amplified *inv*A gene (284 bp).

isolates exhibited typical biochemical characteristics of *Salmonella* on the basis of IMViC reaction, gas production and sugar fermentation as per standard techniques. All the isolates tested negative oxidase, positive catalase, positive indole in typtone broth, positive methyl red, negative Voges-Proskauer, urease negative and citrate positive tests and fermented glucose, fructose and maltose.

Today, the frequencies of bacterial strains resistant to antimicrobial agents have increased dramatically in the environment as a consequence of the wide spread use of drugs [14]. A significant public health concern and the possibility of transfer of resistant genes between bacteria in the natural habitats have attracted attention. In the present study, antimicrobial susceptibility pattern against 5 antibiotics was carried out for 40 *Salmonella* isolates of which ciprofloxacin was 77.5% susceptible, whereas kanamycin exhibited (75%) susceptibility, nalidixic acid (65%), ampicillin (62.5%) and norfloxacin (52.5%). It was observed that norfloxacin which was routinely used in present prognosis was the least susceptible antimicrobial agent found in the present study. Also, this is the first report on the prevalence of *Salmonella*, determining its antimicrobial-resistance pattern along with detection of *inv*A gene of *Salmonella* from poultry farms and slaughter houses of this geographical area. In the current study, the overall prevalence of *Salmonella* and fecal shedding of *Salmonella* was lower in organized farms than the slaughter houses. Despite the use of selective broth, we recovered higher proportions of *Salmonella* (43%) from the both clinical as well as post-mortem samples.

The present study supports the ability of these specific primer sets to confirm the isolates as *Salmonella*. All PCR products of isolates included positive control, screened by PCR, resulted in 284 bp amplified fragment. No amplified DNA fragments were obtained from non-Salmonella species. The ability of *Salmonella* specific primers to detect *Salmonella* species rapidly and accurately is primarily due to the primer sequences that are selected from the gene *inv*A [15]. The *inv*A gene codes for protein in inner membrane of bacteria, which is necessary for invasion to epithelial cells [16]. The PCR assay carried out in this study for the detection of the *inv*A genes in 40 *Salmonella* isolates from poultry meat samples was present in 22 isolates (55%). These findings are in agreement with the earlier reports [17-20]. Furthermore it has been observed that this gene is involved in the invasion of the cells of the intestinal epithelium and is present in pathogenic *Salmonella*. Therefore for salmonellosis to occur it is important that a gene responsible for invasion must be present. This gene is essential for full virulence in *Salmonella* and is thought to trigger the internalization required for invasion of deeper tissue [21].

Conclusion

High levels of Salmonellosis infections in poultry farms has raised an eye amongst the poultry management personnel's to consider various effective control programs to prevent the economic loss resulting from mortality and spreading of infection. In the present work, PCR based methods with genus-specific primers belonging to *inv*A, proved to be quick, specific and sensitive and certain in identification and confirmation of *Salmonella* isolates.

Acknowledgment

We are grateful to the Department of Microbiology, Assam University in regard to financial assistance for completion of this investigation.

References

1. Amavisit P, Browning GF, Lightfood D, Anderson CS (2001) Rapid PCR detection of Salmonella in horse faecal samples. Vet Microbiol 79: 63-74.

2. Antunes P, Reu C, Sousa JC, Peixe L, Pestana N (2003) Incidence of Salmonella from poultry products and their susceptibility to antimicrobial agents. Int J Food Microbiol 82: 97-103.

3. Bauer AW, Kirby WM, Sherris JC, Turck M (1966) Antibiotic susceptibility testing by a standardized single disk method. Am J Clin Pathol 45: 493-496.

4. Baumgartner A, Heirnann P, Schmid H, Liniger M, Simmen A, et al. (1992) Salmonella contamination of poultry carcasses and human salmonellosis. Archiv fuer Lebensmittelhygiene 43: 123-124.

5. Darwin KH, Miller VL (1999) Molecular basis of the interaction of Salmonella with the intestinal mucosa. Clin Microbiol Rev 12: 405-428.

6. Jamshidi A, Bassami MR, Afshari-Nic S (2009) Identification of Salmonella spp and Salmonella typhimurium by a multiplex PCR-based assay from poultry carcasses in Mashhad-Iran. Int J Vet Res 3: 43-48.

7. Jerngklinchan J, Koowatananukul C, Daengprom K, Saitanu K (1994) Occurrence of salmonellae in raw broilers and their products in Thailand. Journal of Food Protection 57: 808-810.

8. Khan AA, Cheng CM, Khanh TV, Summage-West C, Nawaz MS, et al. (2006) Characterization of class 1 integron resistance gene cassettes in Salmonella enterica serovars Oslo and Bareilly from imported seafood Journal of Antimicrobial Chemotherapy 58: 1308-1310.

9. Kinsella KJ, Prendergast DM, McCann MS, Blair IS, McDowell DA, et al.(2009) The survival of Salmonella enterica serovar Typhimurium DT104 and total viable counts on beef surfaces at different relative humidities and temperatures. J Appl Microbiol 106: 171-180.

10. Lampel KA, Orlandi PA, Kornegay L (2000) Improved template preparation for PCR-based assay for detection of food-borne bacterial pathogens. Appl Environ Microbiol 66: 4539-4542.

11. Malorny B, Hoorfar J, Bunge CR, Helmuth R (2003) Multicenter validation of the analytical accuracy of Salmonella PCR: towards an international standard. Appl Environ Microbiol 69: 290-296.

12. Nagappa K, Tamuly S, Brajmadhuri, Saxena MK, Singh SP, et al. (2007) Isolation of Salmonella Typhimurium from poultry eggs and meat of Tarai region of Uttaranchal. IJBT 6: 407-409.

13. Nashwa MH, Mahmoud AH, Sami SA (2009) Application of multiplex polymerase chain reaction (MPCR) for identification and characterization of Salmonella Enteritidis and Salmonella Typhimurium. Journal of Applied Sciences Research 5: 2343-2348.

14. Panisello PJ, Rooney R, quantick PC, Stanwell-Smith R (2000) Application of foodborne disease outbreak data in the development and maintenance of HACCP system. International Journal of Food Microbiology 59: 221-234.

15. Rahn K, De Grandis SA, Clarke RC, Curtiss R, Gyles CL, et al. (1992) Amplification of an invA gene sequence of Salmonalla typhimurium by polymerase chain reaction as a specific method of detection of Salmonella. Molecular and Cellular Probes 6: 271-279.

16. Shanmugasamy M, Velayutham T, Rajeswar J (2011) invA gene specific PCR for detection of Salmonella from broilers. Veterinary World 4: 562-564.

17. Trafny EA, Kozlowska K, Szpakowska M (2006) A novel Multiplex PCR assey for the detection of Salmonella enteric serovar Enteritidis in human faeces. Letters in Applied Microbiology 43: 673-679.

18. Uyttendaele MR, Debevere CM, Lips RM, Neyts KD (1998) Prevalence of Salmonella in poultry carcasses and their products in Belgium. International Journal of Food Microbiology 40: 1-8.

19. Whyte P, Meglli K, Collins D, Gormely E (2002) The prevalence and PCR detection of Salmonella contamination in raw poultry. Veterinary Microbiology 89: 53-60.

20. Xu J, Moore JE, Murphy PG, Millar BC, Elborn JS, et al. (2004) Early detection of Pseudomonas aeruginosa-comparison of conventional versus molecular (PCR) detection directly from adult patients with cystic fibrosis (CF). Annals of clinical Microbiology and Antimicrobials.

21. Zahraei T, Mahzoonae MR, Ashrafi A (2006) Ampllification of invA gene of Salmonella by polymerase chain reaction (PCR) as a specific method for detection of Salmonella. Journal of the Faculty of Veterinary Medicine University of Tehran 6: 195-199.

4

Effect of Almond and Pistachio Juices Processing By-Products on Physicochemical, Sensorial and Textural Properties of Cookies

Assaâd Sila[1,2]*, Nadia Bayar[1], Imen Ghazala[1], Nadhem Sayari[1], Ellouz-Chaabouni S[1,3], Ali Bougatef[1,4] and Ellouz-Ghorbel R[1]

[1]*Unité Enzymes et Bioconversion, Ecole nationale d'Ingénieurs de Sfax, Université de Sfax, Tunisie*

[2]*Institut Régional de Recherche en Agroalimentaire et Biotechnologie: Charles Viollette, Equipe ProBioGEM, Université Lille 1, France*

[3]*Unité de service commun bioréacteur couplé à un ultrafiltre, Ecole nationale d'Ingénieurs de Sfax, Université de Sfax, Tunisie*

[4]*Institut Supérieur de Biotechnologies de Sfax, Département de Technologies Alimentaires, Tunisie*

Abstract

The aims of the present work were to study the composition of almond and pistachio juices processing by-products and their use in cookies formulation. These by-products have relatively high carbohydrate, protein, fat, calcium and magnesium contents. There were slight difference ($p > 0.05$) in essential amino acid contents between almond by-products and pistachio by-products and also both juices processing by-products showed contains polyunsaturated fatty acids (22.43% in almond juices by-products processing (ABP) and 20.83% in pistachio juices by-products processing (PBP). The incorporation of ABP and PBP in the formulation of cookies did not affect physical and textural parameters. This study suggested that almond and pistachio juices processing by-products may be alternative additives in food, pharmaceuticals and cosmetic preparations.

Keywords: Almond; Pistachio; By-products; Nutritional quality; Cookies formulation

Introduction

Nowadays, there is great political and social pressure to reduce the pollution arising from industrial activities. Almost all developed and undeveloped countries are trying to adapt to this reality by modifying their production processes and often contain undesirable contaminants so that their residues can be recycled. During the last ten years there has been an increased consciousness of environmental protection. Agro-industrial by-products present potential air and water pollution problems. High-moisture wastes are also difficult to burn [1]. Indeed, Agro-industries are major contributors to the worldwide industrial pollution problem. With the tremendous pace of technology development, substantial research is devoted to cope with wastes of ever increasing complexity generated by agro-industries [1,2]. Therefore, agro-industries more than any other industrial sector in this field, require a dynamic and comprehensive approach for appropriate waste management.

Juice processing generates large amounts of wastes, which create burdensome disposal problems and environmental concerns. The utilization of these by-products, which has a wide range of nutritional values, may be economically worthy [3]. This biowaste contains, however, several biomass materials that can be exploited to produce useful marketable products. It contains large amounts of biomolecules with high added-value that can be used for the formulation and development of various pharmaceutical, medicinal, and nutritional products. In fact, agro-waste has traditionally been transformed into flour which is used as fertiliser or animal feed [3]. The low value-added flour, from agro-industrial by-products, led scientists to search for new alternatives to add value to these by-products.

The aims of this work were to characterize the almond and pistachio juices processing by-products and to evaluate their physicochemical and textural effects on characteristics of cookies.

Materials and Methods

Reagents

All chemicals and reagents used were of analytical grade. Water was obtained from a Culligan system; the resistivity was approximately 18 MΩ. Commercial ingredients were used for preparation of cookies.

Materials

The pistachio (*Pistacia vera*) and bitter almond (*Prunus amygdalus var amara*) by-products were obtained in fresh condition from a juices processing plant located in Sfax, Tunisia. The samples were packed in polyethylene bags, placed in ice with a sample/ice ratio of approximately 1:1 (w/w) and transported to the research laboratory within 30 min.

Almond and pistachio juices extraction

Almond and pistachio fruits were supplied from a local market in Tunisia. Fruits were washed with water and grinding. Almond and pistachio fruits powder were soaked in water for 1 or 2 hours and finally filtered with a 50-mesh sieve. The juice samples were then transferred into 200 ml hermetically capped glass bottles. The residue (almond and pistachio juices processing by-products) was stored in sealed plastic bags at -20°C.

Determination of chemical composition

The moisture and ash content were determined according to the AOAC standard methods 930.15 and 942.05, respectively [4]. Total nitrogen content was determined by using the kjeldahl method [5]. Crude protein was estimated by multiplying total nitrogen content by the factor of 5.18 for bitter almond (F.A.O) and 6.25 for pistachio [6]. Crude fat was determined gravimetrically after soxhlet extraction of dried samples with hexane [5]. The total sugars were determined by the

---

*Corresponding author: Assaâd Sila, Unité Enzymes et Bioconversion, Ecole nationale d'Ingénieurs de Sfax, Université de Sfax, 3038 Sfax, Tunisie
E-mail: assaadsila@gmail.com

phenolesulphuric acid method [7]. All measurements were performed in triplicate.

Determination of mineral contents

Analyses of calcium (Ca), magnesium (Mg), sodium (Na) and potassium (K) contents in freeze-dried samples were carried out using the inductively coupled plasma optical emission spectrophotometer (ICP-oes) (Model 4300 dv, Perkin elmer, Shelton, Ct, USA) according to the method of AOAC [4]. Sample (1 g) was mixed well with 1 ml of 70% nitric acid. The mixture was heated on the hot plate until digestion was completed. The digested sample was transferred to a volumetric flask and the volume was made up to 10 ml with deionized water. The solution was then subjected to analysis. The solution was then subjected to analysis. The concentration of minerals was calculated and expressed as mg/kg sample.

Determination of dietary fibers

Insoluble and soluble dietary fibres were determined according to the AOAC enzymatic-gravimetric method of Prosky et al. [8]. Briefly, the defatted samples were gelatinized with heat stable alpha amylase (100°C, pH 6) for 15 min and then enzymatically digested with protease (60°C, pH 7.5) for 30 min, followed by incubation with amyloglucosidase (60°C, pH 4.5) for 30 min, to remove protein and starch. Then, the samples were filtered, washed (with water, 95% ethanol and acetone), dried and weighted to determine insoluble fibre. Four volumes of 95% ethanol (preheated to 60°C) were added to the filtrate and to the water washings. Then, the precipitates were filtered and washed with 78% ethanol, 95% ethanol and acetone. After that, the residues (soluble fibre) were dried and weighted. The obtained values were corrected for ash and protein. Total dietary fibre was determined by summing insoluble dietary fibre and soluble dietary fibre.

Determination of starch contents

After removing sugars with ethanol (80%), starch was isolated by extraction with perchloric acid reagent (52%) twice, from a sugar-free residue according to the method described by McCready et al. [9]. Starch in the extract was determined using the anthrone reagent and colorimetric measurement at 630 nm.

Total phenolic content

Total phenolic contents were determined according to Escarpa and González [10]. 5 g of samples were extracted for 1 h with 10 ml of methanol at room temperature and in darkness. The mixture was centrifuged at 4186 g for 15 min. The pellet was extracted for 30 min with 10 ml of methanol and then with 5 ml for 30 min. The extracts were adjusted to a final volume of 25 ml using methanol. The total concentration of phenol in the extract was determined according to the Foline-Ciocalteu method [11]. Results were expressed as milligrams of gallic acid equivalents (GAE).

Amino acid analysis

Samples were dissolved (1 mg/ml) in ultrapure water and the boiled with 6 N HCl containing 0.1% phenol and norleucine (Sigma–Aldrich, Inc., St. Louis, MO, USA) as internal standard. HCl was removed under vacuum after 24 h of hydrolysis 110°C. Dried samples were reconstituted in application buffer and injected in application buffer and injected onto a Biochrom 20 amino acid analyser (Pharmacia, Barcelona, Spain).

Fatty acids analysis

Samples were dissolved in 0.5 ml of hexane. Then, 0.2 ml of potassium hydroxide in methanol (2N) was added for the fatty acid methylation process. The mixture was vortexed then centrifuged and the upper phase containing fatty acid methyl esters were analyzed by Gas Chromatography (GC). GC analyses were performed on a Shimadzu, GC 17 A chromatograph, equipped with a flame ionization detector and a capillary column (50 m × 0.32 mm × 0.5 mm, PERICHROM Sarl, France). The oven temperature was programmed as follows: the initial temperature (100°C) was raised to 150°C at a rate of 30°C/min and held at this temperature for 5 min. The temperature was then increased to 190°C (at 10°C/min) and maintained for 14 min before being increased (at 5°C/min) to 255°C and held for 10 min. The injector and detector temperatures were 255 and 70°C, respectively. Nitrogen was the carrier gas with a flow rate of 1.13 ml/min. The identification of fatty acids was achieved by comparing retention times with those of authentic standards analysed under the same conditions. Peak areas were measured with an HP computing integrator. Results which are means of triplicates were expressed as w/w percentage of total fatty acids by Sagdiç et al.

Preparation of cookies

Formulation: All ingredients used in the cookies' formulation were commercially available (Table 1). All cookies were prepared according to AACC International Method 10-53.01 [12]. After baking at 200°C for 11 min, cookies were removed from the oven and kept on the baking sheet for 5 min. Cookies were then removed from the baking sheet and placed on a wire cooling rack for 25 min. Once the cookies reached room temperature, they were used in subsequent experimental analysis.

Textural analysis: For one-cycle compression, a Texture Analyzer (LLOYD instruments, Fareham, UK) was used. All experiments were conducted in a controlled temperature room at 25°C. A cylinder probe (diameter 19 mm) compressed samples, placed in a plastic food container at fixed quantity. The crosshead was allowed to descend at the rate of 40 mm/min to a total compression at 20 mm in all experiments. The instrument automatically recorded the forceetime curve. All operations were automatically controlled by the texture "Nexygen Lot" software connected to the instrument Firmness (g) is the peak force of the compression of the product and was measured by the Nexygen MT machine software [13]. Three replicates were conducted at each sample. Volume and propagation rate of cookies were determined according to the AOAC standard methods [12]. Weight loss was determined by weight difference before and after cooking.

Sensory analysis: The products were evaluated by an untrained panel consisting of thirty panellists (11 males and 19 females) from the students and the staff members of the National School of Engineer

Ingredient	Weight (g)				
	Control	Cookie A	Cookie B	Cookie C	Cookie D
Flour	150	100	75	100	75
Sugar	85	85	85	85	85
Shortening	85	85	85	85	85
Sodium bicarbonate	1.3	1.3	1.3	1.3	1.3
Salt	1	1	1	1	1
Water	15	15	15	15	15

Cookie A + B: Substitution of flour by almond by-products.
Cookie C + D: Substitution of flour by pistachio by-products.

Table 1: Cookies baking formulation.

(Sfax, Tunisia). Their ages ranged from 22 to 50 years. The cookies were evaluated based on a six-point hedonic scale, where one represented "disliked extremely" and seven represented "liked extremely". Each consumer was given 7 samples labelled with random 3-digit codes simultaneously and asked to evaluate the cookies for appearance, taste, color, odour and texture. The mean value of these sensory proprieties was evaluated as overall acceptability

Microbiological analysis: Specimen of each sample were suspended in sterile 0.1% (w/v) peptone-water solution and homogenized at 350 rpm for 2 min at room temperature. Total floral bacteria were determined after 48 h of incubation at 30°C using a Plate Count Agar (PCA) medium. Fecal coliforms were determined on a desoxycholate citrate agar lactose (DCL) medium after 48 h of incubation at 44°C. Staphylococci were determined on a Baird–Parker agar medium after 48 h of incubation at 37°C. Moulds were determined after cooking as reported by Guiraud and Galzy (1980). All measurements were performed in triplicates.

Statistical analysis

All experiments were carried out in triplicate, and average values with standard deviation errors are reported. Mean separation and significance were analyzed using the SPSS software package (SPSS, Chicago, IL). Correlation and regression analysis was carried out using EXCEL program.

Results and Discussion

Proximate composition

The composition of almond juices processing by-products (ABP) and pistachio juices processing by-products (PBP) is shown in Table 2. There were not significant differences ($p<0.05$) in moisture and fat contents between almond by-products (25.92 and 45.56%, respectively) and pistachio by-products (26.41 and 43.87%, respectively). Fat represent the major component for both juices processing by-products, similar results were previously reported by Prosky et al. [8] for Tunisian almond (45.56%) and by Ghrab et al. for Tunisian pistachio (50%). The PBP showed higher protein content (19.68 ± 3.51%) when compared ($p<0.05$) to the ABP (12.81 ± 2.1%) that indicates their possible use in different food formulations for improving functional properties. As expected, the contents in total sugars the PBP (21.59 ± 2.35%) was significantly lower ($p<0.05$) than that of the ABP (32.01 ± 0.76%).

Mineral contents

The ash content in ABP and PBP (Table 2) was comparable to the other by-products (washed orange bagasse, washed peach bagasse, oat bran, rice bran and wheat bran) which have relatively low ash content (between 2.6% and 8%) [14-17]. Human, as well as animal, studies

Compositions	ABP	PBP
Moisture (%)	25.92 ± 0.37[a]	26.415 ± 0.25[a]
Protein (%)	12.81 ± 2.10[a]	22.68 ± 3.51[b]
Fat (%)	45.56 ± 0.72[a]	43.87 ± 0.30[a]
Ash (%)	3.12 ± 0.035[a]	3.11 ± 0.21[a]
Magnesium (mg/100 g)	1366.96 2.51[a]	1071.87 1.95[b]
Calcium (mg/100 g)	717.92 ± 1.88[a]	516.20 ± 1.46[b]
Sodium (mg/100 g)	452.08 ± 1.2[a]	362.91 ± 0.97[b]
Potassium (mg/100 g)	29.75 ± 0.33[a]	26.53 ± 1.03[a]
Total Sugars (%)	32.01 ± 0.76[a]	21.59 ± 2.35[b]

Values expressed are means of three independent determinations. [a,b]Different superscripts in the same row indicate the significant differences ($p<0.05$).

Table 2: Proximate composition of almond and pistachio juices processing by-products. Physico-chemical composition was calculated basis on the dry mater.

Compositions	ABP	PBP
Starch (%)	0.38 ± 0.04[a]	2.32 ± 0.05[b]
Total fibers (%)	26.3 ± 0.21[a]	21.49 ± 0.10[a]
Insoluble fibers (%)	23.83 ± 1.57[a]	17.80 ± 0.92[a]
Soluble fibers (%)	2.45 ± 0.14[a]	3.68 ± 0.28[a]
Total phenolic (mg of gallic acid/100 g)	54.30 ± 2.66[a]	480.48 ± 9.95[b]

Values expressed are means of three independent determinations.
[a,b]Different superscripts in the same row indicate the significant differences ($p<0.05$).

Table 3: Proportion of starch, dietary fibers and phenolic compounds in almond and pistachio juices processing by-products.

originally showed that optimal intakes of elements such as sodium, potassium, magnesium, calcium, manganese and iodine could reduce individual risk factors, including those related to cardiovascular disease [18,19]. Both juices processing by-products were rich in mineral elements. Macroelements (Ca and Mg) play a crucial rule for enzymes and proteins physiological activities. Magnesium is the most abundant element presented in ABP and PBP. The potassium concentration was similar to that in pistachio juice processing by-products (26 mg/100g) and in almond juices processing by-products (29 mg/100g). These results may be useful for the evaluation of dietary information.

Starch and dietary fibers contents

Dietary fibers and starch contents were also determined (Table 3). The starch content of the pistachio juices processing by-products (2.32%) was significantly ($p<0.05$) higher than that of almond juices processing by-products (0.38%). Starch properties and interactions with other constituents, particularly water and lipids, are of interest to the food industry and for human nutrition [20]. There are no significant differences ($p<0.05$) in dietary fibers contents between ABP (26.3 ± 0.21%) and PBP (21.49 ± 1.1%), with a predominance of insoluble fibers. Epidemiologic support that dietary fiber intake preventing obesity is strong and that fiber intake is inversely associated with body weight and body fat [21]. Furthermore, almond and pistachio juices processing by-products can be exploited into dietetic foods.

Total phenolic content

As shown in Table 3, the PBP showed higher total phenolic content (480.48 mg of gallic acid/100 g) when compared ($p<0.05$) to the ABP (54.30 mg of gallic acid/100 g). The pistachio juices processing by-products may be an alternative additive in food, pharmaceuticals and cosmetic preparations instead of many toxic synthetic antioxidants.

Amino acid composition

The amino acid compositions of almond and pistachio juices processing by-products expressed as residues per 1000 total amino acid residues were shown in Table 4. Significant differences ($p<0.05$) were observed for the amino acid proportions between almond and pistachio juices processing by-products.

Glutamic acid is the most dominant amino acid in ABP and PBP (116 and 202 residues, respectively). Alanine and leucine concentrations were significantly ($p<0.05$) higher in ABP than in PBP. Nevertheless, pistachio juices processing by-products contained higher levels of serine, glutamic acid, cysteine and arginine. There were slight difference in essential amino acid contents between almond by-products and pistachio by-products. Therefore, ABP and PBP show a high nutritional value, based on these amino acid profiles, and could be a good dietary protein supplements to poorly balanced dietary proteins. Furthermore, amino acid composition analyses of almond and pistachio juices by-products processing help to study their antioxidant effect and mechanism. As presented in Table 4, The hydrophobic

Amino acids	Number of residues/1000	
	ABP	PBP
Asp	106[a]	103[a]
Thr	54[a]	33[b]
Ser	58[a]	72[b]
Glu	116[a]	202[b]
Gly	91[a]	88[a]
Ala	119[a]	66[b]
Cys	4[a]	15[b]
Val	61[a]	55[a]
Met	19[a]	12[a]
Ile	43[a]	36[b]
Leu	91[a]	74[b]
Tyr	25[a]	22[a]
Phe	42[a]	43[a]
His	21[a]	20[a]
Lys	52[a]	35[b]
Arg	53[a]	77[b]
Pro	45[a]	47[a]
TEAA	387[a]	310[a]
THAA	455[a]	354[b]
Essential/non-essential ratio	0.63	0.45

TEAA = total essential amino acids
THAA = total hydrophobic amino acids
[a,b]Different superscripts in the same row indicate the significant differences ($p<0.05$).

Table 4: Amino acid composition of almond and pistachio juices processing by-products.

amino acids content of the almond by-products (455 residues per 1000 residues) was significantly ($p<0.05$) higher than that of pistachio juices processing by-products (354 residues per 1000 residues). Amino acids in ABP and PBP are possibly involved in antioxidative activity. Several amino acids, such as Tyr, Met, His, Lys, and Trp, may significantly contribute to the antioxidant activity [22].

Comparison of the amino acid content and almond juices processing by-products to the reference values recommended by the FAO/WHO/UNU33 showed that ABP would meet the range of amino acids requirements for children and adults. The ratios of essential to non-essential amino acids in ABP and PBP were 0.63 and 0.45, respectively.

Fatty acid composition

The fatty acid compositions of almond and pistachio juices processing by-products lipids were determined (Table 5).

Two fatty acids are highly represented in the lipidic composition of almond and pistachio juices by-products processing: oleic ($C_{18:1}$) and linoleic ($C_{18:2}$) acids. These fatty acids account for 91% and 90% of the total fatty acids of ABP and PBP lipids, respectively. Similar results were previously reported by [23-25]. Oleic acid has a fundamental role in cardiovascular disease prevention and is indispensable for the healthy growth of human skin [26]. Monounsaturated fatty acids (MUFA) were the major fatty acids, representing 69.56% and 70.70% of the total fatty acids in the ABP and in PBP, respectively. It has been recognized that a diet rich in MUFA may be an alternative choice to a low-fat diet, which may lower blood cholesterol levels, modulate immune function, decrease susceptibility of LDL oxidation and improve the fluidity of HDL [2,27,28]. Both juices processing by-products showed contains polyunsaturated fatty acids (PUFA) (22.43% in ABP and 20.83% in PBP). PUFA are described as having various health benefits (Ruxton et al.). The American Heart Association (AHA) has currently endorsed

the use of n-3 PUFA at a dose of approximately 1 g/day of combined eicosapentaenoic acid (DHA) and docosahexaenoic acid (EPA), either in the form of fatty fish or fish oil supplements (in capsules or liquid form) for patients with documented coronary heart disease [29].

Incorporation of juices processing by-products in the formulation of cookies

Almond and pistachio juices by-products processing were used in the formulation of cookies as the substituent of the wheat flour at levels of 33% (Cookie A) and 50% (Cookie B) almond by-products and 33% (Cookie C) and 50% (Cookie D) pistachio by-products.

Textural analysis: Textural analyses of different cookies were shown in Table 6. Volume, weight loss and propagation rate of cookies decrease with the increase in the rate of substitutions. There were not significant differences ($p>0.05$) between control, ABP and PBP. Hardness for control cookies was found to be 63.59 ± 3.35 while for cookies D it was 77.13 ± 5.14. Cookies A and B were found to be harder than control and cookies C and D, the difference made by the structural strength provided by physic-chemical composition of almond and pistachio by-products processing. Based in the statistic test, the level of substitution exerted a main effect on peak force; the addition of all by-products resulted in cookies with force peak that were significantly ($p<0.05$) different from that of the wheat control. Cookies with almond by-products have a peak force higher ($p<0.05$) than that for the wheat control and cookies with pistachio by-products. No significant differences ($p>0.05$) in peak force were found between degrees of substitution that are less than 50%.

Sensory analysis: The sensory analysis scores (Data not shown) indicated that no significant differences were observed in odour between cookies. However, significant differences ($p<0.05$) were observed in taste, appearance, and texture. In fact, the consumer rather appreciates the texture of cookies C and D, the taste of the control without by-products and the appearance of cookies A and B.

Microbiological quality: All the cookies samples were free of aerobes, mould, coliforms and staphylococcus. These results could be related to the richness in polyphenolic compounds of the almond and

Fatty acid (%)	ABP	PBP
SFA	8.015[a]	8.474[a]
MUFA	69.561[a]	70.695[a]
PUFA	22.425[a]	20.831[b]
UFA	91.986[a]	91.526[a]
UFA/SFA	11.48[a]	10.80[a]
PUFA/MUFA	0.32[a]	0.29[a]
$C_{14:0}$	0.015[a]	0.025[b]
$C_{16:0}$	6.192[a]	6.713[a]
$C_{16:1}$	0.828[a]	1.149[b]
$C_{17:0}$	0.025[a]	0.024[a]
$C_{17:1}$	0.028[a]	0.040[a]
$C_{18:0}$	1.755[a]	1.650[a]
$C_{18:1}$	68.705[a]	69.367[a]
$C_{18:2}$	22.384[a]	20.747[b]
$C_{18:3}$	0.041[a]	0.084[b]
$C_{20:0}$	0.028[a]	0.062[b]
$C_{20:1}$	-	0.139

Values expressed are means of three independent determinations.
[a,b]Different superscripts in the same row indicate the significant differences ($p<0.05$).

Table 5. Fatty acid composition of almond and pistachio juices processing by-products (% of total fatty acids).

	Control	Cookie A	Cookie B	Cookie C	Cookie D
Volume	$33.96 \pm 0.59^{a,A}$	29.72 ± 0.84^b	27.27 ± 0.52^b	29.60 ± 0.22^B	27.12 ± 0.39^C
Propagation rate	$1.67 \pm 0.09^{a,A}$	0.34 ± 0.09^b	0.24 ± 0.09^b	0.39 ± 0.05^B	0.18 ± 0.08^C
Weight loss (g)	$2.16 \pm 0.05^{a,A}$	2.14 ± 0.06^b	2.11 ± 0.05^b	2.13 ± 0.10^B	1.82 ± 0.04^C
Peak force (N)	$63.59 \pm 3.35^{a,A}$	105.24 ± 2.4^b	116.64 ± 2^b	67.00 ± 4.59^B	77.13 ± 5.14^C

Values expressed are means of three independent determinations.
[A,B,C,a,b] Different letters indicates significant differences among samples.

Table 6: Physical and textural properties of different cookies.

pistachio juices processing by-products. The phenolic compounds are well-known for their antimicrobial activity [30]. Also Microbiological quality of cookies could be related to the heat treatment during the cooking.

Conclusion

The aims of this work were to characterize the almond and pistachio juices processing by-products and to evaluate their nutritional quality. Results revealed that almond and pistachio juices processing by-products have relatively high carbohydrate, fat, calcium magnesium and essential amino acid, contents. Indeed, both juices processing by-products showed high nutritional value, based on their amino acid contents. Monounsaturated fatty acids were the major fatty acids of both by-products. ABP and PBP can be incorporated into formulations of cookies. Therefore, they represent a potential source of oil and protein for the human diet.

Acknowledgement

This work was funded by the Ministry of Higher Education and Scientific Research, Tunisia.

References

1. Petruccioli M, Raviv M, Di Silvestro R, Dinelli G (2014) Agriculture and agro-industrial wastes, byproducts, and wastewaters: origin, characteristics, and potential in bio-based-compounds production reference module in earth systems and environmental sciences. Comprehensive Biotechnology 6: 531-545.

2. Da Silva AC, Jorge N (2014) Bioactive compounds of the lipid fractions of agro-industrial waste. Food Res Inter 66: 493-500.

3. Pushpa S, Murthy M, Naidu M (2012) Sustainable management of coffee industry by-products and value addition-A review. Resour Conserv Rec 66: 45-58.

4. AOAC (2000) Official methods of analysis (17thedn). Association of Official Analytical Chemists, Washington.

5. AOAC (1995) Official methods of analysis (15thedn). Association of official Analytical Chemists, Washington DC.

6. Mahmoudabadi SK, Panahi B, Agharahimi J, Salajegheh F (2012) Determination of compounds existing in fruits of three Pistachios (Pistacia Vera L) cultivar in Kerman Provaince. J Biol Envir Sci 6: 81-86

7. Dubois M, Gilles K, Hamilton J, Rebers P, Smith F, et al. (1956) Colorimetric method for determination of sugars and related substances. Anal Chem 28: 350-356.

8. Prosky L, Asp NG, Schweizer TF, DeVries JW, Furda I, et al. (1988) Determination of insoluble, soluble, and total dietary fiber in foods and food products: interlaboratory study. J Assoc Off Anal Chem 71: 1017-1023.

9. Mccready RM, Guggolz J, Silviera V, Owens HS (1950) Determination of starch and amylose in vegetables. Anal Chem 22: 1156-1159.

10. Escarpa A, González MC (2001) Approach to the content of total extractable phenolic compounds from different food samples by comparison of chromatographic and spectrophotometric methods. Anal Chim Acta 427: 119-127.

11. Maksimovia Z, Malencia D, Kovacevia N (2005) Polyphenol contents and antioxidant activity of Maydis stigma extracts. Bioresour Technol 96: 873-877.

12. AACC (2010) Baking quality of cookie flour (Method 10-50-05). American Association of Cereal Chemists, International Approved Methods of Analysis.

13. Abbès F, Bouaziz M, Blecker C, Masmoudi M, Attia H, et al. (2011) Date syrup: Effect of hydrolytic enzymes (pectinase/cellulase) on physicochemical characteristics, sensory and functional properties. LWT - Food Sci Technol 44: 1827-1834.

14. Grigelmo-Miguel N, Martina-Belleso O (1999) Characterization of dietary fibre from orange juice extraction. Food Res Inter 31: 335-361.

15. Grigelmo-Miguel N, Martina-Belleso O (1999) Comparison of dietary fibre from by-products of processing fruits and green from cereals. Lebenson Wiss Technol 32: 503-508.

16. Grigelmo-Miguel N, Gorinstein S, Martina-Belleso O (1999) Characterization of peach dietary fibre concentrate as a food ingredient. Food Chem 65: 175-181.

17. Abdul-hamid A, Luan YS (2000) Functional properties of dietary fibre prepared from defatted rice bran. Food Chem 68: 15-19.

18. Anke M, Groppel B, Kronemann H (1984) Significance of newer essential trace elements (like Si, Ni, As, Li, V,...) for the nutrition of man and animals. In: Bratter P, Schramel P (eds.) Trace elements-analytical chemistry in medicine and biology. Berlin, pp: 424-464.

19. Sanchez- Castillo CP, Dewey PJS, Aguirre A, Lara JJ, Vaca R, et al. (1998). The mineral content of Mexican fruits and vegetables. J Food Comp Anal 11: 340-356.

20. Copeland L, Blazek J, Salman H, Chiming-Tang M (2009) Form and functionality of starch. Food Hydrocolloid 23: 1527-1534.

21. Slavin JL (2005) Dietary fiber and body weight. Nutrition 21: 411-418.

22. Chen HM, Muramoto K, Yamauchi F, Nokihara K (1996) Antioxidant activity of designed peptides based on the antioxidative peptide isolated from digests of a soybean protein. J. Agric. Food Chem 44: 2619-2623.

23. Kodad O, Socias I, Company R (2008) Variability of oil content and of major fatty acid composition in almond (Prunus amygdalus Batsch) and its relationship with kernel quality. J Agric Food Chem 56: 4096-4101.

24. Nanos GD, Kazantzis I, Kefalas P, Petrakis C, Stavroulakis GG (2002) Irrigation and harvesting on almond quality and composition. Sci Hor 96: 249-256.

25. Safari M, Alizadeh H (2007) Composition of the oil major Iranian nuts. J Agric Sci Technol 9: 251-256.

26. Bruckert E (2001) Les phytosterols, place dans la prise en charge du patient hyperlipidémique. OCL 8: 312-316.

27. Hargrove RL, Etherton TD, Pearson TA, Harrison EH, Kris-Etherton PM, et al. (2001) Low fat and high monounsaturated fat diets decrease human low density lipoprotein oxidative susceptibility in vitro. J Nutr 131: 1758-1763.

28. Villa B, Calabresi L, Chiesa G, Risè P, Galli C, et al. (2002) Omega-3 fatty acid ethyl esters increase heart rate variability in patients with coronary disease. Pharmacol Res 45: 475.

29. Kris-Etherton PM, Harris WS, Appel LJ (2003) Omega-3 fatty acids and cardiovascular disease: new recommendations from the American Heart Association. Arterioscler Thromb Vasc Biol 23: 151-152.

30. Pereira JA, Pereira AP, Ferreira IC, Valentão P, Andrade PB, et al. (2006) Table olives from Portugal: phenolic compounds, antioxidant potential, and antimicrobial activity. J Agric Food Chem 54: 8425-8431.

5

Effect of Incorporating Whey Protein Concentrate on Chemical, Rheological and Textural Properties of Ice Cream

El-Zeini Hoda M[1]*, Moneir El-Abd M[1], Mostafa AZ[2] and Yasser El-Ghany FH[2]

[1]*Dairy Science and Technology Department, Faculty of Agriculture, Cairo University, Cairo, Egypt*

[2]*Food Technology Research Institute, Agriculture Research Center, Giza, Egypt*

Abstract

Effect of whey protein concentrate as a partial substitution of milk solids not fat in ice cream formula was investigated by replacing 1, 2, 3 and 4% of mix solid not fat. Mixes and resultant ice cream samples were evaluated for their chemical, physicochemical, and rheological properties as well as the sensory quality attributes. Implementing whey protein concentrate in ice cream recipes increased total protein, freezing point and consistency, while, ash, lactose, specific gravity decreased significantly (p<0.001). Apparent viscosity as well as flow time of mixes was significantly (p<0.001) increased with substitution of milk solid not fat by whey protein concentrate. Flow behavior was also affected showing higher yield stress. The consistency coefficient (k) was more affected by the presence of whey protein concentrate in the recipe than the flow behavior index (n). However, increasing whey protein concentrate decreased hardness, cohesiveness, gumminess and chewiness values in texture profile analysis, while there were increase in adhesiveness, springiness and hesion values in fresh ice cream samples over that of stored. The ice cream became smoother and highly acceptable for the panelists by replacing milk solid not fat with whey protein concentrate up to 3%. From the data obtained, it could be recommended that ice cream can be produced with high quality by substituting milk solid not fat with whey protein concentrate up to 3%.

Keywords: Ice cream; Whey protein concentrate; Rheological properties; Texture profile analysis

Introduction

Ice cream is a complex colloidal food that composes of four distinct phases: a continuous serum phase known as a matrix, and three distinct dispersed phases corresponding to fat droplets, ice crystals and air cells [1]. Whey protein concentrate (WPC), a by-product from cheese production, is used in ice cream production to develop a better quality. Because of its good water binding property, WPC delays development of coarseness and increases ice cream mix viscosity. Moreover, the presence of WPC in ice cream could enhance fine dispersion of air cells and lower the ice crystal size in ice cream owing to its foaming property. Furthermore, WPC also lowers surface tension, stabilizes the fat emulsion, controls fat destabilization and enhances partial coalescence due to the emulsifying properties of protein [2]. Patel et al. [3] study the effect of adding WPC to vanilla ice cream and found that WPC could improve physical properties and sensory quality of ice cream by resisting changes in ice cream during storage, enhancing its nutritional value, and increasing consumers' acceptance.

Rheology reflects the manner in which food materials respond to an applied stress or strain and it generally relates to the flow properties of food materials. Rheological properties are important in determining the quality of finished goods; also, they are related to sensory perception. Elucidation of relationships between structural changes during processing, rheology, and sensory perception is very important. For dairy foods, the rheology of fluid and plastic products is relevant to consumer acceptability. Texture profile analysis (TPA) method is widely used for texture evaluation of food products. Human eating action normally consists of several bites. In order to better describe the eating actions of human, the TPA test performs two bites every bite includes loading and unloading cycle. Many research groups have conducted instrumental texture profile analysis (TPA) for assessing the textural properties of food [4].

So, the aims of this study were, to evaluate the effect of different substitution levels of WPC on the chemical composition, texture, rheological properties of ice cream, and to determine the changes of the rheological model as a result of the substitution of MSNF with different WPC%. Moreover, to establish the relationships between chemical composition and rheological behavior of ice cream to set the best substitution level of WPC.

Materials and Methods

Materials

Fresh buffalo's skim milk (90.9% moisture, 0.1% fat, 3.4% protein, 4.9% lactose and 0.7% ash) and fresh concentrated cream (29.4% moisture, 67% fat, 1.3% protein, 1.7% lactose and 0.6% ash) were obtained from the herd of Faculty of Agriculture, Cairo University and used as an ingredient for preparing the ice cream mixes. Low heat skim milk powder (3.8% moisture, 0.8% fat, 33.4% protein, 54.1% lactose and 7.9% ash) was obtained from Abou El-Hool-Import/Export Co., Cairo, Egypt. Whey protein concentrate powder (4.7% moisture, 5.9% fat, 77.7% protein, 9.1% lactose and 2.6% ash) was supplied by Davisco Foods International, Inc, USA. Commercial grade sugar cane was obtained from the local market, Sodium carboxymethyl cellulose (CMC) as a stabilizer was obtained from Mifad Company, Giza, Egypt. Vanilla was obtained from the local market and used to flavour final ice cream.

*Corresponding author: El-Zeini Hoda M, Dairy Science and Technology Department, Faculty of Agriculture, Cairo University, Cairo, Egypt E-mail: dr_hodazeini@yahoo.com

Manufacture of ice cream

Ice cream mix contained 8% fat, 12% milk solid not fat, 15% sucrose, 0.25% stabilizer. Skimmed milk powder was substituted with WPC at 1.0, 2.0, 3.0 and 4.0% of dried milk solids not fat in the base mix (Table 1).

Methods

Total solids, total protein content and ash were determined according to AOAC [5]. Titratable acidity of mixes was determined according to Lawrance [6]. Lactose content was determined according to Arbuckle [7]. Fat content was determined according to Divide [8]. Values of pH were measured using a digital laboratory pH meter (HI 93 1400, Hanna instruments). The specific gravity of mix was measured using a bottle pyconometer as described by Winton [9] at 20°C. The weight per gallon (lb) of ice cream mixes and the final frozen products were calculated according to Kessler [10] by multiplying the specific gravity by the factor of 8.34. Freezing point of ice cream mix was measured as described in FAO [11].

Consistency of the mix was measured as the time (sec) to empty a 50-ml pipette according to Lawrance [6]. Apparent viscosity of mix was determined according to Petersen et al. [12] using a Brookfield viscometer (Brookfield DVIII Ultra Programmable Rheometer equipped with a spindle No. SC4-21 in 250 ml cup. Approximately 24 hours after preparation of the ice cream mixes, viscosity and shear stress were measured periodically at shear rates ranging from 18.6 to 186 sec^{-1}. At each shear rate, shear stress was recorded after two minutes of spindle rotation to ensure a steady reading. All samples were adjusted at 20 ± 1°C before loading in the viscometer device. Collected shear stress/shear rate data was exported to a Microsoft Excel spreadsheet and used to generate Power Law equation and the Herschel-Bulkley equation. Power Law fluids are characterized by a straight line log-log plot of shear stress vs. shear rate, according to the Power Law equation. A straight line plot of shear stress vs. the shear rate allows application of the Herschel-Bulkley equation which usually used to determine the yield stress. Overrun of ice cream samples was calculated by using the method given by Benezech [13].

$$Overrun = \frac{weight\ of\ mix - weight\ of\ the\ same\ volume\ of\ ice\ cream}{weight\ of\ the\ same\ volume\ of\ ice\ cream} \times 100$$

Texture profile analysis

Texture profile analysis test of ice cream samples was done using a Universal Testing Machine (TMS-Pro) equipped with 1000 N (250 lbf) load cell and connected to a computer programmed with Texture Pro™ texture analysis software (program, DEV TPA With hold). A flat rod probe (49.95 mm in diameter) was used to uniaxial compress the ice cream samples to 50% of their original height. The texture profile

analysis test set condition was adjusted to a test speed 50 mm/sec, trigger force 1N, deformation 40% and holding 2 sec between cycles. Test was carried out on samples which left in refrigerator until the temperature became -12°C.

Sensory evaluation of resultant ice cream was judged by 10 staff members of the Food Technology and Research Center, Dairy Department. The evaluation comprised of flavor (45 points), body and texture (35 points) melting properties (10) and colour (10) [6].

Statistical analysis

Data were analysed statistically using the MSTAT-C (ver 2.10, MSU, USA.) package on a personal computer. All experiments were carried out in triplicates. Differences were considered significant at p<0.05.

Results and Discussion

Properties of ice cream mixes

Chemical composition of ice cream mixes with whey protein concentrate as a substitution of MSNF in base formula is shown in Table 2. Fat was adjusted in all mixes to almost 8% for recipe formula during the procedures.

The average value of protein contents in different ice cream treatments was stated in Table 2. A proportional replacement of MSNF with WPC resulted in a significant increase (p<0.001) of protein contents of ice cream mixes. The increase mainly due to the higher protein content in added WPC (77.7%) compared with SMP (33.4%). Among treatments with WPC, the protein content was significantly affected by the ratio added (LSD= 0.084 at 0.05 α level). The total protein content increased in an ascending order with increasing the ratio of WPC substitution being the highest at 4% WPC with a high correlation (0.999). Regression analysis student T test showed a significant difference for the treatment of ice cream mix with variable contents of WPC (p<0.001). These data agreed with the findings of Awad [14] and Suneeta [15].

The usage of WPC as a MSNF replacer leads to a significant reduction (p<0.001) in ash content as shown in Table 2. Ash contents decreased in ice cream mixes with the addition of WPC ratio. This decrease could be due to the differences in ash contents of WPC (2.6%) and SMP (7.9%). Coefficient correlation (-0.985) showed a tight inverse relationship between WPC and ash%. The obtained results are in a harmony with the findings of Awad [14] and Patel et al. [3]. Lactose values decreased by increasing the substitution level of WPC in the mixes which due to a less content of lactose in WPC (9.1%) than in SMP (54%). However, WPC with low lactose content can be safely used at higher levels without concerning of sandiness development defect in ice cream [16]. The proportional use of WPC as a MSNF replacer leaded to a significant differences in lactose content (p<0.001) with a high negative correlation (-0.991). Among treatments, significant differences were found as a result of increasing WPC% (LSD= 0.1975 at α= 0.050).

Physicochemical properties

Acidity and pH values: Titratable acidity values of ice cream mix increased gradually from 0.21 to 0.26% (Table 3). The results showed that within treatments the acidity values differed insignificantly for the control, T$_1$, T$_2$ and T$_3$, while T$_4$ significantly differed from the rest of the treatments (LSD= 0.01883 at 0.05 α level). Incorporating WPC in the ice cream formula increased the acidity values significantly

Ingredients	Control	Level of substitution (g/kg mix)			
		T$_1$	T$_2$	T$_3$	T$_4$
Sugar	150	150	150	150	150
Stabilizer	2.5	2.5	2.5	2.5	2.5
Fresh skim milk	670.38	670.38	670.38	670.38	670.38
Cream	117.69	117.69	117.69	117.69	117.69
Dried skim milk	59.43	48.94	38.16	27.08	17.10
WPC 80	0.00	11.18	22.17	33.05	43.04
Total	1000	1000.69	1000.9	1000.7	1000.71

T$_1$, T$_2$ T$_3$, T$_4$: Corresponding to 1, 2, 3 and 4% WPC substitution of milk solid not fat

Table 1: Formulation of different ice cream mixes with WPC as a substitute of milk solid not fat (g/ kg mix).

Treatments	Total solids	Fat	Total protein	Ash	Lactose content
Control	36.96	8.23	4.37[e]	1.073[a]	7.98[a]
T$_1$	36.37	8.23	4.94[d]	0.943[b]	7.24[b]
T$_2$	36.24	8.20	5.65[c]	0.883[c]	6.51[c]
T$_3$	36.16	8.23	6.29[b]	0.821[d]	5.81[d]
T$_4$	36.29	8.20	6.92[a]	0.750[e]	5.42[e]

a-e: Means with different letters within a column are significantly different from each other at α=0.05 as determined by Duncan's multiple range tests.

Table 2: Chemical composition (%) of ice cream mixes with different ratios of WPC as a partial substitution of milk solid not fat.

Treatments	Acidity	pH	SG (-)	W/gal (lb)	FP (°C)	Con.(sec)	AV(c.p)
Control	0.21[c]	6.72[a]	1.108[a]	9.244[a]	-2.47[a]	44.13[e]	45.65[e]
T$_1$	0.21[c]	6.72[a]	1.093[a]	9.037[b]	-2.41[ab]	51.66[d]	47.8[d]
T$_2$	0.22[c]	6.55[b]	1.082[a]	9.012[b]	-2.35[ab]	60.50[c]	55.57[c]
T$_3$	0.23[bc]	6.47[c]	1.080[a]	9.011[b]	-2.27[ab]	74.59[b]	60.2[b]
T$_4$	0.26[a]	6.41[d]	1.078[a]	9.004[b]	-2.19[b]	82.67[a]	65.65[a]

a-e: Means with different letters within a column are significantly different from each other at α=0.05 as determined by Duncan's multiple range tests

Table 3: Physicochemical and rheological properties of ice cream mixes with different ratios of WPC.

(p < 0.001). The differences in acidity values among treatments are due to the differences in chemical composition and mainly the protein content. Natural titratable acidity in milk and milk products is dependent on casein, albumin, phosphates, citrates and carbon dioxide [17]. Obtained results were in agreement with those of Tirumalesha et al. [18] and Patel et al. [3]. The pH values of ice cream mixes with WPC in base formula are presented in Table 3. The pH values of ice cream mixes decreased significantly by substituting MSNF by WPC in the base formula (p<0.001). The differences in pH values of ice cream mixes are related to the original composition and acidity. Student T test confirmed the significancy of increasing the WPC% on the pH values of ice cream mixes with a tight correlation between both factors (0.964). The decline of pH values were attributed to the reduction of skimmed milk powder (SMP) which contains milk proteins and buffering salt system of phosphate and citrate [19]. The obtained results were in harmony with those obtained by Patel et al. [3] and Castro et al. [20].

Specific gravity and weight/gallon: The effect of replacing MSNF with WPC at different ratios on specific gravity (SG) of ice cream mixes was shown in Table 3. The specific gravity values of ice cream mixes decreased by substituting MSNF with WPC in the recipe (p<0.001). An inverse relationship was found between SG and the ratio of WPC added to the ice cream mix. Within treatments, no significant differences was found (LSD value=0.1575 at α=0.050). Awad and Metwally [14] and Awad [21] mention that SG values of ice milk mix decreased by substituting SMP with total milk protein or rice flour in the recipes.

Weight/gallon is a reflection of SG as it is the product of multiplying SG by 8.34. Values of weight/gallon are presented in Table 3. The control ice cream had the highest weigh/gallon and differed significantly (p<0.01) from that of WPC. A negative correlation (-0.779) indicated that the substitution of MSNF with WPC at ascending ratios decreased weight/gallon.

Freezing point (°C): Whey proteins, lactose and mineral salts in any given ingredient can be taken into account to efficiently manage water-to-ice freezing performance and transitioning. This, in turn, affects freezing conditions, mix performance and finished product qualities such as body (chew, bite) and texture (smoothness). Freezing point of ice cream mixture is directed proportionally to the number of

particles in solution, type and molecular weight of the solutes in the mix. The more dissolved solids in the solution, the lower the freezing point. The freezing point varies with the composition of the mix and concentration of the soluble constituents within the mix. The freezing point of ice cream mixes was significantly affected by adding WPC to ice cream recipes (p<0.001). The mixes showed higher freezing points with substituting MSNF by WPC. Control treatments showed descending freezing point among all treatments (LSD= 0.2306 at α= 0.05), while that with 4% WPC substitution (T$_4$) showed the highest. The high freezing point in treatments with WPC could be due to its lower lactose and other true solutions solutes with high protein contents [22]. The results obtained are in line with those of Patel et al. [3] and Awad [21].

Rheological properties of ice cream mixes

Consistency: The mix consistency was expressed as the flow time which defined as: the time (sec.) required to empty a constant volume of ice cream mix. The flow time was significantly increased (p < 0.001) by incorporating WPC in ice cream formula as shown in Table 3. The high flow time was a sign for improvement in product consistency. The flow time increase or consistency improvement correlated with the ratio added of WPC to the ice cream formula (0.994). The increase of flow time of ice cream mix could be related to the high viscosity of the mixes containing WPC. The results are in agreement with Haque and Ji [23] and Ruger et al. [2].

Apparent viscosity: Viscosity defined as the internal resistance of a substance to flow when a shear stress is applied. Viscosity behavior is influenced by the complex hydrodynamic properties (i.e., size, shape, and hydration potential) and independent on the shear rate and time, while, resistance to flow is caused essentially by molecular or ionic cohesion. Viscosity data are often derived from single-point measurements. Since such measurements give no information about the flow behavior (dependence on shear rate), data may be contradictory, especially if the flow curves (shear stress vs. strain) under consideration actually cross each other. The viscosity of a fluid may increase in a linear or non linear fashion with a transition from Newtonian to non Newtonian behavior as the total solids concentration is increased.

Table 3 presented the viscosity values of ice cream mixes contain WPC in base formula. The data cleared that the viscosity of ice cream mix was significantly affected (p ≤ 0.001) by adding WPC as a part of total milk solids in base recipe. The increase in viscosity values of mixes contain WPC could be due to the higher protein contents and /or to the nature and type of protein added into formula. Several investigations showed that proteins and especially WPC increases the product viscosity if added to the formula [2,20,24-26] and Sadar [26].

Flow behavior model: To elucidate the basic flow behavior combination of ice cream mixes containing WPC, the changes in shear stress induced by changing the shear rate were investigated. A set of WPC% ice cream was tested under increasing shear rate with constant strain and shear stress/shear rate relationship was shown in Figure 1. A typical Herschel-Bulkley behavior was observed, as linear best fitting lines of the shear stress/shear rate data of the control and WPC% mixes crossed the y axe at 0 shear rates with yield stresses exhibited in all treatments. Higher substitution levels of WPC in ice cream mixes resulted in an upward shifting of the flow curve (building up of structure leading to increase in the sample viscosity). The shear stress increased in the treatments with increasing the ratio of WPC added (p<0.001) significantly [19,27].

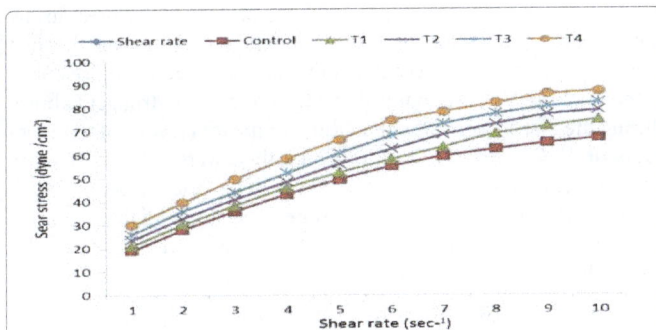

Figure 1: Flow behavior of ice cream mixes with different concentrations of WPC.

Yield stress values: Yield stress is defined as the minimum force required initiating flow. This means that subjecting a material to stress less than the yield stress will lead to a nonpermanent deformation or a slow creeping motion over the time scale of the experiment. Yield stress is related to the internal structure of the material which must be destroyed (overcome) before flow has occurred. Now, yield stress is routinely measured and used in the food industry not only for basic process calculations and manufacturing practices, but also as a test for sensory and quality indices and to determine the effect of composition and manufacturing procedures on structural and functional properties. The values of yield stress were calculated by fitting the shear stress and shear rate data to the Herschel-Bulkley equation:

$$\sigma = \sigma_0 + \eta_a \gamma$$

Where σ, Shear stress; σ_0, Yield stress=(shear stress at zero shear rate; η_a, Apparent viscosity [mPa.s]; γ, Shear rate.

The 'best fit' routine was used and the Herschel-Bulkley model was fitted with high regression coefficients, (Table 4) for all the samples, indicating a fitting model.

Power low equation: Flow behavior parameters of ice cream mixes as affected by WPC substitution were evaluated by fitting the shear stress/shear rate data to the power low equation [13]:

$$\sigma = K\gamma^n$$

Where σ, shear stress [dyne/cm^2]; K, consistency index [mPa.s]=viscosity at 1sec^{-1}; γ, shear rate [s^{-1}]; n, a dimensionless number that indicates the closeness to Newtonian flow. For a Newtonian liquid n is 1; for a dilatant fluid n is greater than 1; and for pseudoplastic fluid n is less than 1.

Power law constants (k and n) were obtained using double logarithmic plot [28]. Power law constant and R^2 values for log-log plots for all treatments are given in Table 5. Flow behavior index (n) increased by the substitution of MSNF with WPC. The flow behavior index measures the degree of departure from Newtonian flow and results were consistent with pseudo-plastic flow for which "n" is less than 1. However, addition of WPC had a little effect on "n" value and maintained between 0.5-0.7 showing slight deviation from Newtonian flow. The consistency coefficient (k) was more affected by the presence of WPC in the recipe than the flow behavior index (Table 5). The plots of the consistency index (K) and the yield stress (σ_0) were shown in Figure 2. While, the K and n derived from the model were given in Table 5. The results were consistent with the observed plot, a shear thinning line that had a decreasing gradient (n<1), their η decreased with increasing

shear. However, the presence of a yield stress values denotes a minimal stress which must be exceeded prior to flow occurring due to shear. This has been interpreted as the existence of a network structure, the bonds of which must be broken to allow flow.

Apparent viscosity-shear rate properties: Apparent viscosity values at a different shear rate (from 18.6 up to 186 sec^{-1}) are presented in Figure 3. Obtained results revealed a considerable decrease (p<0.001) in viscosity with increasing shear rate at all concentration ranges of WPC. At the shear rate of 37.2 s^{-1} the curves leveled off and then decreased with the same constant until the maximum applied shear rate of 186 s^{-1}. Such behavior is typical for a shear thinning system, so ice cream mix with different replacing ratio of WPC can be characterized as Herschel-Bulkley with shear thinning fluids, with the flow curves lacking a linear characteristic A rapid breakdown of the structure occurs on initial shearing followed by much slower changes at higher shear rates. This behavior can be attributed to combined effects of breakdown of weak links between the proteins and/or between the proteins and stabilizer, and of reformation of such links as a result of Brownian motion and molecular collisions [29].

Physicochemical properties of ice cream: Ice cream is a complex system with many ingredients in its formulation that can interact. The formulation as well as the actual processing conditions used in its manufacture can affect the final properties of the ice cream. However, the effect of WPC as a substitution of MSNF on some properties of resultant ice cream is presented in Table 6.

Specific gravity: Specific gravity SG is one of the important physical properties of ice cream. It gives some information about the quality of the resultant ice cream such as body and texture, incorporated air and melting quality of ice cream. These data indicated that the SG of resultant ice cream decreased with incorporating WPC in the formula. The decrease in SG values correlated proportionally (-0.966) with the added ratio of WPC. Specific gravity depends on the formula components as well as the ability of the mix to retain air cells in ice cream matrix. Ice cream treatment with 4% WPC in the formula T$_4$ showed the lowest SG as the total protein was the highest (6.92%) in comparison to the control mix (4.37%), which meant more air incorporation in the body of ice cream with more protein membranes constructed. WPC% affected the SG significantly (p<0.001) with high R^2. Within treatments, there was no significant differences between T3

Treatments	Liner equation	Herschel-Bulkley equation	
		Yield stress	R^2
Control	y = 0.5526x + 1.32	1.32	0.995
T$_1$	y = 0.5643x + 1.39	1.39	0.998
T$_2$	y = 0.5465x + 1.54	1.54	0.996
T$_3$	y= 0.5279x + 1.69	1.69	0.995
T$_4$	y = 0.4909x + 1.95	1.95	0.993

Table 4: Effect of WPC on yield stress of ice cream mixes.

Treatments	Liner equation	Flow parameters		
		Consistency coefficient K (dyne.S^2/cm^2)	Flow behavior index (n) (-)	Power law R^2
Control	y= 0.5844x + 1.19	3.29	0.584	0.994
T$_1$	y= 0.580 x+ 1.27	3.53	0.580	0.998
T$_2$	y= 0.567x+ 1. 42	4.14	0.567	0.997
T$_3$	y= 0.5482x + 1.57	4. 81	0.548	0.995
T$_4$	y= 0.5104x + 1.83	6.23	0.510	0.993

Table 5: Effect of WPC as a partial substitution of MSNF on the consistency (K) and flow behavior index (n) of ice cream.

and T4 (LSD= 0.059 at α= 0.05). The results obtained are in line with those of [14].

Values of weight per gallon are also shown in Table 6. The weight per gallon (lb) values was calculated of SG values obtained for resultant ice cream formula; therefore, they follow the same trend as SG.

Overrun: Some well-known foods such as ice cream are produced by incorporating air into liquid and producing foam. The foam is stabilized by surface-active agents which is collected at the interface. The amount of air incorporated is expressed in term of overrun. An increase in volume of ice cream over that of ice cream mix by incorporating air is called overrun. Overrun percent in ice cream was significantly affected by adding WPC as MSNF substitution in the treatments (p<0.001). The more the protein, the more the protienious bubbles trapping air inside and resulting in high overrun (R^2= 0.952, that was stated by Patel et al. [3]. The high overrun percentages in treatments containing WPC could be related to the high foaming ability of ice cream mixes. There are several factors that affect overrun in ice cream includes total solids in the mix and type of freezer used. The highest the total solids content, the greater the possible overrun [18,30-32]. Who pointed out that there was a significant improvement in both whipping rate and overrun by increasing the replacement of skim milk solids with the admixture of butter milk powder and WPC. Akalin [24] mentioned that the presence of WPC may facilitate the initial stabilization of newly formed air bubbles in the freezer better than UF retentates of NDM.

Texture profile analysis: Texture profile analysis method is widely used for texture evaluation of food products. Human eating action normally consists of several bites. In order to better describe the eating actions of human, the TPA test performs two bites every bite includes loading and unloading cycle. Texture profile analysis attributes of resultant ice cream as influenced by different replacing ratio of WPC and storage period are given in Figure 4. Fresh ice cream samples had hardness values of 66.77, 60.85, 50.30, 44.23 and 39.95 (N) for control, T_1, T_2, T_3 and T_4 treatments with 1, 2, 3 and 4% WPC, respectively. Hardness of resultant ice cream was affected by replacing MSNF with WPC at different ratios and storage for 14 days at -18 ± 1°C (p<0.001). The hardness values were lower in all treatments with WPC than control when fresh or after 14 days of storage, may be due to the higher overrun percentages. During storage, the hardness increased over that of fresh ice cream but decreased with storage proceeding and reached 70.37, 66.98, 62.50, 58.45 and 51.66 (N) for control, T_1, T_2, T_3 and T_4, respectively, may be due to shrinkage (air cell collapsing) results from a loss of discrete air bubbles in ice cream samples. Negative correlation (-0.942) was obtained between WPC ratio and ice cream hardness, while low positive (0.491) was found between storage period and hardness. Obtained results are in harmony with those represented by Tirumalesha and Jayaprakasha [18].

Adhesiveness is recognized as the work required to overcome the attractive forces between surface of the ice cream and surface of other materials with which the ice cream contacts. Data showed that the adhesiveness values were higher in all treatments with not only WPC but also storage period and the interaction of the two factors (p<0.001) than control with a correlation of 0.903. During storage, adhesiveness values decreased gradually and significantly (p<0.001) to reach 8.56, 12.37, 16.84, 20.63 and 25.53 (mj) for control, T_1, T_2, T_3 and T_4 treatments with 1, 2, 3 and 4% WPC, respectively (Figure 4).

Cohesiveness defined as the strength of internal bonds making

up the body of the product, it is the ratio of the positive area during the second compression to that of the first peak during the first compression. results suggested that the internal structure of ice cream without added WPC was bonded and the bonds were stronger to break during the first compression than those of the ice cream with different ratios of WPC (p<0.001). During storage, there were little increasing in cohesiveness values (Figure 4) over those of fresh ice cream (p<0.01). The increasing of obtained values may be correlated with the increasing of hardness in resultant ice cream. Cohesiveness negatively correlated with WPC% (-0.767) and positive (0.0336) with storage period.

Hesion compares the strength between the internal bonds in a material to the strength of sticking of the material to any surface. If the former is higher than the later, the sample keeps its structure intact with less sticking force. If the reverse took place, the sample will rupture when pulled from the sticking surface which indicates loose structure. As indicated in Figure 4, the hesion was negative in all ice cream treatments which indicated a loose structure with weak internal bonds. WPC% brought about a loose structure in favored of ice cream. Storage increased the hesion values leading to stronger and more compact structure which is unfavorable in ice cream.

Springiness (referred to as "elasticity") which is the rate at which a

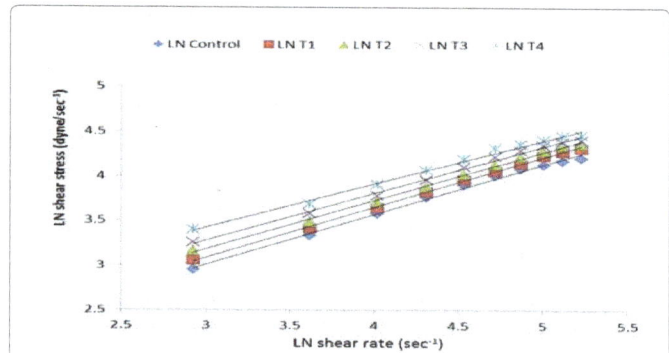

Figure 2: Double logarithmic plot for ice cream mixes with (WPC).

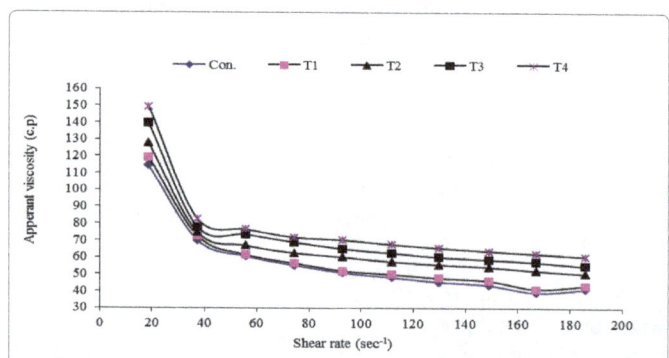

Figure 3: Effect of shear rate on viscosity of ice cream mixes with different ratios of WPC.

Treatments	Specific gravity (-)	Weight per gallon (lb)	Overrun (%)
Control	0.9133[a]	7.6216	47.64[d]
T_1	0.8243[b]	6.8784	54.23[c]
T_2	0.7355[c]	6.1379	61.95[b]
T_3	0.6914[cd]	5.7694	67.31[a]
T_4	0.6740[d]	5.6252	69.94[a]

a-e: Means with different letters within a column are significantly different from each other at α=0.05 as determined by Duncan's multiple range tests
Table 6: Effect of WPC% on specific gravity and overrun ice cream properties.

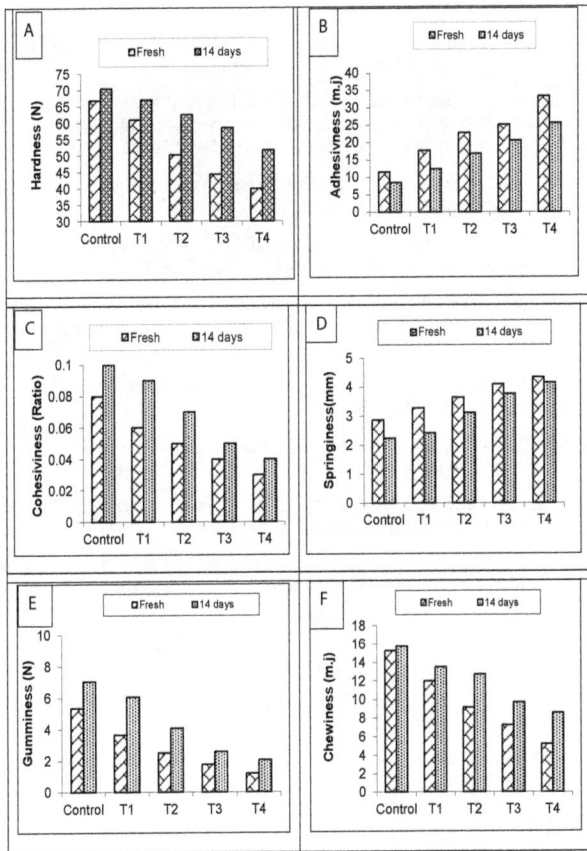

Figure 4: (a) Hardness, (b) Adhesiveness, (c) Cohesiveness, (d) Springiness, (e,f) Gumminess and Chewiness values of fresh and stored (14 days) ice cream with different replacing ratio of WPC.

deformed material returns to its original shape on removal of deforming force. Obtained data revealed that the springiness increased (Figure 4) proportionally ($p<0.001$) with increasing WPC%. The springiness correlated positively (0.823) with the quantity of WPC added to the ice cream formula and negatively with the storage period (-0.333) with high R^2 (0.889) for all treatments.

Gumminess, defined as "the energy required for disintegrating a semisolid food product to a state ready for swallowing," and is related to the primary parameters of hardness and cohesiveness (Figure 4). Ice cream gumminess decreased by increasing substitution level of WPC. There was a little increase of gumminess values during storage of ice cream samples at -18 ± 1°C for 14 days, which due to the increasing values of hardness and cohesiveness during the storage. Chewiness gives an indicator about the energy required to masticate the ice cream product to a state ready for swallowing. Chewiness values of ice cream decreased by increasing substitution levels of WPC. The chewiness values were lower (Figure 4) in all treatments with WPC than control which brought about by reducing hardness and cohesiveness values with the presence of WPC.

Sensory evaluation: Texture is a property difficult to evaluate with the use of a machine because it can only quantify the textural parameters in terms of a few specific characteristics, as it is a multi parameter characteristic, detected by several senses and derived from the food structure, texture is evaluated well by individuals who can

perceive and describe all attributes of a product's texture [33]. Sensory panel evaluation is an important indicator of potential consumer preferences.

Sensory attributes evaluated were presented in Table 7. Panelists scored the T_4, the least flavor. T_3 was the most acceptable flavor among the ice cream (LSD= 0.4859 at 0.05 α level). Totally adding WPC enhanced the flavor significantly ($p<0.001$). All WPC ice cream received flavor ratings higher than control except T_4, while T_3 received the highest score for flavor. On the other hand, increasing the ratio of WPC lowered the flavor score. Incorporating WPC in ice cream formula up to 3% resulted in rich flavor, while with the higher ratio of WPC a slight unpleasant flavor was detected. Panelists did not observe any significant difference between T_1 and T_2 samples. Similarly, body and texture scores for ice cream showed no significant difference (LSD= 0.4834 at 0.05 α level) between the control and T_1. T_1 with 1% WPC had more desirable scores, than T_2, T_3 and T_4 with 2, 3 and 4% WPC, whereas control scored the highest (33.5), most desirable compared with all other treatments. WPC% significantly decreased the obtained score ($p<0.001$) for overall texture acceptance compared with the control, that was insured by the negative correlation between WPC% and body of the ice cream.

Ice cream containing 1%WPC was smoother and gummier than ice cream containing higher ratios of WPC. It was likely due to the increase in protein content in ice cream mix as MSNF was substituted with WPC. Whey proteins had good water binding capacity and could be adsorbed onto ice crystals surface [3]. Hence, less free water was able to flow and form ice crystals, subsequently preventing large ice crystals formation and resulting in smoother texture in the ice cream containing WPC. Body and texture of ice cream scored had the same trend of flavour when judging body and texture. The differences among all treatments in body and texture are related to the effect of substitution level and the nature of proteins in WPC. At higher replacing ratio of WPC the texture appeared as a fluffy in resultant ice cream due to great incorporation percentage of air in relation to the percentage of total solids in the mix.

The meltdown properties of ice-cream constitute a critical performance parameter for the product so much, so that, in some cases, these properties contribute towards the formation of a quality judgment as important as the sensory properties of the product [34]. WPC% played a central role in affecting meltdown behavior ($p<0.001$). The scores of melting properties for ice cream treatment with WPC as substitution of MSNF is shown in Table 7. The results indicated that there were significant improvements in melting quality of ice cream with adding WPC up to 3%. Ice cream samples with WPC became slightly less melted and needed more time to melt which preferred by panelists (LSD value=0.2919 at α=0.050), while with WPC higher ratio the ice cream became more susceptible to melting which is not

Properties		Treatments†				
		Control	T_1	T_2	T_3	T_4
Flavour	(45)	42.11^c	42.84^b	43.16^b	43.86^a	41.40^d
Body and texture	(35)	33.50^a	33.14^a	32.43^b	31.84^c	30.79^d
Melting properties	(10)	8.16^c	8.51^b	8.89^a	9.01^a	7.57^d
Appearance	(10)	8.2^c	8.5^b	8.57^b	8.9^a	8.5^b
Total	(100)	90.97	92.99	94.18	95.61	88.26

$^{a-e}$: Means with different letters within a column are significantly different from each other at α=0.05 as determined by Duncan's multiple range tests

Table 7: Sensory quality attributes of ice cream samples with WPC as a partial substitution of milk solid not fat.

targeted. Negative low correlation (-0.102) was obtained as a result of distorted trend associated with increasing WPC%. The appearance sensory response to the ice cream samples was affected by the variation in WPC%. Although these attributes are similar, the sensory panel found differences (p<0.001) for each percentage of WPC when scoring the samples for textural appearance.

Conclusion

Nutritional and functional properties can be enhanced by incorporating Whey protein concentrate as ingredient contains biologically active proteins as a substitute of the skimmed milk powder at 10, 20, 30 and 40 per cent levels in the ice cream preparation. The WPC incorporated ice cream samples improved the melting quality of ice cream with adding WPC up to 3%. The results reveal that the resultant ice cream has a higher sustainability in the mouth. Hence, it could be concluded that whey protein concentrate could be incorporated in the ice cream replacing skimmed milk powder with improved sensory properties besides improving the protein content of the ice cream.

References

1. Goff HD (1997) Colloidal aspect of ice cream, a review. Int Dairy J 7: 363-373.

2. Ruger PR, Baer RJ, Kasperson KM (2002) Effect of double homogenization and whey protein concentrate on the texture of ice cream. J Dairy Sci 85: 1684-1692.

3. Patel MR, Baer RJ, Acharya MR (2006) Increasing the protein content of ice cream. J Dairy Sci 89: 1400-1406.

4. Tabilo G, Flores M, Fiszman SM, Toldra F (1999) Postmortem meat quality and sex affect textural properties and protein breakdown of dry-cured ham. Meat Sci 51: 255-260.

5. AOAC (2006) Official Methods of Analysis of the Association of Official Analytical Chemists. Association of Official Analytical Chemists, Arlington, Virginia, USA.

6. Arbuckle WS (1986) Ice cream. The AVI Publishing Company Inc, Westport, USA.

7. Lawrance AJ (1968) The determination of lactose in milk products. Aust J Dairy Tech 23: 103-106.

8. Divide CL (1977) Laboratory Guide in Dairy Chemistry Practical. Dairy Training and Research Institute, Univ. of Philippines, Los Banos.

9. Winton AL (1958) Analysis of Foods, 3rd edn. P. 6. John. Wiley and Sons. Inc., New York, USA.

10. Kessler HG (1981) Food Engineering and Dairy Technology. Kessler, Freising, Germany.

11. FAO (1977) Regional Dairy Development and Training Center for the Near East Laboratory Manual Spring.

12. Petersen BL, Dave RI, McMahon DJ, Oberg CJ, Broadbent JR, et al. (2000) Influence of capsular and ropy exopolysaccharide-producing Streptococcus thermophilus on Mozzarella cheese and cheese whey. J Dairy Sci 83: 1952-1956.

13. Benezech T, Maingnnat JF (1994) Characterization of the rheological properties of yoghurt. A review. J. Food Engineering 21: 447-472.

14. Awad RA, Metwally AI (2000) Evaluation of total milk proteinate as a milk solids source in ice cream manufacture. Annals Agric Sci, Ain Shams Uaniv, Cairo 45: 603-618.

15. Suneeta P, Prajapati JP, Patel AM, Patel HG, Solanky MJ, et al. (2007) Studies on the effect of whey protein concentrate in development of low-fat ice cream. Journal of Food Science and Technology 44: 586-590.

16. Parsons JG, Dybing ST, Coder DS, Spurgeon KR, Seas SW, et al. (1985) Acceptability of ice cream made with processed whey and sodium caseinate. J Dairy Sci 68: 2880-2885.

17. Atherton VH, Newlander JA (1977) Chemistry and Testing of Dairy Products. AVI Publ Co, Inc, Westport, CT.

18. Tirumalesha A, Jayaprakasha HM (1998) Effect of admixture of spray dried whey protein concentrate and butter milk powder on physico-chemical and sensory characteristics of ice cream. Indian J Dairy Sci 51: 13-19.

19. Kerdchouay P, Surapat S (2012) Effect of skimmed milk substitution by whey protein concentrates in low-fat coconut milk ice cream.

20. Castro E, Silva C, Osorio F, Miranda M (2000) Characterization of caramel jam using back extrusion technique. Latin Amer Appl Res 30: 227-232.

21. Awad RA (2007) Performance of rice flour in ice cream manufacture. Proceedings of the 10th Egyptian Conference on Dairy Science & Technology 517-534.

22. Muse MR, Hartel RW (2004) Ice cream structure elements that affect melting rate and hardness. J Dairy Science Association.

23. Haque ZU, Ji T (2003) Cheddar whey processing and source. Effect on non-fat ice cream and yoghurt. Inter J of Food Sci and Tech 38: 463-473.

24. Alvarez VB, Vodovotz WY, Ji T (2005) Physical properties of ice cream containing milk protein concentrates. J Dairy Sci 88: 862-871.

25. Akalin AS, Karagozlu C, Unal G (2008) Rheological properties of reduced-fat and low-fat ice cream containing whey protein isolate and inulin. Eur Food Res Technol 227: 889-895.

26. Herald TJ, Aramouni FM, Abu- Ghoush MH (2008) Comparison study of egg yolks and egg alternatives in french vanilla Ice cream. J Text Stud 39: 284-295.

27. Sadar LN (2004) Rheological and textural characteristics of copolymerized hydrocolloid solutions containing curdlan gum.

28. Parnell-Clunies EM, Kakuda JM, Deman (1986) Influence of heat treatment of milk on the flow properties of yoghurt. J Food Sci 51: 1459-1462.

29. Tang Q, Munro PA, McCarthy OJ (1993) Rheology of whey protein concentrate solution as a function of concentration, temperature, pH and salt concentration. J of Dairy Research 60: 349-361.

30. Tomer V, Kumar A (2013) Development of high protein ice-cream using milk protein concentrate. IOSR Journal of Environmental Science, Toxicology and Food Technology (IOSR-JESTFT) 6: 71-74.

31. Pandiyan C, Kumaresan G, Annal Villi R, Rajarajan G (2010) Incorporation of whey protein concentrate in ice cream. Int J Chem Sci 8: 563-567.

32. Alfaifi MS, Stathopoulos CE (2010) Effect of egg yolk substitution by sweet whey protein concentrate (WPC), on physical properties of Gelato ice cream. International Food Research Journal 17: 787-793.

33. Mazaheri-Tehrani M, Yeganehzad S, Razmkhah-sharabiani S, Amjadi H (2009) Physicochemical and Sensory Properties of Peanut Spreads Fortified with Soy Flour. World Applied Sciences Journal 7: 192-196.

34. Tharp B, Gottemoller T, Kilara A (1992) The role of processing in achieving desirable properties in health responsive frozen dessert. Proceedings of Pennsylvania State Ice Cream Centennial Conference, Pennsylvania State University 227-246.

6

Broiler Meat Quality Evaluation Created in Simulated Conditions of Heat

Santos Vaz AB[1*], **Aline G Ganecco**[1], **Juliana Lolli MM**[1], **Mariana P Berton**[1], **Cássia RD**[1], **Greicy Mitzi BM**[2], **Marcel M Boiago**[3], **Luciana Miyagusku**[4], **Hirasilva Borba**[1] and **Pedro A de Souza**[1]

[1]Faculty of Agrarian and Veterinary Sciences, University Estadual Paulista, Jaboticabal, São Paulo, Brazil
[2]Federal University of Alagoas, Arapiraca, Maceió, Brazil
[3]State University of Santa Catarina, Chapecó, Santa Catarina, Brazil
[4]Federal University of Mato Grosso do Sul, Campo Grande, Mato Grosso do Sul, Brazil

Abstract

The effect of different periods (0, 24, 48 and 72 h) of condition heat on the physical and chemical qualities of broiler meat was evaluated. Five hundred Cobb 500® chicks were used, of which 100 were reared at a thermoneutral temperature, ideal for every rearing stage, constituting the control group. The other 400 animals were reared in a climate chamber at 32 ± 2°C, simulated conditions heat for birds. The physical and chemical qualities of the meat was evaluated at 21, 35, and 42 days. This experiment was carried out using a completely randomized design with a factorial of 2 × 4 (temperature and periods of conditions heat, respectively) and four replicates. The means were compared by Tukey`s test at the 5% significance level. It was found that the heat affected the qualitative properties of the meat, particularly its lipid oxidation, water retention capacity, shear force, r value, and pH. Microbiological assessment was carried out on days 21, 35, and 42. The temperature treatments were not found to be associated with the occurrence of associated with the occurrence of any of the microbial species considered.

Keywords: High temperature; Physiological alterations; Qualitative properties

Introduction

Carcass quality is related to variations in conviction rate in the slaughterhouse, and primarily by factors such as age, sex, nutrition, handling, transportation, ambient temperature, time of fasting, and the method of harvesting birds on the farm [1]. Birds are homeothermic animals that have a thermoregulatory center located in their hypothalamus, which is constituted of neurons that are activated in response to temperature changes, triggering behavioral reactions and adaptive mechanisms for thermoregulation. This center is responsible for maintaining and controlling homeothermy through heat exchange with the environment [2]. When in high environmental temperatures, the optimum strategies adopted by the bird for heat dissipation are: increased respiratory rate hyperventilation and peripheral vasodilation, which do not promote evaporative heat loss [3].

Climatic aspects are the major limiting factor in the development of industrial production system in warm regions, because most modern commercial broiler strains were genetically improved for breeding in temperate countries. Therefore, heat stress is a particularly serious concern for the poultry industry in the tropics and during the summer in temperate countries, since it causes stunted growth, immunosuppression, and high mortality rates, resulting in significant economic losses in production [4].

The higher susceptibility of birds to heat stress is directly related to the relative humidity of the air and the environmental temperature because, when exposed to heat, compensatory physiological responses occur to enable a return to the thermal comfort zone [5].

The problems arising from increased susceptibility to heat stress causes changes in the chicken meat, seriously damaging the industrialization of meat products as well as causing increased rejection by the consumer due to the resulting physico chemical and organoleptic changes to the meat. An animal's defense against foreign agents such as viruses, bacteria, protozoa, and other parasites occurs due to the fundamental functioning of the immune system. The balance between immune system function and environmental challenges is a determining factor for animal health [6].

The main factors that negatively interfere with the immune system are stress caused by management, mycotoxins, and low levels of vitamins and minerals in the diet. Ensuring that these factors are controlled may improve animal health and productivity, as well as directly affect the cost of animal production. Over the past few years, the influence of stress on neuroimmune function has been widely studied, due to the notoriously damaging effects on individuals. Animals exposed to adverse situations suffer the effects of stress, including delays in growth, reproductive damage, and even death in animals grown for meat production [7]. It is also important to note that thermal stress, over a prolonged period, causes lesions to the mucosa of the gastrointestinal tract and hinders the absorption of nutrients [8,9].

Salak-Johnson and McGlone [10] review studies of the effect of stress on immunity and particularly the suppression of the cellular and humoral immune responses. These authors concluded that stressful conditions trigger the release of glucocorticoids and this makes animals more susceptible to infectious diseases.

Thaxton and Siegel [11] and Miller and Quershi [12] showed that birds exposed to different types of environmental stress show reduced immune functions. When hens are exposed to temperatures varying from 32.2°C to 43°C for short periods, or cycles of constant high temperatures, the immune response is significantly reduced. Heckert et al. [13] showed that broilers subjected to overcrowding (20 birds/m²), show reduced immune activity.

According to Fuller [14], any factor that causes an imbalance in

*Corresponding author: Santos Vaz AB, Faculty of Agrarian and Veterinary Sciences, University Estadual Paulista, Jaboticabal, São Paulo, Brazil
E-mail: alinebuda@zootecnista.com.br

intestinal microbiota may allow pathogenic microorganisms to invade and multiply; therefore, an imbalance of intestinal microbiota directly reflects the health of the host. According to Courrier [15], acute infections caused by *Salmonella* serotype Enteritidis usually occur in young birds or birds reared under stressful conditions.

The most significant pathogenic bacteria in aviculture are *Salmonella* spp.; *Campylobacter jejuni,* involved in gastroenteritis outbreaks; and Listeria monocytogenes, associated with meningitis and meningoencephalitis [16]. In addition to affecting animal productivity, these bacteria are hugely important to public health, since they are intimately related to infections from food. Bacteria such as *Salmonella* spp. are more frequent in human cases of food-borne illnesses from meat and chicken products. These bacteria are found in the intestinal tracts of humans and other animals and can multiply in culture and produce visible colonies at 37°C in 24 h. The optimal pH for the development of *Salmonella* spp. is approximately neutral, with values above 9.0 and below 4.0 considered bactericidal Gast [17].

Another important group of microorganisms are *Campylobacter* spp., especially C. jejuni, which acts as a pathogenic agent or forms the normal microbiota of the gastrointestinal tract of animals such as cattle, birds, sheep, dogs, and cats [18]. Hence, meat is the greatest source of intestinal campylobacteriosis. *L. monocytogenes* is the only species of the genus Listeria that is pathogenic to humans. Listeriosis is the food-borne disease that causes the greatest number of deaths and hospitalizations (91% of cases), particularly in pregnant women, newborns, and immunocompromised individuals [19].

Thus, conditions of heat is relevant in aviculture, since high temperatures may compromise bird immune systems resulting in increased proliferation of pathogenic bacteria and decreased populations of beneficial bacteria in the gastrointestinal tract. In this sense, stress may interfere with the behavior of intestinal microbiota, which protects the host against pathogens and opportunistic infections. Due to the issues described above, the aim of this work was to evaluate the impact of heat condition on the physical, chemical and microbiological qualities of chicken meat.

Materials and Methods

Animals, initial management, and experimental conditions

The work was submitted to the Ethics Committee on the Use of Animals of the Faculty of Agrarian and Veterinary Sciences, University Estadual Paulista and was approved under protocol number 4207/2010. The experiment was conducted in climate chambers in the experimental aviary facility at Faculty of Agrarian and Veterinary Sciences, University Estadual Paulista - UNESP, Câmpus de Jaboticabal – Brazil.

For the experiment, 500 male Cobb strain 500 broilers were used. The animals received water and feed ad libitum during the 45 days of husbandry. The diets were formulated according to the ages of the birds and were based on their nutritional requirements Rostagno et al. [20] The animals were vaccinated against Gumboro disease (intermediate strain Lukert) for 7 and 19 day old birds, and against New castle disease (strain Ulster) for 12 and 24 day old birds.

Two climatic chambers were used to house the birds, with a density of 10 birds/m², distributed in boxes of 2.5 × 1 m dimensions, containing 25 birds each. One hundred birds were kept in a chamber at a thermoneutral temperature, ideal for every stage of husbandry, according to the recommendations of Cobb. The other 400 birds were

housed in another climatic chamber with heating and cooling systems, and were subjected to different periods of heat (0 h, 24 h, 48 h and 72 h), at which the camera's internal temperature rose to 32 ± 2°C. A 24 h light period was used throughout the experimental period in both chambers.

At 21 days of age the birds were subjected to different periods of conditions simulated heat. After maintaining condition heat for 72 h the heaters were turned off and the animals were kept in a thermoneutral environment until the next phase of simulated condition heat, at 35 and 42 days. After each condition heat simulated period, 12 birds from each chamber were slaughtered in the experimental abattoir in the aviary sector at FCAV / UNESP using conventional slaughtering procedures. The carcasses were then packed in plastic bags and placed in boxes with crushed ice until the time of analysis. The remaining birds stayed in their boxes until the next period of simulated conditions of heat.

Evaluation of meat quality

The carcasses were sent to the laboratory of Animal Products Technology at FCAV/UNESP, where physical and chemical analysis of the pectoralis major muscle was performed. The pH in the muscle was measures 24 h after slaughter using a Testo 205-digital pH meter, coupled to a Digimed glass probe.

The color was assessed in three parts of the muscle sample using a Minolta Chroma Meter CR-300 colorimeter, and the CIELab system to evaluate the L* (lightness) parameters, ranging from black (0) to white (100), a* (red content) ranging from green (- 60) to red (60), and b* (yellow content), ranging from blue (- 60) to yellow (60). The water holding capacity (whc) was evaluated according to the methodology described by Hamm [21,22], by measurement of the water released when a 10 kg pressure is applied for five minutes to 0.50 g samples of muscle tissue. The percentage of water lost was calculated from the difference in sample weight before and after undergoing pressurization. To determine the weight loss by cooking (wlc). The samples were weighed and packed in plastic bags and cooked in a water bath at 85°C for 30 min until the internal temperature reached 75°C. Next, the samples were removed from the bags, left at ambient temperature, and reweighed to calculate the weight loss [23].

The softness was evaluated from the shear strength of the samples after firing, taken perpendicular to the muscle fiber orientation, using Warner-Bratzler blade adapted with a Stable Micro Systems TA-XT2i texturometer, and the results were expressed as the maximum shear force in kgf/cm² [23]. The thiobarbituric acid reactive substances (tbars), resulting from lipid oxidation of the chicken meat samples, were determined according to the method described by Pikul et al. [24]. Triplicate samples, weighing around 10 g were homogenized with 50 mL of 7.5% trichloroacetic acid solution (tca) The supernatant was filtered and 4 mL aliquots of the filtrate were treated with 5 mL of thiobarbituric acid solution (tba), placed in boiling water, cooled, and analyzed with a spectrophotometer at 538 nm. The results were expressed in milligrams of the tbars per 1 kg of sample. The value of r was determined according to the methodology described by Honikel and Fischer [25]. This assessment was based on the extraction of nucleotides through homogenization with 1 M perchloric acid using a ratio of, 1:10 v / m. After filtering, a 0.1 mL aliquot was diluted with 4.9 mL of 0.1M phosphate buffer at pH 7.0 and was analyzed at 250 nm (inosine monophosphate) and 260 nm (adenosine triphosphate) in a spectrophotometer. The r value was determined from the ratio of the two absorbance wavelengths.

Preparation of swab dilutions

Swab dilutions were prepared according to the methods of Apha (2001). From each sample, 1 mL of the swab transport solution was removed aseptically and placed into 9 mL of 0.1% sterilized peptone water, which was then homogenized in a Stomacher for one minute. An initial dilution of 10:1 was obtained, and decimal dilutions were prepared (up to 10:5) using the same dilutant.

Detection of *Campylobacter* spp.

To detect *Campylobacter* spp., the SimPlate method was used according the manufacturer specifications Biocontrol [26,27]. Initially, 1 mL of the sample, 9 mL sterile distilled water, 0.025 mL of rifampicin and 0.04 mL of Hemin were added to each tube containing the substrate to hydrate it. This mixture was then added to the center of the plates and distributed over all cavities using circular movements. The plates were incubated in an inverted position for 48 h at 42°C in the dark under microaerophilic conditions (5% O_2, 10% CO_2, and 85% N_2), in anaerobiosis jars. After incubation, the cavities were observed and those with color changes from yellow to red were considered presumptive positive samples. These plates that showed colored cavities were observed using a fluorescence camera under 365 nm UV light and the colored cavities that did not fluoresce were considered to be positive for *Campylobacter*. To confirm the positive results, the difference between the numbers of red and fluorescent cavities was estimated and the SimPlate conversion table was used to obtain the total *Campylobacter* count per plate.

Detection of *Escherichia coli*

For the *Escherichia coli* assay, the most probable number method described by Hunt and Rice [27] was used. A set of three tubes containing lauryl sulphate tryptose broth with 4-methylumbelliferyl-beta-D-glucuronide (LST-MUG) by dilution were used, and 1 mL of the dilution was added to each tube of 10 mL LST-MUG. Next, the tubes were incubated at 35°C ± 0.5°C for 24 h to observe the development and/or production of gas. In the samples with a positive result, i.e., those with growth and/or gas production, the tubes were observed under ultraviolet (6 W) and long-wave (365 nm) lamps. Those with blue fluorescence were confirmed as positive for *E. coli*. The result was based on the most probable number (MPN) table.

Determination of the total enterobacteria

For the total enterobacteria assays, the plate count method described by Kornacki and Johnson [28] was used. Each dilution (1 mL) was inoculated onto empty sterile petri dishes and violet red bile glucose agar was added. After complete solidification of the medium, it was covered with an additional layer of the same medium. The plates were incubated in an inverted position at 35°C ± 1°C for 24 h. After the incubation period, the typical colonies of total enterobacteria were counted.

Detection of lactic bacteria

For the lactic bacteria assay, the plate count method of Hall and Yousef Hall and Yousef [29] was used. Each dilution (1 mL) was inoculated onto empty sterile petri dishes, and the Man, Rogosa and Sharpe culture media was subsequently added. The plates were incubated in an inverted position at 35°C ± 1°C for 72 h. After the incubation period, the plates with colonies were counted and those with at least five colonies present were selected for subsequent Gram staining and the catalase test. Isolation, selection, and identification of

Listeria monocytogenes and *Salmonella* spp. by real time polymerase chain reaction (RT-PCR).

To detect the presence of the pathogenic *Salmonella* spp. and *L. monocytogenes*, the real-time polymerase chain reaction was used (RT-PCR) Biocontrol [26], which currently is accepted for use by the Ministry of Agriculture, Livestock and Supply (MAPA) due to its specificity.

RT-PCR was performed using SYBR Green dye following the manufacturer's guidelines using Assurance GDS Rotor-Gene®. If *Salmonella* spp. were confirmed, samples were subjected to a pre-enrichment, selective enrichment, differential plating, preliminary confirmation of typical colonies and a polyvalent somatic serological test.

Pre-enrichment

A portion of the 1 mL cloacal swab sample was homogenized in 9 mL of 0.1% buffered peptone water and incubated at 37°C for 18 h.

Selective enrichment

The pre-enrichment tube was carefully shaken and 0.1 mL was transferred to 10 mL of Rappaport Vassiliadis broth and between 1 mL and 10 mL tetrathionate broth. The Rappaport Vassiliadis broth was incubated at 41.5°C ± 1°C for 24 h and the tetrathionate broth at 37°C ± 1°C for 24 h.

Differential plating

The selective enrichment tubes were shaken in a "vortex" agitator and a sample of the tetrathionate broth was streaked onto bismuth sulfite agar and xylose lysine deoxycholate agar. The same procedure was repeated using the Rappaport Vassiliadis broth. The plates were incubated in an inverted position at 35°C for 24 h to verify the development of typical colonies of *Salmonella* spp.

Preliminary confirmation of typical colonies of *Salmonella* spp.

When a suggestive colony was found, part of the cell mass was removed from the center of the typical colony using an inoculation needle and inoculated in inclined tubes containing iron lysine agar and triple sugar iron (TSI) Agar. Two typical colonies from each plate were selected, and inoculation was performed by stabbing and streaking on the slant, using the same loop to inoculate both tubes. The tubes were incubated at 35°C ± 1°C for 24 h and the typical reaction for *Salmonella* spp. was observed.

Polyvalent somatic serological test

Two squares of approximately 2 cm^2 were marked onto a glass slide, using a glass hydrophobic marker pen. A sample from the culture grown for 24 h in TSI was transferred with a loop to each of the squares, placing the culture sample in the upper part of each square. A drop of physiological saline solution was added to the lower part of one of the squares and the culture was thoroughly emulsified. A drop of polyvalent somatic serum anti-*Salmonella* spp. was added to the lower part of the other square. Holding the slide against a well-lit dark background, the slide was gently inclined and rotated in order to agitate the emulsion to observe the occurrence of agglutination in the square containing the serum. This result was compared with the result of the emulsion performed with the saline solution on the other square (the negative control), so that there was no confusion between cloudy appearance and an agglutination reaction.

Statistical analysis

The experimental design was completely randomized in a factorial scheme of 2 x 4 (temperatures thermoneutral and simulated conditions of heat) duration of heat (0, 24, 48, and 72 h) respectively, with four replications. The averages were compared by Tukey`s test at a 5% significance level, using the SAS [30] statistical program.

Results and Discussion

The results obtained for the thiobarbituric acid reactive substances (tbars), water holding capacity (whc), shear force (sf) and weight loss by cooking (wlc) of the broiler meat at 21 days are shown in Table 1.

There was a significant correlation between the temperature (thermoneutral and heat) and the duration of heat with the tbars (p<0.01), whose breakdown is shown in Table 2. Through the breakdown of this correlation, it is found that the lipid oxidation degree of the meat from birds that were subjected to heat was higher. This is expected because a high body temperature can cause the release of steroids, initiating the peroxidation of membrane lipids, and thus oxidation of the meat from the birds that were subjected to heat at 32 ± 2°C is increased. Heat stress can also cause disturbances in the balance between oxidant and antioxidant defense systems, causing lipid peroxidation and oxidative injury to proteins and DNA [31].

With regards to the results of the whc and sf, it is noteworthy that the duration of heat significantly alters these characteristics (p<0.01). The whc of meat was lower for birds that were subjected to 24 h of heat, as indicated by statistical difference (p<0.01) with those that experienced 0 h of heat. However, the lower whc is consistent with the notion that the pH drops during maturation, for birds that were subjected to 24 h of heat, resulting in denaturation of myofibrillar proteins. Thus, this type of meat can cause irreparable disorder during industrialization providing income disabled during processing due to its difficulty in retaining water compared to regular meat. These results are similar to those reported by Fischer et al. [32] who found there was a lower whc in the breast meat of broilers reared under heat conditions.

Smoothness is a very important factor in the consumer's perception of meat quality. It is closely related to the amount of intramuscular water and, therefore, to the whc of the meat, so that the higher the water content in the muscle set, the greater the tenderness of the meat. According to the results obtained, it appears that the 72 h duration of heat caused more tenderness in the meat compared to other periods of 0-24 h, coinciding with a high whc value.

Table 3 shows the results obtained for the analysis of lightness (L*), redness (a*), yellow intensity (b*), pH, and r value of broiler chicken meats at 21 days of age. It can be observed that there was only a significant correlation between the temperature (thermoneutral and heat) and the duration of heat with the pH variable (p<0.01) the breakdown of wich is shown in Table 4. Notably, the pH value of the meat at 21 days after heat for 24 h was significantly lower (p<0.05) than that for the control group. The low pH value in animals that were subjected to heat may be related to the acceleration of glycolysis reactions post mortem. Muscle tissues with pH values below 5.8 about 15 min after slaughter, when the carcass is still at near-physiologic temperatures, usually undergoes partial protein denaturation, impairing its functional properties by Tankson et al. [33].

The lightness (l*), intensity of yellow (b*), and redness (a*) of broiler chicken meat at 21 days, showed no significant differences between the temperatures (thermoneutral and heat) tested and were not influenced

by the period of exposure to high temperature (p>0.05). Regarding the red content (a*), a significant increase (p<0.05) in the chest of the birds that remained in a hot environment for 48-72 h compared to birds that were not exposed to heat is observed. Most authors found a higher red content in the muscles of birds reared in cold weather, reporting that the environmental temperature has a greater influence on the development of muscle fiber type.

The r value differed (p<0.05) between temperature (thermoneutral and heat) and duration of the exposure. This value represents an indirect measure of the depletion of ATP in the muscle during the development of rigor mortis, with r becoming greater as the ATP is

	Tbars (mg TMP/Kg am)	whc (%)	sf (kgf/cm²)	wlc (%)
Temperature (T)				
Thermoneutral	0.021	70.90A	0.90A	30.44A
Heat	0.266	69.86A	0.99A	31.12A
F test	190.00**	1.78NS	0.88NS	0.80NS
Duration in hours of heat (D)				
0	0.077	73.71A	1.17A	30.97A
24	0.109	68.37B	1.02A	31.44A
48	0.148	68.49AB	0.86AB	31.27A
72	0.243	70.95AB	0.74B	29.44A
F test	34.04**	10.62**	3.38**	1.45NS
F Int. TxD	30.39**	0.72NS	0.35NS	0.18NS
CV (%)	11.37	3.77	16.71	8.55

Averages in the same column followed by the same letter do not differ by Tukey`s test (5%); **(p<0.01); CV = Coefficient of Variation; NS = Not Significant.

Table 1: Averages obtained for thiobarbituric acid reactive substances (tbars), water holding capacity (whc), shear force (sf), and weight loss by cooking (wlc) of broiler meat at 21 days of age.

Temperature (T)	Duration in hours of heat (D) – tbars (mg TMP/kg am)			
	0 h	24 h	48 h	72 h
Thermoneutral	0.026Ba	0.027Ba	0.009Aa	0.019Ba
Heat	0.085Ac	0.538Aa	0.163Ab	0.277Ab

Averages in the same column followed by the same capital letter do not differ by Tukey`s test (5%); averages in the same row followed by the same lowercase letter do not differ by Tukey`s test (5%).

Table 2: Breakdown of the interaction between temperature and duration (in hours) of heat for the means of the thiobarbituric acid reactive substances (tbars) of broiler meat at 21 days of age.

	l*	a*	b*	pH	r value
Temperature (T)					
Thermoneutral	50.19A	3.64A	2.76A	5.71	0.86B
Heat	49.54A	3.39A	2.69A	5.58	0.96A
F-test	0.56NS	1.13NS	0.07NS	13.69**	9.37**
Duration in hours of heat (D)					
0	48.85A	2.91B	2.67A	5.79	0.73B
24	50.26A	3.22AB	2.87A	5.57	0.88A
48	50.96A	4.01A	2.53A	5.62	0.95A
72	49.44A	3.92A	2.82A	5.61	1.08A
F test	1.10NS	5.21**	0.33NS	7.00**	18.74**
F Int. TxD	0.17NS	0.14NS	0.25NS	6.62**	1.69NS
CV (%)	6.08	18.22	10.45	2.17	12.35

Averages in the same column followed by the same letter do not differ by Tukey`s test (5 %); ** (p < 0.01); CV = Coefficient of Variation; NS = Not Significant.

Table 3: Averages obtained for lightness (l*), redness (a*), yellow intensity (b*), pH, and r value of broiler meat at 21 days of age.

Temperature (T)	Duration in hours of heat (D) – pH			
	0	24	48	72
Thermoneutral	5.58Aa	5.99Ab	5.63Aa	5.63Aa
Heat	5.54Aa	5.61Ba	5.60Aa	5.60Aa

Averages in the same column followed by the same capital letter do not differ by Tukey's test (5%); averages in the same row followed by the same lowercase letter do not differ by Tukey's test (5%).

Table 4: Breakdown of the interaction between temperature and duration of heat (in hours) for the average pH of broiler meat at 21 days of age.

	tbars (mg TMP/kg am)	whc (%)	sf (kgf/cm^2)	wlc (%)
Temperature (T)				
Thermoneutral	0.008	73.18A	1.63A	27.94A
Heat	0.065	70.07B	1.28B	27.87A
F test	179.01**	12.50**	17.72**	0.01^{NS}
Duration in hours of heat (D)				
0	0.021	72.08A	1.54AB	27.65A
24	0.031	71.83A	1.79A	27.98A
48	0.041	71.47A	1.29BC	29.08A
72	0.059	71.15A	1.19C	26.89A
F test	19.67**	0.23^{NS}	10.07**	1.76^{NS}
F Int. TxD	19.37**	1.96^{NS}	0.42^{NS}	1.24^{NS}
CV (%)	14.35	4.21	19.67	8.08

Averages in the same column followed by the same letter do not differ by Tukey's test (5%); **(p < 0.01); CV = Coefficient of Variation; NS = Not Significant.

Table 5: Averages obtained for thiobarbituric acid reactive substances (tbars), water holding capacity (whc), shear force (sf), and weight loss by cooking (wlc) of broiler meat at 35 days of age.

consumed. According to these results, it can be concluded that birds subjected to heat for 24, 48, and 72 h showed a faster decrease of ATP in the muscle, resulting in a higher r value and accelerated rigor mortis. These results were similar to those reported by Marchi et al [34] who found higher r values, for birds subjected to heat at 35°C.

The tbars, whc, sf and wlc results for broiler meats at 35 days of age are presented in Table 5. It can be seen that there is a significant correlation between temperature (thermoneutral and heat) and the duration of the heat with the tbars (p<0.01), whose breakdown is described in Table 6. In analyzing the breakdown of this correlation, an increased lipidoxidation is observed in the meat of the birds that were subjected to heat. These data are consistent with those found by other authors, who report that heat leads to increased generation of free radicals, resulting in higher rates of lipid oxidation in the meat. Lin et al. [35] investigated the exposure of chickens to simulated conditions heat (32°C for 6 h) and the results of this study suggest that elevated body temperature can induce metabolic disorders, causing oxidative stress and raising the tbars levels in the plasma and liver compared with the control chickens.

There was no correlation (p>0.05) between temperature (thermoneutral and heat) and the duration of heat for the whc, sf, and wlc at 35 days of age. There was a statistical difference (p<0.05) between the whc values obtained for birds at different temperatures (thermoneutral and heat).

The ability to retain water, one of the main contributors to the yield during product processing, was lower in birds stressed by heat, wich causes a reduced juiciness Fletcher [36] and lower product life Barbut [37]. The data obtained are consistent with those in the literature Woefel et al. [38] which identified a trend of reduced whc in meat originating

from animals that were stressed before slaughter. The results show that the shear strength at the different temperatures (thermoneutral and heat) and the duration of heat influence the tenderness of the meat (p<0.05).

The softness increased in birds that were exposed to high temperatures for a heat period of 72 h, differing significantly from that at 0-24 h. The wlc of broiler meats at 35 days, showed no significant difference (p>0.05) for all temperatures (thermoneutral and heat) and was not affected by different periods of exposure to high temperature. The results of this study agree with the data obtained by Oba et al. [39], where the husbandry temperatures of 32 (hot), 26 (thermoneutral), and 18°C for animals at 47 days did not affect this characteristic.

Table 7 show no significant correlation (p>0.05) between the temperature (thermoneutral and heat) and the duration of heat for the color (l*, a*, and b*), pH and r value of chicken meat at 35 days.

The lightness (l*) was influenced (p<0.05) by the temperature (thermoneutral and heat) with a lower l* value being observed after heat. The length, 24 and 72 h, for which the birds were subjected to high temperature, significantly affected the yellow intensity (b*) of the meat compared to the period of 0 h heat (p<0.05). As seen from Table 7, there was no statistical difference (p>0.05) for the pH and r value, for the birds at different temperatures (thermoneutral and heat), in contrast to what was seen for the animal meats at 21 days. According to the data obtained, a lower incidence of pale breast meat was found in birds reared under stressful conditions. These results are similar to those previously described in the literature. Fischer et al. [15] found darker coloration reduced l* and less yellow reduced b*) in the meat of birds subjected to heat stress at 35°C. Bianchi et al. [40] when studying the influence of the seasons on the quality of broiler meat, observed that birds slaughtered during the summer also exhibited a pale chest which contradicts the results obtained in the present work.

Temperature (T)	Duration in hours of heat (D) – tbars (mg TMP/kg am)			
	0	24	48	72
Thermoneutral	0.004Ba	0.013Ba	0.007Ba	0.007Ba
Heat	0.021Ac	0.061Ab	0.069Ab	0.111Aa

Averages in the same column followed by the same capital letter do not differ by Tukey's test (5%); averages in the same row followed by the same lowercase letter do not differ by Tukey's test (5%).

Table 6: Breakdown of the interaction between temperature and duration of heat in hours for the average of the thiobarbituric acid reactive substances (tbars) of broiler meat at 35 days of age.

	l*	a*	b*	pH	r value
Temperature (T)					
Thermoneutral	49.92A	3.23A	0.71A	5.53A	0.97A
Heat	48.27B	3.48A	0.68A	5.59A	0.96A
F test	7.73**	0.91^{NS}	0.06^{NS}	1.69^{NS}	0.00^{NS}
Duration in hours of heat (D)					
0	49.17A	3.22A	1.38A	5.51A	1.05A
24	48.98A	3.24A	0.35B	5.66A	0.99A
48	49.05A	3.21A	0.77AB	5.56A	0.94A
72	49.18A	3.76A	0.29B	5.52A	0.88A
F test	0.03^{NS}	1.10^{NS}	8.62**	2.77^{NS}	1.73^{NS}
F Int. TxD	0.39^{NS}	0.04^{NS}	0.16^{NS}	0.47^{NS}	1.72^{NS}
CV (%)	4.07	14.05	9.03	2.53	13.49

Averages in the same column followed by the same letter do not differ by Tukey's test (5%); ** (p <0.01); CV = Coefficient of Variation; NS = Not Significant.

Table 7: Average obtained for lightness (l*), redness (a*), yellow intensity (b*), pH, and r value of broiler meat at 35 days of age.

The results obtained for tbars, whc, sf, and wlc for broiler meat at 42 days of age are shown in Table 8. There was a significant correlation ($p<0.01$) between the temperature (thermoneutral and heat) and the duration of heat for tbars, the breakdown of wich is shown in Table 9. The breakdown of this correlation, shows higher lipid oxidation in meat from birds that were subjected to heat for periods of 24, 48, and 72 h of heat compared to that of 0 h. These results suggest that the high temperature induced metabolic disorders involving oxidative stress, as was seen previously at 21 and 35 days.

There was no correlation ($p>0.05$) between temperature (thermoneutral and heat) and the duration of heat for the whc, sf, and wlc at 42 days of age (Table 8). Of the evaluated parameters only the whc was not significantly different ($p>0.05$) at all temperatures (thermoneutral and heat) and durations of heat, contrary to what occurred at 21 and 35 days.

It is observed that the meat of the birds that remained in high temperatures for 24, 48, and 72 h was stiffer, than that at 0 h ($p<0.05$), resulting in a decreased quality for this type of meat. However, higher rates of sf indicate that there was less post mortem proteolytic potential, leading to a decrease in softness.

The results indicate that the wlc in the breast muscle was affected ($p<0.05$) by the duration of the heat. Greater wlc was observed in the breasts of birds kept at $32 \pm 2°C$ for 48 h, averaging 30.64%, than in the breasts of birds subjected to $32 \pm 2°C$ for 0 h with an average of 26.81%. Some authors report that a faster rigor mortis leads to greater weight losses in cooking, but this hypothesis is not consistent with the results obtained in this work since the value of r was not statistically different ($p>0.05$) for the duration of heat (Table 10). Bressan and Beraquet [41] reported that the wlc in the chest muscles was influenced by the preslaughter environmental temperatures. The birds that were maintained at 30°C provided meat with wlc values (28.7%) greater than birds reared at 17°C (27.2%).

There was no correlation ($p>0.05$) between temperature (thermoneutral and heat) and the duration of heat for the color variables (l^*, a^*, and b^*), pH, and r at 42 days of age (Table 10). The results of color (l^*, a^*, and b^*) and r, show that the temperature (thermoneutral and heat) and the duration of the heat does not significantly alter these features ($p>0.05$). A significant increase in the pH of meat from birds

	tbars (mg TMP/kg am)	whc (%)	sf (kgf/cm²)	wlc (%)
Temperature (T)				
Thermoneutral	0.010	72.32A	2.15B	28.88A
Heat	0.081	73.92A	2.74A	29.15A
F test	145.94**	1.67NS	16.88**	0.11NS
Duration in hours of heat (D)				
0	0.023	73.47A	1.87B	26.81B
24	0.047	72.69A	2.90A	29.60AB
48	0.049	73.63A	2.49A	30.64A
72	0.063	72.68A	2.52A	26.98AB
F test	8.26**	0.16NS	8.94**	3.87*
F Int. TxD	5.19**	2.00NS	2.54NS	1.49NS
CV (%)	10.52	5.80	15.05	9.80

Averages in the same column followed by the same letter do not differ by Tukey's test (5%); * (p < 0.05); ** (p < 0.01); CV = coefficient of variation; NS = Not significant

Table 8: Averages obtained for thiobarbituric acid reactive substances (tbars), water holding capacity (whc), shear force (sf), and weight loss by cooking (wlc) of broiler meat at 42 days of age.

Temperature (T)	Duration in hours of heat (D) – tbars (mg TMP/Kg am)			
	0	24	48	72
Thermoneutral	0.015Ba	0.005Ba	0.011Ba	0.009Ba
Heat	0.041Ab	0.083Aa	0.089Aa	0.112Aa

Averages in the same column followed by the same capital letter do not differ by Tukey's test (5%); averages in the same row followed by the same lowercase letter do not differ by Tukey's test (5%).

Table 9: Breakdown of the interaction between temperature and duration (in hours) of heat for the average of the thiobarbituric acid reactive substances (tbars) of broiler meat at 42 days old.

	l*	a*	b*	pH	r value
Temperature (T)					
Thermoneutral	49.48A	2.86A	0.59A	5.59A	0.91A
Heat	48.84A	2.57A	0.89A	5.64A	0.94A
F test	0.61NS	2.19NS	1.89NS	0.72NS	0.88NS
Duration in hours of heat (D)					
0	48.92A	2.62A	0.86A	5.45B	0.94A
24	48.83A	2.55A	0.84A	5.48B	0.96A
48	48.07A	3.12A	0.80A	5.52B	0.87A
72	49.83A	2.57A	0.76A	6.01A	0.94A
F test	1.08NS	1.95NS	0.61NS	30.51**	1.49NS
F Int. TxD	0.14NS	0.81NS	2.64NS	0.26NS	0.18NS
CV (%)	5.73	15.01	9.57	2.96	11.66

Averages in the same column followed by the same letter do not differ by Tukey's test (5%); ** (p < 0.01); CV = Coefficient of Variation; NS = Not Significant.

Table 10: Averages obtained for lightness (l*), redness (a*), yellow intensity (b*), pH, and r value of broiler meat at 42 days of age.

that remained in the heat for 72 h, compared to periods of 0, 24, and 48 h ($p<0.05$) was observed. The pH of the muscle has been associated with other meat quality attributes, including softness, whc, wlc, juiciness, and microbial stability. The final pH of the meat is intimately connected with the glycogen concentration in the muscle moments before slaughter, as this will significantly affect the pH reduction Roça [42]. According to the results, the highest pH value results from a depletion of glycogen *in vivo*, preventing reduction of the pH post mortem. However, it can be concluded that the rigor of the muscle was not sufficient for meat processing.

The results for *Salmonella* spp., *L. monocytogenes*, and *Campylobacter* spp. in the bird cloacal samples are shown in Table 1. Note that all results were negative during all experimental phases.

The results for the lactic bacteria, *E. coli*, and total enterobacteria counts on the cloacal swab samples are shown in Figures 1-3. The assay results of the 16 cloacal samples obtained using a swab from birds during their first few days of life were used to monitor the microbiological conditions that precede thermal heat in birds.

Note that there was no significant difference (Figures 1-3) in the incidence of lactic bacteria, *E. coli*, and total enterobacteria between the samples from animals that were grown at a thermoneutral temperature and those that were subjected to heat at $32°C \pm 2°C$. In all samples, there was a significant increase in log UFC/mL or log NMP/mL over the 21 days; however, a correlation with acute heat could not be confirmed, since both the control samples and the samples with treatment showed the same profile.

Conclusion

Although determining the population of total enterobacteria was important, as it is a group of microorganisms frequently present in the gastrointestinal tract of birds. Lactic bacteria are associated with

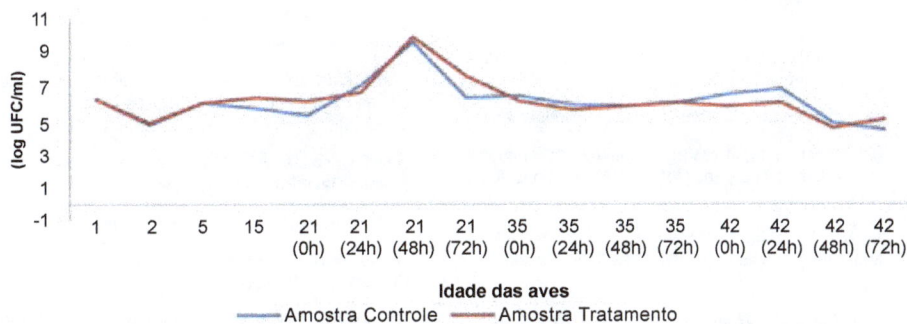

Figure 1: Lactic bacteria counts in cloacal swab samples obtained from the broilers on the following experimental days: 1, 2, 5, 15, 21, 35, and 42 (at 0 h, 24 h, 48 h, and 72 h).

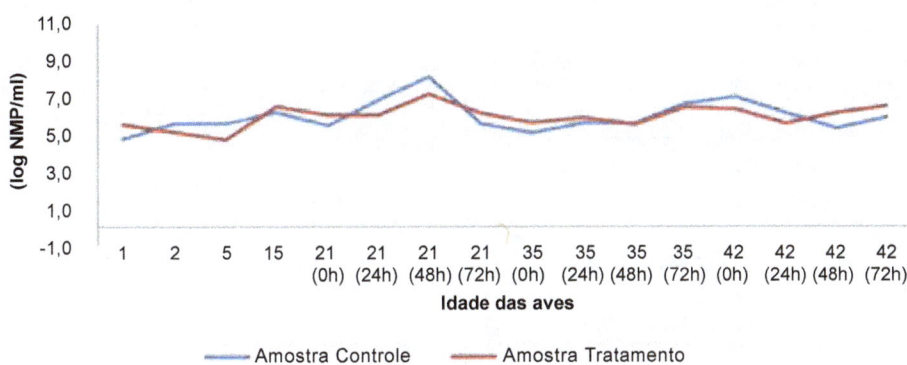

Figure 2: Results obtained for the Escherichia coli screening in cloacal swab samples from birds on the following experimental days: 1, 2, 5, 15, 21, 35, and 42 (at 0 h, 24 h, 48 h, and 72 h).

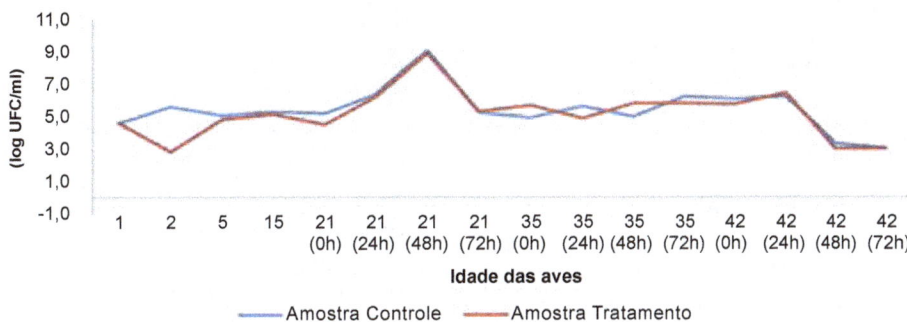

Figure 3: Results obtained from the total *Enterobacteriaceae* screening in cloacal swab samples from the broilers on the following experimental days: 1, 2, 5, 15, 21, 35, and 42 (at 0 h, 24 h, 48 h, and 72 h).

protection against pathogenic microorganisms, and determining this population was important since in conditions of heat, the number of lactic bacteria may decrease, which contributes to an increase in pathogenic bacteria in the gastrointestinal flora. Thus, decreases in the zootechnic and physiological indices of animals are expected, but this was not observed in our study using heat.

In the available literature, we did not find articles assessing the relationship between heat and the incidence of lactic bacteria, *Campylobacter* spp., *E. coli*, total enterobacteria, and *L. monocytogenes*. However, some authors suggest that after heat, there may be a greater incidence of pathogenic microorganisms, since the increase in temperature decreases the immune response of birds, leading to a greater colonization of the intestine by pathogens in these animals.

Quinteiro-Filho et al. [43] verified that heat is capable of increasing *Salmonella* spp. colony forming units (UFC/g) in the spleen of animals and causes a decrease in weight gain and food consumption and an increase in mortality [44-48].

References

1. Mendes AA, Komiyama CM (2011) Management strategies aimed at broiler carcass and meat quality. Revista Brasileira de Zootecnia 40: 352-357.

2. Borges SA, Maiorka A, Da Silva AVF (2003) Fisiologia do estresse calórico e a utilização de eletrólitos em frangos de corte. Ciência Rural 33: 975-981.

3. Lavor CTB, Fernandes AAO, De Sousa FM (2008) Effect of thermal insulating materials in aviaries on broiler performance. Revista Ciência Agronômica 39: 308-316.

4. Renaudeau D, Collin A, Yahav S, De Basilio V, Gourdine JL (2012) Adaptation

to hot climate and strategies to alleviate heat stress in livestock production. Animal 6: 707-728.

5. Caires CM, De Carvalho AP, Caires RM (2008) Nutrição de frangos de corte em clima quente. Revista Eletrônica Nutritime 5: 577-583.

6. Santin E (2012) Modulating the immune system of birds to increase productivity.

7. Moberg GP (2000) Biological response to stress: Implications for animal welfare. The biology of animal stress: basic principles and implications for animal welfare, Wallingford, England.

8. Kornacki, JL, Johnson JL (2001) Enterobacteriaceae, coliforms, and Escherichia coli as quality and safety indicators. Public Health Association, Washington.

9. Sifri M (2006) Informal nutrition symposium: dynamics of the digestive system- Introduction. J App Poul Res 15: 122.

10. Salak-Johnson JL, Mcglone JJ (2007) Making sense of apparently conflicting data: Stress and immunity in swine and cattle. J Anim Sci 85: 81-88.

11. Thaxton JP, Siegel HS (1982) Immuno-depression in young chickens by high environmental temperature. Poultry Sci 42: 202-220.

12. Miller L, Qureshi MA (1991) Introduction of heat shock proteins and phagocytic function of chicken macrophage following in vitro heat exposure. Vet Immunol Immuno 37: 34-42.

13. Heckert RA, Estevez I, Russek-Cohen E, Pettit-Riley R (2002) Effects of density and perch availability on the immune status of broilers. Poultry Sci 81: 451-457.

14. Fuller R (1989) Probiotics in man and animals: A review. J Appl Bacteriology 66: 365-378.

15. Courrier DE (1991) Mycotoxicosis mechanisms of immunosuppression. Vet Immunol Immunop 30: 73-87.

16. Byrd JA, Anderson RC, Callaway TR, Moore RW, Knape KD, et al. (2003) Effect of experimental chlorate product administration in the drinking water on Salmonella typhimurium contamination of broilers. Poultry Sci 82: 1403-1406.

17. Gast RK (2003) Salmonella infections. Dis Poultry 11: 567-583.

18. Scarcelli E, Genovez ME, Cardoso MV, Souza MCAM, Grasso LMPS, et al. (1998) Evaluation of the potential spread of Campylobacter spp. by different animal species. Arq Ins Biol 65: 55-61.

19. European Food Safety Authority (2006) The community summary report on trends and sources of zoonoses, zoonotic agents, antimicrobial resistance and foodborne outbreaks.

20. Rostagno HS, Albino LFT, Donizete JL, Gomes PC, De Oliveira RF, et al. (2005) Brazilian tables for poultry and swine: Food composition and nutritional requirements, Viçosa.

21. Hamm R (1960) Biochemistry of meat hydration. Adv Food Res 10: 335-443.

22. Cason JA, Lyon CE, Papa CM (1997) Effect of muscle opposition during rigor on development of broiler breast meat tenderness. Poultry Sci 76: 785-787.

23. Wheeler TL, Cundiff LV, Koch RM, Crouse JD (1996) Characterization of biological types of cattle (Cycle IV) carcass traits and longissimus palatability. J Animal Sci 74: 1023-1035.

24. Pikul J, Leszczynski DE, Kumerow FA (1989) Evaluation of three modified TBA methods for measuring lipid oxidation in chicken meat. J Agriculture and Food Chemistry 37: 1309-1313.

25. Honikel KO, Fischer C (1977) A rapid method for detection of PSE and DFD porcine muscles. J Food Science 42: 1633-1636.

26. Biocontrol (2010) Assurance GDS Genetic Detection System. USA.

27. Hunt ME, Rice EW (2005) Microbiological examination. American Public Health Association, American Water Works Association & Water Environment Federation, Washington DC 21: 949-958.

28. Kornacki JL, Johnson JL (2001) Enterobacteriaceae, coliforms, and Escherichia coli as quality and safety indicators. Public Health Association, Washington 4: 69-82.

29. Hall ST, Yousef AE (2005) Tests for groups of microorganisms. Stand Met Examin Dairy Products. 17: 234-247.

30. SAS Institute (2002) SAS user's guide: statistics.

31. Droge W (2002) Free radicals in the physiological control of cell function. Physiological Reviews 82: 47-95.

32. Fischer PC, Brossi C, Golineli BB, Castillo C (2005) Chicken breast quality subjected to acute heat stress. Piracicaba, São Paulo, Brazil.

33. Tankson JD, Vizzier-Thaxton Y, Thaxton JP, May JD, Cameron JA (2001) Stress and nutritional quality of broilers. Poultry Sci 80: 1384-1389.

34. Marchi DF, Oba A, Dos Santos GR, Soares AL, Shimokomaki M (2010) Evaluation of halothane as a stressor in chickens. Ciências Agrárias 31: 405 - 412.

35. Lin H, Decuypere E, Buyse J (2006) Acute heat stress induces oxidative stress in broiler chickens. Comparativ Biochem Physiol 144: 7-11.

36. Fletcher DL (2002) Poultry meat quality. World's Poultry Sci J 58: 131-145.

37. Barbut S (1998) Estimating the magnitude of the PSE problem in poultry. J Muscle Food 9: 35-49.

38. Woelfel RL, Owens CM, Hirschler EM, Martinez-Dawson R, Sams AR (2002) The characterization and incidence of pale, soft, and exudative broiler meat in a commercial processing plant. Poultry Sci 81: 579-584.

39. Oba A, De Souza PA, De Souza HBA, Leonel FR, Pelicano ERL, et al. (2007) Meat quality of broilers to diets supplemented with chromium, created in different ambient temperatures. Acta Scientiarum Animal Sci 29: 143-149.

40. Bianchi M, Petracci M, Sirri F, Folegatti E, Franchini A, et al. (2007) The influence of the season and market class of broiler chickens on breast meat quality traits. Poultry Sci 86: 959-963.

41. Bressan MC, Beraquet NJ (2002) Effect of pre-slaughter factors on the quality of chicken breast meat. Ciências Agrotecnicas 26: 1049-1059.

42. Roça RO (2002) Meat technology and related products. Faculdade de Ciências Agronômicas UNESP, Botucatu 45-51.

43. Quinteiro-Filho WM (2010) Heat stress impairs performance parameters, induces intestinal injury, and decreases macrophage activity in broiler chickens. Poultry Sci 89: 1905-1914.

44. Rostagno HS (2005) Brazilian tables for poultry and swine food composition and nutritional requirements. Publisher UFV, Viçosa MG.

45. Martindale RG (2005) Contemporary strategies for the prevention of stress- related mucosal bleeding. Am J Hea Sys Phar 62: 511-517.

46. Furlan RL, Macari M (2002) Termorregulação. Avian physiology applied to broilers, Jaboticabal.

47. NPAH (1995) Accreditation Standards for Monitoring and Diagnosis of avian Salmonella (S. enteritidis, S. gallinarum, S. pullorume, S. typhimurium) Laboratories. Ministry of Supply and Agricutura. Agriculture Defense Department, National Plan for Avian Health, Brazil.

48. American Public Health Association (2001) Committee on Microbiological Examination of Foods. Comp met microbe examination of foods, American Public Health Association, Washington.

7

Effect of Spent Yeast Fortification on Physical Characteristics of Cassava-Wheat Composite Bread

Amaraegbu A, Adewale P and Ngadi M*

Department of Bioresource Engineering, McGill University, Canada

Abstract

In this study, the effect of substituting wheat flour with 10%, 20%, and 30% cassava flour (CF) and the addition of 5, 10, and 15 g SBY (protein concentrates) on cassava-wheat composite bread was studied. Density and specific volume of the bread were influenced ($p<0.05$) by the percentage composition of CF. The breadcrumb total color difference was significantly ($p<0.05$) influenced by the addition of SBY. In the crust, only the brightness remained significantly different between the control and SBY fortified composite bread. The parameters a*, b* and L* of the fortified bread crusts were not significantly different. The desirability score of the two major variables: % composition of CF was 20% and SBY concentration was 10 g as optimum values of the fortified cassava–wheat composite bread. The Microstructure of 5 and 10 g of fortified bread were similar but different from that of 15 g fortified bread. Findings from this study revealed suitability of SBY in composite bread making to improve physical properties of the bread.

Keywords: Protein fortifier; Cassava flour; Composite bread; Physical properties; Microstructure

Introduction

Nutritive facts of food products have been the focus of most consumers. In order to improve the quality of wheat flour based bread, composite bread is produced from wheat flour by the addition of low protein flours. Due to an apparent reduction in protein content of composite bread, the effects of different protein fortifying materials have interested many researchers. However, finding a readily available fortifying ingredient that is economical and sustainable is a challenge. Spent brewer's yeast is one of the major by-products from the brewing industry which is readily available. It is a valuable source of cheap fiber and very rich in β-glucans [1-3]. β-glucans are natural cell wall polysaccharides which constitute a major structural component of bakers and brewing yeast [4]. Thammakiti et al. [2] reported a successful procedure to extract high purity β-glucans from spent brewer's yeast and equally examined the suitability of its functional properties for food products. The authors found that the β-glucan obtained from their study had a higher apparent viscosity, water-holding capacity, and emulsion stabilizing capacity, but very similar oil binding capacity as compared with commercial β-glucan from baker's yeast. The authors suggested that the β-glucan obtained from brewer's yeast could be used in food products as a thickening, water-holding, or oil-binding agent and emulsifying stabilizer. For the modification of nutritive value, texture and stability, β-glucan has been reported as an interesting hydrocolloid because of its positive effects on human and animal health such as cholesterol-lowering, antifibrotic, anti-inflammatory, antimicrobial, and antidiabetic [4-6]. Therefore, investigation of the effects of incorporation of spent brewer's yeast in the cassava-wheat flour bread formulations is a worthwhile investment.

Most of the supplement flours (e.g. cassava flour (100 g of cassava = 1.2 g of protein), maize flour (100 g of maize flour = 6.9 g of protein), etc.) to wheat flour to form composite bread are of low protein content Begum et al. [7]. For instance, the effect of the addition of protein concentrates from natural and yeast fermented rice bran on the rheological and technological properties of wheat bread was investigated by Chinma et al. [8]. The authors reported that the springiness, cohesiveness and resilience values of wheat bread were not significantly different from composite bread. Their results further revealed that composite bread surface had embedded granules of protein-like deposits with small spores. Pasting characteristics of wheat-chia blends for bread formulation was examined by Švec et al. [9]. Similarly, Begum et al. [7] studied protein fortification and use of cassava flour for bread formulation. The authors observed a strong correlation between sensory scores and soy protein fortification techniques. Their results equally showed that wheat flour could be replaced partially by cassava flour in the composite bread preparation by applying fermentation and soy protein fortification techniques. However, low-cost protein fortifiers such as spent brewer's yeast (SBY) have not been well researched in the literature for the production of composite bread. Recently, Martins et al. [6] studied the effect of spent yeast fortification on physical parameters, volatiles and sensorial characteristics of home-made bread. The fortification of home-made bread using spent yeast was reported to have an influence on the crumb darkened and increased crumb and crust springiness and had an impact on the volatile profile of the home-made bread. In an attempt to improve gluten-free bread quality, Blanco et al. [10] enriched their composite bread from rice flour and hydroxypropyl methylcellulose with acidic food additives. The authors concluded that addition of monosodium phosphate to the dough increased the volume of the gluten-free bread, which subsequently improved the appearance, odor, taste and texture of the composite bread. The objectives of this study were to enhance the protein content of composite bread produced from cassava-wheat flour blends using SBY and to investigate the effects of SBY on the physical properties of the composite bread.

Materials and Methods

Materials

Commercial wheat flour with the nutritive fact of 4 g of protein in

---

*Corresponding author: Ngadi M, Department of Bioresource Engineering, McGill University, 21111 Lakeshore Rd., Ste-Anne-de-Bellevue, QC, H9X 3V9, Canada
E-mail: michael.ngadi@mcgill.ca

30 g of wheat flour was purchased from Walmart store in Montreal, Canada. Cassava flour with 7.75% moisture, 1.57% ash, 2% protein and 91% starch was obtained from a local tropical market. Cassava flour was mixed with wheat flour at three levels (10%, 20% and 30%) for the preparation of dough [11]. The spent brewer's yeast of 45g of protein per 100g of spent yeast was procured from Lallemand Bio- Ingredient, Montreal, Canada. The inactive spent brewer's yeast contained less than 300 colonies forming unit (CFU/g) which is far below the standard plate count (< 10000 CFU/g). The amount of spent brewer's yeast used was determined by calculating the amount of protein lost from the cassava-wheat blend. Other required ingredients (table salt, granulated sugar, shortening, Fleishmann's active dry yeast) were purchased from a Walmart store in Montreal. For the control sample, 20% cassava flour blend was used to make the bread without the addition of spent brewer's yeast.

Methods

Preparation of bread with composite flour

The composite bread was produced modifying a method described by Taofik et al. [9]. Cassava flour was added to wheat flour at three levels (10%, 20%, and 30%) w/w flour basis (Table 1). The dough was baked in a pre-programmed bread maker (black and decker B3200, Middleton, USA). The machine was pre-programmed as follows: (1) all ingredients (cassava-wheat flour % composition, 4% sugar, 1% salt, 1.3% active dry yeast and 4% shortening) w/w flour basis (Table 2); (2) Kneaded/mixed for 30 min; (3) fermented for 120 min at 40 °C; (4) baked at an incremental temperature to a maximum temperature of 105 °C in the fermentation pan for 45 min as monitored by Hotmux thermocouple (DCC Corporation, Westfield Ave., Lower Level Pennsauken, NJ, USA). The baking process which included, mixing, kneading, leavening and baking took approximately 3 hours 15 min to make a complete cycle. After cooling, the bread samples were packed in polyethylene bags and. Samples were stored in Ziploc bags at room temperature (25 °C) in preparation for further analysis.

Composite bread physical properties

Evaluation of bread mass, volume, and specific volume: The mass of the bread was determined by weighing the whole baked bread (sample) with a sensitive electronic weighing balance (0.001 g accuracy). The bread dimensions were measured using a digital caliper and a ruler. The caliper was used to determine the height (thickness) of the bread. Due to the variability of the bread's height, length, and breadth, these three dimensions were measured at three different locations of the bread to estimate the mean value. Specific volume was calculated as cm^3/g by dividing the volume of the bread loaf by its weight. The bread density is the reciprocal of specific volume.

Level	Cassava four (%)	Spent yeast required to replace the protein lost (g)
1	10	5
2	20	10
3	30	15

Table 1: Ingredients used for preparing composite bread (Levels of experimental design for dough formulation).

Dough component	Amount (g)	Amount (%)
Water	232.00	61.8
Salt	3.75	1.0
Shortening	15.00	4.0
Sugar	15.00	4.0
Active dry yeast	5.00	1.3

Table 2: Ingredients used for preparing composite bread (Constant ingredients in dough formulation).

Determination of bread crumb and crust color: A Minolta spectrophotometer CM 3500d (Minolta, Ramsey, NJ, USA) was used to determine the color parameter lightness (L), redness (+a), greenness (-a), yellowness (+b), Blueness (-b) of the baked loaves crust and crumb. The colorimeter was calibrated using a white calibration plate and zero calibration black boxes before use. The machine was then standardized with 100% wheat flour bread before the measurement of bread crust and crumb. The crumb color was determined after cutting the bread to expose the crumb in triplicate at different locations. Similarly, the crust color was determined in triplicate by taking the average result of three measurements (the sides were excluded). The colorimetric results were used in equation 1 to determine the brown index (BI) and total color difference (ΔE) of the crust and crumb (equation 2) which was estimated according to Maskan [12]:

$$BI = \frac{[100x - 0.31]}{0.17} \tag{1}$$

$$\text{Where, } x = \frac{a + 1.75L}{5.645L + a - 3.012b}$$

$$\Delta E = \sqrt{\left(L_o - L\right)^2 + \left(a_o - a\right)^2 + \left(b_o - b\right)^2} \tag{2}$$

Textural analysis - crumb softness: The softness of the bread crumb was analyzed using a tensile testing machine (Instron-4502, Instron Corporation, USA) controlled by a computer software (Instron Series IX, version 8.25). A portion of each breadcrumb [2.5 (length) × 2.5 (breadth) × 2.5 (thickness)] cm was cut from the central portion of loaves with a bread knife on which the test was carried out. Care was taken during cutting in order to obtain a clean and undistorted surface. The breadcrumb softness was examined using a compression test by a cylinder plunger with a diameter of 5 mm to a depth of 12.5 mm to reach a load of 50 N. The slices were placed on a flat surface beneath the probe. The probe was brought down until it touched the surface of the slice. The slices were compressed at a deformation speed of 50 mm/min, the compression curves were recorded. The amount of force exerted on the bread slice was as a function of the distance traveled by the probe before it returned its initial position at the same speed. The results obtained were calculated using Instron Series IX software, version 8.25 (Instron Corporation, USA), which recorded the plunger force as a function of time, to determine softness of the breadcrumb. The compression data of each breadcrumb sample was collected in triplicate.

Scanning electron microscopy

The microstructure of the composite bread enriched with three levels of SBY (5, 10 and 15%) and a control (Bread without SBY) were examined through a scanning electron microscope (SEM) (JEOL JSM-6460LV, Tokyo, Japan) operated at 20.0 kV. Composite bread samples (cassava-wheat) of 20% cassava flour fortified with SBY were prepared for scanning electron microscopy. Bread samples were cut into approximately 5 × 5 × 2 mm^3 cubes with a sharp razor. Each sample was mounted on aluminum stubs, coated with a thin layer of gold before being scanned and digital images were acquired at 50 × magnification over regions of 1280 × 1040 pixels. Three images were captured on each sample over square regions of 5 × 5 mm^2.

Statistical analysis

A full factorial design was used to verify effects of % composition of cassava flour and SBY on the quality of composite bread production from cassava-wheat flour (Table 1). The bread produced was analyzed separately and was allowed to cool down in a desiccator for a maximum of 4 h at room temperature prior to analysis. Results were reported as averages of each of the three replications (all treatments were evaluated in three batches). Statistical analysis was carried out to investigate factor effect leverage and desirability using JMP® 11.2.0 (Statistical Analysis Systems, SAS Institute Inc., Cary, NC, USA). Leverage Plot reports the effect of each factor on the response variable and gives insight on possible multicollinearity observations. Desirability functions were used to find the optimal value for each factor.

Results and Discussion

Bread mass, volume, and specific volume

The loaf size parameters of the cassava-wheat composite bread (volume, mass, density, and specific volume) were successfully estimated in this study. Figures 1A-1D shows the results of some physical properties of the bread loaf. The results of bread mass, volume, density, and specific volume range from 506 to 587 g, 1540 to 2038 cm³, 0.250 to 0.391 g/cm³, and 2.55 to 3.90 cm³/g, respectively as shown in Figure 1. All these parametric values of the bread varied significantly

($p<0.05$) based on percentage composition of cassava flour but not in the case of SBY. The bread mass is not significantly ($p>0.05$) affected by SBY concentrations (Figure 1A). The correlation analysis explained a high correlation coefficient ($r \leq 0.98$) correlation between the cassava flour and mass of loaves. Increase in cassava flour concentration caused an increase in the mass; this is as a result of insufficient carbon dioxide, and water binding capacity of cassava flour due to its high concentration of damaged starch [13-15]. A reverse case was observed for volume. There was a downward trend in the volume (Figures 1B and 1C), and specific volume as the concentration of cassava flour (%) composition increased (Figure 1D). Both CF and SBY concentrations significantly ($p<0.05$) influenced the volume and specific volume of the bread. This is as a result of the attenuation of gluten content of the wheat flour with the CF which contains no gluten. Decrease in gluten concentration leads to an early offset of gelatinization [5], plasticized the starch- protein network, causing the gas cell walls to become firmer and tougher, making it difficult for expansion to occur [13-15].

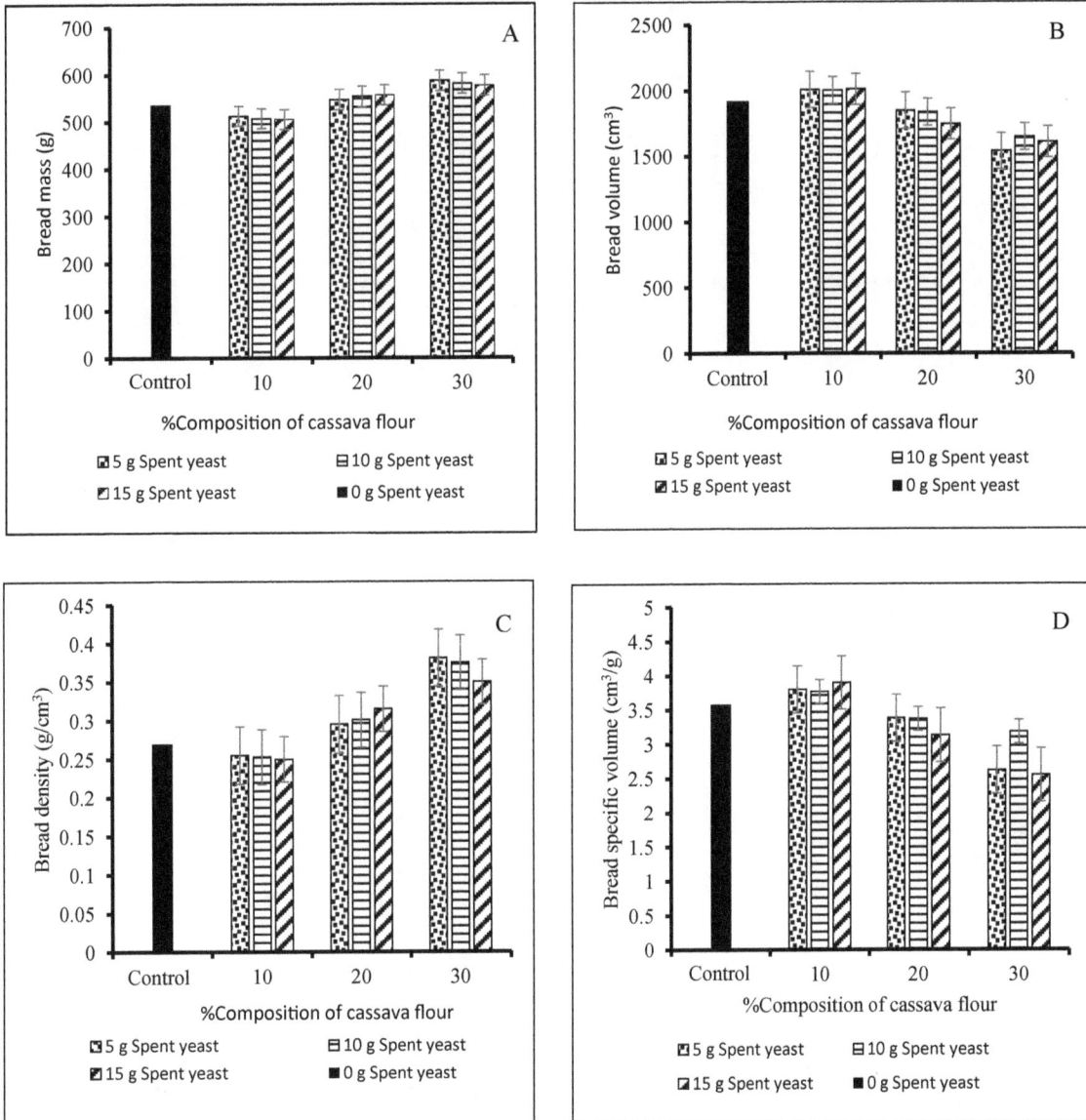

Figure 1: Bread size parameters (A) Bread mass (B) Bread volume (C) Bread density (D) Bread specific volume.

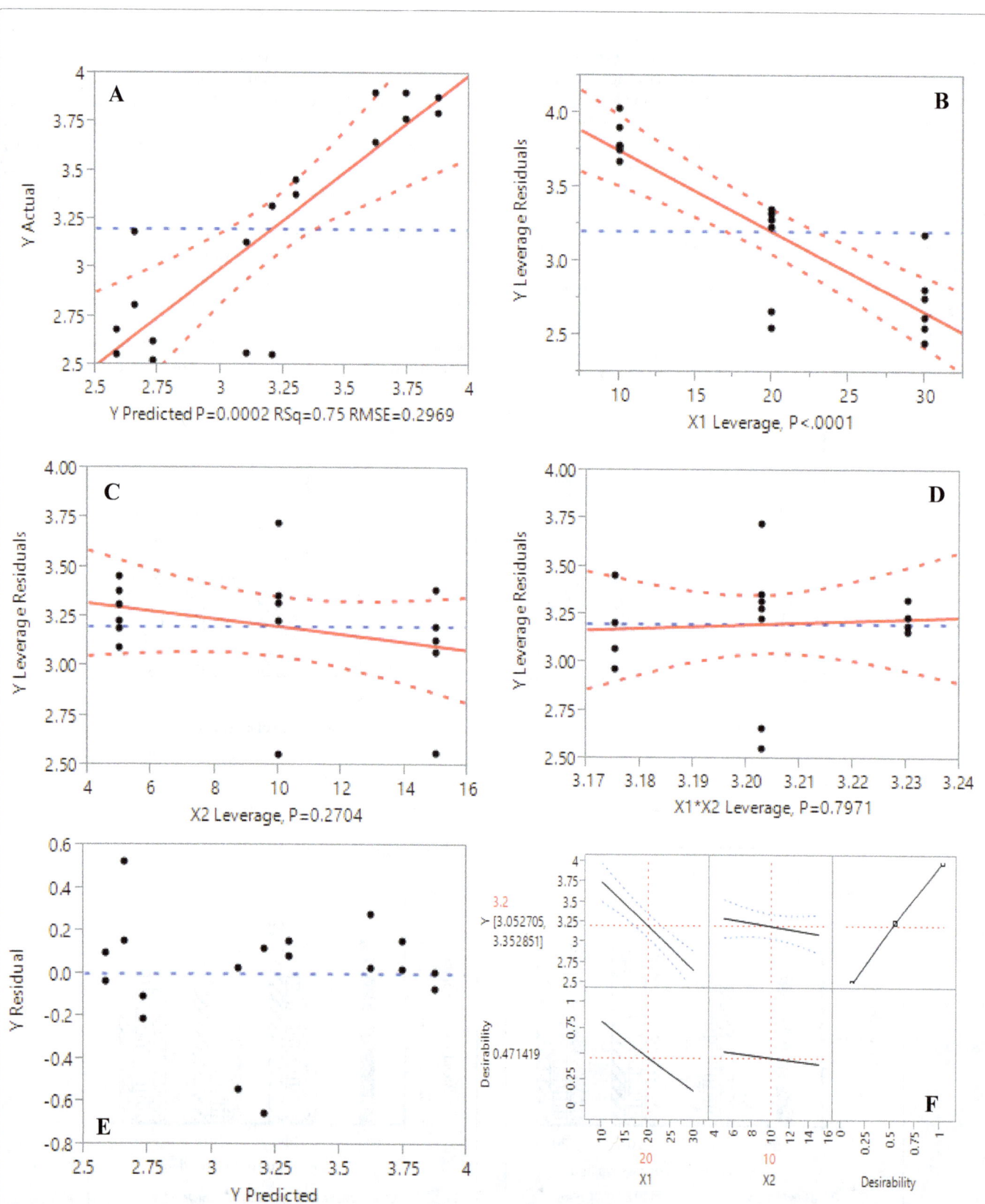

Figure 2: Effects leverage and desirability plots of bread specific volume based on (A) Yield Prediction (B) percentage composition of cassava flour (C) Spent yeast concentration (D) Factors interaction (E) Specific volume residual plot (F) Desirability of factors.

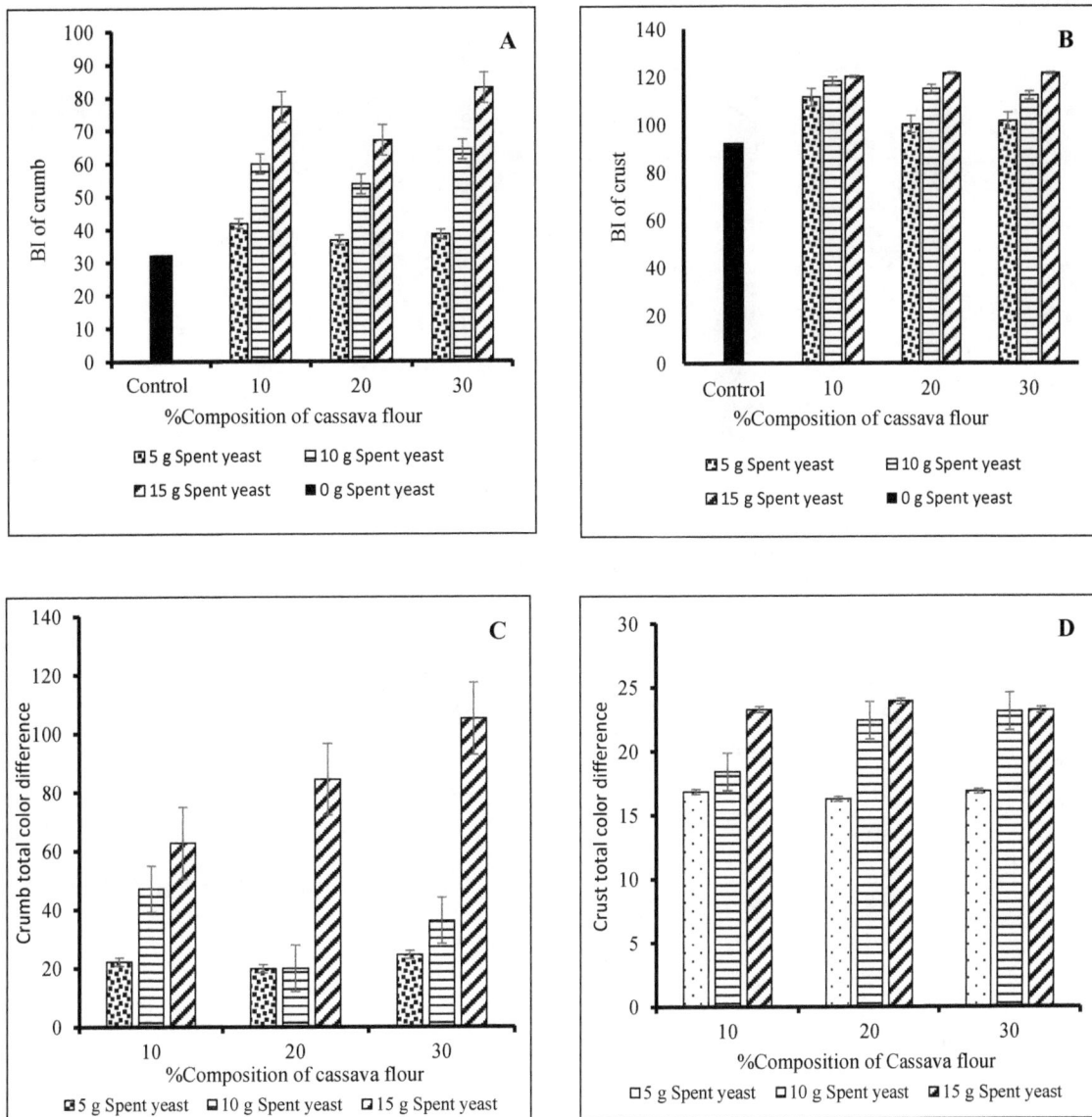

Figure 3: Crust and crumb color (A) Brown index of crumb (B) Brown index of crust (C) Crumb total color difference (D) Crust total color difference.

The higher the bread density the lower the specific volume and the bread density was a reciprocal of bread the specific volume. The expression to determine the specific volume of bread combines both mass and volume of the bread. Thus, further statistical analysis was carried out to understand the effects of CF and SBY concentration on the bread specific volume and their desirability towards the bread production. Figures 2A-2F shows the effect leverage and desirability plots of the percentage composition of CF and SBY on the bread properties based on the specific volume. The plot of actual versus predicted values, which shows the observed data against the predicted data of the bread specific volume, is shown in Figure 2A. The figure shows the leverage plot for the whole model with regression coefficient $R^2 = 0.75$, and RMSE = 0.2969 ($p<0.05$). The residual plot shows a random pattern around a horizontal axis, indicating a good fit of the model (Figure 2E). The effect leverage plot, (Figure 2B) shows the percentage composition of CF (X1) to significantly ($p<0.05$) influence the specific volume of the bread. The effect leverage plot, (Figure 2C) shows the concentration of SBY (X2) and interaction between

percentage composition of CF and SBY (Figure 2D) to be insignificant ($p>0.05$) on the specific volume of the bread. This can be ascribed to the inactive nature of the SBY used. SBY is added to enhance the protein content of the bread. So, it has no effect on the risen of the bread dough. Similar results were reported by Martins et al. [6], that two-way ANOVA indicated no significant differences due to β-glucan fortification by dry spent yeast addition ($P>0.05$) on the bread volume. Figure 2F shows the statistical test carried out to investigate the actual effect of these factors using desirability function. The desirable values for percentage composition of CF and SBY concentration are 20% and 10 g, respectively with a corresponding desirable value for specific volume to be 3.2 cm³/g.

Color properties of the bread crust and crumb

The effect of SBY on the color properties of composite bread crust and crumb were studied using two proximate parameters (brown index and total color difference) as shown in Figures 3A-3D. The addition of SBY significantly ($p<0.05$) influenced the color of the breadcrumb

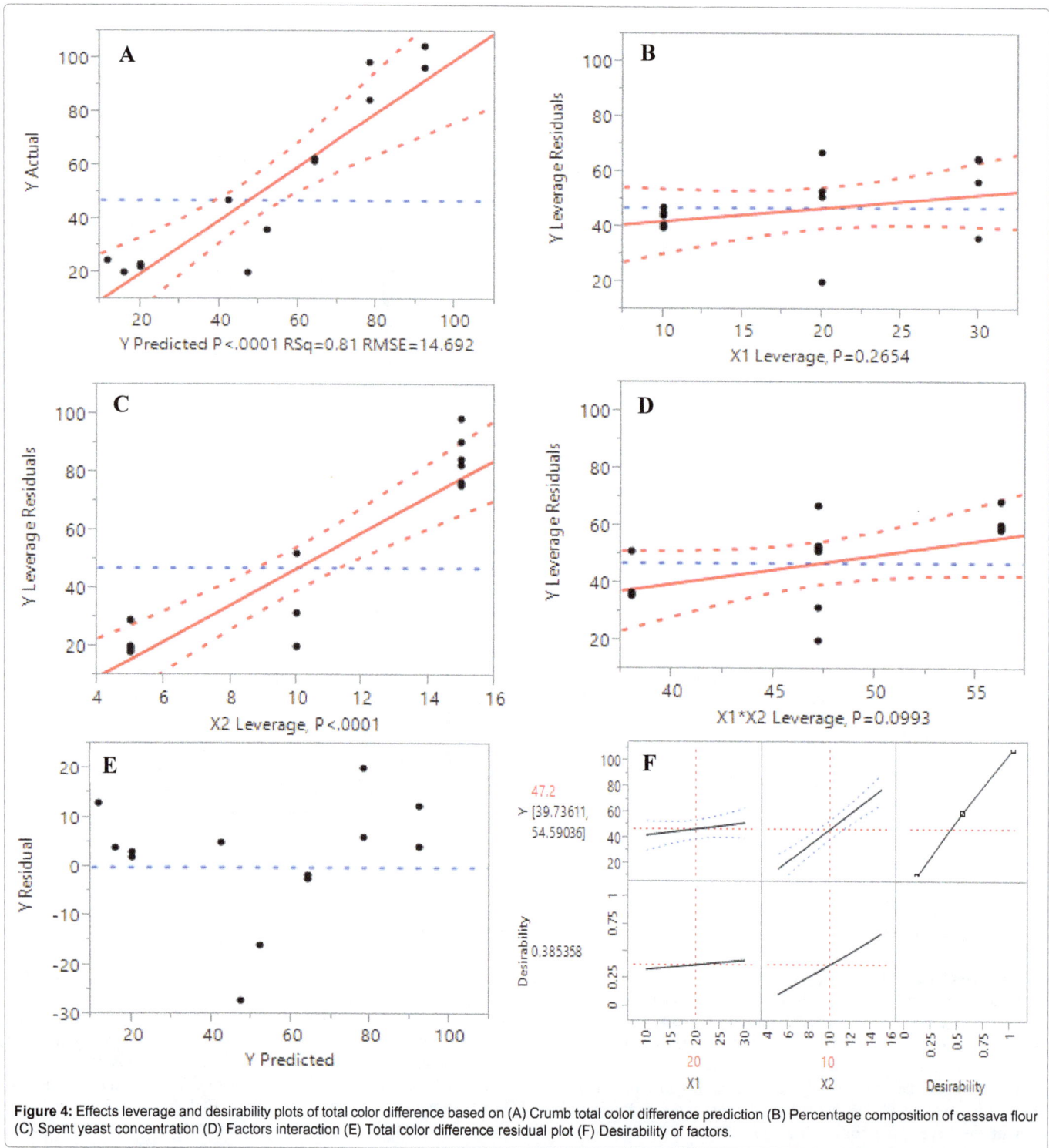

Figure 4: Effects leverage and desirability plots of total color difference based on (A) Crumb total color difference prediction (B) Percentage composition of cassava flour (C) Spent yeast concentration (D) Factors interaction (E) Total color difference residual plot (F) Desirability of factors.

but no significant influence was observed for the crust (Figures 3A and 3B). The L values decreased with the addition of SBY in bread while a, b, BI, and ΔE values increased. Total color difference parameter was used as a key analytic response to discuss the effect of SBY on the bread color. In ΔE, each of the bread samples was compared with the control bread (20% cassava flour-80% wheat flour without SBY). The control bread was lighter (L) in the crumb and crust with lower red color (a) than composite bread samples, while bread containing 15 g SBY had

the lowest (L) and highest (a) values (Figure 3C). In addition, control bread was less yellow (b) than composite bread samples. The lower L and higher (a) values of composite bread may be due to their higher protein contents due to the addition of SBY which perhaps caused increased Maillard reaction during baking. The Maillard reaction is a chemical reaction between amino acids and reducing sugars that induce browning of food products and their desirable flavor. Bread color is a very good indicator of bread quality, how the individual ingredients affect the bread formulation, the storage conditions as well as the process variable [15-17]. The brown index shows less significance of the crust color than the crumb color, this is because the crust color was also affected by the heat (direct contact of the oven pan) than the crumb (Figures 3C and 3D). Figures 4A-4F) shows the effect leverage and desirability plots of the percentage composition of CF and SBY on the bread properties based on the total color difference of the crumb. The plot of actual versus predicted values, which shows the observed data against the predicted data of the bread specific volume is shown in Figure 4A. The figure shows the leverage plot for the whole model with regression coefficient $R^2 = 0.81$, and RMSE = 14.692 (p<0.05). The residual plot shows a random pattern around a horizontal axis, indicating a good fit of the model (Figure 4E). The effect leverage plot, (Figure 4B) shows the percentage composition of CF (X1) to not significantly (p>0.05) influence the total color difference of the bread. The effect leverage plot, (Figure 4C) shows the concentration of SBY (X2) as a significant (p<0.05) factor that influenced the total color difference of the breadcrumb. The interaction between percentage composition of CF and SBY (Figure 4D) had no influence (p>0.05) on the total color difference of the breadcrumb. This can be ascribed to the inactive nature of the SBY used. SBY was added to enhance the protein content of the bread. So, it has no effect on the risen of the bread dough. Desirability function was used to investigate the actual effect of independent variables on the composite bread quality. The desirable values for percentage composition of CF and SBY concentration are 20% and 10 g, respectively with a corresponding desirable value for the total color difference was 47.2 (Figure 4F).

Texture characteristics: crumb hardness

Figure 5 shows the breadcrumb hardness per percentage composition of CF and concentration of SBY. The crumb hardness is directly proportional to the percentage composition of CF and the concentration of SBY added. The 30% CF-wheat bread had the highest hardness and lowest springiness values while the lowest hardness and highest springiness was observed in 5 g SBY fortified bread. Figures 6A-6F shows the effect leverage and desirability plots of the percentage composition of CF and SBY on the bread properties based on the crumb hardness. The plot of actual versus predicted values, which shows the observed data against the predicted data of the bread specific volume, is shown in Figure 6A. The figure shows the leverage plot for the whole model with regression coefficient $R^2 = 0.99$, and RMSE = 0.197 (p<0.05). The residual plot shows a random pattern around a horizontal axis, indicating a good fit of the model (Figure 6E). The effect leverage plot, (Figure 6B) shows the percentage composition of CF (X1) to significantly (p<0.05) influenced the hardness of the breadcrumb. The addition of SBY (X2) to cassava-wheat flour in bread making significantly (P<0.05) increased bread hardness, which led to a slight reduction in its cohesiveness value (Figure 6C). A similar finding was reported by Totosaus et al. [18] in their investigation on the effect of lupinus (*Lupinus albus*) and jatropha (*Jatropha curcas*) protein concentrates on wheat dough texture and bread quality. The authors reported that added proteins in wheat dough compete with gluten for available water which results in increased hardness but decreased cohesiveness. In contrary, Martins et al. [6] reported that the effect of β-glucan fortification by dry spent yeast addition revealed no significant effects on texture and color parameters of the bread. The interaction between percentage composition of CF and SBY (Figure 6D) had no influence (p>0.05) on the hardness of the breadcrumb. However, the confidence level was almost at the borderline. It might be because of poor extensibility of the wheat-cassava network and the dilution of the gluten contributed to this effect. This could also be as a result of SBY activation during baking and kneading period. In the study of Gómez

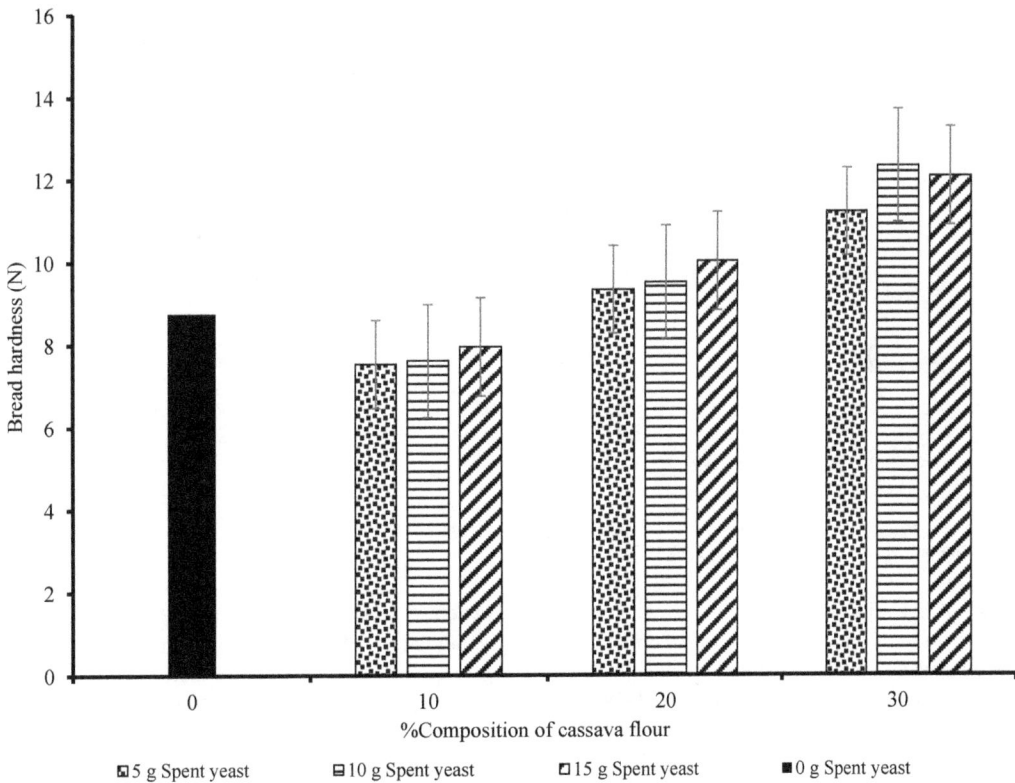

Figure 5: Bread crumb hardness.

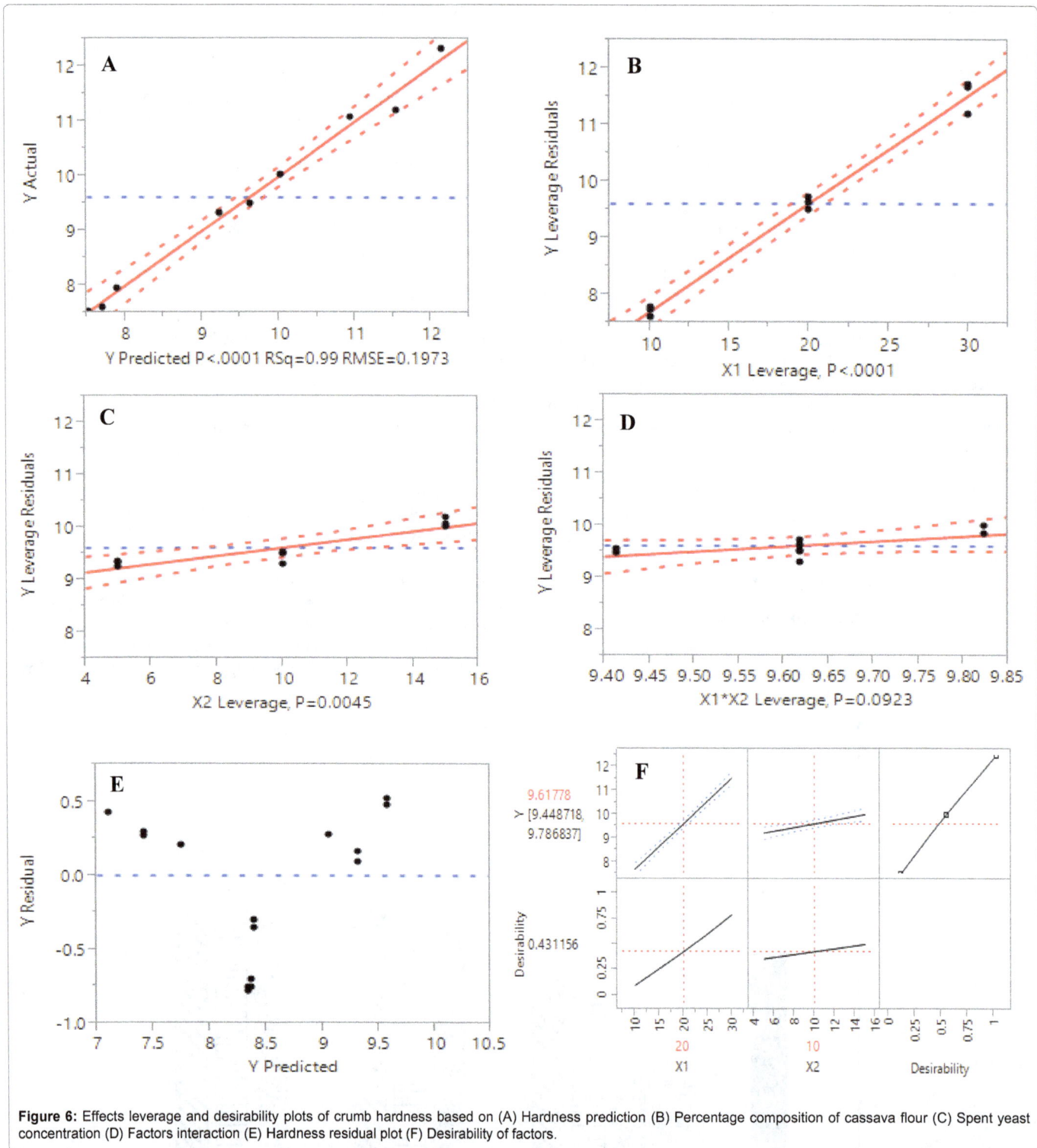

Figure 6: Effects leverage and desirability plots of crumb hardness based on (A) Hardness prediction (B) Percentage composition of cassava flour (C) Spent yeast concentration (D) Factors interaction (E) Hardness residual plot (F) Desirability of factors.

et al. [19] on the effect of extruded wheat bran on dough rheology and bread quality, the authors reported that hardness of bread crumb was related to the gumminess, chewiness parameters and hydration effect which partly accounted for the hard texture of composite bread samples. Jensen et al. [13] results showed that, depending on the type of cassava flour, up to 30% of the wheat flour could be replaced without any significant differences from control bread and the addition of fibers resulted in a lower volume and harder and more cohesive crumb structure. In this study, a statistical test carried out to investigate the actual effect of independent variables using desirability function showed the desirable values for percentage composition of CF and SBY concentration to be 20% and 10 g, respectively with a corresponding desirable value for bread hardness as 9.62 N (Figure 6F). Thus, the best cassava-wheat composite bread can be made by the addition of 20% CF and 10 g of SBY.

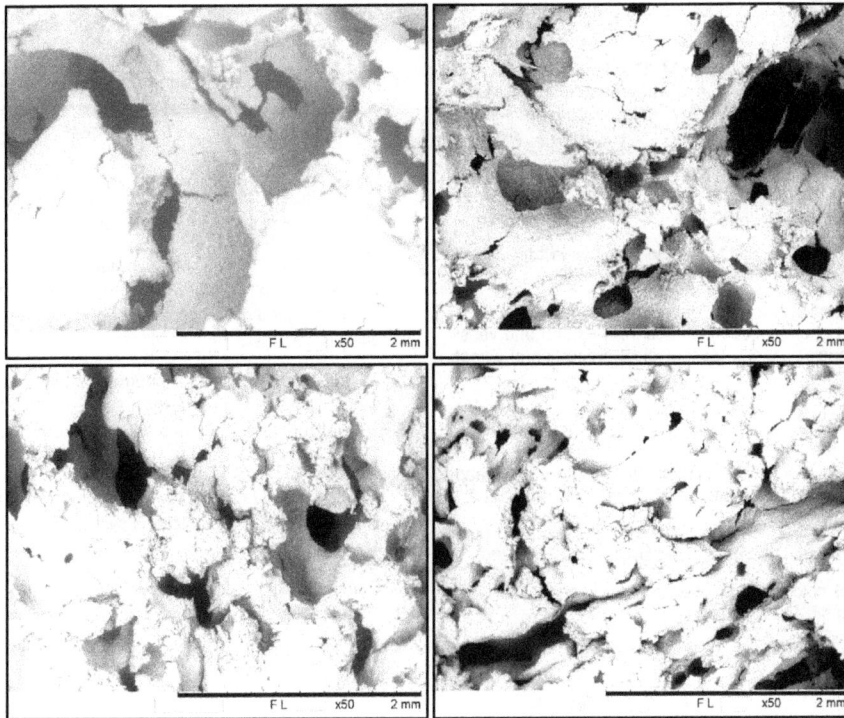

Figure 7: Scanning electron micrographs of crumb of three SBY-fortified composite bread (cassava-wheat) and control (without SBY) at the magnification × 50 (a) Control (without SBY) (b) 5% SBY, (c) 10% SBY (d) 15% SBY.

Microstructure of cassava-wheat bread

The microstructure of composite bread samples was examined after 24 h of storage in a desiccator to keep the samples dry before digital imaging on SEM. The effect of different concentrations of SBY in bread formulation on the morphological features of crumb structure was visualized. Figures 7A-7D shows the SEM micrographs of breadcrumb section. A qualitative information on pore structure from the SEM showed that all the bread samples were characterized by heterogeneously distributed pores of different sizes. In addition, SBY, the pores appeared to be slightly smaller as compared with control samples. The pores of bread fortified by 5 g SBY appeared to be the largest out of the levels of treatment for SBY concentrations (Figure 7B). The pore size decreased as the SBY concentration increased while the smallest pores distribution was predominant at 15 g SBY (Figure 7D). The microstructure of 5 g SBY fortified bread was similar to that of 10 g SBY while 15 g SBY was distinct from either concentration of SBY. The higher the concentration of SBY in the formulation the more compact the breadcrumb. This result affirms our findings in section 4.3, desirable value for SBY concentration to be 10 g. Control bread apparently differ from the SBY fortified bread (Figure 7A) with more opened structure, rigid and fragile features. Crumb firmness was observed in SBY fortified samples. The crumb firmness increased with increase in the concentration of SBY. This is one of the positive impacts of SBY on the breadcrumb apart from its nutritive values. Similar findings were reported by Gomez et al. [19] from their study of dough properties and bread quality of wheat–barley composite flour as affected by β-glucanase. The authors reported that the macroscopic characteristics (firmness and springiness) and microstructure of the dough displayed a remarkable improvement when β-glucanase was added; however, a higher level of β-glucanase (0.06% and 0.08%) could result in sticky dough and consequently lower final product quality.

Conclusion

This study has been able to show that varying spent yeast and cassava flour concentration to produce a cassava-wheat composite bread leads to a significant difference in the physical properties of composite bread produced. The study clearly showed that spent yeast up to 10 g of protein replacement will cause no negative changes in the bread quality or marketability of the bread but rather adds to the richness in brownness. The bread quality such as volume, mass, softness gradually diminished as the cassava flour increased. This study showed SBY (10 g of protein) and CF (20%) stood a good chance to compete with whole wheat in the market in terms of nutrition, quality, and health. The protein content was calculated based on the percentage composition of the flours and spent yeast used for bread making. There was no protein analysis done to determine the protein content of the composite bread. Therefore, further studies are required to accurately determine the response of sensory and protein content of composite bread as they for optimizing consumer acceptability.

Acknowledgement

The authors are acknowledging the International Fund for Agricultural Development (IFAD) and National Science and Engineering Research Council of Canada (NSERC).

References

1. Kittisuban P, Ritthiruangdej P, Suphantharika M (2014) Optimization of hydroxypropylmethylcellulose yeast β-glucan, and whey protein levels based on physical properties of gluten-free rice bread using response surface methodology. LWT-Food Sci Technol 57: 738-748.

2. Thammakiti S, Suphantharika M, Phaesuwan T,Verduyn C (2004) Preparation of spent brewer's yeast β-glucans for potential applications in the food industry. Int J Food Sci Technol 39: 21-29.

3. Xu X, Pu Q, He LIU, Na Y, Wu, et al. (2009) Rheological and SEM Studies on the interaction between spent brewer's yeast B-glucans and K-carrageenan. J Texture Stud 40: 482-496.

4. Hamed A, Ragaee S, Marcone M, Abdel-Aal ESM (2015) Quality of bread and cookie baked from frozen dough and batter containing β-glucan-rich barley flour fraction. J Food Qual 38: 316-327.

5. Kittisuban P, Ritthiruangdej P, Suphantharika M (2014) Optimization of hydroxypropylmethylcellulose yeast β-glucan, and whey protein levels based on physical properties of gluten-free rice bread using response surface methodology. LWT-Food Sci Technol 57: 738-748.

6. Martins ZE, Erben M, Gallardo AE, Silva R, Barbosa I, et al. (2015) Effect of spent yeast fortification on physical parameters, volatiles and sensorial characteristics of home-made bread. Int J Food Sci Technol 50: 1855-1863.

7. Begum R, Rakshit SK, Rahman SMM (2011) Protein fortification and use of cassava flour for bread formulation. Int J Food Prop 14: 185-198.

8. Chinma CE, Ilowefah M, Shammugasamy B, Mohammed M, Muhammad K, et al. (2015) Effect of addition of protein concentrates from natural and yeast fermented rice bran on the rheological and technological properties of wheat bread. Int J Food Sci Technol 50: 290-297.

9. Švec I, Hrušková M, Jurinová I (2015) Pasting characteristics of wheat-chia blends. J Food Engi 172: 25-30

10. Blanco CA, Ronda F, Pérez B, Pando V (2011) Improving gluten-free bread quality by enrichment with acidic food additives. Food Chem 127: 1204-1209.

11. Shittu TA, Aminu RA, Abulude EO (2009) Functional effects of xanthan gum on composite cassava wheat dough and bread. Food Hydro 23: 2254-2260.

12. Maskan M (2001) Kinetics of color change of kiwi fruit during hot air and microwave drying. J Food Eng 48: 169-176.

13. Jensen S, Skibsted LH, Kidmose U, Thybo AK (2015) Addition of cassava flours in bread-making: Sensory and textural evaluation. LWT - Food Sci Technol 60: 292-299.

14. Nindjin C, Amani G, Sindic M (2011) Effect of blend levels on composite wheat doughs performance made from yam and cassava native starches and bread quality. Carbohydr Polym 86:1637-1645.

15. Shittu TA, Raji AO, Sanni LO (2007) Bread from composite cassava-wheat flour: I Effect of baking time and temperature on some physical properties of bread loaf. Food Res Int 40: 280-290.

16. Motrena SG, Carvalho MJ, Canada J, Alvarenga NB, Lidon FC, et al. (2011) Characterization of gluten-free bread prepared from maize, rice and tapioca flours using the hydrocolloid seaweed agar-agar. Recent Research in Science and Technology 3: 64-68.

17. Pasqualone A, Caponio F, Summo C, Paradiso VM, Bottega G, et al. (2010) Gluten-free bread making trials from cassava (*Manihot Esculenta Crantz*) flour and sensory evaluation of the final product. Int J Food Properties 13: 562-573.

18. Totosaus A, López H, Güemes-Vera N (2013) Effect of lupinus (*Lupinus albus*) and Jatropha (*Jatropha curcas*) protein concentrates on wheat dough texture and bread quality: optimization by a D-Optimal mixture design. J Texture Stud 44: 424-435.

19. Gómez M, Jiménez S, Ruiz E, Oliete B (2011) Effect of extruded wheat bran on dough rheology and bread quality. LWT-Food Sci Technol 44: 2231-2237.

8

Effect of Extrusion Variables on the Extrudate Properties of Wheat-plantain Noodle

Sobowale SS[1]*, Bamgbose A[1] and Adeboye AS[2]

[1]*Department of Food Technology, Moshood Abiola Polytechnic, Abeokuta, Ogun State, Nigeria*
[2]*Department of Food Science, University of Pretoria, Hatfield, Pretoria, South Africa*

Abstract

Matured green plantain (*Musa paradisiaca*) was processed into plantain flour and composted with wheat flour prior to extrusion. In this study, the effect of extrusion cooking variables [Barrel Temperature (BT, Feed Moisture Content (FMC) and Screw Speed (SS)] on the extrudate properties of whate-plantain noodles were investigated, while predictive models were also developed. Results obtained showed that changing the cooking variables significantly ($p \leq 0.05$) affected all the extrudate properties studied. An increase in the screw speed (6.3-8.4 m/s) and feed moisture content (40-50%) resulted in a substantial increase in the expansion ratio (48%), residence time (62%) and specific mechanical energy (83.6%) Likewise, an increase in barrel temperature increased the mass flow rate (64.5%) of the extrudates significantly ($p \leq 0.05$). Regression analysis also revealed that the screw speed and the feed moisture content were the major extrusion cooking variables, as they showed significant ($p \leq 0.05$) linear quadratic and interaction effects on mass flow rate, residence time, specific mechanical energy and expansion ratio. The study showed that optimization of the combined effects of a 50% feed moisture content, 6.3 m/s screw speed of 6.3 m/s and a 100°C barrel temperature gave the optimum processing conditions for expanded wheat-plantain noodles. These variables are important considerations for commercial and mass production of wheat-plantain noodles.

Keywords: Noodle; Extruder; Feed moisture content; Screw speed; Barrel temperature

Introduction

Plantain is an important food crop in Nigeria as well as in all humid tropical zones of Africa. Over 80% of the harvested plantain is obtained during the period of September to February of the year, and there is much wastage of this crop during this peak period as a result of glut resulting in seasonal availability and limitations of use. Wheat flour is the main raw material for noodles production. Its cultivation thrives best in temperate regions worldwide [1]. It is one of the major raw material for wide range of food products ranging from baked foods to extruded products, such as pasta and noodles.

Due to health benefits, nutritional improvement and improved utilization of local crops, attention is shifting towards substituting wheat in noodles production with a locally cultivated crop such as plantain. This fruit is an excellent source of nutrient when eaten. Furthermore, it possesses a high carbohydrate content, low fat content, they are good source of vitamins and minerals particularly iron, potassium, calcium. The sodium content is low in dietary terms hence the recommendation is made for low sodium diets. It is recommended to produce plantain flour from green fruits since it has high starch content of about 35% on wet basis [2].

Extrusion cooking process is a high temperature short time (HTST) process in which moist food material is fed into the extruder where the desired temperature and pressure are obtained over the required period of residence time [3]. For cooking of the product generally, external heat is not applied but it is achieved through shear and friction in the extruder. Extrusion cooking is used worldwide for the production of expanded snacks, starch, modified ready to eat cereals baby foods, pasta and pet foods [4,5]. This technology has many distinct advantages including versatility, low cost, better product quality and no process effluents [6]. Recently, research into the field of manufacturing and processing of pasta eliminates the needs for cooking for as long as it results in the production of noodles [7]. This therefore necessitates the need to examine the influence of extrusion cooking variables on plantain-wheat noodles and its extrudates. The objectives of this work are to evaluate the influence of some extrusion cooking variables (feed moisture content, screw speed and barrel temperature) on wheat-plantain noodles and to predict the optimum cooking temperature-time for the production of wheat-plantain noodles using response surface methodology (RSM).

Materials and Methodology

Fresh matured green plantain (*Musa paradisiaca*) and wheat flour were respectively obtained from Siun and Kuto markets in Abeokuta (7.15°N, 3.35°E), Ogun state, Nigeria.

Processing of fresh matured green plantain to flour

The plantain fruits were washed with distilled water, peeled and sliced to about 5 mm diameter using a stainless steel knife. The slices were subsequently steamed for 5 minutes to inactivate enzymes. Subsequently, the pulp was drained and dried in an oven drier (Gellamp, England) at 60°C for 24 hours, after which the dried plantain slices were milled into flour using attrition mill (Figure 1). The flour were then screened through a 0.25 mm sieve and packed in high density polythene (HDPE) bags, using the methods of Ojure and Quadri [8].

*****Corresponding author:** Sobowale SS, Department of Food Technology, Moshood Abiola Polytechnic, Abeokuta, Ogun State, Nigeria
E-mail: sobowale.sam@gmail.com

Extrusion process of wheat-plantain noodles

Extrusion process of wheat-plantain noodles (Figure 2) was done as described by Oke, et al. [9], using a locally fabricated laboratory scale single screw extruder. The flours were mixed at a ratio of 70% wheat and 30% plantain flour. Based on the most stable product expansion and stability of the extruding product, the following extrusion conditions were used; feed moisture content varied at 30, 40 and 50%; barrel temperature of the extruder was set at 80, 100 and 120°C; screw speeds 6.3, 7.3 and 8.4 m/s, while a 2.5 mm restriction die was used. Constant feeding rate was kept throughout the experiments.

Determination of moisture content

The moisture content of the plantain was determined according to AOAC [10] method. 5 g of the samples were accurately weighed into an evaporating dish and dried in an oven until constant weight.

Measurement of extrudate properties

Torque (T): This was determined by reading directly from the extruder operation panel during extrusion runs according to Oke, et al. [9]. A value of 2.0 A was subtracted from the total obtained being the motor driving force. Means values of torque were expressed in Nm^{-1}. Hence, torque was calculated as follows:

$$T = r \times f$$

$$T = r \times f \times Sin\ \theta$$

Where; r = displacement vector (a vector from the point from which torque is measured to the point where force is applied); F = force vector; × = denotes the cross product; θ = angle between the force vector and the lever arm vector.

Mass flow rate (MFR): Mass flow rate was determined using the methods of Iwe, et al. [11] and Oke, et al. [9]. A constant torque and extrusion conditions were reached (steady state operation). Subsequently, a stopwatch was used to time the entry of the samples (beginning) until they were flowing out of the extruder die orifice at 60s interval. Mean weight of triplicate collections were collected for each run, as the means flow rate for that run in kilogram per second.

Hence;

Figure 1: Flow chart for production of plantain flour (Source: Ojure and Quadri [8]).

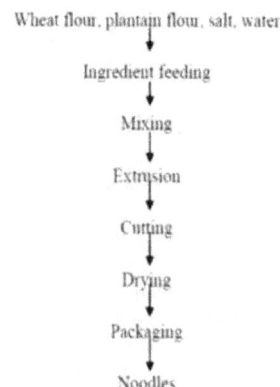

Figure 2: Flow chart for the production of wheat-plantain noodles using extruder (Source: Oke, et al [9]).

Mass flow rate = weight/time (kg/s)

Specific mechanical energy (SME): Specific mechanical energy is defined as the total mechanical energy input to obtain 1g of extrudate (Jkg^{-1}). The SME was thus determined using the methods of Rosentrater, et al. [12] and Oke, et al. [9].

$$SME = T = \omega \times 60/M_{feed}$$

where; SME = specific mechanical energy consumption (Jkg^{-1}); T = Net torque exerted on the extruder drive (Nm); ω = angular velocity (rpm); M_{feed} = mass flow rate of the raw sample (kg/s)

Therefore, M_{feed} was calculated using:

$$M_{feed} = M_{prod} \times 1-M_{cf}/1-M_{ci}$$

where; M_{prod} = mass flow rate of the extrudate (kg/s); M_{cf} = moisture content of the collected extrudate (% w.b.); M_{ci} = moisture content of the raw sample before entering the extruder (% w.b.)

Residence time (RT): Residence time during extrusion was determined using the methods of Oke, et al. [9]. The RT (s) was expressed as the time taken for a print of red colour to show up at the die orifice.

Expansion ratio (ER): Expansion ratio was determined as described by Conway and Anderson [13] and Rosentrater, et al. [12]. The ratio of diameter of extrudate and diameter of die was used to express the expansion of the extrudate. The diameter of extrudate was determined as the mean of 10 random measurements made with a vernier caliper. The expansion ratio was calculated as

Expansion ratio = diameter of extrudate (mm)/die diameter (mm)

Response surface methodology (RSM)

All results obtained were subjected to RSM. This was used to build up mathematical models that will facilitate the qualitative interpretation and description of the relationships existing between the selected dependent extrusion variables selected (torque, MFR, RT, SME, ER) and the independent extrusion variables (FMC, SS and BT). The extrusion cooking was carried out adopting a five variable response surface analysis using a central composite design (CCD) which was nearly orthogonal.

The generalized regression model fitted was $Y = Bo + b_1FMC + b_2SS + b_3BT + b_{11}FMC^2 + b_{22}SS_2 + b_3BT^2 + b_{12}FMC*SS + b_{13}FMC*BT$

+ b_{23}SS*BT + ϵ. Where Y = objective response, FMC = feed moisture content, SS = Screw speed, BT = barrel temperature and ϵ = random error in which the linear, quadratic and interaction effects were involved.

The resulting models were tested for significance using Analysis of variance (ANOVA) and coefficient of determination (R^2). This was determined by SAS 9.1 (SAS Inc. USA). Significant terms were accepted at $p > 0.05$. The R^2 of 0.6 was accepted for predictive purposes [9].

Results and Discussion

The wheat-plantain flour extrudates properties which include torque, residence time, mass flow rate, specific mechanical energy and expansion ratio are shown in Table 1. The effect of FMC, SS and BT on the extruder torque, MFR, RT, ER and SME are shown in Figures 3-7, respectively. The extruder torque (T) ranged between 68.75 and 91.66 Nm, the mass flow rate (MFR) ranged between 2.07 and 3.04 kg/s, the residence time (RT) ranged between 49.0 and 62.0 s, the expansion ratio (ER) ranged from 2.4 to 6.0 and the specific mechanical energy (SME) ranged between 10459.23 and15505.61 kJ/s. According to Oke, et al. [9], the estimated torque of an extruder is related to the power consumed and approximately 98% of this power input is utilized for shearing and less than 5% is consumed for pumping. The results were similar with the observed extrudate that increase in feed moisture content leads to increase in torque values [9,11,13]. Increase in screw speed at constant feed moisture content and barrel temperature leads to decrease in torque value, residence time and specific mechanical energy, while mass flow rate and expansion ratio of the extrudate increases [5,14].

The regression equations presented in Table 2 were used to generate surface plots shown in Figures 3-7. The response surfaces of all the responses measured were vertically displaced to significantly value (p < 0.05). The low R^2 value observed for the responses, ER (0.2725) and SME (0.2067) indicated that they were not influenced by the FMC and SS as suggested by RSM. The coefficient of determination was quite high for b-values. Specific mechanical energy input (SME) has been

described as a good quantitative descriptor of extrusion processes, as it allows for the direct comparison of different combinations of extrusion conditions such as rates, MFR and torque [15]. The quantity and amount of mechanical energy delivered to the extruded material directly relates to the extent of macromolecular transformations which takes place. This includes starch conversion and changes in the rheological properties of the dough [15]. As stated by Mitchell and Areas [16], increased SME results in a lower viscosity which thus promotes mobility may lead to an increase in rate bubble growth rate. Residence time and temperature relation is important in maillard reactions. An increase in temperature for longer time leads to an increase of the reactivity between the sugar and the amino group. The combined effects of the feed moisture content of 50%, screw speed of 6.3 m/s and extrusion temperature of 100°C could be subsequently used for the production of expanded wheat-plantain noodles [17,18]. These composite blends (wheat-plantain flour) could be of significance in the formulation of diets for diabetic patients and also for use as binders or disintegrants in tablet formulations.

Conclusion

This study showed that the wheat and plantain flour could be useful to produce noodles of good quality using extrusion cooking. It was observed that changes in the barrel temperature, feed moisture content and screw speed significantly (p ≤ 0.05) affected the torque, residence time, specific mechanical energy, expansion ratio and mass flow rate of all the extrudates studied. Increase in the feed moisture content (40-50%) and screw speed (6.3-8.4 m/s) resulted in a substantial increase in the expansion ratio (48%), residence time (62%) and specific mechanical energy (83.6%), while increase in the barrel temperature also significantly affected the mass flow rate (64.5%) of the extrudates. The study showed that optimization of the combined effects of the feed moisture content of 50%, screw speed of 6.3 m/s and extrusion temperature of 100°C gave the optimum processing conditions for wheat-plantain noodles. These variables are important considerations for commercial and mass production of expanded wheat-plantain noodles and for health conscious individuals and also

Run	Factor 1 FMC (%)	Factor 2 S.S (m/s)	Factor 3 BT (°C)	Response 1 Torque (Nm)	Response 2 MFR (kg/s)	Response 3 RT (s)	Response 4 ER	Response 5 SME (kJ/s)
1	30	5.23	80	91.66	2.07	62	2.4	15505.61
2	50	5.23	80	91.66	2.87	52	4.8	10459.23
3	40	6.28	100	78.57	2.48	56	4	12933.67
4	40	5.23	100	91.66	2.33	60	4.4	13014.89
5	40	6.28	100	78.57	2.48	56	4	13305.69
6	40	6.28	100	78.57	2.47	57	4	12984.72
7	50	5.23	120	91.66	2.94	51	5.2	13316.16
8	40	6.28	100	78.57	2.47	56	4	13396
9	40	6.28	100	78.57	2.43	57	2.8	12984.72
10	30	6.28	100	78.57	2.19	59	2.8	10924.41
11	50	6.28	100	78.57	2.92	50	5.6	10804.8
12	30	5.23	120	91.66	2.08	61	3.2	16068.66
13	40	6.28	80	78.57	2.4	58	4	13041.92
14	40	6.28	120	78.57	2.49	56	3.6	14097.7
15	30	7.33	120	68.75	2.16	60	3.2	16986.65
16	30	7.33	80	68.75	2.08	62	2.4	16891.75
17	40	7.33	100	68.75	2.68	52	3.6	13378.87
18	50	7.33	80	68.75	2.99	50	4.8	10798.66
19	50	7.33	120	68.75	3.04	49	6	11108.18

FMC: Feed Moisture Content; SS: Screw Speed; BT: Barrel Temperature; T: Torque; SME: Specific Mechanical Energy; MFR: Mass Flow Rate; RT: Residence Time; ER: Expansion Ratio.

Table 1: Table of results for extrusion runs.

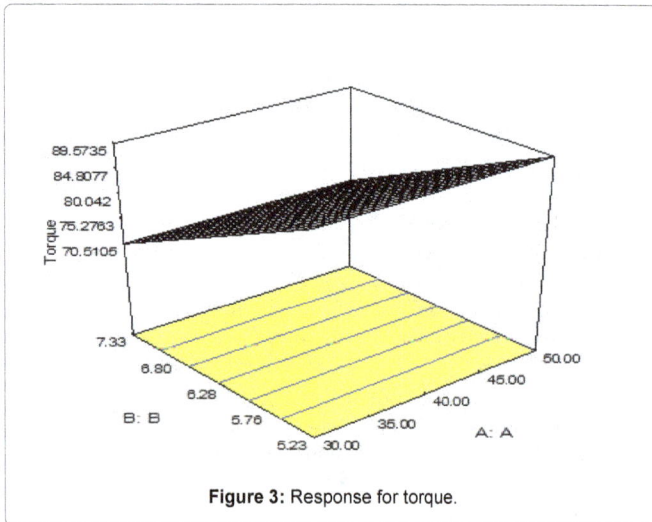

Figure 3: Response for torque.

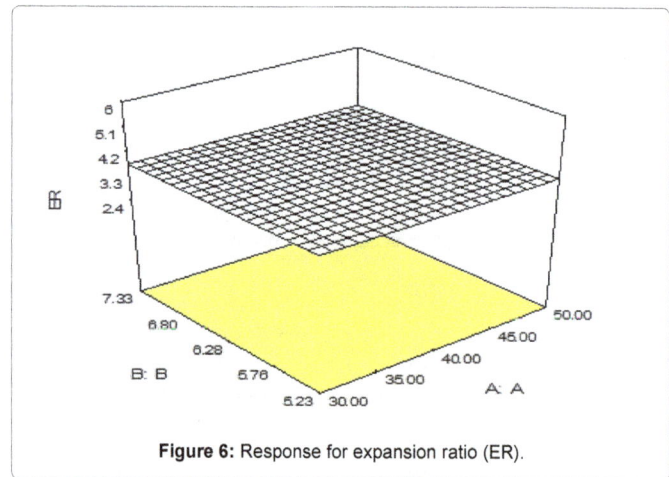

Figure 4: Response for mass flow rate (MFR).

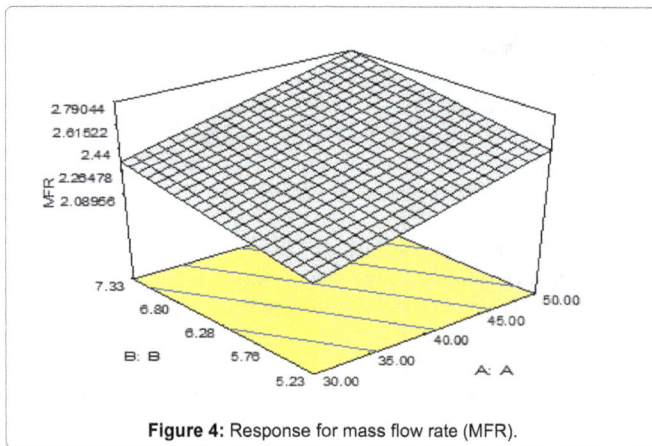

Figure 5: Response for residence time (RT).

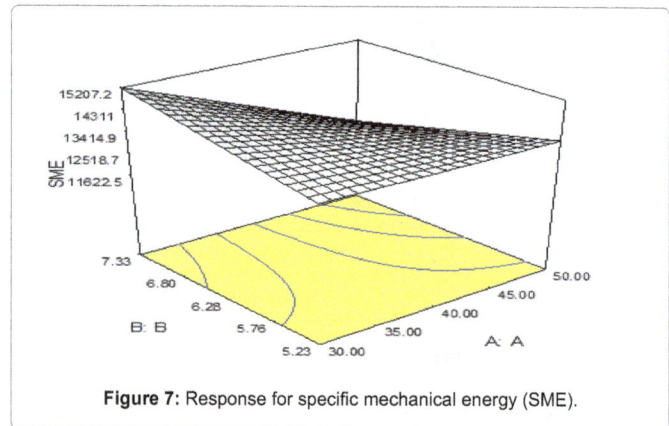

Figure 6: Response for expansion ratio (ER).

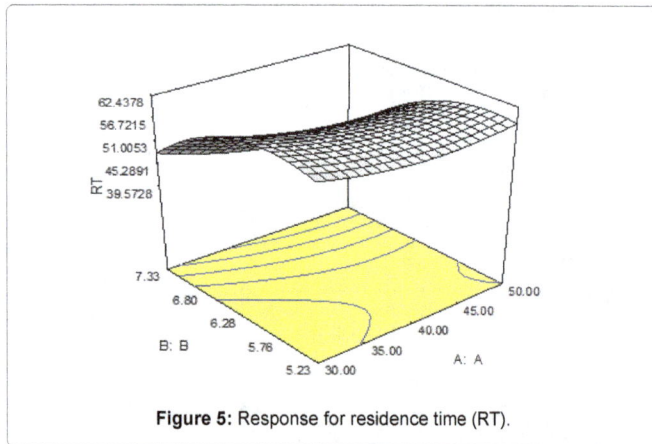

Figure 7: Response for specific mechanical energy (SME).

	TORQUE	MFR	RT	ER	SME
Intercept	80.04	2.44	55.90	3.78	13539.70
A	0.00	0.22	2.66	-	926.68
B	9.53	0.13	7.85	-	-124.83
C	0.00	0.03	0.12	-	611.60
A^2	-	-	2.41	-	-
B^2	-	-	6.57	-	-
C^2	-	-	2.06	-	-
AB	-	-	1.63	-	865.85
AC	-	-	-1.38	-	657.83
BC	-	-	1.38	-	-721.01
R^2	0.8458	0.4672	0.6381	0.2725	0.2067
Adjusted R^2	0.8169	0.3672	0.3184	0.1361	0.0579
Predicted R^2	0.7893	0.1525	-1.5949	-0.1345	0.3590
Lack of fit	0.27 NS	0.86 NS	13.29 NS	0.97 NS	2.26 NS

R^2: Coefficient of Determination; SME: Specific Mechanical Energy; MFR: Mass Flow Rate; RT: Residence Time; ER: Expansion Ratio

Table 2: Regression equation coefficient, presented as actual terms and statistical analysis for response of extrudates.

for use as binders or disintegrants in tablet and diet formulations. It is therefore, recommended that further studies on the effect of different ratios of wheat-plantain flour and other ingredients for the production of large scale snacks and others pasta should be investigated to provide a broader understanding of the feed moisture content and extruder variables.

References

1. Oladunmoye OO, Akinoso R, Olapade AA (2010) Evaluation of some physical-chemical properties of wheat, cassava, maize and cowpea flours for bread making. Journal of Food Quality 33: 693-708.

2. Simmond NW (1976) Food Uses of Banana. Macmillan Publishers, Ibadan, Nigeria.

3. Kitabatake N, Doi E (1992) Denaturation and Texturization of Food protein by Extrusion Cooking. Food Extrusion Science and Technology, New York.

4. Malleshi NG (1986) Processing of Small Millets for Food and Industrial uses.

Small Millets in Global Agriculture, Oxford and IBH Publishing Company, New Delhi.

5. Bhattacharya S, Prakash M (1994) Extrusion cooking of blends of rice and chickpea flour: A response surface analysis. Journal of Food Engineering 21: 315-330.

6. Camire ME, Camire A, Krumhar K (1990) Chemical and nutritional changes in foods during extrusion. Critical Reviews in Food Science and Nutrition 29: 35-57.

7. Kim CH, Maga JA (1993) Influence of starch type, starch/protein composition and extrusion parameters on resulting extrudate expansion. Developments in Food Science 32: 957-964.

8. Ojure MA, Quadri JA (2012) Quality evaluation of noodles produced from unripe plantain flour. International Journal of Research and Reviews in Applied Science 13: 740-752.

9. Oke MO, Awonorin SO, Workneh TS (2013) Effect of varieties on physicochemical and pasting characteristics of water yam flours and starches. African Journal of Biotechnology 12: 1250-1256.

10. AOAC (2000). Official Methods of Analysis, Association of Official Analytical Chemists, Virginia, USA.

11. Iwe MO, Vanzuilichem DJ, Ngoddy PO (2001) Extrusion cooking of blends of soyflour and sweet potato flour on specific mechanical energy (SME), extrudate temperature and torque. Journal of Food Science 53: 450-454.

12. Rosentrater KA, Muthukumarappan K, Kannadhason S (2009) Effects of ingredients and extrusion parameters on aqua feeds containing DDGS and Tapioca Starch. Journal of Aquaculture and Feed Science Nutrition 1: 6-21.

13. Conway HF, Anderson RA (1973) Protein fortified extruded food products. Cereal Science Today 18: 94-97.

14. Choudhury GS, Gautam A (1999) Screw configuration effects on macroscopic characteristics of extrudates produced by twin-screw extrusion of rice flour. Journal of Food Science 64: 479-487.

15. Moraru CI, Kokini JL (2003) Nucleation and expansion during extrusion and microwave heating of cereal foods. Comprehensive Reviews in Food Science and Food Safety 2: 147-165.

16. Mitchell JR, Areas JA (1992) Structural Changes in Biopolymers during Extrusion. Food Extrusion Science and Technology, Marcel Dekker, New York.

17. Anounye JC, Jigam AA, Ndaako GM (2012) Effects of extrusion cooking on the nutrients and anitnutrients composition of pigeon pea and unripe plantain blends. Journal of Applied Pharmaceutical Science 2: 58-162.

18. Deshpande HW, Poshadri A (1999) Physical properties and sensory characteristics of extruded snacks prepared from Foxtail millet based composite flours. International Food Research Journal 18: 10-16.

9

Effect of Substitution of Sucrose with Date Palm (*Phoenix dactylifera*) Fruit on Quality of Bread

Nwanekezi EC*, Ekwe CC and Agbugba RU

Department of Food science and Technology, Faculty of Agriculture and Veterinary Medicine, Imo State University, Owerri, Sudan

Abstract

The proximate, physical and sensory analysis of bread samples produced by substituting the level of sucrose with date palm fruit pulp (50:0 g, 37.5:12.5 g, 25:25 g, 12.5:37.5 g and 0.50 g) was investigated. The proximate analysis revealed that the protein, moisture, ash, crude fiber and fat contents increased with increase in the level of the date palm fruit pulp. There were increases in the levels of nutrients which ranged from 15.19-19.43% (protein), 1.65-4.43% (crude fiber), 2.44-4.11% (ash) and 28.19-28.92% (moisture). However, there was decrease in the level of carbohydrate content from 45.39 to 35.13% as the level of date palm pulp increased. The specific volume also decreased as the level of the date palm fruit increased ranging from 3.12 cm^3/g to 2.93 cm^3/g; the addition of date palm fruit pulp had no effect on the loaf volumes which ranged from 1920.1 to 1925.0 cm^3. The sensory analysis, using a 25-man panel of judges revealed that all the loaf samples were acceptable organoleptically. However, the substitution of sucrose with date palm fruit pulp powder increased the nutritional value of the bread samples.

Keywords: Sucrose; Date palm; Carbohydrate; Bread samples

Introduction

Bread is a food product that is universally accepted as a very convenient form of food that has desirability of all population, rich and poor, rural and urban. Its origin dates back to the neolithic era and is still one of the most consumed and acceptable stable in all parts of the world [1]. In Nigeria, bread has become the second most widely consumed and non-indigenous food product after rice and has become an important source of food to Nigerians. It is consumed extensively in most homes, restaurants and hotels [2].

One of the ingredients for bread-making is sugar. Sugar is the primary food for the yeast [3]. In the course of bread-making, the wheat flour dough is fermented with yeast. Fermentation is a process by which yeast acts, on sugar and changes them into carbondioxide gas and alcohol (Gisslen). The release of carbondioxide gas produces the leavening action in breads. Other roles of sugar in bread making according to Bali [3] are: It helps to improve the crust color through browning reaction. Sugar acts as preservative as it is anti-staling agent. It helps bread to retain moisture by keeping the bread moist. It acts as bread improver and imparts flavor to bread. Sugar also tenderizes the bread by preventing gluten formation [4].

Fermentable sugar in bread dough comes from two sources. It is added to the dough by the baker and it is produced from flour by enzymes that break down the wheat starch into sugar (Gisslen). The concentration of added sugar in bread dough varies from none to about 8% or a little higher [5].

Date palm fruit *(Phoenix dactylifera)* is a delicious fruit with a sweet taste and a fleshy mouth feel. The major component of date fruits are carbohydrates (mainly the sugars; sucrose, glucose and fructose), which may constitute about 70% [6]. The sugars are easily digested and can immediately be moved to the blood after consumption and can quickly be metabolized to release energy for various cell activities. Date fruits are a good source of fiber and contain very important vitamins and minerals, including significant amounts of calcium, iron, fluorine and selenium [6]. The date fruit can be used as a practical supplement for iron deficiency without any side effects. At least six vitamins (thiamine, riboflavin, niacin, ascorbic acid, pyridoxine and vitamin A) have been reported to be present in dates in visible consideration [7].

Nigeria spends most of its foreign exchange on importation of sugar. This foreign currency spent on sugar importation depletes the country's foreign exchange reserve. Complete replacement of sucrose with date palm fruit in bread making will not only save substantial fraction of foreign exchange expended on importation of sugar but will also uplift the nutritional profile of bread in view of numerous nutrients in date palm fruit.

Therefore, the aim of this study is to evaluate the quality of bread made with date palm fruit in place of sugar (sucrose).

Origin and distribution of date palm

Phoenixdactylifera L., date palm is among the most important species in the palm family (*Arecaceae*), which encompasses about 200 genera and more than 2,000 species [8,9] and includes *P. Canasiensis*(Canary Islands date palm) *P. Rechinata*(Senegal date palm) and *P. Slyvestris*(India sugar palm). The species name was inspired by the finger like shape of the fruit and the genus from the legendary bid of ancient Greece.

Historically, date palm cultivation was practiced by ancient civilizations and in nowadays considered one of the oldest domesticated fruit-bearing trees. Remains of date palms were found in Jericho-the oldest site town to date to be the origin of agriculture. Date palm cultivation gained socio-economic importance among tribes and countries due to its ecological plasticity and high adaptation to and conditions where the annual precipitation rarely exceed 250 mm combined with hot summers up to 50°C and cold writers down to 10°C [8].

*Corresonding author: Nwanekezi EC, Department of Food science and Technology, Faculty of Agriculture and Veterinary Medicine, Imo State University, Owerri, Sudan, E-mail: chibyzaps@gmail.com

Phoenix dactylifera is a widely distributed species occurring in diverse geographic soil and climate areas [10]. The vast majority of the trees are located in the Middle East and North Africa although the crop has been established in California, Arizona and Mexico in the Americas. The common requirement among all date palm growing areas is the high temperature (35°C) necessary for an optional development of pollen and the low relative humidity for fruit setting and ripening. Such desert-adapted tree large quantities of water drawn from deep in the soil through a well-established root system or from surface irrigation. Date palm grows in nearly rainless regions at 9-39°C North latitude, which are represented by the Sahara and Southern fringe of the Near East (Arabia Peninsula, Southern Iraq, and Jordan).

FAO estimates that the harvested area of date growing was 1.3 million ha in 2009 (FAO statistics).

Edible dates go through four ripening stages termed kimri, khalal, rutab and tamr(Fayadh and AL-Showimann). These represent the immature astringent green, mature full, coloured, soft brown and hard raising-like stages of development, respectively. In first 4-5 weeks, the dates are full green and become kimri at this stage, average fruit size is 27.5 mm long x 17.8 mm in diameter and weighs 5.8 g an average, quickly increasing due to the accumulation of carbohydrates and moisture content. Acidity is quite high at this stage with an average protein level of 5.6%, fat 0.5%, and ash 3.7% [11]. In the Khalal stage, the fruit colour changes from green to yellowish/reddish tone depending on the colour, over a period of 3 to 5 weeks. Sugar and moisture content decreases from values recorded at the kimri stage along with a decrease in acidity. In this stage, the fruit averages 32.5 mm in length and 21 mm in diameter while the fruit weight increases to 8.7 g [11].

The percentage of protein, fat and ash decrease to 2.7, 0.3 and 2.8% respectively. At the rutab stages, dates begin to soften (2-3 weeks period) due to an increased loss of moisture content and an increase of enzymatic activities of pectinases and poly galactinosases. The protein, fat and ash percentage in this stage decrease to 2.6, 0.3 and 2.6% respectively. At the tamr stage, dates and drier and rather firm in texture while their colour turns to a darker one.

The date palm commonly known as "Dabino" by the Hausa tribe is believed to have been introduced into Nigeria in the early 17th century through the trans-Sahara trade made from North Africa and Muslim Pilgrims in Pilgrimage to the Holycities of Mecca and Medina [12].

Date palm fruit is grown in Northern Nigeria including Kaduna, Katsina, Kano, Sokoto, Kebbi, Jigawa, Yobe, Bonu, Gombe, Bauchi and Adamawa States. Other states including Plateau, Taraba, Nassarawa, Southern Kaduna and Niger State could be classified as marginal areas for date palm cultivation in the country [13]. Date production in Nigeria has two fruiting seasons (Dry and Wet season fruits), but only the dry season fruits is economically useful. Date palm cultivation has remained restricted to compound, homesteads and few orchards in the Northern part of the country. The statistical of the annual date production in the country from the studied states deduce so far is over 21,000 metric tonnes from the available data shown in Table 1.

Harvest and post-harvest handling of dates

Dates are harvested at or near maturing; harvest is generally by hand with access to crown of the tree being by way of climbing or mechanical lifts. Completely mechanized harvest shakers used in some perennial crops is not developed enough for routine commercial use at this time [14]. In many traditional areas of date production, where the bulk of production is by small farms with limited resources, dates are usually transported directly to open air markets. Because of their low moisture content, dates can be successfully stored for sometimes without specialized storage conditions.

Physiological and pathological factors which and lower the fruit quality include: black nose, associated with high humidity during the Khalal storage, black scald, associated with abnormally high temperature and humidity storage conditions may be promote fruit defects such as darkening of the skin and sugar spots. In addition, dates are sometimes attacked by various pathogens including *Aspegillus, Alternaria, Penicillum.*

Diet contribution and uses of date fruits

Besides the use of fresh fruits for human consumption a number of by-products derived from dates also have various uses. These include Jam, Jelly, Juice, Syrup and fermented beverage [15]. Cull dates from grading and sorting, as well storage and conditioning are often utilized as animal feed. Several reports show that a number of bacteria compounds can be extracted from these by-products, thereby adding industrial value which could compensate for the economic loss from under-grading and/or deterioration. Various, metabolites also are reported to be produced from dates or their by-products, such as citric acid, oxytetrachine and ethanol [15].

Dates represent an important nutritional element in the diet population where the trees are grown. Dates contain a high percentage of carbohydrate (total sugars, 44-88%), protein (2.3-5.6%), fat (0.2-9.3%) and essential fibre (6.4-11.5%) and the seed (7.7-9.7%). Dates are known for numerous other nutritional properties due to their redness in non-starch polysaccharide and liquid [16]. Other nutritional properties include a number of the anti-oxidant molecules such as polyamine, phenolics, tannins and glutathione known for their health enhancement attributes.

Bread

According to Pomeranz and Shellenbergerthe history of bread is almost as old as the history of making, since it has long been used as a sacred symbol in religious ceremonies. In ancient times, the Egyptians used it as both sacrifice and a tribute to their gods. Today Christianspotray the body of Christ at communion with bread. Bread

S.NO	States	Annual Production in Metric Tonnes (MT)
1	Adamawa	200
2	Bauchi	6,000
3	Borno	1,000
4	Gombe	1,500
5	Jigawa	5,000
6	Kano	6,000
7	Plateau	Insignificant
8	Taraba	Insignificant
9	Yobe	2,000
10	Kaduna	-
11	Nassarawa	-
12	Katsina	-
13	Zamfara	-
14	Kebbi	-
15	Sokoto	-
	Total	**21,700 (MT)**

Source: Abdul – Qadir et al.,

Table 1: Statistics and annual date's production in Nigeria.

has many variations depending on the shape, size, texture, colour and taste [17].

Bread products are well accepted worldwide because of their low cost, ease of preparation, versatility, sensory attributes, convenience and nutritional properties. Bread in human nutrition is not only a source of energy, but also supplier of irreplaceable nutrients for the human body. It provides little fat, but high qualities of starch and dietary fibre as well as cereal protein. Apart from that, bread contains to B group vitamin and minerals which are mostly magnesium, calcium and iron [18].

Types of bread

According to Serrem et al. [19], there are different types of bread which include:

α) **White bread:** This is made from flour, contains only the central core of the grain (endosperm).

β) **Bran bread:** This is made with endosperm and about 10% brain. It can also be referred to as white bread with added colouring often caramel colouring to make it broom, this is commonly labeled in America as wheat bread (as opposed to whole wheat bread). It contains the whole grain (endosperm and bran). It is also referred to as "whole grain" or "whole-wheat bread", especially in Non-America.

χ) **Wheat germ bread:** This has added wheat germ for flavoring whole-grain bread. It can also be referred to as whole meat breads or wheat bread with added white grains to increase its fiber content as in whole grain bread.

δ) **Unleavened bread:** This is used for the Jewish feast Passover; it does not include yeast which is responsible for rising.

ε) **Crisp bread:** Crisp bread is a flat and dry type of bread or cracker containing mostly rye flavor.

ϕ) **Flat bread:** Flat bread is often simple, made with flour, water and salt and then formed into flattened dough; most are unleavenedmade without yeast or sour dough culture, though some are made with yeast.

Baking ingredients

There are several ingredients involved in the production of bread. Some of the ingredients are mandatory (flour, water, yeast, salt), while others are optimal (sugar, milk, salt, flavor, fat, emulsifiers) [20].

- **Flour:** Contains starch, proteins, and lipids. Approximately 80-90% of the total wheat proteins are storage proteins and they play major role in bread production because of its essential functions in bread structure. The gluten network forms when flour is combined with water and some energy input [21]. It is crucial for the retention of air and carbohydrate during bread-making and this gives bread its structure.

- **Water:** Water acts as a solvent during the formation of bread dough. When all the ingredients are mixed together for dough formation, water hydrates the flour, the yeast is dispersed [21]. Secondly, water acts as a plasticizer during mixing and after baking [21].

- **Yeast:** Produces CO_2 for leavening the bread [22]. Yeast fermentation produces reducing sugars, which interacts with the dough proteins in the surface, under the influence of heat. This process is known as the Millard reaction which causes browning of the bread crust and contributes greatly to bread flavor [23].

- **Salt:** Sodium chloride (salt) is included at levels of about 2% in bread making [24]. The main reasons for adding salt are to:

  - Develop flavour: Without salt bread is tasteless, salt also intensifies the bread flavor developed by other ingredients [17].

  - Retard fermentation: Salt is used to control fermentation [22].

  - Strength of the Gluten: By suppressing the repulsion charges and increasing the molecular interaction between protein chains [25].

  - Affects the colour crust.

- **Sugar:** Sugar is the basic source of energy which yeast converts into CO_2 during dough proofing. Sugars are usually used by yeast during the early stages of fermentation. Later more sugars are released by the action of enzymes in the flour and then used for gas production. The concentration of sugar used in dough depends on the type of the product and desired crust characteristics. Sugar is added to provide pleasant flavor and to develop a desired crust colour [26]. The reason why most bread recipes call for sugar is to impart sweetness. It is also a source of fermentable carbohydrate for yeast especially when flour is low in amylolytic activity [27]. Other functions of sugar in bread include the improvement of texture of the crumb, retention of the moisture in the crumb and adding to the nutritional value of the bread. Some natural alternatives of white sugar for bakery products are: raw honey, maple, syrup, molasses, corn syrup, steria, xylitol, agara, nectar, brown rice syrup, evaporated cane juice, black strap molasses, date sugar and organic sugar have been listed by Anon [28], Phillips [29] and Khan [30] reported that one cup of date sugar is equivalent to one cup of granulated or brown sugar. Dates are dried and then grind into a powder to make date sugar. This sugar is high in fibre, vitamins and minerals. The substitute with white sugar is 1:1 [31].

- **Lipids:** Lipids can be used in bread-making either in the form of fats or oils and are usually referred to as shortening. Lipids also improve the keeping quality, softness and moistness and contribute to bread texture. Lipids embedded into the protein matrix are essential as they interact with proteins during dough mixing and contribute to the visco-elastic properties of the gluten network, required for expansion and gas retention during proofing [32].

Bread production process

α) **Mixing:** According to Dendy and Dobraszizky [33] mixing serves the purpose of blending and hydrating the dough ingredients, developing and aerating the dough.

β) **Proofing:** Proofing is known to be cardinal step in the bread making process. This step is necessary to produce the highly leavened structure with bread [33]. The bubble structure which was formed while mixing expands during proofing and will be set during baking. Thus, during proofing the visco-elastic gluten complex is transformed into a continuous three dimensional dough network [34].

χ) **Baking:** Baking results in series of physical, chemical and bio-chemical changes in bread. The reaction includes volume expansions, evaporation of water, formation of a porous structure, denomination of protein, gelatinization of starch, crust formation and "reaction" protein cross linking, method of fat crystals and their incorporation into the surface of air cells, rupture of gas cells and sometimes fragmentation of cell walls [35].

δ) **Cooling:** Heat transfer and evaporation are the characteristics methods by which a loaf of bread is cooled.

ε) **Packaging:** One major purpose of packaging is to provide accurate nutritional information to consumers. The bread package also functions to keep the bread together and maintain product quality (Tables 2 and 3).

Nutritional value of bread

Like all other foods of cereal origin, bread is eaten mainly as a cheap source of energy. It contains about 40-45 percent available carbohydrate and has an energy value of 900 – 1000 kg/100 g. Because considerable amount of bread are eaten, its other constituents also contribute substantially to the daily intake of nutrients. It contains 8 – 9 percent protein and significant amounts of minerals and vitamins [36].

Materials and Methods

Material collection

The date palm fruit was brought at Ama-Hausa, Douglas, Owerri Imo State. Other ingredients like fat, wheat, flour, sugar, milk and salt was brought at Eke-ukwu market, Owerri, Imo State.

Sample preparation

Date palm powder was produced first by removing the seeds of the fruit manually with the aid of knife and weighing the dried palm fruit. The date palm fruit was washed with water to remove adhering dirts. The de-seeded fruit was then oven dried at 65°C for 8 h and subsequently milled using hand milling machine (Figure 1).

Bread production process

The bread was produced using the straight dough method. Date palm fruit pulp (DPFP) was used as a replacement for granulated sugar at the following ratios: 100:0, 75:25, 50:50, 25:75, 0:100 (sugar: DPFP), which was at these percentages 0, 25, 50, 75, and 100% of sugar with DPFP. The dry ingredients were measured in the required quantities and mixed; water was added to form the dough and mixed thoroughly. The dough was kneaded until smooth and the air spaces became small. The dough was allowed to proof at room temperature 30-45 minutes. The dough was baked in a hot oven (230°C) until golden brown (Figure 2).

Analysis

Ash was determined by the furnace maceration gravimetric method [37]. The moisture content was analyzed using the method described by AOAC [38]. Crude fibre was determined using the Weendemethod. James [39]. Crude protein was determined using the kjeldahl method Chang [40]. Fat content was determined using the soxlet continuous extraction gravimetric method by Min and Boff [41]. Carbohydrate was calculated by difference as the Nitrogen free extractions (Bemiller) (Figures 3 and 4).

Sensory evaluation

Sensory evaluation was conducted for the bread samples by a Twenty-five member semi-trained sensory panel. They were asked to access for taste, texture, smells, crumb colour, crust colour and general acceptability using 9-point hedonic scale. The scale ranged from nine for "extremely like" to "dislike extremely" one (1). They were instructed to take one or two bites, slowly masticate the product before rating the sample and take over all acceptances [42].

S/N	Characteristics	Requirements
a	Specific Volume (min)	4.0
b	Moisture (%) max)	40.0
c	Total solid content (%) (min)	60.0
d	Protein (%) (min)	10.0
e	pH of the aqueous extracts	5.3 – 6.0
f	Ash content (%) (max)	0.6
g	Acid insoluble Ash (max)	0.5
h	Fat content (%) (max)	2.0
i	Crude fibre (%)(max)	0.5
j	Carbohydrate (%) (max)	48.0
k	Energy (kj/100g): on dry basis	900 - 1000

Source: SON (2004)

Table 2: Analytical values of white bread using standard formula.

Ingredient	(g) S100/0	S75/25	S50/50	S25/75	S0/100 (%)
Sugar	50	37.5	25	12.5	-
Date palm fruit powder	-	12.5	25	37.5	50
Wheat flour	500	500	500	500	500
Baking yeast	50	50	50	50	50
Baker's yeast	20	20	20	20	20
Common salt	9	9	9	9	9
Water (ml)	275	275	275	275	275
Ascorbic Acid	0.25	0.25	0.25	0.25	0.25

$S^{100}/_0$ = Bread with 100% sucrose
$S^{75}/_{25}$ = Bread the sucrose was substituted with 25% date palm fruit pulp
$S^{50}/_{50}$ = Bread the sucrose was substituted with 50% date palm fruit pulp
$S^{25}/_{75}$ = Bread the sucrose was substituted with 75% date palm fruit pulp
$S^{0}/_{100}$ = Bread the sucrose was completely replace with date palm fruit pulp.

Table 3: Recipe of different substitution of sucrose with date palm (phoenix dactylifera) fruit on quality of bread.

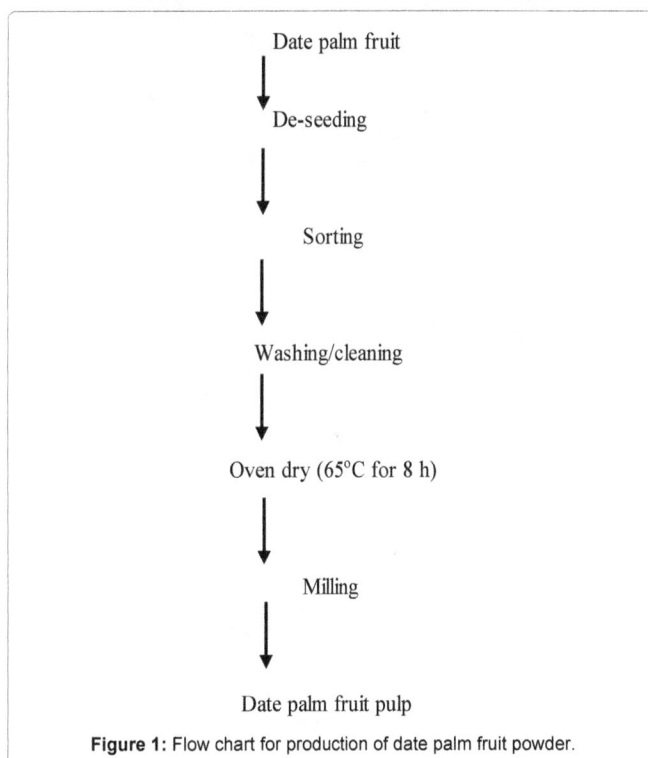

Date palm fruit
↓
De-seeding
↓
Sorting
↓
Washing/cleaning
↓
Oven dry (65°C for 8 h)
↓
Milling
↓
Date palm fruit pulp

Figure 1: Flow chart for production of date palm fruit powder.

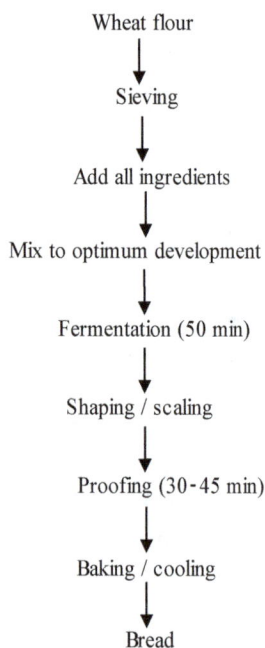

Figure 2: Flow process for Bread Production.

Figure 3: Date palm fruit pulp.

Figure 4: Bread samples during baking.

Loaf volume and specific volume determination

The volume of the loaf was determined by seed displacement method [38] with slight modification. Rice grains were used in place of rape seed. The grains were layered in the box of known dimension. Bread loaf was inserted and the grains poured to fill the remaining space till the grains were running over. A straight edge rule was used to level the grain. The volume of the grain was measured using a measuring cylinder and the volume of bread loaf calculated as the volume of the rectangular box less the volume of rice grain. Loaf specific volume was calculated as the ratio of loaf volume of its weight.

Statistical analysis

The data obtained from the study was statistically analyzed using one factor randomized design, analysis of variance (ANOVA). The means were separated using fishers Least significant difference (LSD) [43].

Results and Discussion

Proximate composition

The carbohydrate content of the bread samples, their sugars (sucrose) were substituted at varying levels with date palm *(Phoenix dactylifera)* fruit pulp which ranged from 45.39 to 35.13%. The carbohydrate content of the loaves was found to decrease significantly ($P < 0.05$) with increase in the level of date palm fruit pulp (Table 4). This is due to replacement of sucrose, a carbohydrate with date palm fruit pulp which contains other nutrients apart from sugar. Also the protein, moisture, ash, crude fibre and fat increased significantly ($P < 0.05$) as the level of date palm fruit pulp increased. This agreed with work of Sadiq et al. [44] that the increase in protein, presence of significant quantity of these nutrients in the bread could greatly improve the nutritional quality of bread and this is highly beneficial to consumers.

However, significant difference ($P < 0.05$) existed between 100% sucrose bread sample $S^{100}/_0$, 25% ($S^{15}/_{25}$), 50% ($S^{50}/50$), 75% ($^{25}/_{75}$) and 100% ($S^0/_{100}$) in carbohydrate and crude fibre. There were no significant differences between 75% ($S^{25}/_{75}$) and 100% date palm fruit pulp ($S^0/_{100}$) in protein, 100% sucrose bread sample ($S^0/_{100}$) and 25% ($S^{75}/^{25}$) in ash and between 50% ($S^{50}/_{50}$) and 75% ($S^{25}/_{75}$) in fat content (Table 5).

Physical analysis

The physical characteristics of the five loaves samples are similar, most especially the volumes which showed no significant difference. The bread volumes ranged from 1920.1 to 1925 cm³ in terms of specific volume, loaf sample with 100% sucrose had the highest specific volume of 3.12 cm³/g which was only significantly higher ($P < 0.05$) than the loaf sample with 100% date palm fruit pulp but was similar with the other three loaf samples their sucrose were substituted 25%, 50% and 75% date palm fruit pulp. This result agreed with Obiegbuna et al. [45] stating that specific volume is a function of hydration and the reduced volume with increasing date palm fruit pulp could be as a result of minimal hydration of 50%, having used 275 ml of water for 500 g of flour to form a dough.

Sensory evaluation of bread samples

The results of the comparative sensory evaluation of the different substitution of sucrose with date palm fruit pulp are indicated in Table 6.

There were no significant differences ($P < 0.05$) between the loaf samples in taste, texture, aroma, crumb colour, crust colour and general acceptability.

Conclusion

This study has shown that bread can be produced adequately using date palm fruit pulp with variations in the sucrose levels. The proximate

Nutrients%	$S^{100}/_0$	$S^{75}/_{25}$	$S^{50}/_{50}$	$S^{25}/_{75}$	$S^{0}/_{100}$	LSD
Carbohydrate	45.39 ± 0.29^a	43.53 ± 0.42^b	39.10 ± 0.16^c	36.74 ± 0.13^d	35.13 ± 0.12^e	1.03
Protein	15.19 ± 0.10^d	16.3 ± 0.10^c	18.08 ± 0.10^b	19.07 ± 0.18^a	19.3 ± 0.18^a	0.37
Moisture	28.19 ± 0.32^b	28.51 ± 0.27^{ab}	28.79 ± 0.23^{ab} 28	$.87\pm0.03^a$	28.92 ± 0.09^a	0.655
Ash	2.44 ± 0.02^d	2.61 ± 0.02^d	3.17 ± 0.01^c	3.51 ± 0.04^b	4.11 ± 0.15^a	0.20
Crude fibre	1.65 ± 0.02^e	1.99 ± 0.10^d	3.32 ± 0.06^c	4.09 ± 0.13^b	4.43 ± 0.04^a	0.22
Fat	6.34 ± 0.06^d	7.02 ± 0.09^c	7.53 ± 0.12^b	7.71 ± 0.03^b	7.98 ± 0.05^a	0.24

* Mean values in the same row with the same superscript are not significantly different at P < 0.05
The means were separated using least significant.
$S^{100}/_0$ =   Bread with 100% sucrose
$S^{75}/_{25}$ =   Bread the sucrose was substituted with 25% date palm fruit pulp
$S^{50}/_{50}$ =   Bread the sucrose was substituted with 50% date palm fruit pulp
$S^{25}/_{75}$ =   Bread the sucrose was substituted with 75% date palm fruit pulp
$S^{0}/_{100}$ =   Bread the sucrose was completely replace with date palm fruit pulp.

Table 4: Mean values for the proximate composition of bread as affected by different substitution of sucrose with date palm (phoenix dactylifera) fruit pulp.

Bread samples	Specific volume (cm³/g)	Loaf volume (cm³)
$S^{100}/_0$	3.12^a	1920.3
$S^{75}/_{25}$	2.08^{ab}	1923
$S^{50}/_{50}$	3.06^{ab}	1925.3
$S^{25}/_{75}$	3.02^{ab}	1922.2
$S^{0}/_{100}$	2.93^b	1920.1
LSD	0.16	-

* Mean values in the same row with the same superscript are not significantly different at P < 0.05
The means were separated using least significant
$S^{100}/_0$ =   Bread with 100% sucrose
$S^{75}/_{25}$ =   Bread the sucrose was substituted with 25% date palm fruit pulp
$S^{50}/_{50}$ =   Bread the sucrose was substituted with 50% date palm fruit pulp
$S^{25}/_{75}$ =   Bread the sucrose was substituted with 75% date palm fruit pulp
$S^{0}/_{100}$ =   Bread the sucrose was completely replace with date palm fruit pulp.

Table 5: Mean values of the specific volume and loaf volume of different substitution of sucrose with date palm (phoenix dactylifera) fruit pulp on quality of bread.

Bread samples %	Taste	Texture	Aroma	Crumb colour	Crust colour	General Acceptability
$S^{100}/_0$	8.04	7.96	7.28	6.8	6.52	8.24
$S^{75}/_{25}$	7.89	7.91	7.18	6.72	6.28	7.92
$S^{50}/_{50}$	7.8	7.82	7.12	6.64	5.96	7.75
$S^{25}/_{75}$	7.55	7.77	7.02	6.58	5.32	7.58
$S^{0}/_{100}$	7.24	7.68	7.04	6.4	5.56	7.44
LSD	-	-	-	-	-	-

* Mean values in the same column are not significantly different at P < 0.05
$S^{100}/_0$ =   Bread with 100% sucrose
$S^{75}/_{25}$ =   Bread the sucrose was substituted with 25% date palm fruit pulp
$S^{50}/_{50}$ =   Bread the sucrose was substituted with 50% date palm fruit pulp
$S^{25}/_{75}$ =   Bread the sucrose was substituted with 75% date palm fruit pulp
$S^{0}/_{100}$ =   Bread the sucrose was completely replace with date palm fruit pulp.

Table 6: Mean values of the sensory evaluation of bread as affected by substitution of sucrose with date palm (phoenix dactylifera) fruit pulp.

analysis indicated that the nutritional value of the bread increased as the date palm fruit pulp level increased.

Recommendation

Author thereby suggest public enlightenment on the nutritional importance of date palm fruit and also recommend that sucrose can be substituted with date palm fruit.

References

1. Mannay S, Shadaksharaway CM (2005) Facts and principles. New Age International Ltd. Publishers.

2. Shittu TX, Raji AO, Sani LO (2007) Bread from composite cassava-wheat flour: I. Effect of baking time and temperature on some physical properties of bread loaf, food research International 40: 280-290.

3. Bali PS (2009) Bread fabrication. In: Food production operations. Oxford University Press, USA.

4. Gusba J (2008) Sugar, sugar!:A look at the functional role sugar in baking.

5. Campbell AM, Penfield MP, Gris world RM (1979) Yeast breads and Quick breads. In: The experimental study of foods. Houghton Mifflir Company, Boston.

6. Al-Shahib W, Marshall RJ (2003) The fruit of date palm: its possible use as best food for the future. Internal Journal of Food Science and Nutrition 54: 247-259.

7. Al-Hooti S, Jiuan S, Quabazard H (1995) Studies on the physics-chemical characteristics of date fruits of fine UAEA cultivars at different stages of maturity. Arab gulf J 13: 553-569.

8. El-hadrami I, El-Hadrami A (2009) Breeding date palm. In: Jani SM, Priyadarshan PM. Breeding plantation Tree crops, Springer, New York, USA.

9. Jain SM, Al-Khayi JM, Johnson DV (2011) Date Palm Biotechnology. Springer, USA.

10. El- hadrami A, Daayf F, El- Hadrami I (2011) Date palm genetics and breeding. In: Jani SM, Al – Khayi JM and Johnson DV. Date Palm Biotechnology, Springer Netherlands, USA.

11. Al- Hooti S, Sidu JS, Quabazard H (1995) Studies on the physics-chemical characteristics of date fruits of fine UAEA cultivars at different stages of maturity. Plant food for human nutrition 50: 101-113.

12. Omamor IB, Aisagbonli CI, Oruade Dimain EA (2000) Present status of Date palm diseases, Disorders and pests in Nigeria. Proceedings of the date palm international symposium, NnidHoek, Namibia.

13. Abdul-Qadir IM, Garba ID, Esiegbe E, Omofonmwan EI (2011) Nutritional components of date palm and its production states in Nigeria. International Journal of Agricultural Economics and Rural Development 4: 83-89.

14. Glasner B, Botes A, Zaid A, Emmens J (2002) Date harvesting, parking house management and marketing aspects. In: The Encyclopedia of fruits and Nuts Jawick J and Pall ER Cambridge University Press, Cambridge, USA.

15. El- hadrami A, Al Khayri JM (2012) Socio-economic and traditional importance of date palm. Emirates Journal of food and Agriculture.

16. Elleuch M, Berbes S, Rosiseux O, Blecker C, Deroame C, et al. (2008) Date flesh: chemical compositions and characteristics of the dietary fibre. Food chem 111: 676-682.

17. Cauvain S (1998) Technology of bread-making. In.Cauvain SP and young LS. Food Science and Nutrition.

18. Isserilykska D, Karadjor G, Agelor A (2001) Mineral composition of Bulgarian wheat bread European Food Research and Technology 213: 244-245.

19. Serrem CA, Kock HLD, Taylor JRN (2011) Nutritional quality, sensory quality and consumer acceptability of Sorghum and bread wheat biscuits fortified with defatted soy flour. International Journal of food science and Technology 46:74-83.

20. Ukpabi A (2008) Cassava processing utilization. NRCRI, Umuahia, Nigeria.

21. Cauvain SP, Young LS (2008) Bakery Food Manufacture and Quality: Water Control and Effects,(2ndedn). Blackwell science, oxford,l USA.

22. William T, Pullen G (1998) Functional ingredients.In. Cauvain S.P and young LS (eds.) Technology in bread-making. Blackie Academics and Professional, New York.

23. Kent NL, Evers AD (1994) Kent's Technology of Cereals: An Introduction of Students of food science and agriculture. Elsener science, Oxford, USA.

24. Hoseney RC (1998) Principles of cereal Science and Technology. American Association of cereal chemists, Inc. Miniesota, USA.

25. Stauffer CE (1998) Principles of dough formation. In: Cauvain SP and Young LS. Technology in bread-making. Blackie Academic and Professional, New York, USA.

26. Salhlstrum S, Park W, Shelton DR (2004) Factors influencing yeast formation and effect LMW sugars and yeast fermentation on health bread quality. Cereal chem 81: 328-335.

27. Okaka JC, Ikegwu F (2011) Dietary fiber and encouraging the use of monogastric nutrition. Journal of Science and Techn.

28. Anon (2010) Natural sugar substitute: 10 healthier alternatives to refined sugar.

29. Phillips S (2010) Dry sweeteners.

30. Khan F (2010) Natural ingredient you can use to replace sugar. Demand Media Inc.

31. Anon (1998) Substitute Natural and Artificial sweeteners.

32. Demiralp H, Celik S, Koksel H (2000) Effect of oxidizing agents and defatting on the electrophoretic patterns of flour proteins during dough mixing. Eur Food Res Technol 211: 322-235.

33. Dendy DAV, Dobrasczyk BJ(2001) Cereals and Cereal Products: Chemistry and technology. Food science and nutrition, Springers Book Archives, USA.

34. Atwell WA (2007) wheat flour. Fagan Press handbook series, Minnesota, USA.

35. Salbni SS, Baik OD, Marcotte M (2012) Neural network for predicting thermal conductivity of bakery products. J Food Eng 52: 299-304.

36. Lean ME (2006) Fox and Cameron's food science, nutrition and health. Nutrition in health and disease, CRC press, Taylor and Francis Group, USA.

37. AOAC (2010) Official methods of analysis. (18thedn). Association of official analytical chemists.

38. AACC (2000) Approved methods of American association of cereal chemists. (10thedn). AACC International.

39. James CS (1995) The adverse effects of long term cassava (Manihotesculenta Ganta).

40. Chang (2003) Protein analysis in food analysis. Klewer Academic or plenium Publishers, New York.

41. Min DB, Boff JM (2008) Crude fat analysis. In: Neilson SS. Food Analysis. Kluwer Academic/Plenium Publishers, New York, USA.

42. Annette-Cannetar BA (2013) Department of food science and technology. Rutgers University, Newyork,USA.

43. Steel RG, Torrie JH, Dickey DA (1996) Principles and procedures of statistics. A Biometric Approach. MC Graw-Hill companies.

44. Sadiq IS, Izugaize T, Shuaibu M, Dogoyaro AI, Garba A, et al. (2013) The Nutritional Evaluation and Medicinal Value of Date Palm (Phoenix dactylifera). International Journal of Modern Chemistry 4: 147-154.

45. Obiegbuna JE, Akubor PI, Ishiwu CN, Ndife J (2013) Effect of substituting sugar with date palm pulp meal on the physiochemical, organoleptic and storage properties of bread. African Journal of Food Science 7: 113-117.

10

Development and Quality Evaluation of Ready to Use Sweet Potato and Mushroom (*Pleurotus ostreatus*) Tikki mix

Virendra Singh* and Lakshmi BK

Department of Food Process Engineering, Vaugh School of Agricultral Engineering and Technology, Sam Higginbottom Institute of Agriculture, Technology and Sciences, Allahabad, Uttar Pradesh, India

Abstract

Experiment was conducted at Research Laboratories of Department of Food Process Engineering, SHIATS, Allahabad to utilize the Sweet potato flour and oyster mushroom powder for preparation of Tikki mix. The sweet potato tubers were brought from local market of Allahabad and oyster mushroom obtained from Department of Plant protection, SHIATS Allahabad and then tubers with sorted washed, peeled, sliced, blanched, dried and milled into flour form. Sweet potato flour contains less amount of protein, although rich in dietary fiber content and carbohydrate, so a successful combination with oyster mushroom powder for Tikki mix production would be nutritionally advantageous. In this experiment Sweet potato flour was blended with oyster mushroom powder in ratio of 94%, 88%, 82% with control 100% Sweet potato flour. These were evaluated for sensory analysis that included color, flavor, taste, texture and overall acceptability of SPF and mushroom powder Tikki mix and analyzed for chemical properties viz; moisture, protein, fiber, fat and ash contents. On the basis of nutritional value T_2 was found high fiber content than other samples and high score for over-all acceptability with containing of 88% sweet potato flour. Thus, the product was found to be acceptable after the storage period.

Keywords: Sweet potatoes flour; Oyster mushroom powder; Standardization; Physic-chemical analysis; Sensory evaluation; Tikki mix

Introduction

The sweet potato (*Ipomoea batatas* L.) belong to the Convolvulaceae or morning glory family. Sweet potatoes are good sources of vitamins C and E as well as dietary fiber, potassium, and iron, and they are low in fat and cholesterol. Sweet potato, either fresh, grated, cooked and mashed, or made into flour, could, with high potential for success, replace the expensive wheat flour in making buns, chapattis (flat unleavened bread) and mandazis (doughnuts) [1]. Sweet potato flour is used as a raw material for processing into other products.

Mushroom is fleshy, spore-bearing reproductive structures of fungi grown onorganic substrates and for a long time, have played an important role as a human food due to its nutritional and medicinal properties [2]. Mushrooms are a good source of protein, vitamins and minerals and are known to have a broad range of uses both as food and medicine. A high nutritional value of oyster mushrooms has been reported with protein (25-50%), fat (2-5%), sugars (17-47%), mycocellulose (7-38%) and minerals (potassium, phosphorus, calcium, sodium) of about 8-12%. Edible mushrooms are also rich in vitamins such as niacin, riboflavin, vitamin D, C, B1, B5 and B6.

Sweet potato flour contains less amount of protein, although rich in dietary fiber content and carbohydrate, so a successful combination with oyster mushroom powder for Tikki mix production would be nutritionally advantageous. Shelf life encompasses both safety and quality of food. Safety and spoilagerelated changes in food occur by three modes of action: biological (bacterial/enzymatic), chemical (autoxidation/pigments) and physical. The approach in the present study was to use sweet potato flour in different proportions and important mushroom powder, analyze nutritional attribute and shelf life of Ready to use Sweet potato and mushroom Tikki mix [3].

Materials and Methods

The experimental studies were carried out in Research Laboratories of Department of Food Process Engineering, SHIATS, Allahabad. The methodology adopted has been described under the below.

Procurement of raw material

Good quality of sweet potatoes without any bruises and other major ingredients that is fresh oyster mushrooms, spices, corn flour and salt etc were purchased from local market of Allahabad.

Experimental plan

The experimental plan used for the present research is given in Tables 1 and 2. Figure 1 shows the flow chart for the preparation of sweet potato flour, Figure 2 shows preparation of Mushroom powder flour and Figure 3 shows the flow chart for the preparation of sweet potato and mushroom Tikki mix. Table 2 shows the different combination of sweet potato flour and mushroom powder for tikki mix preparation [4].

Preparation of sweet potato flour: Figure 1 represents Process flow chart for preparation of sweet potato flour

Preparation of Mushroom powder flour: Figure 2 represents Process flow chart for preparation of mushroom powder

Preparation of sweet potato and mushroom Tikki mix: Figure 3 represents Process flow chart for sweet potato and mushroom Tikki mix

Physico-chemical analysis for sweet potato and mushroom Tikki mix

Moisture content: The moisture content of the developed Tikki mix was determined by the method described AACC (200), Method no-44- 15 A,

*Corresonding author: Virendra Singh, Department of Food Process Engineering, Vaugh School of Agricultral Engineering and Technology, Sam Higginbottom Institute of Agriculture, Technology and Sciences, Allahabad, Uttar Pradesh, India, E-mail: virendra.singh788@gmail.com

S. No.	Parameters	Description
1.	Product	Sweet potato flour and Mushroom *Tikki* Mix
2.	Variable	Sweet potato flour and mushroom powder
3.	Storage conditions	Ambient
4.	Packaging material	HDPE
5.	Mode of packaging	Sealing Machine
6	Sample size	100 g.

Table 1: Experimental Plan.

Treatments	Sweet potato flour (%)	Mushroom powder (%)	Corn flour (%)
T_0	100.00	0.00	0.00
T_1	94.00	4.00	2.00
T_2	88.00	8.00	4.00
T_3	82.00	12.00	6.00

Table 2: Treatment Combination.

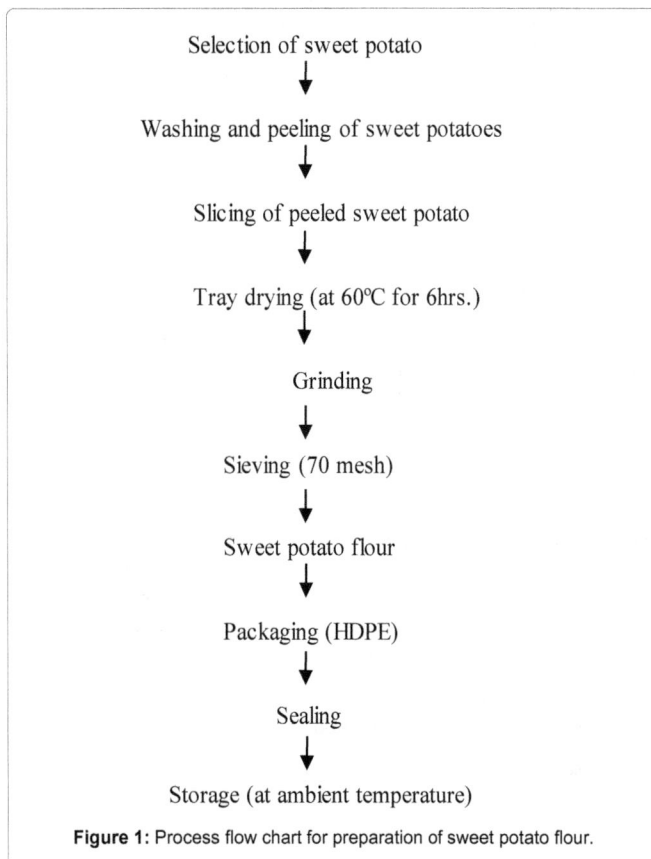

Figure 1: Process flow chart for preparation of sweet potato flour.

$$\% \text{ Moisture } = \frac{w1 - w2}{w1 - w} \times 100 \tag{1}$$

Where,

W = weight in gram of the empty moisture dish.

W1= weight in gram of the moisture dish with the material before drying.

W2= weight in gram of the moisture dish with the material after drying.

Ash content (%): The residue remaining after the incineration of sample at 550°C – 600°C is regarded as ash. The ash content was determined by the method described in AACC (2000), Method no. 08-01.

$$\% \text{ Ash } = \frac{W2 - W1}{\text{Wt. of sample}} \times 100 \tag{2}$$

Where,

W = weight in gram of the empty dish.

W1= weight in gram of the dish with dried material taken for test.

W2 = weight in gram of the dish with the ash.

Protein content (%): It was determined by the Method no. 32-10.

$$\% \text{ Nitrogen } = \frac{W2 - W1}{W} \times \text{N of NaOH} \times 1.4 \tag{3}$$

Where

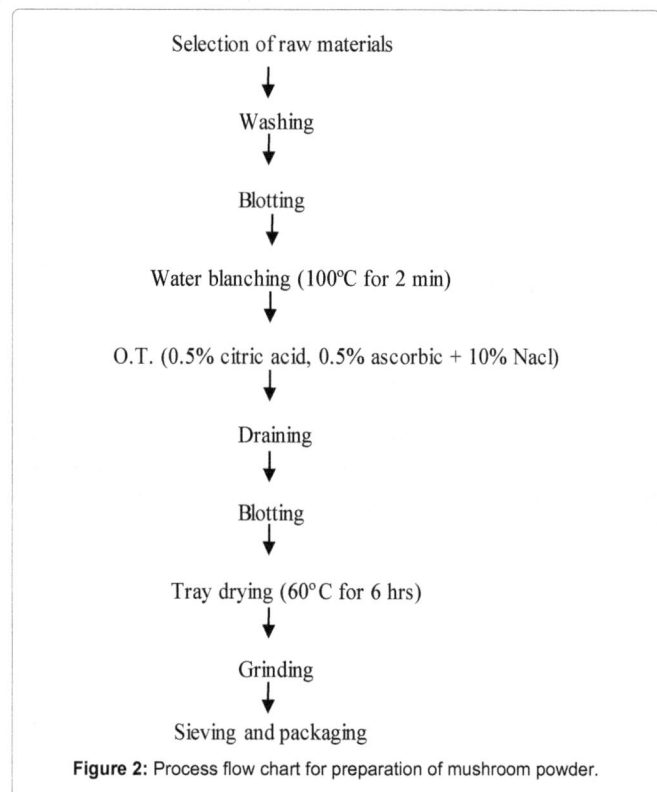

Figure 2: Process flow chart for preparation of mushroom powder.

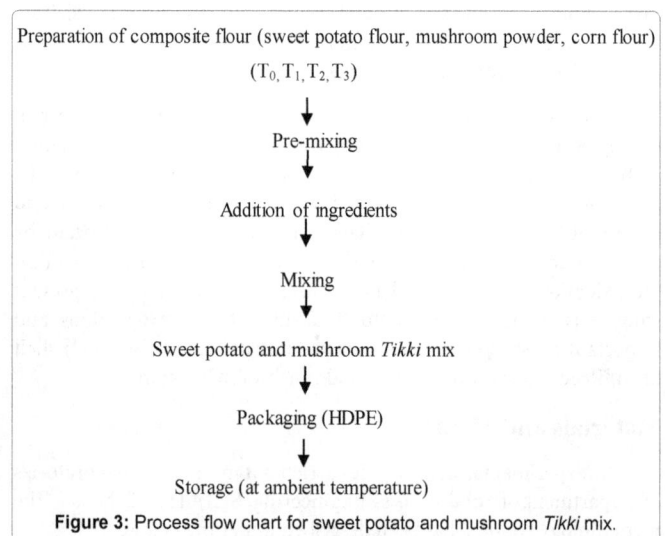

Figure 3: Process flow chart for sweet potato and mushroom *Tikki* mix.

N = Normality of NaOH

W = Wt. of sample

B = Blank titer value (ml)

S = Sample titer value (ml)

% Protein = % Nitrogen x Empirical factor

Empirical factor: 5.8

Fiber content (%): The fiber content was determined by the method described in Ranganna.

$$\% \text{ Fiber content} = \frac{\text{Loss in weight}}{\text{Weight of sample}} \times 100 \qquad (4)$$

Carbohydrate content (%): It was determined by Duboi's Method.

Absorbance corresponds to 0.1 ml of the test = x mg of glucose

100 ml of the sample contains =0.1 × 100 mg of glucose = % of total carbohydrate content.

Fat content (%): The fat content was determined by the method described in AACC, Method no. 30-25.

$$\% \text{ Fat content} = \frac{M2 - M1}{M} \times 100 \qquad (5)$$

Where; M_1 = Initial wt. of round flask

M_2 = Final wt of flask + fat

M= Weight of sample

Sensory analysis of developed tikki mix

The samples were evaluated on the basis of color, taste, flavor, texture and overall acceptability by a panel of judges using 9-point hedonic scale [5] (Figure 4).

Results and Discussion

Physico-chemical analysis of developed sweet potato and mushroom tikki mix

Proximate analysis of sweet potato and mushroom tikki mix is presented in Table 3.

The ash content of sweet potato and mushroom tikki mix was 2.94% and the moisture content was 8.7%. The carbohydrate content in sweet potatoes was 78.06%. So, it can be analyzed that sweet potato has good nutritional quality [6]. Physical parameters of developed tikki mix (Tables 4 and 5) shows that there was a significant decrease in the diameter of control (T0) and different treatments (T1, T2, T3,) after incorporating tikki mix with Sweet potato flour. Physico-chemical analysis of sweet potato and mushroom tikki mix During the present investigation no significant difference was found in the percent moisture content, and ash content was observed on increasing the incorporation of SPF in the treatments (i.e. T0, T1, T2, and T3) [7]. Whereas, there was a significant change in the values of fiber and fat content. This was because in the present study the formulation was based on 100:00:00%, 94:04:02%, 88:08:04%, 82:12:06%, of sweet potato flour, mushroom powder and corn flour [8]. Sweet potato flour has a lower moisture content but high carbohydrate and fiber content whereas, corn flour is rich in moisture, protein and carbohydrate content. Therefore, a significant difference was observed between the samples. The moisture content of control and the sample tikki mix was between the range of 1.33% to 2.86%. But the ash content of control and the sample tikki

mix was between the range of 2.41% to 3.15%. The fiber content was between the range of 2.02% to 7.36%. Fat content varied from 4.97% to 6.53% which was within the acceptable limits for tikki mix. Figure 5 shows the physico-chemical quality of sweet potato and mushroom tikki mix for various treatments (Table 5). The moisture content of tikki mix increased linearly with increase in concentration of sweet potato [9], this is attributed to high water binding capacity of sweet potato which retained higher moisture content in ultimate products. The results for moisture content of the tikki mix were similar with the results obtained by, who incorporated corn flour in preparation of the tikki mix.

The ash content of tikki mix increased significantly due to higher ash content of sweet potato and due to externally added fat during tikki mix preparation. Both refined mushroom powder and sweet potato flour were having lower fat content and hence the total fat content in samples were similar where as there was slight reduction in fat content with increase in sweet potato flour incorporation [10]. The results of proximate composition of sweet potato based tikki mix are similar with the results obtained by the fiber content of tikki mix increased significantly, due to higher fiber content of sweet potato flour. As fiber absorbs large amount of water, it gives a sensation of fullness (having an appetite completely satisfied). Sensory evaluation

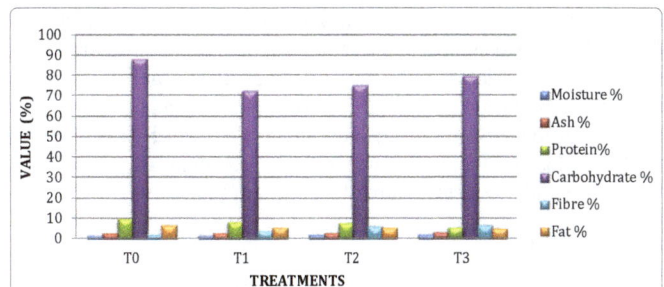

Figure 4: Physico-chemical analysis of standardized sweet potato flour and mushroom tikki mix.

S. No.	Constituents	Percentage (%)
1.	Ash	2.94 ± 0.6
2.	Moisture	8.70 ± 0.2
3.	Protein	3.15 ± 0.04
4.	Carbohydrate	78.06 ± 0.05
5.	Fiber	6.57 ± 0.03
6.	Fat	0.58 ± 0.2

Table 3: Physico-chemical analysis of Sweet Potato Flour and mushroom tikki mix.

Treatment	Moisture %	Ash %	Protein%	Carbohydrate %	Fibre %	Fat %
T_0	1.85	2.58	9.85	88.42	2.03	6.47
T_1	1.94	2.86	8.38	72.19	4.05	5.74
T_2	1.95	2.88	7.67	75.23	5.97	5.59
T_3	2.14	3.07	5.77	79.55	6.45	5.05

Table 4: Physico-chemical analysis of Standardized Sweet Potato Flour and mushroom tikki mix.

Treatments	Color	Taste	Flavor	Texture	O.A.A
T_0	8.66	8.66	8.00	8.66	9.00
T_1	8.00	8.00	7.33	8.33	8.00
T_2	8.33	8.66	8.33	8.66	8.66
T_3	7.66	7.33	6.66	7.33	7.33

Table 5: Sensory evaluation for ready to use sweet potato flour and mushroom *Tikki* mix.

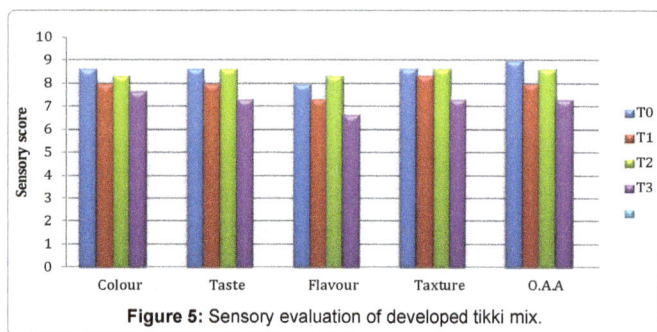

Figure 5: Sensory evaluation of developed tikki mix.

for product standardization tikki mix supplemented by different levels of substitutions of sweet potato flour were sensory evaluated and compared with control tikki mix 100% sweet potato flour. Data indicated that the percent score of tikki mix containing 82% sweet potato flour were found to be the most acceptable [11]. At 82% level of incorporation, all the attributes scored highest level. The color scores of tikki mix with 82% sweet potato flour reached maximum than to the rest of the proportions similar to the control sample. Thus, incorporation of sweet potato flour at 82% level improved the sensory attributes namely texture, flavor, color and over all acceptability. The nutritional quality of the developed biscuits was enhanced due to the addition of Sweet potato flour. Thus, the sensory evaluation (Table 5) depicts that highest amount of sweet potato flour that can be incorporated to develop acceptable biscuit was 82%, i.e. sample T2 was the best regarding all sensory attributes [12]. The score of colour reduced significantly to 8.33, this was due to increasing level of sweet potato flour which gave a color to tikki mix which was not liked much by the panelist. Similarly, the score of taste, flavor and texture also reduced significantly (Tables 3-5).

Conclusion

It can be concluded that T_2 found satisfactory after testing of Chemical analysis and Sensory analysis of developed Sweet potato flour and mushroom Tikki mix were found 2 month of shelf life on ambient temperature with HDPE packaging. There was significant increase in ash, moisture, carbohydrate and fiber content. It is strongly recommended that highly nutritious may be prepared by replacing upto the level of 82% of sweet potato without adversely affecting the overall acceptability of these snacks. Shelf life analysis was done at an interval of 20 days and the Tikki mix was found acceptable for period of 60 days. This study indicates that feasibility of developing such products.

References

1. Hagenimana V, Carey EE, Gichuki ST, Oyunga MA, Imungi JK (1998) Carotenoid content after drying and processing sweetpotato products. Ecol Food Nutr 37: 455-473.

2. Benjamin (2007) Rheology and chemistry of dough. In: Pomeranz Y (ed) Wheat chemistry and Technology. (3rd edn), American Association of Cereal Chemists, Minnesota.

3. Ajlouni SD, Beelman RB, Thompson DB, Mou JL (1992) Stipe trimming at harvest increases shelf life of fresh mushroom. J Food Sci 57: 1361-1363.

4. Dove AS, Coote SA (2008) Drying of sweet potato at different temperature. Pol. Journal Food Nutr Sci 57: 487-496.

5. FAO (1972) Food composition tables for use in East Asia. Part- 1, food policy and nutrition division, FAO Rome.

6. Jones (1965) Genetic variation in color of sweet potato flour related to its use in wheat-based composite flour products. Cereal Chemistry 74: 681-686.

7. Peters D, Wheatley C (1997) Small scale agro-enterprises provide opportunities for income generation: Sweet potato flour in East Java, Indonesia. Quarterly Journal of International Agriculture.

8. Suismono (1995) Sweetpotato processing for flour and noodles. Msc thesis, Bogor Agricultural University (IPB), Bogor, Indonesia.

9. Van Hal M (2000) Quality of sweet potato flour during processing and storage. Food Rev Int 16: 1-37.

10. Woolfe JA (1992) Sweet potato: An untapped resource. Cambridge, Cambridge University Press, UK.

11. Walde SG, Velu V, Jyothirmayi T, Math RG (2006) Effect of pretreatment drying methods on dehyration of mushroom. Journal of food engineering 74: 108-115.

12. Zhang M, Tang J, Mujumdar AS, Wang S (2006) Trends in micro-wave related drying of fruits and negatives. Trends in food science and technology 17: 524-534.

11

Applying Light, Histochemical and Scanning Histological Methods for the Detection of Unauthorized Animal and Herbal Content in Street Meat Sandwich: What is in the Sandwich We Eat?

Abdel Hafeez HH[1]*, Zaki RS[2] and Abd El-Magiud DS[3]

[1]Department of Anatomy, Embryology and Histology, Faculty of Veterinary Medicine, Assuit University, Assuit, Egypt
[2]Department of Food Hygiene, Faculty of Veterinary Medicine, Branch of New Valley, Assiut University, Assuit, Egypt
[3]Department of Forensic Medicine and Toxicology, Faculty of Veterinary Medicine, Branch of New Valley, Assiut University, Assuit, Egypt

Abstract

Samples of the total of 105 different meat sandwiches products were examined (Kofta, Hawawshi, and shawerma sandwich, 35 sandwiches from each type of product were collected from New Valley City from different restaurants during the year 2016 and analyzed by light and scanning electron microscope for detection of meat adulteration. Select half samples from each group for light and histochemical microscopic examination and the rest of the same group for scanning electron microscopic examination. The sections were stained using hematoxylin and eosin, PAS, Wigert's and Crossman's trichrome, bromophenol blue and ATPase. The histological examination revealed that a variety of tissue types besides skeletal muscle were observed including connective tissue fibers, Lung, ruminant stomach, Large elastic blood vessels, heart muscle, adipose tissue, cartilage (hyaline and white fibrocartilage) and spongy bone, lymphatic tissue (spleen), plant materials, in addition to sand particles. With use ATPase enzyme staining can suspect fetal tissue in Hawawshi meat with abundant dark (slow contracted) muscle fiber than light (fast contracting) muscle fibers. The findings of the present research suggest the histological technique as an effective method for qualitative evaluations of street meat sandwich adulteration.

Keywords: Street meat sandwich; Meat adulteration; Histochemical microscopic examination

Introduction

From a socioeconomic point of view, street food plays a great role in conflicting food and nutritional provisions consumed in different cities at cheap payment to the lower and middle livelihood people chiefly meat products which are well known to complement at least half of marketable meat that enclose heat cooked or processed meat [1]. Intentional or economically-motivated adulteration of food has recently been defined as "the fraudulent addition of nonauthentic substances or removal or replacement of authentic substances without the purchaser's knowledge for economic gain of the seller" [2]. Food adulteration is considered by the USA are conceivable food safety issues on the authority that adulteration of food is simulated that people have narrow or no awareness of any possible food safety indications of their actions. In order to present higher conservation to society, strategic regulations must be established on the hypothesis that food safety may be settled by adulteration is deserved [2]. Alongside from the adulteration form, some animal tissues such as central nervous tissues could be a way to infect human beings [3] inform of bovine spongiform encephalopathy (BSE) which is known to be set by feeding cattle with scrapie infected sheep tissues [4]. It has been recorded that both human new variant Creutzfeldt-Jakob disease (nvCJD) and BSE belong to the family of fatal TSE diseases, both contribute the same way of infection, namely the abnormal prion protein (PrPsc), that bundles in cytoplasmic vesicles in the infected individuals' and animals' brains and is very resistant to heat [5]. The meat of great economic value so the usage of fraudulent tissues in meat products is unwanted but probable. Formerly, many research works have reported that histological examination as an efficient technique to detect unauthorized tissues in some meat products [6]. This study aimed to use the histological technique as a simple and inexpensive method for determination of unauthorized and herbal content in street meat sandwich. This study aimed to use histological techniques as simple and inexpensive methods for determination of unauthorized animal and herbal content in street meat sandwich.

Materials and Methods

Collected samples

Samples of the total of 105 different meat sandwiches products were examined (Kofta, Hawawshi, and shawerma sandwich, 35 sandwiches from each type of product were collected from New Valley City from different restaurants during the year 2016 and analyzed for detection of meat adulteration.

Sample selection

Select half samples from each group for light microscopic examination and the rest of the same group for scanning electron microscopic examination. In addition to unfixed selected frozen sections (10 μm) were obtained.

Sample preparations for light microscopic examinations

Six different areas were obtained from each sample from different parts (size of the sample about 1 cm long and 0.5 cm thickness) and finally, each sample has six blocks represented different parts. During trimming of samples found hard structure and very difficult in cutting, then selected sample were put in decalcifying agent (neutral buffer formalin-nitric acid) composed of the following constituents for 20

*Corresponding author: Abdel Hafeez HH, Department of Anatomy, Embryology and Histology, Faculty of Veterinary Medicine, Assuit University, Assuit, Egypt
E-mail: hhmmzz91@gmail.com

days: 20 ml nitric acid-80% 20 ml-10% neutral buffer formalin, 160 ml distilled water [7]. Then Specimens were washed by 0.1 M Na-phosphate buffer (pH 7.2-7.4), then immediately fixed in Bouin's fluid for 2 h. Bouin`s fixed samples were extensively washed in 70% ethanol (3 × 24 hours) to get rid of the fixative before the subsequent steps of tissue processing for preparation of paraffin blocks. Fixed samples were dehydrated in ascending grades of alcohols at 80%, 90% for 3 hours at each concentration and 100% for two hours. The sampled were cleared using methyl benzoate. Dehydrated samples were then impregnated and embedded in Paraplast (sigma Aldrich). Serial sections of 5-7 μm were cut using a Richert Leica RM 2125 Microtome, Germany and mounted on glass slides. Sections were kept in an incubator at 40°C for dryness and stained with Hematoxylin and eosin stain [8] and used for general histological examination.

Histochemical investigations

The following staining methods were used: Crossmon's trichrome for collagenous fibers and to differentiate between the different tissue constituents [9]; Weigert [10] for staining of elastic tissue; the periodic acid-Schiff reaction (PAS) according to McManus [11] and *Representative sections were stained with bromophenol blue stain for detection of proteins [12].* All staining were cited by Bancroft et al. [7]. Stained sections were then; examined using DMLS light microscope (Leica, Germany) outfitted with MC120 HD camera (Leica, Germany).

Adenosinetriphosphatase (ATPase) histochemical staining: The adenosine triphosphates (ATPase) stain is to use as a histochemical stain for to distinguish between muscle fiber types. Individual muscle sections were stained with either acidic myofibrillar ATPase, alkaline myofibrillar ATPase, for assessment of primary fibers and secondary fibers [13]. ATPase is method based on contractile type alone, as revealed by myofibrillar [14]. Different researchers have used this method for classifying muscle fiber types in different species, Peter et al. [15] in Pork, Picard et al. [14] and Crosier et al. [16] in bovine, De Freitas et al. [17] in ovine, Francisco et al. [18] in buffalo. Histochemical analysis of ATPase was performed on unfixed selected frozen sections (10 μm) were obtained in Leica cryostat CM 1900-6-1 (Richert, Germany) and stained with by varying pH (acidic pH 4.2 and alkaline pH 9.4). The procedure was done as the description in Bancroft et al. [7]. During the staining procedure, different muscle fiber, type I (FOG, fast-twitch-oxidative-glycolytic, dark fiber) and type II (Light fiber, SO, slow-twitch oxidative) may be differentially stained [7]. Histochemically, fast fibers display high mATPase activity under alkaline conditions and low activity under acid conditions (alkali-stable, acid-la- bile), whereas slow fibers exhibit the inverse (alkali-labile, acid-stable [19].

Morphometrical and statistical analysis of ATPase enzyme: Dark and light muscle fibers were counted using Image J in the different meat product. Fiber counts were performed to estimate fiber number of dark and light muscle cells /500 mm² using 20x objective in each product. All the data are expressed as the mean ± SE.

Sample preparations for scanning electron microscopic (SEM) examinations: Small specimens were Selected from different areas from the rest of Half samples from each group, then washed with 0.1 M Na-phosphate buffer fixed in Karnovsky fixative (10 ml paraformaldehyde 25%, 10 ml glutaraldhyde 50%, 50 ml Phosphate buffer and 30 ml DW) [20] for 4 hours at 4°C, then was used for SEM examination. Thereafter, they were washed in the same buffer used in fixation 5 minutes x 4 times and post-fixed in 1% osmic acid in 0.1 M Na-phosphate buffer for further 2 hours at room temperature. They were washed by 0.1 M Na-phosphate buffer 15 minutes × 4 times. The samples were dehydrated by alcohol 50%, 70%, 90% for 30 min in each concentration and 100% for 2 days with changes many times followed by isoamyl acetate for 2 days and then subjected to critical point drying method with a polaron apparatus. Finally, they were coated with gold using JEOL -1100 E-ion sputtering Device and observed with JEOL scanning electron microscope (JSM – 5400 LV) at KV10 at the Electron Microscopy Unit of Assiut University.

Digital coloring scanning electron microscopic images: To increase the visual contrast between several structures on the same electron micrograph, we digitally colored specific structures either of animal or plant origin (bone, lung tissue, heart, ruminant stomach, blood vessels, fascia and different parts of plants, stem, root, leaves, etc.) to make them more visible to the untrained eye. All the elements were carefully hand colored using the Photo Filtre 6.3.2 program. Coloring images required to change the color balance using the stamp tool to color the objective structures.

Results

In paraffin sections, hawawshi samples were adulterated with lung, hyaline cartilage, Fascia, white fibrocartilage, bone, vascular tissue, tubular organ, spleen, adipose tissue. Lung tissue was distinguished by the alveoli and basophilic cartilage (Figures 1A and 1B). Hyaline cartilage was marked by the presence of chondrocytes embedded in the basophilic cartilage matrix (Figures 1C and 1D). Fascia composed of dense regular collagenous connective tissue (Figure 1E). The white fibrocartilage was characterized by the presence of row aligned chondrocytes parallel to collagen fibers (Figure 1F). The bone tissue could be differentiated by osteocytes (embedded in the bone matrix (Figures 1G and 1H). Vascular tissue was represented by the elastic and muscular artery. The elastic artery was determined by the regularly arranged elastic fibers (Figure 1L). While the muscular artery was identified by the muscular layer in the tunica media (Figures 1I and 1J). The tubular organ was identified by the muscular coat which comprised of smooth muscle fibers (Figure 1M). The spleen was marked by red and white pulps (Figures 1N and 1O). The adipose tissue contained white fat cells (Figure 1P).

Scanned samples of Hawawshi were adulterated with different organs of animal origin. Lung tissues were identified by the presence of alveoli (Figures 2A and 2C). Samples were also contained different organs of the gastrointestinal tract including ruminant stomach, glandular stomach, fascia, bone and vascular tissue. The reticulum was characterized by honey-comb shaped reticular folds (Figures 2D and 2E) and the omasum was recognized by the omasal lamina (Figure 2F). The Glandular stomach was marked by the presence of gastric pits (Figure 2G). Fascia appeared as regularly arranged collagenous fibers (Figures 2H and 2I). Spongy bone was recognized by osteocytes embedded in the bone matrix (Figure 2J). Vascular tissue contained wide blood vessels which filled with blood (Figure 2K).

By light microscopy, hawawshi samples were adulterated with plant tissues including leaves, stem, and root. plant stem was determined by the epidermis (Figures 3B, 3P and 3Q), parenchyma cells (Figures 3B, 3C, 3H, 3I, 3F and 3W), sclerenchyma (Figures 3A-3D , 3G, 3V and 3X), vascular tissue (Figures 3E, 3J and 3L), and collenchyma (Figures 3F, 3V and 3W). Parenchyma cells of the stem of some plant were rich in starch granules (Figure 3O). Plant leaf was identified by mesophyll (Figures 3K-3R and 3S). The root of some plant was recognized in

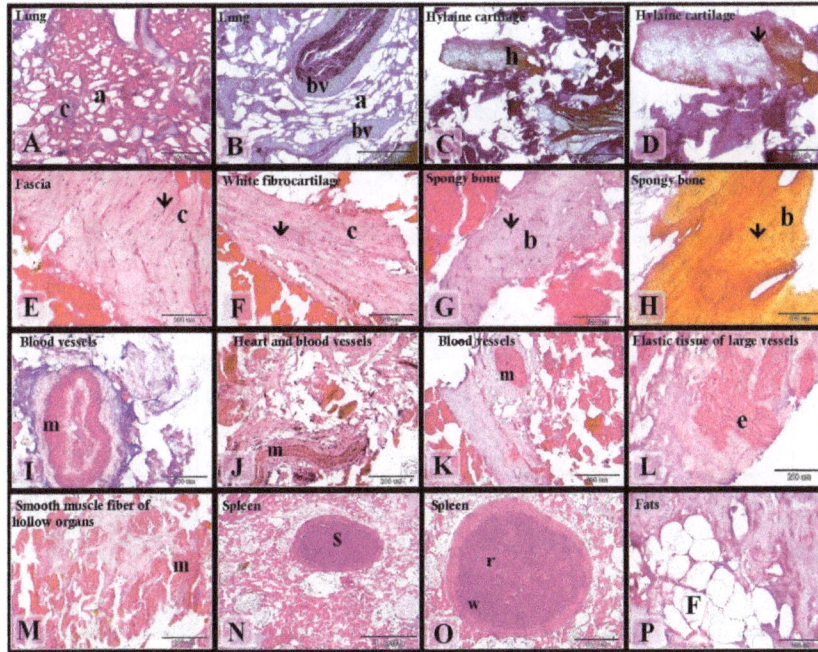

Note: Paraffin sections of samples of Hawawshi stained with H & E (A, E, F, G, I, J, K, L, M, N, O, P) and bromophenol blue (B, C, D, H) showed adulterations with different animal tissues and organs. A, B: lung tissue was identified by Alveoli (a), basophilic cartilage remnants (C). Note blood vessels (Bv). C, D: hyaline cartilage (h) contained chondrocytes (arrow). E: Fascia composed of dense regular collagenous connective tissue. Note, regularly oriented collagen bundles (c), arrow refers to the fibroblasts which had oval to the flattened nucleus. F: white fibrocartilage contained chondrocytes with rounded nuclei (arrow) arrange in rows which aligned parallel to collagen fibres (c). G, H: A part of bone tissue contained osteocytes (arrows) which embedded in the bone matrix (b). I, J, K: muscular arteries identified by the muscular layer (m) in the tunica media. L: a part of the wall of the elastic artery contained regularly arranged elastic fibres (e). M: A part of the muscular coat of the tubular organ which contained smooth muscle fibres (m). N, O: spleen (s) contained red (r) and white (w) pulp. P: a part of the adipose tissue contained white fat cells F).

Figure 1: Adulterations of Hawawshi samples with animal tissues.

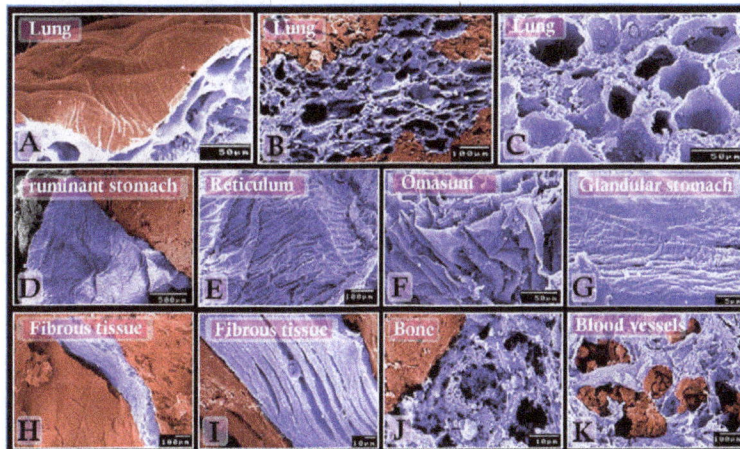

Note: Digitally coloured scanning electron micrographs of animal tissue in Hawawshi samples. A, B, C: lung tissues (blue coloured) contained of alveoli D: low magnification showed ruminant stomach (blue coloured). E: higher magnification of ruminant stomach showed the mucosa of the reticulum which was characterized by honey-comb shaped reticular folds (blue coloured). F: higher magnification of ruminant stomach showed the mucosa of the omasum which was characterized by omasal lamina (blue coloured). G: showed the mucosa of the glandular stomach which was marked by presence of gastric pits (blue coloured). H, I: showed fibrous tissue (fascia) which was composed of regularly arranged collagenous fibres (blue coloured). J: showed bone tissue of spongy type K: showed vascular rich tissue. Note, the wall of the blood vessels (red coloured).

Figure 2: Adulterations of Hawawshi samples with animal tissues.

meat samples (Figures 3T and 3U). The onion was recognized by the epidermal cells (Figure 3P and 3Q).

Scanned samples of Hawawshi were adulterated with leaf and stem of some plants. The stem of the plant was distinguished by the parenchymatous cells (Figures 4A-4C and 4O), Xylem (Figures 4F and 4G), and phloem vessels (Figure 4H). Plant leaf was recognized by outer epidermal and the covering cuticle and vascular tissue such as

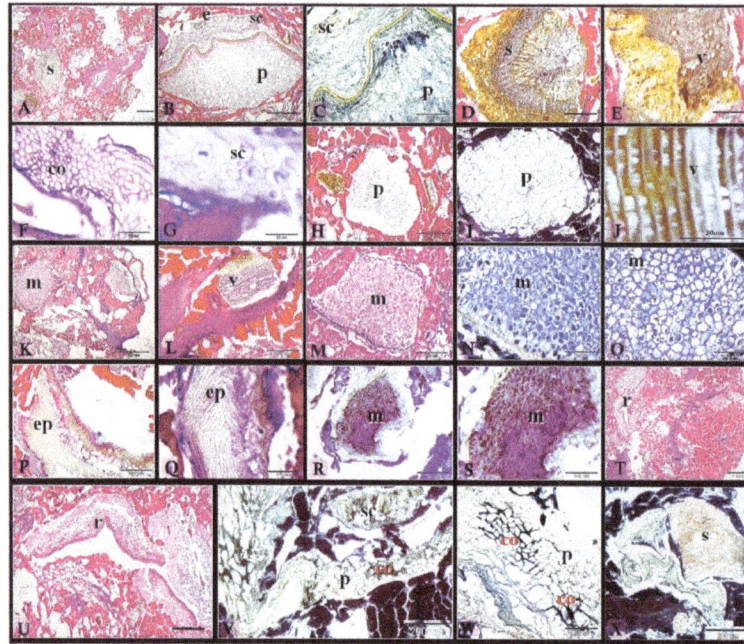

Note: Paraffin sections of samples of kofta stained with H&E (A, B, D, E, G, H, K, M, P, Q, T, U), bromophenol blue (C, I, J, L, O, R, S, V, W, X), PAS stain (F) showed adulterations with the plant. A: Apart of plant stem (s) inside meat sample. B: higher magnification of the plant stem in (fig A) showed epidermis (e), parenchyma cells (P), sclerenchyma (sc). C: plant stem showed parenchyma cells (P), sclerenchyma (sc). D: Plants stem (s). E: A part of plant stem with vascular tissue (v). F: A part of plant stem with collenchyma (co). G: A part of a plant stem contained sclerenchyma (sc). H: parenchyma cells (P) of the plant stem. I: Parts of plant stem with parenchyma cells (P) rich in starch granules. J: Vascular tissue (v) of the plant. K: mat sample adulterated with plant leaf which was identified by mesophyll (m). L: Vascular tissue (v) of the plant. M, N: plant leaf contained mesophyll (m). P, Q: epidermal cells (ep) of the onion. R, S: plant leaf contained mesophyll (m). T, U: meat sample contained the root of the plant (r). V, W: plant stem showed X: meat sample with stem (s) of the plant collenchyma (co) and parenchyma cells (P).

Figure 3: Adulterations of Hawawshi samples with plant tissues.

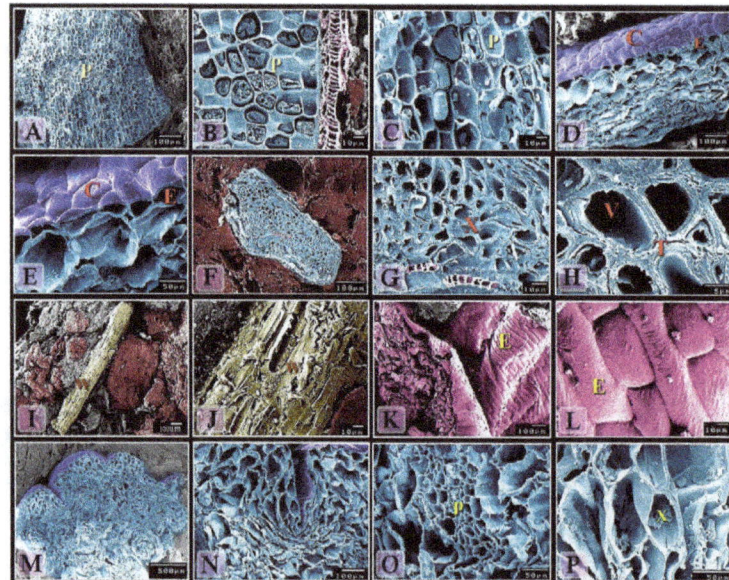

Note: Digitally coloured scanning electron micrographs of plant tissue in Hawawshi samples. A: low magnification showed the stem of the plant. Longitudinal section in the parenchymatous cells (light blue). Xylem vessels (pink). B, C: high magnification of the plant stem showed Xylem vessels (pink) and parenchymatous cells (p). D, E: Cross section in disintegrated plant leaf showed outer epidermal layer (blue coloured) and the covering cuticle (deep blue coloured). F: low magnification of plant leaf showed vascular tissue. Note, longitudinal section in Xylem vessels (pink coloured) and cross section of Xylem vessels (light blue coloured). G, H: higher magnification of the plant leaf showed the vascular tissue of the plant. Note, longitudinal section in Xylem vessels (pink coloured) and cross section of Xylem vessels (light blue coloured). Xylem composed of tracheids (T) and larger vessels (L). I, J: a particle of plant wood (yellow coloured) inside the muscles (red coloured). K, L: showed the surface of the onion epidermal cells (E) which had a characteristic rectangular shape. M-O: showed the part of plant stem (light blue coloured) note xylem (X) and phloem (P).

Figure 4: Adulterations of Hawawshi samples with plant tissues.

Xylem vessels which composed of tracheids and larger vessels (Figures 4D, 4C, 4M and 4N). A particle of plant wood could be observed in meat sample (Figures 4I and 4J). The onion could be identified by their characteristic rectangular shape epidermal cells (Figures 4K and 4L).

In Paraffin sections, kofta samples were adulterated with different animal tissues and organs including heart, vascular tissues, lung tissue, adipose tissue, nerve trunk, fascia, hollow organs, and bone. Heart tissue was distinguished by Purkinje cell fibers (Figure 5H), cardiac muscles (Figures 5E and 5F) and the associated large arterial vessels which had prominent elastic fibers (Figures 5A and 5B) and muscular arteries which were predominated by smooth muscle fibers (Figures 5C, 5D, 5Q and 5R). Meat samples contained adipose tissue which consisted of white fat cells (Figures 5G). Lung tissue was determined by the presence of lung alveoli and basophilic stained hyaline cartilage (Figures 5I-5L). Nerve trunk, which composed of myelinated axons, was observed in meat samples (Figures 5M and 5N). The fascia was identified by regularly arranged dense collagen fibers (Figures 5O and 5P). Meat samples also contained hollow or tubular organs which were recognized by the muscular coat; inner circular and outer longitudinal smooth muscle fibers (Figure 5S). Kofta samples were adulterated by bone tissue (Figure 5T). Scanning samples of kofta showed adulterated with different organs of animal including heart, bone, Vascular tissue, lung tissue, ruminant stomach. The heart was identified by the presence of the papillary D (Figure 6A-6C); a bone tissue of spongy type was recognized in meat samples (Figures 6D and 6E). Vascular tissue contained artery (Figure 6F). Lung tissue characterized by alveoli (Figures 6G-6I). The Reticulum of the ruminant stomach was characterized by honeycomb shaped reticular folds (Figures 6J-6L). By paraffin sections, Kofta samples contained different parts of the plant; leaves and stem. Plant leaves were identified by cuticle, epidermis, mesophyll, parenchyma cells, and vascular tissue (Figures 7D, 7E, 7K-7O and 7Y). Some of the mesophyll cells contained starch granules. Plant stem was recognized by the epidermis, sclerenchyma, collenchyma, and a prominent vascular tissue including Xylem and phloem (Figures 1P-1X, 7A-7C and 7F-7J). Scanning kofta samples showed parts of plant tissue; stem and leaves. Plant stem was distinguished by parenchymatous cells vascular tissue; xylem. Plant leaves were identified by cuticle, epidermal cells, and cortical cells (Figures 8A-8I). Parts of onion were recognized by the outer epidermal cells (Figures 8J-8L).

Scanned kofta samples were contaminated by different types of microbes. The surface of the muscle samples was contaminated with cocci-shaped bacteria, rod-shaped bacteria, thread-like fungi (Figures 9 D-9G). A part of heart tissue was contaminated by cocci-shaped bacteria (Figures 9A and 9B). The mucosa of the reticulum was contaminated by cocci-shaped bacteria (Figures 9H and 9I).

In paraffin sections, samples of shawerma were adulterated with animal and plant tissues. Meat samples were adulterated with adipose tissue which contained fat cells and blood vessels (Figures 10A and 10B). The samples also contained nerve trunk (Figure 10C), cartilage tissue which was distinguished by the basophilic matrix and lacuna (Figure 10C). Meat samples had a part of plant leaf which was recognized by mesophyll and plant stem with parenchyma cells (Figures 10E and 10F).

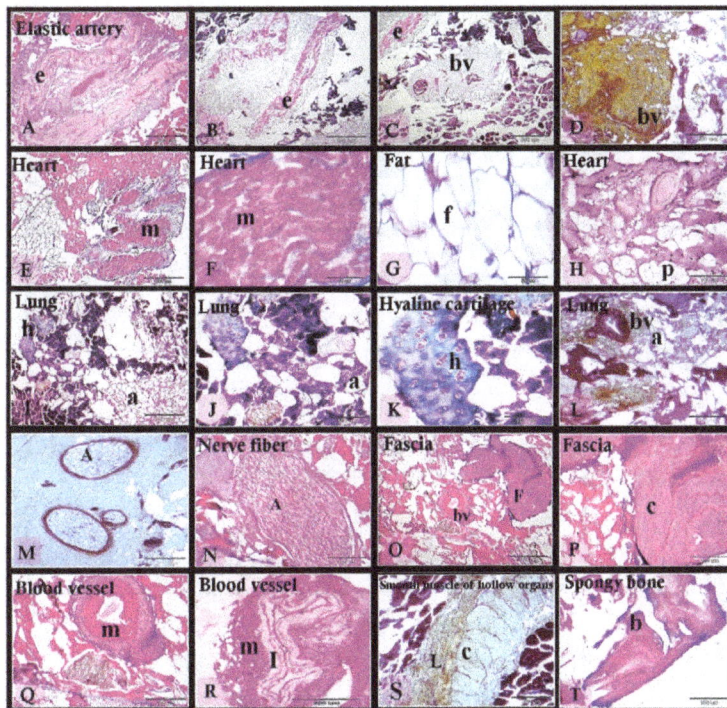

Note: Paraffin sections of samples of kofta stained with H&E (A, E, F, G, N,O, P, Q, T) and bromophenol blue (D, L, M, S), Weigert stain (B, C, H, R), Crossman's trichrome (I, J,K) showed adulterations with different animal tissues and organs. A, B: large elastic artery note tunica media rich in elastic fibres (e). D: large blood vessels associated with the elastic artery (e). D: the wall of a blood vessel (bv). E, F: large blood vessel (bv). E: cardiac muscles (m). G: adipose tissue contained white fat cells (f). H: showed heart with Purkinje cell fibres (p). I- L: lung tissue contained alveoli (a) and hyaline cartilage (h), blood vessel (bv). M, N: showed nerve trunk composed of myelinated axons (A). O, P: meat samples contained fascia (F) which composed of regularly arranged dense collagen fibres . Note, blood vessel (bv). Q: muscular artery had smooth muscle rich tunica media (m). R: large artery note, elastic fibres rich tunic media (m) and prominent lamina elastic internal in tunica intima (I). S. Apart of the muscular wall of the hollow organ. Note inner circular (c) and outer longitudinal (L) smooth muscle fibres . T: spongy bone. Note bone matrix (b).

Figure 5: Adulterations of kofta samples with animal tissues.

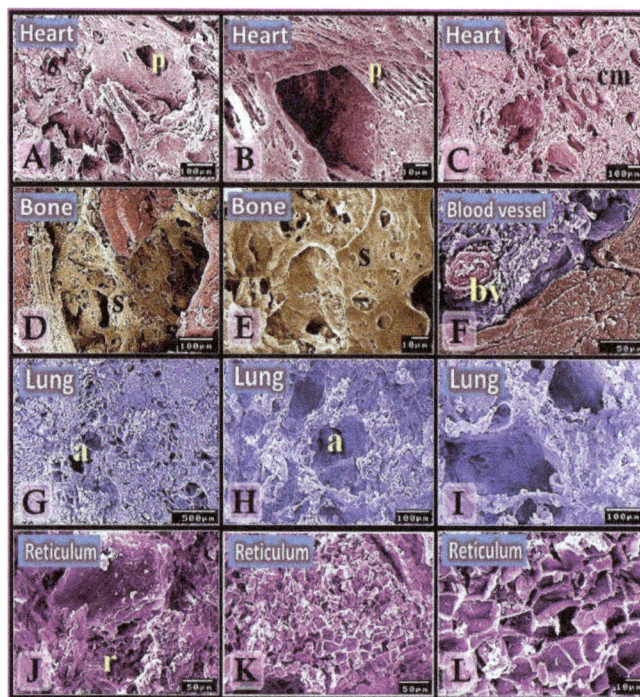

Note: Digitally coloured scanning electron micrographs of animal tissue in kofta samples. A, B, C: Meat samples contained a part of the heart. Note: papillary muscles of the heart (P). Cardiac muscles (cm). D, E: bone tissue (brown coloured) within the muscles (red coloured). Note bone was of spongy type (S). F: An artery (pink coloured) inside the muscular tissue (red coloured). Note peri-arterial connective tissue (violet color). G, H, I: lung tissue (blue coloured) characterized by alveoli (a). J, K, L: Reticulum (r) of the ruminant stomach characterized by honey comb shaped reticular folds.

Figure 6: Adulterations of Kofta samples with animal tissues.

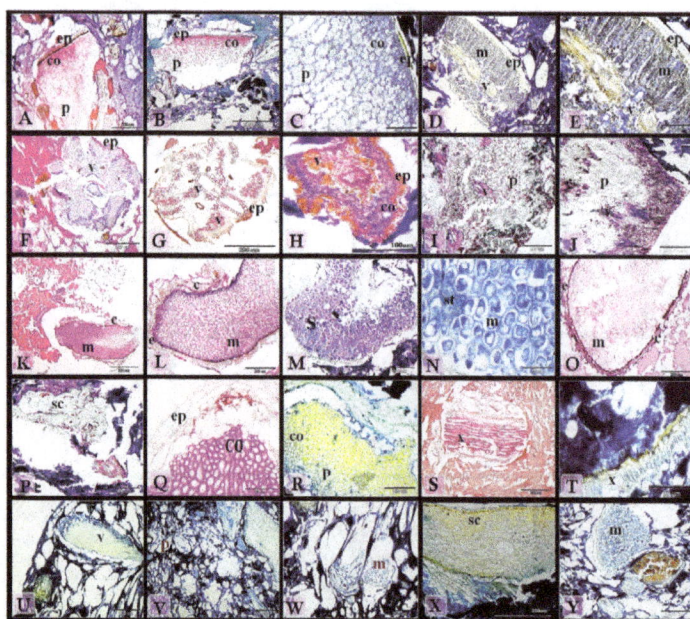

Note: Paraffin sections of samples of kofta stained with H&E (A, F, K, L, S) and bromophenol blue (C, D, E, N, R, T, U, V, W, X, Y), Weigert stain (G, Q, O), Crossman's trichrome (B, H, L, J, M, P) showed adulterations with the plant. A, B, C: cross section in a part of plant stems which was identified by epidermis (ep) parenchyma cells (P), collenchyma (co). D, E: transverse section in plant leaf which was composed of the epidermis (ep), mesophyll (m), vascular tissue (v). F, G: Cross sections in plant stem which had a prominent vascular tissue (v) and epidermis (ep). H, I, J: a part of the plant leaf vascular tissue (v) and epidermis (ep), parenchyma cells (P), collenchyma (co). K, L, M, N, O: plant leaf contained outer epidermis which consisted of cuticle (c) and epidermal cells (e), mesophyll cells contained starch granules (s), sieve tube (st). P, Q, R, S, T, U, V, V and X: A part of plant stem which could be recognized by epidermis (ep), parenchyma cells (P), collenchyma (co), sclerenchyma (sc), vascular tissue (V) including Xylem (X), phloem (p). W, Y: a part of a plant leaf. Note mesophyll (m).

Figure 7: Adulterations of kofta samples with plant tissues.

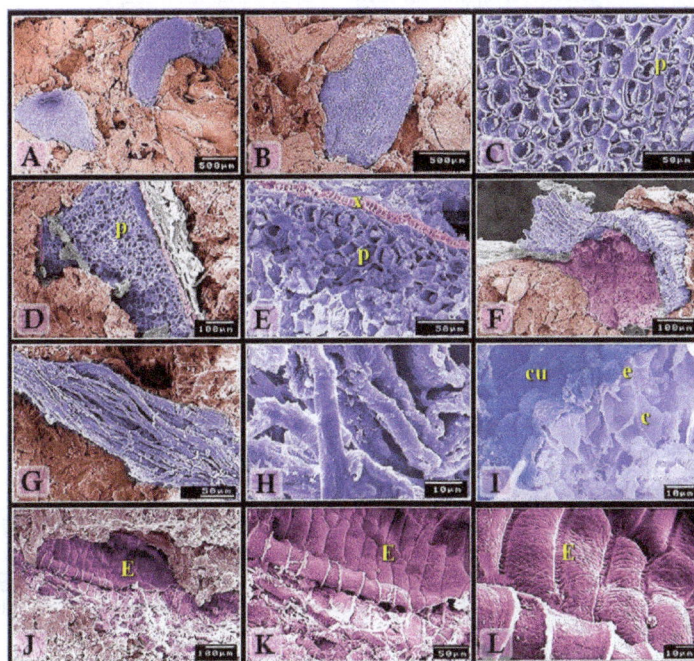

Note: Digitally coloured scanning electron micrographs of plant tissue in kofta samples. A, B: low magnification of plant stems (blue coloured) within the red coloured muscle samples. C, D, E: parenchymatous cells (P) of plant stem. Note, The xylem (pink coloured). F: low magnification of plant tissue (blue coloured) within the meat samples (red coloured). G, H: Low and higher magnification of Plant fibres within meat samples I: plant leaf. Note cuticle (Cu). Epidermal cells (E), cortical cells (C). J, K, L: onion plant .Note, the outer epidermal cells (E).

Figure 8: Adulterations of Kofta samples with plant tissues.

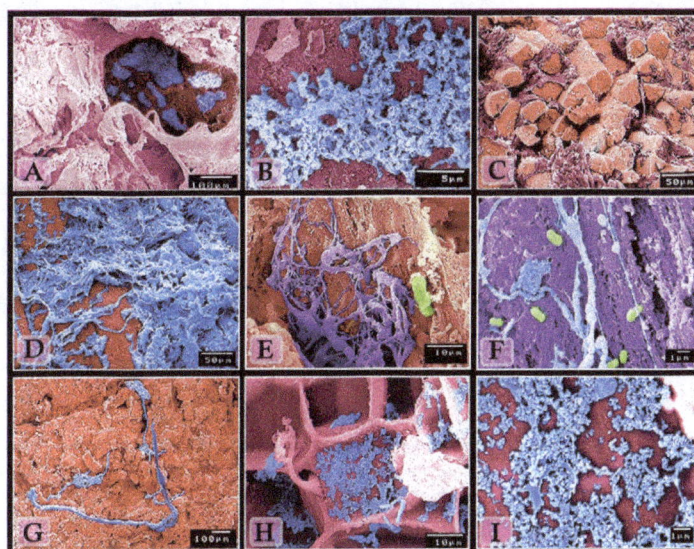

Note: Digitally coloured scanning electron micrographs of kofta samples showed different types of microbes. A, B: the surface of heart tissue (pink coloured) contaminated by cocci-shaped bacteria (blue coloured). C: fibrous tissue (dark red coloured) separated between the muscle cells (bright red coloured). D: contamination of the surface of the muscle samples (red coloured) with cocci-shaped bacteria (blue coloured). E: contamination of the meat samples (red coloured) by rod-shaped bacteria (green coloured). Note Plant fibres (violet coloured). F: contamination of the meat samples by rod-shaped bacteria (green coloured). Note Plant fibres (blue coloured). G: contamination of the meat samples (red coloured) by thread-like fungi (blue coloured). H, I: cocci-shaped bacteria (blue coloured) covering the surface of the reticulum (purple coloured).

Figure 9: Contamination of Kofta samples by microbial cells.

Scanned samples of shawerma samples contained different organs and tissues of the animal. Meat samples contained fibrous tissue which had few fat cells (Figures 11A and 11B). Meat samples was an adulterated by lung (Figures 11C and 11D), spongy bone (Figures 11E and 11F) and different parts of ruminant stomach particularly omasum and glandular stomach (Figures 11M, 11N, 11O and 11P). A large part of intact stomach filled with ingesta which contained herbal contents and Pollen grains (Figures 11I and 11L). Moreover, Scanned samples

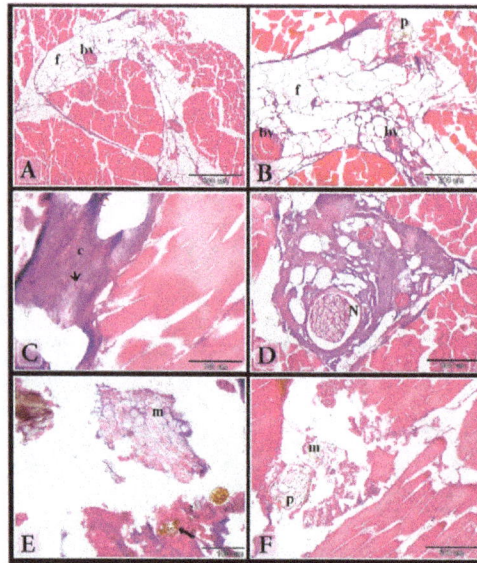

Note: Paraffin sections of samples of shawerma stained with H&E. A, B: meat samples with adipose tissue which contained fat cells (f), blood vessels (bv). C: shawerma eat sample contained cartilage tissue which had a basophilic matrix and lacuna (arrow). D: nerve fibre (N) embedded in meat sample. E: A part of plant leaf was recognized by mesophyll (m). F: A part of plant leaf contained mesophyll (m) and plant stem with parenchyma cells (P).

Figure 10: Adulterations of shawerma samples with animal and plant tissues.

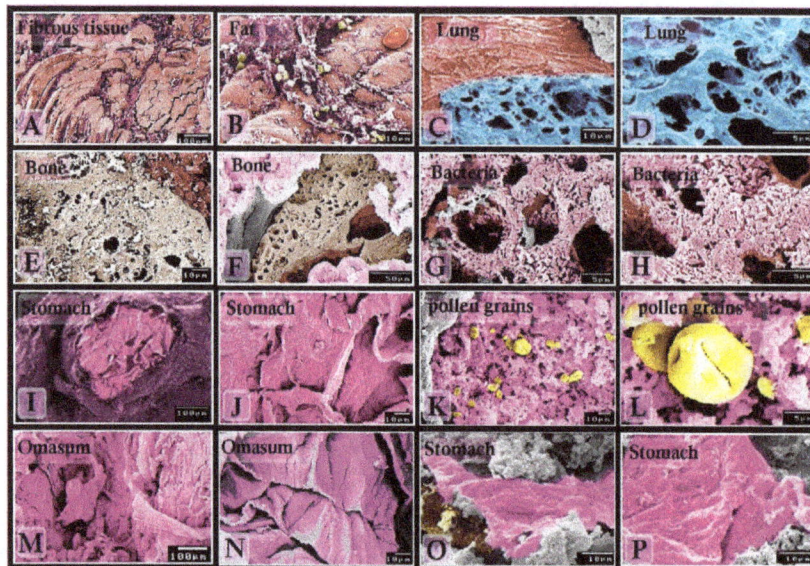

Note: Digitally coloured scanning electron micrographs of shawerma samples adulterated with animal tissue and contaminated with microbial cells. A, B: fibrous tissue (dark brown) with fat cells (yellow coloured) interspersed between muscular cells (light brown coloured). C, D: lung tissue (blue coloured) within the muscular tissue .E, F: bone tissue of spongy type (S, light brown coloured) in the meat samples (dark brown coloured in E and light pink in F). G, H: meat samples were contaminated by rod-shaped bacteria (pink coloured). I, J: meat samples with a part of stomach. Note wall of stomach (dark pink), gastric mucosa (light pink). K, L: Herbal contents of the gastric ingesta. Note, Pollen grains (yellow coloured). M, N: omasal compartment of ruminant stomach in meat samples note omasal lamina (pink coloured). O, P: A part of rumen stomach in the meat samples. Note. Gastric mucosa (pink coloured).

Figure 11: Adulterations of shawerma samples with animal tissues.

of shawerma samples contaminated by microbial cells which were rod-shaped bacteria (Figures 11G and 11H).

Scanning samples of shawerma adulterated with plant tissue and sand particles. Plant stem was identified by the vascular tissue; xylem (Figures 12A-12E). A part of the onion was observed in the meat sample (Figure 12F). The sand particle was recognized in the meat sample (Figures 12G and 12H).

ATPase staining result

Table 1 Reaction for myofibrillar ATPase, in different meat product after preincubation in alkaline medium (pH 9.4) and in acidic media (pH 4.2), to evaluate the contractile ability of muscle fiber types. Slow (dark) and fast (light) fibers. According to the counting of fibers in the different meat product, kofta, shwerma and hwawashi, (Figure 13 and Table 1) found that reactivity in the slow (dark) fiber after

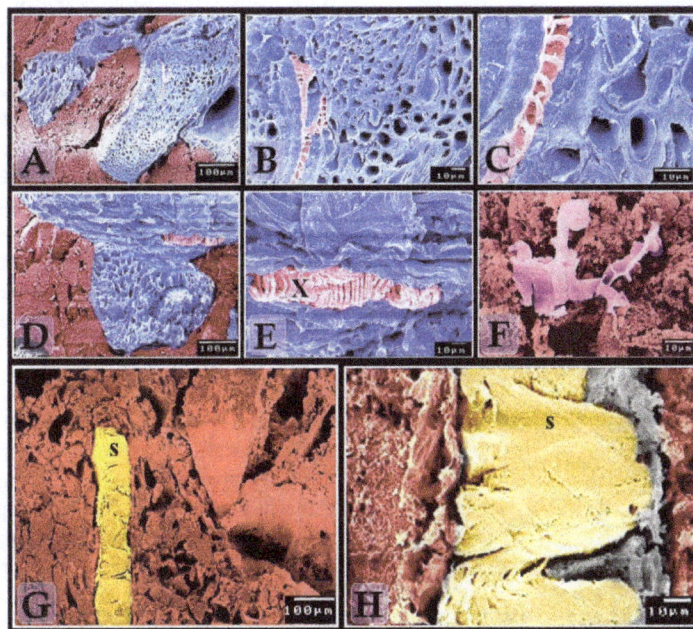

Note: Digitally coloured scanning electron micrographs of shawerma samples adulterated with plant and sand particles. A, B, C, D, E: A part of plant stem showed a cross view of the xylem (blue coloured) and longitudinal view of the xylem (pink coloured). F: A part of the onion (pink coloured) within the meat sample (red coloured). G, H: sand particle (s) within the meat sample (dark brown coloured).

Figure 12: Adulterations of shawerma samples with plant tissues and sand particles.

	Hawawshi	Kofta	Shawerma
PH 4.2	23 + 3.51 dark fibers/500 µm²	22 + 2.0 dark fibers/500 µm²	15 + 2.1 dark fibers/500 µm²
	31+2.3 light fibers/500 µm²	34 + 1.7 light fibers/500 µm²	22 + 1.8 light fibers/500 µm²
PH 9.4	42 + 2.4dark fibers/500 µm²	16 + 3.1 dark fibers/500 µm²	20 + 1.9 dark fibers/500 µm²
	26+ 2.8 light fibers/500 µm²	24 + 3.8 light fibers/500 µm²	22 + 2.4 light fibers/500 µm²

Table 1: Reaction for myofibrillar ATPase, in different meat product after pre-incubation in alkaline medium (pH-9.4) and in acidic media (pH-4.2), to evaluate the contractile ability of muscle fiber types. Slow (dark) and fast (light) fibers.

preincubation in alkaline medium (pH 9.4) was more than fast (light) fiber in hwawashi other product. We suspect that the hwawashi are adulterated with fetal flesh. In addition to the reactivity of ATPase in muscle fiber we are observed the reaction after preincubation in alkaline medium (pH 9.4) and in acidic media (pH 4.2) in different parts of plants (Figure 14).

Discussion

In Egypt, variable meat sandwich sold invariant cities in different governorates as hawawshi (made from minced beef or mutton meat mixed with chopped onion, garlic, black pepper, spices, salt and some fat), Kofta (prepared from cow or sheep meat, salt, black pepper, onion, garlic in addition to amount of fat) and shawerma (made from slices of ground cattle meat or chicken in our study shawerma composed of slices of beef meat, fat tomato, green pepper, onion, black pepper and salts). But our study explores a thunderbolt though you must think before eating. What is the real constituent of street meat sandwich which is a far-reaching maintenance of our health diet?

12 hours of our day life are outdoors subjecting our feed dependence on street meat sandwich at lunch time as quick food and fewer prices especially in developing countries. Fewer prices mean the ability of adulteration as meat is so expensive in developing countries. Evaluation using histological and histochemical methods markedly demonstrated that the constitutions used in preparing these products do not reverence the standard and hygiene food regulation and the products are comprehensively bad quality. Electron microscope and ATPase staining pinpointed the unpermitted tissues. Besides the skeletal muscles, the samples included lung, heart, spleen, bone, cartilage, fibrous connective tissues, hollow tubular organs, plants, sands, fat, pollen grains, suspecting fetal tissues, sand and more illegal tissues illustrated in our results. Skeletal muscle, the normal "meat" of street meat sandwich, virtually composed of one type of cell that is marked in minute fragments, but other tissues or organs detected by some structural arrangement or by the affiliation of certain cell types. The heating and cooking degree does not prevent the histological and histochemical detection of the meat constituent used [21]. Various studies were compiled for disclosure of unauthorized tissues in meat and meat products. The USA conveyed histological examination on eight different brands of hamburgers. Paryson et al. [22] and demonstrated the presence of plant material, adipose tissue, blood vessels, peripheral nerves, bone, and cartilage. Lattore et al. [1] distinguished uncommitted tissues in processed meat products besides the skeletal muscles included gizzard, gland tissue, cartilage, soya, ovary, connective tissue, adipose tissue and lymph node. Histologically smooth muscles and soya tissues identified by Rokni et al. [23]. In Iranian heated sausage salivary gland and nuchal ligament have been investigated that conveys the use of meat achieved from the head of slaughtered animals meat products [24]. Assessment by Sepheri [24] revealed the presence of peritoneal fat, chicken skin, hyaline, cartilage

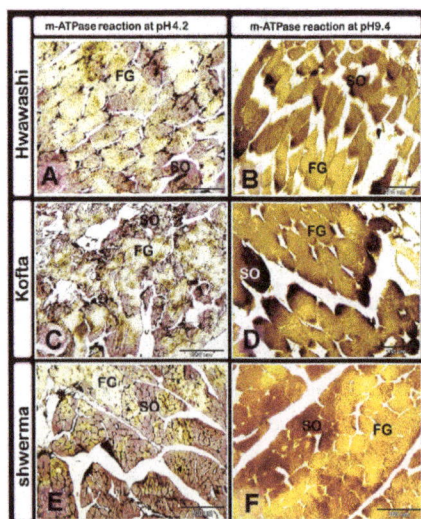

Note: A, C, E Slow contraction fibres (SO) and fast contraction fibres (FG) identified by ATPase reaction at pH 4.2 B, D, F Slow contraction fibres (SO) and fast contraction fibres (FG) identified by m-ATPase reaction at pH 9.4. A and B represent the Hwawashi sections, C and D represent the kofta sections, E and F represent the shwerma sections.

Figure 13: Thick frozen section (10 μm) section stained by m-ATPase enzyme.

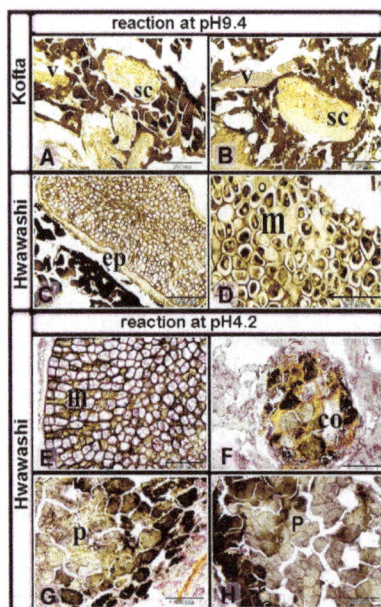

Note: A and B: Kofta samples, C and D: Hwawashi samples adulterated with plant stained with ATPase at pH 9.4. A and B: a part of plant stem contained ATPase –ve sclerenchyma (sc) and vascular tissue (v). C, D: plant leaf had ATPase –ve epidermis and ATPase +ve mesophyll (m). E, F, G, H: Hwawashi samples adulterated with plant stained with ATPase at pH 4.2 E: plant leaf had ATPase +ve mesophyll (m). F: part of plant stem contained weak and strong ATPase +ve collenchyma (co). G, H: part of plant stems showed weak and strong ATPase +ve parenchyma cells (P)

Figure 14: Adulterations of kofta and hwawashi samples with plant tissues and stained with ATPase at pH-9.4 and 4.2.

and kidney in sausage. Fraud authentication detected by Sadeghi et al. [25] counting esophagus, heart muscles, cartilage, bone, lymph nodes and spleen in sausage samples as shown in kofta and hawawshi samples.

Agreement to the histological estimation of hamburger, minced meat, and kabab loghme demonstrated the residence of fraud tissues such as adipose tissue, plant material, nerves and cartilages, blood vessels that in correspondence with our observation. Sadeghinezhad et al. [25] scoped on qualitative and quantitative accuracy of histological examination for the dedication of herbal content and unauthorized animal in minced beef meat as 5%, 10%, 15% and 20% of soya and chicken gizzard were composed. The presence of ruminant stomach in hawawshi, kofta and shawerma settled with Baskaya et al. [26] as they examined 27 ready to sell minced meat samples in Ankara and found cartilage tissue (11.1%) of the samples, and organs related to alimentary canal in another 3 (11.1%) of the samples tested. Little inedible tissue traces identified by Yildiz et al. [27] of 75 ready to sell meatball samples in Istanbul.

Citizen et al. [28] showed the presence of non-meat materials such as mechanically deboned meat and ground bone fragments to processed meat. The presence of all morphological characters and structure of the plants as a whole with presence of pollen grains in shawerma indicate that all of the contents of these meat sandwich not only minced meat or slices of meat, additives or spices but also the food contents of the stomach of animals in which its meat muscles are used in the sandwich which led to contamination also it may be due to unhygienic measures during preparation as it appeared with presence of different microbial cells whether rods, cocci bacteria which may lead to food spoilage and food poisoning in addition to the presence of thread- like fungi in kofta in spite perfectly an internal temperature of 80°C should be attained to distinct in raw meat mixes. Nevertheless, not only adulteration by unauthorized tissues but also a source of specific species of bacteria likes *Staphylococcus aureus, Campylobacter jeguni, Escherichia coli* O157H7 and *Salmonella.* Because of its great level and its liability to detent pathogenic bacteria, minced meat is an important product to make a threat to human health [28]. The existence of hollow organs or tubular organs in kofta and hawawshi suspect the presence of the intestine or uterus, the especially uterus is prohibited and forbidden to be used in according to religious factor. Different researchers were observed the similar result of the unauthorized organ by histological methods. Rezain and Rokni [29] were observed that the udder, salivary gland, lymph node, skin and its accessories are unauthorized organs authenticated by the histological study of heated meat products in Iran. Citizen et al. [28] observed in processed meat products foreign tissues originated from poultry as well as Kidneys, skin, heart, bones, cartilage, gizzards heart and lung. TFC [30] itemized the structural elements required to be added in the composed meat mixtures (Skeletal muscles and a small amount of fat). As the presence of unauthorized tissues as inedible parts with high amounts of fat horns, bones, cartilage and nail are prohibited; the tissues antithetically alter the structural element of processed and manufactured meat products and originate biased antagonism and profit.

The high presence of fascia (dense fibrous connective tissue), cartilage either hyaline or white fibrocartilage, in an excessive amount in all samples in accordance with Citizen et al. [28] as they implicated their addition reduce the quality and nutritional value of meat used. Knowing the differences between skeletal muscle fiber types as type I (dark-slow fiber) and type II (Light-fast fiber) is important for researchers and of particular interest for the food industry because meat tenderness depends in part on the proportion of the different types of fibers [14], In addition for detection of fetal and adult flesh according to the percentage of dark and light fibers. The fetal skeletal muscle fiber in bovine contains dark fibers (primary fiber) more than light fibers (secondary fiber) [16] and in ovine muscle [17,31]. ATPase analysis in this investigation is suspected the presence of fetal flesh more in hawawshi than other samples, due to a high number of dark fiber. Animal fetal flesh is not allowed to be used in food due to not only poor quality of meat as it contain high moistures, the sodden appearance of the meat but also it appeared as a result of abortion which may be due to bacteriological, viral or pathological disease which leads to severe contamination of meat causing infection to the consumers. It is rejection during the inspection in the abattoir [32]. Economically, from a religious point of view and ethics related to human health. It is important to identify animal tissues in meat and meat products and processed meat in order to have healthier consuming from both illegal competition and undesirable adulteration. Histological methods could play a role for routine examination for authentically and quality to all types and forms of meat-based products to save consumers counterfeit and fraudulent practices of meat substitution.

Acknowledgment

Authors wish to thank Mr. Ahmed Ibrahim, the technician in electron microscope unit in Assuit University for providing the facilities to achieve this study.

The authors would like to express deep appreciation and thanks to to Dr Soha Soliman, lecturer of histology ,Department of Histology, Faculty of Veterinary Medicine, South Valley University, Qena　and Dr Doaa Mokhtar,Department of Anatomy, Emryology and histology, Faculty for Veterinary medicine, Assuit University, Egypt, for providing us with helpful advises and their kind assistance.

References

1. Latorre R, Sadeghinezha DJ, Hajimohammadi B, Izadi F, Sheibani MT (2015) Application of morphological method for detection of unauthorized tissues in processed meat products. J Food Qual Hazard Control 2: 71-74

2. Moore JC, Spink J, Lipp M (2012) Development and application of a database of food ingredient fraud and economically motivated adulteration from 1980 to 2010. J Food Sci 77: 118-126.

3. Herde K, Bergmann M, Lang C, Leiser R, Wenisch S (2005) Glial fibrillary acidic protein and myelin basic protein as markers for the immunochemical detection of bovine central nervous tissue in heat-treated meat products. J Food Protect 68: 823-827.

4. Wilesmith JW, Wells GAH, Cranwell MP, Ryan BM (1988) Bovine spongi formen cephalopathy: epidemiological studies. Vet Rec 123: 638-644.

5. Irani DN, Johnson RT (2003) Diagnosis and prevention of bovine spongiform encephalopathy and variant Creutzfeldt-Jakob disease. Annu Rev Med 54: 305-319.

6. Prayson BE, McMahon JT, Prayson RA (2008a) Applying morphologic techniques to evaluate hotdogs; what is in the hotdogs we eat? Annal Diagnos Pathol 12: 98-102.

7. Bancroft JD, Layton C, Suvarna SK (2013) Bancroft's theory and practice of histological techniques, ed 7, Churchill Livingstone.

8. Harris HF (1898) A new method of ripening hematoxylin. In: Romeis, B. (ed) Mikroskopische Technik. Munich, Oldenburg.

9. Crossmon G (1937) A modification of Mallory's connective tissue stain with discussion of the principle involved. Anat Rec 69: 33-38.

10. Weigert C (1898) Study on a method for dyeing elastic fibers. ZL Pathol 9: 289-292.

11. McManus J (1948) Histological and histochemical uses of periodic acid. Stain Technol 23: 99-108.

12. Pearse AGE (1985) Histochemical: Theoretical and applied. Churchill, London.

13. Junqueira LC, Carneino J, Kelley RO (1989) Basic histology (6thedn.). New York Prentice Hall International.

14. Picard B, Duris MP, Jurie C (1998) Classifcation of bovine muscle fibres by different histochemical techniques. Histochem J 30: 473-479.

15. Peter JB, Barnard RJ, Edgerton VR, Gillespie CA, Stempel KE (1972) Metabolic profiles of the three fiber types of skeletal muscle in guinea pigs and rabbits. Biochem 11: 2627-2633.

16. Crosier AE, Farin CE, Rodriguez KF, Blondin P, Joseph E, et al. (2002) Development of skeletal muscle and expression of candidate genes in bovine fetuses from embryos produced in vivo or in vitro. Biol Reproduce 67: 401-408.

17. De Freitas CEA, Freitas SBZ, Lopes FS, Pai-Silva MD, Piçarro IC (2009) Skeletal muscles with antagonistic muscular actions: morphological, contractile and metabolic characteristics. Int J Morphol 27: 1173-1178,

18. Franciscoa CL, Jorgea AM, Dal-Pai-Silvab M, Caranib FR, Cabeçob LC, et al. (2011) Muscle fiber type characterization and myosin heavy chain (MyHC) isoform expression in Mediterranean buffaloes. Meat Sci 88: 535-541

19. Guth L, Samaha FJ (1969) Qualitative differences between actomyosin ATPase of slow and fast mammalian muscle. Exp Neurol 25: 138-152.

20. Karnovsky MJ (1965) A formaldehyde-glutaraldehyde fixative of high osmolarity for use in electron microscopy. J Cell Biol 27: 137-138.

21. Hole NH, Meriel PRK, Jones B (1964) The identification of offals in sausages. Analyst.

22. Prayson BE, McMahon JT, Prayson RA (2008b) Fast food ham-burgers: what are we really eating? Annal Diagnos Pathol 12: 406-409.

23. Rokni N (2008) Meat sciences and industries (5thedn). University of Tehran Press, Tehran.

24. Sepehri ES (2008) Histological methods evaluation for detection of adulteration of raw meat products supplied in Tehran. University of Tehran, Tehran, Iran.

25. Sadeghinezhad J, Hajimohammadi B, Izadi F, Yarmahmoudi F, Latorre R (2015) Evaluation of the morphological method for the detection of animal and herbal content in minced meat. Czech J Food Sci 33: 564-569.

26. Başkaya R, Karaca T, Sevinc İ, Cakmak O, Yıldız A, et al. (2004) The histological, microbiological and serological quality of ground beef marketed in İstanbul. Yüzüncü Yıl Üniversitesi Vet Fak Derg 15: 41-46.

27. Yildiz A, Karaca T, Çakmak Ö, Yörük M, Baskaya R (2004) The histological, microbiological and serological quality of meatball marketed in Istanbul. YYU Vet Fak Derg 15: 53-57.

28. Citizen O, Bingol EB, Colak H, Ergun O, Demir C (2011) The microbiological, serological and chemical qualities of minced meat marketed in İstanbul. Turk J Vet Anim Sci 34: 407-412.

29. Rezain M, Rokini N (2003) Histological study of the heated meat products of Mazandaran Province of IRAN.

30. TFC (2012) Turkish food codex: Regulations on fresh meat, prepared meat and prepared meat mixtures. Official Gazette, Ministry of Agricultural and Rural Affairs, Ankara, Turkey.

31. Demirtaş B, Özcan M (2011) A study on ovine muscle development. Istanbul Universitesi veteriner Fakültesi, Dergisi 38: 97-106.

32. Wilson WG (2005) Wilson's practical meat inspection (7thedn). Wilson CIEH.

12

Effects of Adding Different Proportions of Sunflower Seeds on Fatty Acid Composition of Chicken Tissues

Mawada Mahfoudh[1]*, Hajer Trabelsi[1], Khaled Sebei[2] and Sadok Boukhchina[1]

[1]*Biology Department, Lipids Biochemistry Unit, Science Faculty, Tunisia*
[2]*Biology Department, Higher Institute of Applied Biological Sciences, Tunisia*

Abstract

Poultry meat becomes an essential part of our nutrition, our aim is to optimize poultry meat quality to obtain more healthy results. Because of its rich composition of desired fatty acids, we chose sunflower seeds (SS) as chicken's food to search for the best proportion to obtain the best meat composition. A basal diet is given to chickens, with proportions of 25%, 50%, and 75% of sunflower seeds, respectively for each group. The results were very adequate: the more we add SS in chicken's meal, the more its composition is expressed in chickens tissues. So, monounsaturated fatty acids (MUFA) were the major FA, decrease for the favor of polyunsaturated fatty acids (PUFA) in all tissues, linoleic acid become the major FA taking the place of Oleic Acid. Saturated fatty acids (SFA) become, fortunately, lower; it's a target to low the content of SFA, since they have been associated with several human diseases.

Keywords: Fatty acids; Chickens; Sunflower seeds; Poultry meat

Introduction

The quality of meat and animal products has become a major concern for all the poultry sector partners. The lipid content of meat, quality criterion for the consumer, depends on the nutritional characteristics of ingested food, and the quantity and quality of fat added to the diet. In recent years, there has been an increased attention towards the manipulation of the meat lipid amount and composition [1], because of animal sources, such as butter and lard, are characterised by high concentrations of saturated fatty acids. It is well known that a diet rich in saturated fatty acids forms a risk factor for hypercholesterolemia, atherosclerosis and other diseases in humans, in the other side, Many studies have shown the benefits of polyunsaturated fatty acids on human health, especially in relation to heart problems and similarly important diseases [2].

Consequently, poultry meat has become very popular thanks to its nutritional characteristics, Moreover, chicken lipids are characterized by relatively high levels of unsaturated fatty acids, which are considered as a positive and healthy aspect to consumers. So, to obtain a better meat quality, poultry meat is generally manipulated by using a selected food with particular nutritional characteristics. In fact, the ratio of unsaturated to saturated fatty acids in meat should be increased, so to achieve positive health effects [3,4]. Thus, the aim of this work is to manipulate chicken food and evaluate the effect of feeding fat sources on the quality and composition of lipids of thighs, breasts and skin of poultry meat.

As feed additive, we used Sunflower seeds in different proportions to see it's impact on muscle FA deposition, in fact, Sunflower (*Helianthus annuus* L.) is one of the most widely cultivated oilseed in the world and ranks third in importance as a source of vegetable oil [5]. Most studies evaluating effects of seeds FA on poultry meat FA, were experimented by adding the oil of the seed, however, in our experiment, we added the whole seed to the animal (with the bark) and we search for better results. Even though it is an attraction for many consumers, studies on chicken's skin are very low, so, we will try to give details about it. This study provides information on the effects of processing factors on the quality characteristics of chicken's tissues (meat and skin); in economic

practice, it is important to determine how to process chicken's meat to preserve its benefits and to reduce harmful components like SFA, this will be established through special feeding to optimize poultry meat quality and to produce new poultry meat ranges, each one specialized and appointed by the food received.

Material and Methods

Sunflower seeds

Sunflower seed (SS) should be considered rich in energy, since it has from 3.691 to 5.004 kcal of ME.kg^{-1} and from 19.9 to 43.4% of ether extract [5,6]. Varieties with high oil concentration have increased levels of polyunsaturated fatty acids, mainly linoleic [7], and low levels of saturated fatty acids [8].

We chose to give SS to chickens because of its characteristics; sunflower seeds are an excellent source of essential fatty acids, vitamins, minerals and energy. Sunflower kernels actually employed to extract edible oil at commercial levels. Much of their calories come from fatty acids. The seeds are especially rich in poly-unsaturated fatty acids: Linoleic acid (LA), which constitute more than 50% fatty acids in them. They are also good in mono-unsaturated: Oleic acid (OA) that helps lower LDL or "bad cholesterol" and increases HDL or "good-cholesterol" in the blood. Research studies suggest that Mediterranean diet which is rich in monounsaturated fats help to prevent coronary

*Corresponding author: Mawada Mahfoudh, Biology Department, Lipids Biochemistry Unit, Science Faculty, 2092 El Manar 2, Tunisia
E-mail: mawada.mahfoudh@yahoo.fr

artery disease, and stroke by favoring healthy blood lipid profile [9]. Fatty acid profile of SS is presented in Table 1 [10].

Animal rearing

One hundred 1-day old broiler chicks were obtained from a commercial hatchery associated to Regional directorate of Agriculture (Nabeul, Tunisia). They are distributed in four parcels in an animal husbandry, They are placed in favorable conditions to their growth (temperature, humidity, aeration, lighting, troughs, foragers, valid nutrition according to age, vaccination, etc.). The Basal Diet (BD) is a nutritionally balanced diet given to chicks just after birth favorising a good nutrition and a healthy growth; it is composed by some proteins, vitamins, trace of minerals and different kinds of seeds. Chicks of each parcel receive a different proportion of diet (Table 2). After 42 days (the breeding period), chickens were slaughtered at a commercial abattoir (slaughterhouse) by cervical dislocation. Muscle is boneless and divided to breasts, thighs and skins separately, meat is minced using a mincing machine and preserved at -20°C.

Extraction and analysis

Total lipids were extracted from chicken muscles using Folch et al. [11] modified by Bligh and Dyer [12], then by Allen and Good [13]. Methyl esters were prepared by transesterification using cold methanolic potassium hydroxyde solution: it's a transmethylation of the triacylglycerols according to the ISO 5509 method (1978). Fatty acid methyl esters (FAME) were analysed by a gas chromatograph.

Gas chromatography of FAME was performed with an agilent

Fatty acids	Percentage
C16:0	5.8%
C16:1	0.1%
C18:0	3.9%
C18:1	15.9%
C18:2	71.7%
C18:3	0.6%
C18:3	0.1%
C20:0	0.3%
C20:1	0.2%
C22:0	0.7%
C24:0	0.5%

Table 1: Sunflower seeds fatty acids composition.

	Parcel 1	Parcel2	Parcel3	Parcel4
SS	25%	50%	75%	0%
BD	75%	50%	25%	100%

SS: Sunflower Seeds; 25% SS: Chicks fed 25% SS of the total diet; BD: Basal Diet ; 50% SS: Chicks fed 50% SS of the total diet; 75% SS: Chicks fed 75% of the total diet

Table 2: The ration given to each parcel.

6890N gas chromatography equipped with a flame ionization detector (FID). The operating conditions for the gas chromatograph were as follows; we began by a temprature of 75°C, increasing by 5°C/min to 148°C, from 148 to 158°C, the temperature was increased by 2.5°C/min, from 158 to 225°C per min, the temperature was increased at the rate of 5°C/min. The temperature of the injector and the detector remained stable at 280°C. The column head pressure of the conductor gas (helium) was $1.3g/cm^2$.

Statistical analysis

The experimental data were analyzed using the analysis of variance (ANOVA) and the Statistical Analysis System (XLSTAT 2013) (Addinsoft USA). Differences at $p \leq 0.05$ were considered statistically significant by Duncan's new multiple range test. All values were expressed as means ± standard deviation of at least triplicate repetitions. Each value is a mean ± standard deviation (SD) of a triplicate analysis performed on different samples Means with different letters were significantly different with $p \leq 0.05$.

Results

Effects on breast's fatty acids

The effects of experimental diets (SS), in its different proportions, on breast FA profile is in Table 3; Ten fatty acids(FA) were found in the analyse of FAME. Most of values Were considered statically significant by Duncan's new multiple range test. The major FA in standard breast is oleic acid, same result for chicks fed 25% SS and 50% SS, then it's reversed at75% SS, major FA become linoleic acid, the difference is significant ($p \leq 0.05$). Palmitic acid exist too with a sufficient percentage, but, it decrease significantly ($p \leq 0.05$) with the consumption of SS, like all other saturated fatty acids (Myristic, stearic, arachidic, béhénic).

Breast's TFA (total fatty acids) evolution are shown in Figure 1. It shows that SFA of breast decrease with consumption of sunflower seeds whatever the proportion, their most decrease is at 75% SS and this proportion represent the top of PUFA (40.730%) and the best propotion of UFA (77.744%) values. Concerning MUFA (C16:1, C18:1, C22:1), they are concentrated in the breast of standard chickens with the highest value, same for SFA, with consumption of SS, they decrease for the favor of PUFA which are in remarkable evolution: from 24.07% to 40.73%, the highest level of UFA is in 75% SS chicks group. The polyunsaturated to saturated fatty acid ratio (P/S) is around 1.1% but it has a peak in chick's breast of 75% SS it reaches 1.828%.

Effects on thigh's FA

Table 4 shows the effects of Sunflower Seeds on FA profile of Thigh. In first steps of minimum added diet, (standard and 25% SS) thigh's predominant FA is oleic acid; the more chicks consume SS the more oleic acid decrease, the more linoleic acid increase, so, in 50% and 75% proportions, the predominant FA becomes linoleic acid (C18:2n-6).

	C14:0	C16:0	C16:1	C18:0	C18:1 (n-9)	C18:2 (n-6)	C18:3 (n-3)	C20:0	C22:0	C22:1	SFA	MUFA	PUFA	UFA	P/S
Standard(BD)	0.460[c]	21.880[d]	4.860[c]	6.385[b]	39.940[c]	24.070[a]	0.000[a]	1.845[c]	0.180[b]	0.380[c]	30.750	45.180	24.070	69.250	1.124
25% SS	0.440[bc]	17.785[b]	2.370[b]	7.140[d]	38.180[bc]	30.925[b]	0.570[d]	2.000[d]	0.150[b]	0.440[c]	27.515	40.989	31.495	72.483	1.144
50% SS	0.410[b]	19.380[c]	4.060[c]	7.140[d]	36.935[b]	31.860[b]	0.190[b]	1.490[b]	0.075[b]	0.320[c]	28.495	41.313	32.050	73.363	1.124
75% SS	0.235[a]	14.220[a]	1.575[a]	6.575[c]	35.075[a]	40.365[c]	0.365[c]	1.045[a]	0.200[b]	0.365[b]	22.275	37.014	40.730	77.744	1.828

Values are mean ± SD. Values that are followed by different letters within each row are significantly different ($p \leq 0.05$)
SFA: Saturated Fatty Acids; MUFA: Monounsaturated Fatty Acids; PUFA: Polyunsaturated Fatty Acids; UFA: Unsaturated Fatty Acids; P/S: Polyunsaturated to Saturated Fatty Acids Ratio; STD: Standard Chicks Fed Basal Diet; 25% SS: Chicks fed 25% Sunflower Seeds of the total diet; 50% SS: Chicks fed 50% Sunflower Seeds of their total diet; 75% SS: Chicks fed 75% Sunflower Seeds of their total diet.

Table 3. Fatty acids profile of chicken's breast.

The effect of treatments on concentration of FA (C14:0, C16:0, C16:1, C18:0, C18:1, C18:2) was noticeable (p ≤ 0.05). TFA evolution is shown in Figure 2. Thigh's SFA decrease with the consumption of SS; it begins by 32.695% for standard ration and decreases proportionally with the increase of SS propotion, it becomes 21.715% for 75% SS. The same experience showed a significant effect on MUFA in thigh, it decreases proportionally with consumption of SS. While, PUFA grow increasingly with consumption of Sunflower seeds and reaches 46.685% at 75% of SS of total alimentation, so, UFA develop too with the uptake of Sunflower seeds.

Effects on skin's FA

Skin tissue is also analysed. It was affected by Sunflower seeds nourishment. The results has been shown in Table 5. Even for Skin, the same results are found; oleic acid, is the major FA for standard group and it stays the same in 25% SS group, then it's reversed in 50% SS and 75% SS ration to the linoleic acid. SFA and MUFA decrease with the SS uptake to the favor of PUFA in fat tissue too (Figure 3).

Discussion

The breast's major FA is oleic acid at 25% and 50% SS. Our major FA is reversed at 75% SS, it becomes linoleic acid with a very important percentage (40.365%). Crespo and Esteve Garcia [14] proved that Birds fed sunflower oil showed higher values of linoleic acid (c18:2 n-6) than those fed tallow, olive, or linseed oil. Oleic acid was the second FA (since 75% SS) like in SSFA (sunflower seeds FattyAcids) (Table 2) and in chicken's meat FA: Table 2, showing SSFA composition, demonstrate that this composition is reflected on FA profile of chicken's meat. The major FA (linoleic acid(LA)) of SS (71%) influenced LA in chicken tissue, it becomes 40%. However it beomes the second FA (at 75% SS), oleic acid have anyway an important value staying relatively high and beneficial.

MUFA and SFA are concentrated in the breast and thigh of standard chicken with high percentages, they decrease for the favor of PUFA. Similar trends are reported by Grau et al. [15] who determined

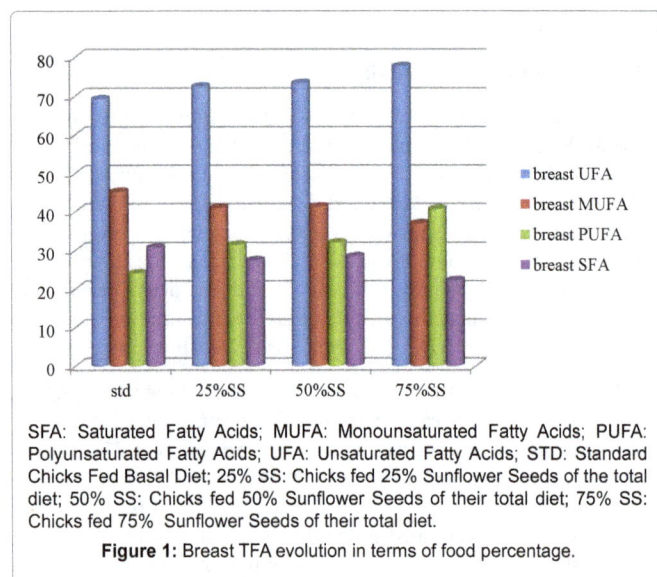

SFA: Saturated Fatty Acids; MUFA: Monounsaturated Fatty Acids; PUFA: Polyunsaturated Fatty Acids; UFA: Unsaturated Fatty Acids; STD: Standard Chicks Fed Basal Diet; 25% SS: Chicks fed 25% Sunflower Seeds of the total diet; 50% SS: Chicks fed 50% Sunflower Seeds of their total diet; 75% SS: Chicks fed 75% Sunflower Seeds of their total diet.

Figure 1: Breast TFA evolution in terms of food percentage.

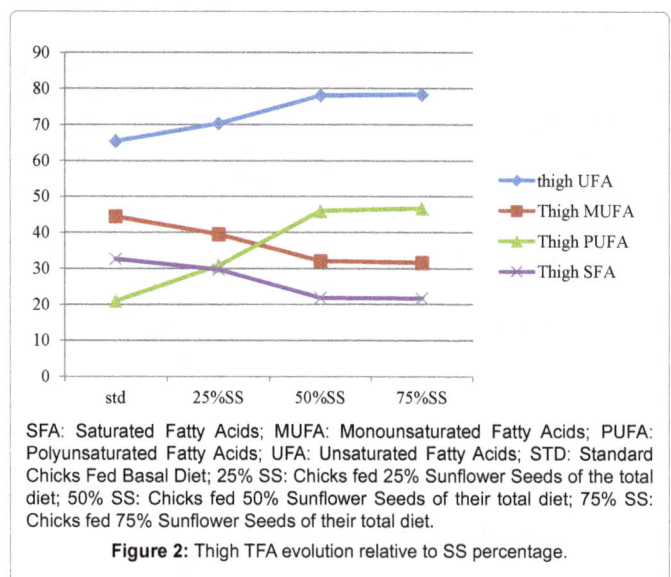

SFA: Saturated Fatty Acids; MUFA: Monounsaturated Fatty Acids; PUFA: Polyunsaturated Fatty Acids; UFA: Unsaturated Fatty Acids; STD: Standard Chicks Fed Basal Diet; 25% SS: Chicks fed 25% Sunflower Seeds of the total diet; 50% SS: Chicks fed 50% Sunflower Seeds of their total diet; 75% SS: Chicks fed 75% Sunflower Seeds of their total diet.

Figure 2: Thigh TFA evolution relative to SS percentage.

	C14:0	C16:0	C16:1	C18:0	C18:1 (n-9)	C18:2 (n-6)	C18:3 (n-3)	C20:0	C22:0	C22:1	SFA	MUFA	PUFA	UFA	P/S
Standard(BD)	0.570[c]	23.645[c]	4.995[c]	6.620[b]	39.125[c]	20.850[a]	0.155[a]	1.725[a]	0.135[a]	0.320[a]	32.695	44.440	21.005	65.445	0.642
25% SS	0.430[b]	20.025[b]	3.130[a]	7.340[c]	36.080[b]	30.625[b]	0.190[a]	1.815[a]	0.125[a]	0.295[a]	29.735	39.505	30.815	70.320	1.036
50% SS	0.285[a]	13.770[a]	1.900[a]	6.325[a]	29.860[a]	45.880[c]	0.140[a]	1.335[a]	0.165[a]	0.335[a]	21.880	32.095	46.020	78.115	2.100
75% SS	0.325[a]	13.860[a]	2.060[a]	6.470[ab]	29.230[a]	46.505[c]	0.180[a]	0.920[a]	0.140[a]	0.325[a]	21.715	31.615	46.685	78.300	2.149

Values are mean ± SD. Values that are followed by different letters within each row are significantly different (p ≤ 0.05)
SFA: Saturated Fatty Acids; MUFA: Monounsaturated Fatty Acids; PUFA: Polyunsaturated Fatty Acids; UFA: Unsaturated Fatty Acids; P/S: Polyunsaturated to Saturated Fatty Acids Ratio; STD: Standard Chicks Fed Basal Diet; 25% SS: Chicks fed 25% Sunflower Seeds of the total diet; 50% SS: Chicks fed 50% Sunflower Seeds of their total diet; 75% SS: Chicks fed 75% Sunflower Seeds of their total diet.

Table 4: Fatty acids profile of thigh meat.

	C14:0	C16:0	C16:1	C18:0	C18:1 (n-9)	C18:2 (n-6)	C18:3 (n-3)	C20:0	C22:0	C22:1	SFA	MUFA	PUFA	UFA	P/S
Standard (BD)	0.460[b]	23.060[d]	5.110[d]	7.360[c]	39.730[c]	21.715[a]	0.000[a]	2.295[c]	0.120[c]	0.150[a]	33.295	44.990	21.715	66.705	0.652
25% SS	0.455[b]	20.475[c]	2.630[c]	8.250[d]	35.165[b]	31.590[b]	0.120[b]	1.210[b]	0.000[a]	0.110[a]	30.390	37.905	31.710	68.615	1.043
50% SS	0.310[a]	15.365[b]	1.990[a]	7.140[b]	31.290[a]	40.105[c]	0.180[c]	3.090[d]	0.100[bc]	0.430[b]	26.005	33.710	40.285	73.995	1.549
75% SS	0.365[a]	14.720[a]	2.230[b]	5.675[a]	32.330[a]	43.510[d]	0.325[d]	0.645a	0.070[b]	0.130[a]	21.475	33.710	43.835	78.525	2.041

Values are mean ± SD. Values that are followed by different letters within each row are significantly different (p ≤ 0.05)

SFA: Saturated Fatty Acids; MUFA: Monounsaturated Fatty Acids; PUFA: Polyunsaturated Fatty Acids; UFA: Unsaturated Fatty Acids; P/S: Polyunsaturated to Saturated Fatty Acids Ratio; STD: Standard Chicks Fed Basal Diet; 25% SS: Chicks fed 25% Sunflower Seeds of the total diet; 50% SS: Chicks fed 50% Sunflower Seeds of their total diet; 75% SS: Chicks fed 75% Sunflower Seeds of their total diet.

Table 5: Fatty acids profile of skin.

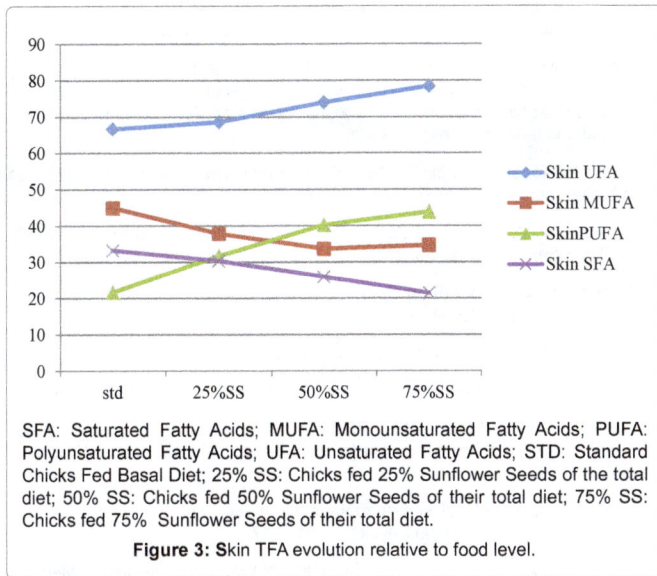

SFA: Saturated Fatty Acids; MUFA: Monounsaturated Fatty Acids; PUFA: Polyunsaturated Fatty Acids; UFA: Unsaturated Fatty Acids; STD: Standard Chicks Fed Basal Diet; 25% SS: Chicks fed 25% Sunflower Seeds of the total diet; 50% SS: Chicks fed 50% Sunflower Seeds of their total diet; 75% SS: Chicks fed 75% Sunflower Seeds of their total diet.

Figure 3: Skin TFA evolution relative to food level.

that raw meat from chickens fed with beef tallow had 34.41% SFA, 47.29% MUFA, and 18.34% PUFA, whereas those fed with sunflower oil showed 22.62% SFA, 34.96% MUFA, and 42.47% PUFA [16].

As regards to the skin, like the thigh and breast, skin's linoleic acid grow up proportionally with the consumption of sunflower seeds. Viveros et al. [17] also, showed significant differences in the fatty acid concentrations of abdominal fat among dietary fat sources. The contents of the major fatty acids, palmitic acid (C16:0), stearic acid (C18:0), oleic acid (C18:1n-9), and linoleic acid (C18:2n-6) in abdominal fat reflected the fatty acid profile of the dietary fat. The C18:2n-6 content was significantly greater in Conventional SS diets. The PUFA/MUFA and UFA/SFA ratios were significantly increased (up to 44%). In our experiment, diet's FA are well expressed in fat tissue, even like muscle tissue. Maria et al. [18] showed that it should be stressed that dietary fatty acid composition is best represented by fat tissue rather than muscle tissue. The results of Ortiz et al. [19], showed too a more efficient modification of adipose than intramuscular fat profile, which was probably due to the physiological lipid storage function of adipose tissue.

To synthesize, SFA and MUFA decreases for the favor of PUFA in all tissues (Figures 1-3). Consequently, P/S ratio increase from 0.642% for standard group to 2.149% for 75% SS group. Likewise Ortiz et al. [19] found too that when we feed chickens with a source of fat in C18:1n-9 (HOASS(high oleic acid sunflower seeds) or its oil), or C18:2n-6(CSS(conventional sunflower seeds) or its oil), the proportion of MUFA and PUFA, respectively, increased in abdominal fat as well as in breast and thigh muscles. The proportionality is detected too in the experience of Lopez-Ferrer et al. [20] who used different percentages of flaxseed oil in diets for chicken and obtained different values of Linolenic acid in chicken meat. Increasing the flaxseed oil decreased the SFA and MUFA and increased the PUFA [13]. Breast, thigh and Skin, are affected by food quantity and quality, indeed, the amount added to the diet has changed the quantity (in%) of meat and fat tissue FA, and the quality of FA of the seeds (FASS) is reproduced on the quality of meat FA, and fat tissues.

Comparing breast's and thigh's and skin's FA, we note that thigh and skin are more sensitive to dietry FA than breast; their FA composition is changed and their major FA becomes Linoleic Acid

from 50% SS ration (Breast at 75% SS). In 25% SS ration, results are not well expressed in all tissues, this could be explained by the lower level of fat added, as it has been stated, the dietry fatty acids incorporated into the muscular tissues will be relatively more diluted by de novo synthesized fatty acids at low fat intakes than at high fat intakes, Ortiz et al. [19] and Hardinka et al. [21] Confirm this. It appears that the increased PUFA with proportional decrease of MUFA and SFA are related phenomena, in fact, the concentration of PUFA is resulted from direct deposition of diet fat [22], it is assumed that MUFA and SFA give way to PUFA and degrade in liver, in this direction, Cortinas et al. [23] Showed that with increase in concentration of PUFA in diet, concentration of MUFA decreases in tissue, these results were also obtained by schreiner et al. [24], and Dobrzaneki et al. [25].

Chicks fed sunflower oil showed higher values of linoleic acid (c18:2 n-6), linoleic acid replaced myristic, palmitic, stearic and oleic acids in all tissues [14]. This suggest the same hypothesis that these FA are leaving tissues letting place to other FA derived from diet (essentially PUFA by direct deposition). Where as, Lopez-Ferrer et al. [26] reported that the concentration of SFA and MUFA in tissue have completely dependence on contents of diet. They maintained that this may be dependent on relation between conversion in liver, deposition on tissue and turnover from carbohydrates. Normally in poultry, de novo lipogenesis occurs in the liver from dietary carbohydrates, reactions catalyzed by both: glucose-6-phosphate dehydrogenase and malic enzyme that provide the reduced equivalents required for de novo synthesis. Subsequently, triglycerides are hydrolyzed into glycerol and fatty acids and transported to adipose tissue in chylomicrons and very low density lipoprotein [27] deposited in the fatty organ or oxidized to obtain energy [28-30]. In fact, Doreau and Chilliard [31], show that In birds, the effects of adding dietary fats on adipose tissue metabolism are secondary to changes in lipid metabolism in the liver (the main site of FA synthesis in these species). The reduction of dietary carbohydrates when fat is supplied in the diet decreases the de novo synthesis of FA [32] and VLDL secretion by the liver. The proportion of dietary FA in circulating lipids then increases, so that body fat mainly depends on the direct uptake of exogenous FA. Due to the reduction of de novo FA synthesis, body fat deposition mainly depends on the direct uptake of preformed FA. For this reason, FA composition of body lipids is highly correlated to the nature of FA intake. For example, the proportion of linolenic acid in body fat of chickens fed with linseed oil can reach 34 g/100g instead of 1 g/100g with classical diets [33]. In broilers, lipid classes are unevenly distributed in different tissues; triacylglycerols, the most abundant, and the most affected by food, polar lipids were little affected, triacylglycerols carry with them, so, the dietry lipids to different muscle tissues [19,20]. Since the recommended low value of n-6/n-3 ratio, and knowing that chicken's meat is not a source of n-3 FA, the proportion of 75% which has incured a remarkable transformation and a great increase in n-6 FA, could be discouraged because of the high level of n-6/n-3 ratio (high value of n-6) which can be noxious on the oxidative situation; But we could contradict this hypothesis as our added diet is an excellent source of vitamin E [34], the body's primary fat-soluble antioxidant, it plays an important role in the prevention of cardiovascular disease. It is one of the main antioxidants found in cholesterol particles and helps prevent free radicals from oxidizing cholesterol. Vitamin E has significant anti-inflammatory effects that result in the reduction of symptoms in asthma, osteoarthritis, and rheumatoid arthritis, conditions where free radicals and inflammation play a big role. Vitamin E has also been shown to reduce the risk of colon cancer, help decrease the severity and frequency of hot flashes in women going through menopause, and help reduce the development of diabetic complications [34]. All these benefits of SS are reproduced in chickens and will help to defend against drawbacks of n-6/n-3 (if any). The best proportion in our work could be 50% SS since both FA, Linoleic acid and Oleic Acid are very narrow range giving the desired nutritional balance. A remarquable difference between n-3 and n-6 FA in our experiment, this distribution is not aleatory, the highest values of n-6 fatty acids were found in thighs and breasts of birds fed sunflower seeds, possibly due to the competition for Δ-5 and Δ-6 desaturase between n-3 and n-6 fatty acids in these tissues. For the same reason, the lowest values of n-3 derivatives were found in muscles of birds fed sunflower. These results are in accordance with Crespo and Esteve-Garcia [14]. Differences in tissue fatty acid profiles could be attributed to different roles of fatty acids in these tissues or to their different contents of phospholipids. PUFA are preferentially incorporated into phospholipids [35] and phospholipids are in higher proportion in muscle fat than in adipose tissue fat [36], which could explain different compositions of fat in different tissues.

So, As already reported in literature [37-42] the fat sources used as integration for diets can effectively affect the fatty acid composition of monogastric's lipid tissues [16]. Marion and Woodroof [43] too, found that the fatty acid composition in broiler tissues was easily influenced by dietry lipids.

Conclusion

Adding the whole seed to chicken's diet proved that it is easier and gives significant results. The addition of SS in the poultry diet cause an increase in n-6 FA of their meat. L.A is the major FA, but oleic acid is the second one and its values stay relatively high and beneficial (about 30%). The wealth of meat by UFA, SFA, (n-6) FA, P/S ratio are proportional to consumption of sunflower seeds by chickens. Thigh is more sensitive to dietry FA than breast, it acquires the effects of SS from 50% SS (Breast at 75% SS);the difference is related to richness of each muscle by phospholipids and triacylglycerols essentially, since total fatty acid intake is the same (the same chicken). FA composition of body lipids is highly correlated to the nature of FA intake, and the concentration of PUFA is resulted from direct deposition of diet fat. Poultry meat, a low-fat meat compared with red meat, and rich by PUFA (in our case), is usually recommended by dieticians, but, Some researchers find the large amount of n-6 discouraged, although, in our experience we can say that it is beneficial due to its origin from SS, source of vitamin E, a luxury antioxidant. So, it is concluded that the seeds of conventional varieties of sunflower might be used in poultry feeding in order to increase PUFA contents in both abdominal adipose tissue and intramuscular fat. Adding proportions in ascending order of SS to the parcels, proves that a proportionality is established in the quality and quantity of FA; This will have a great importance in poultry economy by producing types of chicken's meat according to all tastes and needs of consumers; So, we can manipulate chicken meat as we want by adding the food that suits with our needs, it will be reproduced in the meat; we could process by the same technical way to produce: n-3 chickens, n-6 chickens (our study), O.A chickens, SFA chickens, etc. to meet urban demand.

References

1. Wood JD, Richardson RI, Nute GR, Fisher AV, Campo MM, et al. (2004) Effects of fatty acids on meat quality: a review. Meat Sci 66: 21-32.

2. Leskanich CO, Noble RC (1997) Manipulation of the n-3 polyunsaturated fatty acid composition of avian eggs and meat. World's Poultry Science Journal 53: 155-183.

3. Reddy BS (1995) Nutritional factors and colon cancer. Crit Rev Food Sci Nutr 35: 175-190.

4. Søyland E, Drevon CA (1993) The effect of very long-chain n-3 fatty acids on immune-related skin diseases. Eur J Clin Nutr 47: 381-388.

5. Ortiz LT, Alzueta C, Rebolé A, Rodriguez ML, Arija I, et al. (2006) Effect of dietry high oleic acid and conventional sunflower seeds and their refined oils on fatty acid composition of aadipose tissue and meat in broiler chickens. Journal of animal and feed sciences. 15: 83-95.

6. Cheva IB, Tangtaweewipat S (1990) Effect of different levels of sunflower seed in broiler rations. Poultry Science 70: 2284-2294.

7. Senkoylu N, Dale N (1999) Sunflower meal in poultry diets: a review. World's Poultry Science Journal. 55: 153-174.

8. Pelegrini B, Girassol (1989) Uma planta solar que da América conquistou o mundo. São Paulo, Ícone.

9. Umesh R (2015) Sunflower seeds nutrition facts.

10. Esoteric Oils CC, Sallamander Concepts (2010) The world of pure essential oils.

11. Folch J, Lees M, Sloan SGH (1957) A simple method for the isolation and purification of total lipids from animal tissues. Journal of Biological Chemistry 226: 497-509.

12. Bligh EG, Dyer WJ (1959) A rapid method for total lipid extraction and purification. Can J Biochem Physiol 37: 911-917.

13. Allen CF, Good P (1971) Acyl lipids in photosynthetic systems. Methods enzymol 13: 523-547.

14. Crespo N, Esteve-Garcia E (2001) Dietary fatty acid profile modifies abdominal fat deposition in broiler chickens. Poult Sci 80: 71-78.

15. Grau A, Codony R, Grimpa S, Baucells MD, Guardiola F, et al. (2001) Cholesterol oxidation in frozen dark chicken meat: influence of dietary fat source, and Î ± tocopherol and ascorbic acid supplementation. Meat Sci 57: 197-208.

16. Matteo B, Maria FC, Maria TR, Giovanni L (2006) Effect of feeding fat sources on the quality and composition of lipids of precooked ready-to-eat fried chicken patties. Food Chemistry 10: 1327-1337.

17. Viveros A, Ortiz LT, Rodríguez ML, Rebolé A, Alzueta C, et al. (2009) Interaction of dietary high-oleic-acid sunflower hulls and different fat sources in broiler chickens. Poult Sci 88: 141-151.

18. Maria CM, Roseli GP, Carmino H, Nilson ES, Makoto M, et al. (2003) Influence of diets enriched with different vegetable oils on the fatty acid profiles of snail Helix aspersa maxima. Food Chemistry 82: 553-558.

19. Rodríguez ML, Ortiz LT, Alzueta C, Rebolé A, Treviño J, et al. (2005) Nutritive value of high-oleic acid sunflower seed for broiler chickens. Poult Sci 84: 395-402.

20. Sanz M, Carmona JM, Lopez-Bote CJ (2002) Quantitative effect of dietry fatty acids on fatty acid composition and fat firmness in broilers. Arch Geflugelk 66: 211-215.

21. Hardinka C, Zollitsch W, Knaus W, Lettner F (1996) Effects of dietary fatty acid pattern on melting point and composition of adipose tissues and intramuscular fat of broiler carcasses. Poult Sci 75: 208-215.

22. Ali AS, Hossein I, Mohammad KT, Mohammad C (2012) Fatty Acids Profiles in Meat of Broiler Chicks Fed Diet Containing Corn Oil Switched to Fish Oil at Different Weeks of Age. World Applied Sciences Journal 18: 159-165.

23. Cortinas L, Villaverde C, Galobart J, Baucells MD, Codony R, et al. (2004) Fatty acid content in chicken thigh and breast as affected by dietary polyunsaturation level. Poult Sci 83: 1155-1164.

24. Schreiner M, Hulan HW, Razzazi-Fazeli E, Bohm J, Moreira R, et al. (2005) Effect of different source of dietary omega-3 fatty acid on general performance and fatty acid profiles of thigh, breast, liver and portal blood of broilers. J Sci Food Agric 85: 216-226.

25. Dobrzański Z, Jamroz D, Bykowski P, Trziszka T (2002) Effect of fish oil on broiler performance and meat quality, Electronic. Journal of Polish Agriculture Universities 1: 43-51.

26. López-Ferrer S, Baucells MD, Barroeta AC, Galobart J, Grashorn MA, et al. (2001) n-3 enrichment of chicken meat. Use of precursors of long-chain polyunsaturated fatty acids: linseed oil. Poult Sci 80: 753-761.

27. Mourot J, Hermier D (2001) Lipids in monogastric animal meat. Reprod Nutr Dev 41: 109-118.

28. Ferrini G, Manzanilla EG, Menoyo D, Esteve GE, Baucells MD, et al. (2010)

Effects of dietary n-3 fatty acids in fat metabolism and thyroid hormone levels when compared to dietary saturated fatty acids in chickens. Livest Sci 131: 287-291.

29. Žlender B, Antonija H, Vekoslava S, Polak T (2000) Fatty acid composition of poultry meat from free range rearing. Poljopriverda 6: 53-56.

30. Alvarez MJ, Díez A, López-Bote C, Gallego M, Bautista JM, et al. (2000) Short-term modulation of lipogenesis by macronutrients in rainbow trout (Oncorhynchus mykiss) hepatocytes. Br J Nutr 84: 619-628.

31. Doreau M, Chilliard Y (1997) Digestion and metabolism of dietary fat in farm animals. Br J Nutr 78 Suppl 1: 15-35.

32. Hillard BL, Lundin P, Clarke SD (1980) Essentiality of dietary carbohydrate for maintenance of liver lipogenesis in the chick. J Nutr 110: 1533-1542.

33. Edwards HM, Hart P (1971) Carcass composition of chickens fed carbohydrate-free diets containing various lipid energy sources. J Nutr 101: 989-996.

34. The George Mateljan Foundation (2015) The world's healthiest foods.

35. Whitehead CC, Armstrong J, Herron KM (1990) The growth to maturity of lean and fat lines of broiler chickens given diets of different protein content: body composition, plasma lipoprotein concentration and initial egg production. Anim Prod 50: 183-190.

36. Enser M, Wiseman J (1984) The chemistry, biochemistry and nutritional importance of animal fats. Fats in Animal Nutriton, London, UK.

37. Daghir NJ, Raz MA, Uwayjan M (1980) Studies the utilization of full fat sunflower seed in broiler rations. Poultry Science 59: 2273-2278.

38. Bou R, Guardiola F, Grau A, Grimpa S, Manich A, et al. (2001) Influence of dietary fat source, alpha-tocopherol, and ascorbic acid supplementation on sensory quality of dark chicken meat. Poult Sci 80: 800-807.

39. Gray JI, Crackel RL, Ledward DA, Johnston DE (1992) Oxidative flavour changes in meats: their origin and prevention. The chemistry of muscle-based foods. The Royal Society of Chemistry, Cambridge, UK.

40. Lin CF, Gray JI, Asghar A, Buckley DJ, Booren AM, et al. (1989) Effects of dietary oils and a-tocopherol supplementation on lipid composition and stability of broiler meat. J Food Sci 54: 1457-1460.

41. Gallardo MA, Pérez DD, Leighton FM (2012) Modification of fatty acid composition in broiler chickens fed canola oil. Biol Res 45: 149-161.

42. O'Neill LM, Galvin K, Morrissey PA, Buckley DJ (1998) Comparison of effects of dietary olive oil, tallow and vitamin E on the quality of broiler meat and meat products. Br Poult Sci 39: 365-371.

43. Marion JE, Woodroof JG (1963) The fatty acid composition of breast, thigh and skin tissues of chicken broilers as influenced by dietary fats. Poultry Sci 42: 1202-1207.

13

Development of Health Foods from Oilseed Cakes

Sunil L, Prakruthi A, Prasanth Kumar PK and Gopala Krishna AG*

Department of Traditional Food and Sensory Science, CSIR-Central Food Technological Research Institute (CSIR-CFTRI), Mysore, India

Abstract

Four health foods were prepared using copra cake, sesame cake, dried mature coconut kernel without the testa, flattened rice flakes, sugar, coconut water solids, rice bran oil and sesame oil. The nutritional composition of raw materials and health foods were evaluated for various parameters such as moisture, ash, soluble and insoluble fiber (crude fiber), protein, carbohydrates, fats, fatty acid composition, oryzanol, lignans and minerals (sodium, potassium, calcium, iron and zinc). These health foods were also evaluated for sensory acceptance. The health foods had moisture content 2.2% to 3.9%, fat 2.0% to 35.0%, ash 2.1% to 6.2%, protein 9.2% to 12.2%, carbohydrates 42.85% to 83.7%, and crude fiber 2.95% to 6.4%. Among minerals, potassium content was in the range of 39-120.6 mg/100 g, sodium 9.95-49.6 mg/100 g, calcium 7.8-219.6 mg/100 g, iron 3.1-22.0 mg/100 g and zinc 1.9-6.4 mg/100 g. Fatty acid profile was designed to include medium chain, long chain saturated, monounsaturated and polyunsaturated fatty acids. Total phenolics content was in the range of 54.6-105.7 mg/100 g. A product with 0.6% oryzanol and 1% lignans was prepared to provide the hypocholesterolemic effect. These products were acceptable by sensory evaluation. Hence, it can be concluded that these products may be called health foods which provide health benefits to consumers.

Keywords: Coconut water solids, Copra cake; Health foods; Rice bran oil; Sesame oil; Sesame cake

Introduction

Oilseed cakes are the major byproducts obtained after oil extraction in the edible oil industries. Depending upon the extraction methods the percentage of oil in the cake varies. The oilseed cakes have been mainly used for feeding cattle and these oilseed cakes are rich in nutrients like proteins, carbohydrates, antioxidants, vitamins and minerals. These oilseed cakes can also be used for human consumption for proper processing and preparation of low-cost health foods [1,2]. Oilseed cakes are of two types, edible and non-edible grades. Edible cakes are those resulting from edible oil bearing seeds which are being used to meet a part of the nutritional requirements of either animal feed or for human consumption and those which cannot be used as feedstuff due to the presence of toxic compounds and other impurities are differentiated as non-edible [3].

Copra cake has been used for feeding cattle since it is rich in nutrients like protein, fat, nutraceuticals, fibers, vitamins and minerals. The copra cake can be processed and made edible for human consumption by giving certain treatments since it can be used as value-added products in food industries [2]. Coconut water solids may be prepared by drying the tender coconut water, which can be preserved for a longer time and is nutritionally rich in minerals especially sodium and potassium. Sesame cake can be used as a protein supplement in the food industries. Sesame cake also has lignans, a potential antioxidant against lipid oxidation [4]. The sesame oil has high amounts of lignans with health benefits like lowering the cholesterol level [5,6]. Rice bran oil is the only edible oil having an almost balanced ratio of saturated, mono unsaturated and poly unsaturated fatty acids (S:M:P ratio). Rice bran oil is stable especially during frying due to the presence of natural antioxidant called oryzanol which is heat stable antioxidant [7] and oryzanol can also be isolated from rice bran deodorized distillates and incorporated to other vegetable oils to increase the stability during frying [8]. Some oilseed cakes have antinutritional compounds such as tannins, phytates, trypsin inhibitors, pepsin inhibitors, hemaglutinin, etc., and these antinutritional compounds can reduce the bioavailability in our body but fortunately the concentration and activity of these compounds are less and can be inactivated by heat treatment to make palatable for human consumption [9].

Oilseed cakes have been used for feed application for cattle. However, oilseed cakes have never been prepared as food for human consumption due to the presence of antinutritional factors in them. Therefore, the present study aimed to utilize oil seed cakes (copra and sesame) by giving certain heat treatments and coconut water solids as a natural source of minerals and nutrients for the preparation of health foods for human consumption in food industries. With this background and using published research findings, two cakes and two vegetable oils having cholesterol lowering properties viz., rice bran oil and sesame oil were selected for incorporation. Also, to find a use for the new product developed, tender coconut water solids as a natural mineral supplement, food ingredients included improving texture, sensory attributes were also added to get the experimental design was framed in such a way that the raw materials used are safe for human consumption.

Materials and Methods

Collection of raw materials

Copra cake was collected from M/s Nirmal Coconut oil industries Pvt. Ltd., Cochin, sesame cake was collected from a local ghani mill, dried mature coconut kernel powder was collected from M/s Elite Food Industries, Mysore. Sugar, rice bran oil, sesame oil and flattened rice flakes were purchased from the local super market. Coconut water solid was prepared by drying the tender coconut water in the pilot plant using the process developed at the institute (CFTRI Process). Standard gallic acid, hydroxybenzoic acid, chlorogenic acid, vanillic acid, syringic

*Corresponding author: Gopala Krishna AG, Department of Traditional Food and Sensory Science, CSIR-Central Food Technological Research Institute, Mysore-570020, India, E-mail: gopkag11@gmail.com

acid, coumaric acid, caffeic acid, ferulic acid cholesterol, FAME mix and cinnamic acid were procured from Sigma Chemicals Co., St. Louis, USA. All chemicals, solvents, and reagents were of analytical grade.

Oilseed cake treatment

The raw copra cake and sesame cakes were mixed with distilled water (1:4 w/v) and cooked in such a way to get a cooked solid residue and a liquid extract which are separated by decantation of the liquid extract followed by freeze drying ('ScanVac' FREEZE DRIER model: Cool Safe 55-9 Pro with Drying Chamber, Denmark) at −55°C to get two products from each. Cooked solid residues of copra cake and sesame cakes were used as ingredients for the preparation of health foods [2].

Health food preparation

Different health foods (HF1, HF2, HF3, and HF4) were prepared by using the ingredients such as copra cake, sesame cake, mature coconut kernel, flattened rice flakes, sugar, coconut water solids, rice bran oil (RBO) and sesame oil (SESO). The health food HF1 was prepared by blending of mature coconut powder and copra cake along with sugar. HF2 was prepared by blending of a mature coconut, copra cake and sesame cake along with sugar. HF3 was prepared by mixing copra cake, rice flake, and tender coconut water solids along with sugar. HF4 was formulated by using copra cake, sesame cake, rice flake, sugar, rice bran oil, and sesame oil in specific combination (Table 1). Before the formulation of the health foods all the ingredients except TCWS, RBO and SESO were ground into fine powder. The ground ingredients mixed along with other raw materials in a mixer grinder to obtain a uniform distribution of ingredients in health foods. The prepared health foods were transferred to pet jars and kept at 4°C till the time of further analysis.

Proximate composition of health foods

Moisture content: The samples were ground to a fine powder; 10 g of the ground samples were taken in aluminum moisture cups and placed in an oven at 100 ± 1°C for 2 h or till a constant weight was obtained. The moisture content was expressed on a dry basis (Method No. Ac 2-41 1997) [10].

Fat content: The analysis was carried out by AOCS Official Butt-tube Method Ac 3-44 [10]. The samples of raw rice bran, oil cakes, their cooked residues, and extracts were ground to a fine powder, dried in an oven at 100 ± 1°C, packed in 26 mm × 60 mm thimbles and extracted with hexane in Soxhlet apparatus. The extracts were desolventizing by vacuum flash evaporation (Rotavapor RE 121A, Buchi, Switzerland) at controlled temperature and were subjected to various analyses.

Protein content (AOAC Official Method 950.48): The micro-Kjeldahl method was used to determine total proteins using 1 g sample. Sample nitrogen content was calculated using the formula [11]. The method described by the AOAC Official Method 950.02 [11] was used for crude fiber determination.

Dietary fiber content: The estimation of dietary fiber in the samples was done according to the enzymatic-gravimetric method [12]. Briefly, 1 g of ground sample was suspended in 25 ml of 0.1 M phosphate buffer (pH 6), then 0.1 ml of Thermo amylase was added and the mixture was kept in a boiling water bath for 15 min to digest starches. The crucible was kept in an oven (105°C) until the weight became constant and its final weight was determined. The sample-containing crucible was then incinerated at 550°C for 5 h and its weight was determined. To obtain soluble dietary fiber, the volume of the filtrate was adjusted to 100 ml and the soluble fiber was precipitated by adding 4 volumes of warm (60°C) ethanol. The precipitate was filtered through a crucible containing celite, dried at 105°C and the weight of the crucible was determined. The sample-containing crucible was then incinerated at 550°C for 5 h and its weight was determined. Blanks were prepared as above but without the sample.

Ash content: A known weight of the sample was initially charred on a tared silica crucible and placed in a muffle furnace at 550°C for 6 h till the charred material became white. The dish was allowed to cool to room temperature in desiccators and reweighed. The difference in weight was taken as total ash content [11].

Mineral content: Iron, zinc, sodium, potassium and calcium content of raw materials and health foods were analyzed by atomic absorption spectroscopy (AAS) [13].

Total phenolic content (TPC): The phenolics extraction from the samples was carried out according to [14]. The phenolics were extracted from samples using 80% methanol. For CK, 1.0 g of sample was mixed with 5.0 mL of 80% methanol, heated for 5 min on a water bath and vortexed for 2 minutes (twice). The samples were centrifuged at 2500 rpm for 10 minutes at room temperature. The methanol-water layer was collected in another tube and this step was repeated for four times. The extracts pooled and made up to 20 mL with 80% methanol. For CW, it was used in aliquots directly. The TPC was determined using the Folin–Ciocalteu reagent. Different aliquots were mixed with 0.20 mL of Folin–Ciocalteu reagent and were kept for 3 minutes. About 1.0 mL of 15% Na_2CO_3 solution was added and made up to 7 mL with distilled water. The tubes were incubated for 45 minutes and centrifuged at 2000 rpm for 10 minutes at room temperature. The absorbance was determined at 765 nm using a UV-Visible spectrophotometer (Shimadzu corporation, Kyoto, Japan, model UV–1601). The TPC (mg/100) was calculated using gallic acid as a standard compound [15].

Extraction of crude fat and its composition in health foods: The four different health foods were subjected to fat extraction and further this fat was analyzed for various parameters such as fatty acid profile, oryzanol, and lignans content.

Fatty acid composition: Fatty acid methyl esters (FAME) of the oil samples were prepared by transesterification, according to AOCS Method No: Ce 1-62, 1998 [10]. FAMEs were analyzed on a Fisons 8000 series gas chromatograph (Fisons Co., Italy), equipped with a flame ionization detector (FID) and a fused silica capillary column (100 m

Health Foods	Raw materials (%)							
	Sugar	Mature Coconut	Copra Cake	Sesame Cake	Rice flakes	Coconut water solid	Rice bran Oil	Sesame Oil
HF1	40	50	10	-	-	-	-	-
HF2	40	45	7.5	7.5	-	-	-	-
HF3	20	-	20	-	40	20	-	-
HF4	40	15	7.5	7.5	10	-	15	5

Table 1: Composition of health foods.

× 0.25 mm i.d.), coated with 0.20 ml SP2560 (Supelco Inc., Bellefonte, PA) as the stationary phase. The oven temperature was programmed from 140°C to 240°C at 4°C/min with an initial hold at 140°C for 5 min. The injector and FID were at 260°C. A reference standard FAME mix (Supelco Inc.) was analyzed under the same operating conditions to determine the peak identity. The FAMEs were expressed as relative area percentage.

Oryzanol content: Oryzanol content was determined by spectrophotometric method using UV–vis spectrophotometer (model-UV-1601, Shimadzu, Kyoto, Japan) by measuring the optical density at 314 nm of the oil taken in hexane followed by calculation using the extinction coefficient of 358.9 and expressed as g/100 g of oil and reported as milligram/ 100 g of oil [16].

Lignans content: The oil samples (0.01 g), in triplicate, were dissolved in 10 ml of hexane + chloroform mixture (7:3, v/v) and the absorbance at 288 nm was determined [17]. The lignans content was calculated by using the formula:

$$\% \text{ Lignans (as sesamin)} = [(A/W) \times (100/230.1)] \qquad (1)$$

Where,

A is the absorbance of the sample

W is the weight of the sample in gram/100 ml

230.1, E1% 1 cm for sesamin

Sensory attributes: Sensory evaluations of health foods were carried out by 20 untrained taste panelists. They were instructed to taste the samples and to rinse their mouth after each sample taste. They were requested to express their feelings about the samples by scoring the following attributes: appearance, texture, taste, aroma and overall acceptability. Sensory scores were based on a nine-point hedonic scale, where 1 dislikes extremely and 9 are like extremely [18].

Statistical Analysis

The experiment was carried out in quadruplicate. All the quality parameters were analyzed in quadruplicate and the data obtained for each parameter were expressed as a mean ± standard deviation. One-way ANOVA was used to calculate the significant difference in the oils and residues [19].

Results and Discussion

The raw materials used for the preparation of health foods (HF) were dried mature coconut, copra cake, sesame cake, sugar, rice flakes, coconut water solid, rice bran oil (RBO) and sesame oil (SESO) depicted in Table 1. All these raw materials were made into a fine powder before mixing them uniformly in a 10 kg capacity mixer. Four health food products were prepared by standardizing the composition for each product for different health benefits. Cooked residues of copra and sesame cakes were little bitter in taste so sugar was added up to 40% to mask the bitterness in the product. HF1 was based on the utilization of copra cake (10%) and dried matured coconut (50%), these two are rich in protein, fiber, fat, and minerals. HF2 was based on the utilization of copra cake and sesame cake (15%), sesame cake is rich in antioxidant lignans and minerals especially calcium. HF3 was based on copra cake (20%) and coconut water solid (20%), coconut water solid is very rich in minerals especially sodium and potassium. Finally, HF4 was based on the copra, sesame cake, rice bran oil and sesame oil. All the four health foods are sweet in taste and can be eaten as such or can be mixed with water or milk for use as an energy/nutritional/health drink.

The health foods developed were subjected to storage studies. The above health foods were stored for 3 months and analyzed at every 15 days up to 3 months' period for the product's shelf-life characteristics. Table 2 depicts the storage studies of health foods such as moisture, fat, oryzanol and lignans contents. The moisture content of health foods was in the range of 1.75% to 2.0%. Fat content was in the range of 2.0% to 34.9%. Fat content was almost constant during the storage studies this was due to the presence of potent antioxidants oryzanol and lignans. The antioxidants, oryzanol (0.6% to 0.58%) and lignans (0.98% to 1.0%) were stable during the storage studies as they are heat stable antioxidants [5,7] and they may also give synergistic effect for the stability of enhanced shelf-life. The antioxidant oryzanol is a heat-stable compound and it does not degrade even at higher temperatures. Many studies shown the health benefits of oryzanol such as hypocholesterolemic activity [20,21], protective role in lipid peroxidation [22], safety assessment of oryzanol indicates no genotoxic or carcinogenic activity [23], utilization of oryzanol concentrate and purified oryzanol as natural antioxidant for stability and can be used as functional foods [8, 24]. Lignans are natural antioxidants present in sesame oil. Lignans have health benefits, including antioxidant activity [25], anticarcinogenic [26], blood pressure-lowering [27] and serum lipid-lowering [28].

Table 3 depicts the data for the moisture, fat, protein, carbohydrates, ash, crude fiber, oryzanol, and lignans. Carbohydrate content was calculated by difference. The moisture content was ranged from 2.2% to 3.9%. Estimation of lipids is an important factor for nutritional evaluation of any food products. The fat content of health foods ranged from 2.0% to 35.0%. HF1, HF2, and HF4 health foods were oil based foods and HF3 was mainly of coconut water solid and copra cake based product, the fat content was only 2.0% in it. The fat content of the oils

Parameters (%)	Health Foods	Storage period at 38°C/90% RH (days)						
		0	15	30	45	60	75	90
Moisture	HF1	2.0 ± 0.01	1.84 ± 0.08	1.84 ± 0.01	1.77 ± 0.01	1.76 ± 0.02	1.76 ± 0.08	1.75 ± 0.01
	HF2	2.0 ± 0.05	1.98 ± 0.01	1.98 ± 0.05	1.97 ± 0.01	1.97 ± 0.02	1.95 ± 0.01	1.91 ± 0.06
	HF3	2.1 ± 0.01	2.1 ± 0.01	2.1 ± 0.01	2.1 ± 0.02	1.99 ± 0.01	1.99 ± 0.02	1.99 ± 0.05
	HF4	2.3 ± 0.02	2.3 ± 0.01	2.3 ± 0.04	2.3 ± 0.01	2.3 ± 0.01	2.3 ± 0.1	2.3 ± 0.01
Fat	HF1	34.9 ± 0.01	34.9 ± 0.06	34.9 ± 0.04	34.9 ± 0.1	34.8 ± 0.01	34.8 ± 0.02	34.8 ± 0.02
	HF2	32.7 ± 0.01	32.7 ± 0.04	32.7 ± 0.02	32.7 ± 0.05	32.7 ± 0.05	32.7 ± 0.01	32.7 ± 0.01
	HF3	2.0 ± 0.05	2.0 ± 0.05	2.0 ± 0.01	2.0 ± 0.08	2.0 ± 0.02	2.0 ± 0.02	2.0 ± 0.1
	HF4	30.2 ± 0.01	30.1 ± 0.01	30.1 ± 0.02	30.1 ± 0.05	30.1 ± 0.01	30.1 ± 0.05	30.1 ± 0.02
Oryzanol	HF4	0.6 ± 0.02	0.6 ± 0.08	0.6 ± 0.08	0.6 ± 0.01	0.59 ± 0.02	0.59 ± 0.01	0.58 ± 0.01
Lignans	HF4	1.0 ± 0.01	1.0 ± 0.01	1.0 ± 0.01	1.0 ± 0.02	0.98 ± 0.01	0.98 ± 0.05	0.98 ± 0.08

Values are average of triplicate determinations.

Table 2: Storage stability of health foods.

Proximate Composition	HF1	HF2	HF3	HF4
Moisture (%)	2.3 ± 0.1	2.4 ± 0.2	2.2 ± 0.5	3.9 ± 0.2
Fat (%)	35.0 ± 0.08	32.7 ± 0.1	2.0 ± 0.1	30.1 ± 0.08
Protein (%)	11.3 ± 0.4	10.7 ± 0.5	9.2 ± 0.2	12.2 ± 0.4
CHO (%)	42.8 ± 0.08	47.0 ± 0.05	83.7 ± 0.01	50.5 ± 0.2
Ash (%)	2.1 ± 0.3	2.7 ± 0.05	3.3 ± 0.01	6.2 ± 0.02
Crude fiber (%)	6.4 ± 0.05	4.5 ± 0.02	2.9 ± 0.2	3.3 ± 0.4
Oryzanol (mg/100 g)	-	-	-	600 ± 0.1
Lignans (mg/100 g)	-	-	-	1000 ± 0.1
TPC (mg/100 g)	65.52 ± 0.1	54.65 ± 0.2	105.7 ± 0.5	89.64 ± 0.2

Values are average of triplicate determinations.

Table 3: Proximate composition of health foods.

Health Foods	Dietary fiber (%)		Total dietary fiber (%)
	Soluble	Insoluble	
HF1	4.1 ± 0.01	28.2 ± 0.02	32.3 ± 0.06
HF2	2.8 ± 0.06	33.5 ± 0.02	36.3 ± 0.05
HF3	3.6 ± 0.08	52.2 ± 0.06	55.8 ± 0.01
HF4	3.2 ± 0.09	48.0 ± 0.08	51.2 ± 0.02

Values are average of triplicate determinations.

Table 4: Dietary fiber content of health foods.

Health Foods	C8:0	C10:0	C12:0	C14:0	C16:0	C18:0	C18:1	C18:2	SFA	MUFA	PUFA
HF1	9.2 ± 0.2	6.3 ± 0.5	44.0 ± 0.5	21.9 ± 0.1	10.1 ± 0.1	1.4 ± 0.1	5.4 ± 0.5	1.7 ± 0.1	92.9 ± 0.6	5.4 ± 0.4	1.7 ± 04
HF2	8.8 ± 0.5	6.1 ± 0.1	41.6 ± 0.4	21.7 ± 0.8	10.0 ± 0.5	1.8 ± 0.4	7.2 ± 0.2	2.8 ± 0.5	90.0 ± 0.2	7.2 ± 0.6	2.8 ± 0.5
HF3	9.1 ± 0.08	6.4 ± 0.8	44.2 ± 0.2	21.8 ± 0.4	9.9 ± 0.5	1.4 ± 0.6	5.3 ± 0.5	1.9 ± 0.8	92.8 ± 0.8	5.3 ± 0.4	1.9 ± 0.6
HF4	3.0 ± 0.1	2.1 ± 0.1	14.9 ± 0.8	12.6 ± 0.6	12.4 ± 0.1	1.5 ± 0.2	27.8 ± 0.1	25.7 ± 0.4	46.5 ± 0.5	27.8 ± 0.5	25.7 ± 0.8

Values are average of triplicate determinations.

Table 5: Fatty acid composition of health foods.

and cakes depends on the oil extraction method [29]. Oilseed cake is an excellent by-product source of protein for the preparation of health foods in food industries. The protein content of health foods varied from 9.2% to 12.2%. The maximum protein content was observed in the HF4. The ash content varied from 2.1% to 6.2%. There was a wide variation in the ash content of health foods. HF1 had the lowest and HF4 showed the highest ash content. The health foods were also analyzed for crude fiber and the crude fiber was in the range of 2.9% to 6.4%. The HF4 was incorporated with both rice bran and sesame oils to enrich the nutritional value to the product. The rice bran oil is having the antioxidant called oryzanol, a highly stable antioxidant to give stability to the oil. The oryzanol content in HF4 was having 600 mg/100 g of the product. The sesame oil is rich in antioxidant like lignans, the lignans in HF4 had 1000 mg/100 g of the product. The total phenolic contents were also analyzed in the health foods. The oilseed cakes are rich in phenolics especially copra cake, kernel, testa [30,31] and health benefits of phenolics includes antioxidant and anticancer [31,32]. The total phenolic contents were in the range of 54.65-105.7 mg/100 g, the highest phenolic content was found to be in HF3 (105.7 mg/100 g) and lowest was in HF2 (54.7 mg/100 g).

Health foods were analyzed for the total dietary fiber (soluble and insoluble) and are presented in the Table 4. Dietary fibers are best known for its ability to prevent or relieve constipation. The total dietary fibers were ranged from 32.35 to 55.8% and these values were agreed well with those reported for the total dietary fiber of copra and sesame cakes [2]. The soluble and insoluble fibers were ranged from 2.8-4.1 and 28.2-52.2 respectively. The insoluble dietary fiber contents were more compared soluble dietary fiber in all health foods.

Table 5 depicts that fatty acid composition of health foods. Fatty

acid composition is one of the key factors for nutritional evaluation of any foods [33]. HF1 was having coconut oil and its fatty acid composition was mainly the presence of lauric acid (C12: 0 44.0%) and myristic acid (C14: 0 21.9%). HF2 was having both coconut and sesame oil and there was a slight variation in the fatty acid profile compared to HF1. This was due to the presence of sesame oil in the sesame cake of HF2. HF3 health food was based on coconut oil and its fatty acid profile was almost similar to HF1. The HF4 was incorporated with rice bran and sesame oil and its fatty acid composition was well balanced with S:M:P ratio compared to other commercial individual edible oils (Saturated: Mono unsaturated : polyunsaturated fatty acids) i.e., saturated fatty acid was 46.5%, mono unsaturated fatty acid was 27.8 and it had 25.7% of poly unsaturated fatty acids.

Mineral composition is one of the nutritional factors for any foods. Potassium, sodium, calcium, iron and zinc were analyzed from these health foods and provided in Table 6. There was a wide variation in the mineral contents of health foods. The lowest mineral content was found to be zinc in HF3 (1.9 mg/100 g) and highest was found to be calcium in HF2 (219.6 mg/100 g) this was due to the incorporation of sesame cake in the HF2. The RDA for minerals such as potassium, sodium, calcium, iron and zinc for adult men and women are mentioned [34]. The health foods of potassium and sodium content were in the range of 39.0-120.6 mg/100 g (RDA for adult men and women is 4.7 g/day) and 9.9-49.6 mg/100 g (RDA for adult men and women is 1.1-3.3 g/day). Potassium and sodium are essential to maintaining the body's electrolyte balance. The content of calcium was in the range of 7.8-219.6 mg/100 g (RDA for adult men and women is 1000 mg/day). The iron content was in the range of 3.1-22.0 mg/100 g (RDA for adult men and women are 8 mg/day and 18 mg/day). The mineral iron is a very important nutrient

Health Foods	Mineral Composition mg/100 g				
	Iron	Zinc	Sodium	Potassium	Calcium
HF 1	13.1 ± 0.05	3.2 ± 0.1	49.6 ± 0.06	120.6 ± 0.01	95.6 ± 0.1
HF 2	22.0 ± 0.01	6.4 ± 0.05	43.5 ± 005	118.3 ± 0.04	219.6 ± 0.05
HF 3	3.1 ± 0.08	1.9 ± 0.01	16.0 ± 0.01	90.0 ± 0.01	7.8 ± 0.06
HF 4	5.1 ± 0.02	2.3 ± 0.01	9.9 ± 0.12	39.0 ± 0.02	16.2 ± 0.01
Values reported are mean ± SD (n=6).					

Table 6: Mineral composition of health foods.

Health Foods	Appearance	Texture	Taste	Aroma	Overall Acceptability
HF 1	7.3 ± 0.01	6.6 ± 0.05	7.5 ± 0.1	6.8 ± 0.02	7.0
HF 2	7.8 ± 0.02	6.6 ± 0.06	7.5 ± 0.02	7.2 ± 0.05	7.2
HF 3	8.3 ± 0.02	8.2 ± 0.1	8.5 ± 0.01	8.2 ± 0.04	8.3
HF 4	7.5 ± 0.05	7.4 ± 0.01	7.4 ± 0.02	7.0 ± 0.01	7.3
Values reported are mean ± SD (n=6).					

Table 7: Sensory scores of health foods.

because iron deficiency is the most common nutritional disorder and the health foods meet the RDA for adult men and women. The zinc content was in the range of 1.9-6.4 mg/100g (RDA for adult men and women are 11 mg/day and 9 mg/day). The health foods have good sources of minerals, health benefits, and sufficient intakes meet the RDA for adult men and women.

Sensory evaluation of health foods is shown in Table 7. The quality parameters such as appearance, texture, taste, aroma and overall acceptability of products were done by the sensory panelists. All health foods were acceptable by the sensory panelists, among these HF3 was highly acceptable by all the sensory panelists.

Conclusion

From the present study, it can be concluded that the raw materials such as cooked copra cake, cooked sesame cake, and mature coconut kernel can be used as food supplements in the food industries because of their nutrients like protein, dietary fiber (both soluble and insoluble), minerals and potent antioxidants such as lignans in sesame cake. Coconut water solids can be obtained by removing the water content in the tender coconut water, the coconut water solids are a rich source of minerals mainly sodium and potassium and it can be used as food supplements. Rice bran oil and sesame oil were used to enrich the antioxidants like oryzanol in rice bran oil and lignans in sesame oil, rice bran oil gives stability to the products which have got the balanced ratio of saturated, mono unsaturated and poly unsaturated fatty acids (S:M:P ratio). Based on these findings it can be concluded that these can be mixed in particular proportions to make health products. We have prepared four different health foods (HF1, HF2, HF3, and HF4). The results showed that these health foods are having nutritional values since these are rich in health improving nutrients. All these four products had very low moisture content; the protein content was in the range of 9.2% to 12.2%. Phenolics content was in the range of 54.65-105.7 mg/100 g, Mineral contents were also analyzed, potassium content was in the range of 39-120.6 mg/100 g, sodium 9.9-49.6 mg/100 g, calcium 7.8-219.6 mg/100 g, iron 3.1-22.0 mg/100 g and zinc 1.9-6.4 mg/100 g. These products were accepted from the sensory panelists, among four, HF3 was best rated by the panelists. Finally, it can be concluded that oilseed cakes can be used for human consumption by appropriate processing and giving certain treatments.

Acknowledgment

Authors are thankful to Prof. Ram Rajasekharan, Director, CSIR-CFTRI, Mysore for providing infra structural facilities and The Coconut Development Board, Kochi, for funding the project.

References

1. Pandey A, Soccol CR, Nigam P, Soccol VT (2000) Biotechnological potential of agro-industrial residues. Biores Technol 74: 69-80.

2. Sunil L, Prakruthi A, Prasanth Kumar PK, Gopala Krishna AG (2015) Preparation of food supplements from oilseed cakes. J Food Sci Technol 52: 2998-3005.

3. Mitra CR, Misra PS (1967) Amino acids of processed seed meal proteins. J Agri Food Chem 15: 697-700.

4. Ali A, Muhammad NM, Muhammad NA (2014) A review on the utilization of sesame as a functional food. Am J Food Nutr 4: 21-34.

5. Suja KP, John TA, Selvam NT, Jayalekshmy A, Armugham C (2004) Antioxidant efficacy of sesame cake extract in vegetable oil protection. Food Chem 84: 393-400.

6. Gopala Krishna AG, Khatoon S, Shiela PM, Sarmandal CV, Indira TN, et al. (2001) Effect of refining of crude rice bran oil on the retention of oryzanol in the refined oil. J Am Oil Chem Soc 78: 127-131.

7. Sunil L, Srinivas P, Prasanth Kumar PK, Gopala Krishna AG (2015b) Oryzanol as a natural antioxidant for improving sunflower oil stability. J Food Sci Technol 52: 3291-3299.

8. Khalil AH, Mansour EH (1995) The effect of cooking, autoclaving and germination on the nutritional quality of faba beans. Food Chem 54: 177-182.

9. Official methods of the American Oil Chemists Society (1998) AOCS method no. Ac 2–41, The AOCS Official Butt-tube Method Ac 3–44, AOCS method Nos. Ca 5a–40, AOCS method Nos. Cd 8–53. American Oil Chemists Society, Champaign, IL, USA.

10. AOCS (1997) Official Methods of Analysis (16th Edn). Association of official analytical chemists, Washington DC.

11. Asp NG, Johansson CG, Hallmer H, Siljestrom M (1983) Rapid enzymatic assay of insoluble and soluble dietary fiber. J Agric Food Chem 31: 476-482.

12. Raghuramulu N, Nair KM, Kalyanasundram (2003) A manual for laboratory techniques. National institute of Nutrition, Hyderabad, India.

13. Kapila NS, Dissanayake MSD (2005) Effect of method of extraction on the quality of coconut oil. J Sci Univ Kelaniva 2: 63-72.

14. Osawa T, Namiki M (1981) A novel type of antioxidant isolated from leaf wax of eucalyptus leaves. Agri Biol Chem 45: 735-739.

15. Gopala Krishna AG, Hemakumar KH, Khatoon S (2006) Study on the composition of rice bran oil and its higher free fatty acids value. J Am Oil Chem Soc 83: 117-120.

16. Bhatnagar AS, Hemavathy J, Gopala Krishna AG (2013) Development of a rapid method for determination of lignans content in sesame oil. J Food Sci Technol 52: 521-527.

17. Watt BM, Ylimaki GL, Jeffery LE, Elias LG (1989) Basic sensory methods for food evaluation. International Development Research Centre (IDRC), Ottawa, Canada.

18. Steele RGD, Torrie JH (1980) Principles and procedures of statistics. Mc Graw-Hill, New York.

19. Nicolosi RJ, Austrian LM, Hegsted DM (1991) Rice bran oil lowers serum total and low-density lipoprotein cholesterol and apo B levels in nonhuman primates. Atherosclerosis 88: 133-142.

20. VissersMN, Zock PL, Meijer GW, KatanMB (2000) Effect of plant sterols from rice bran oil and triterpene alcohols from sheanut oil on serum lipoprotein concentrations in humans. Am J Clin Nutr 72: 1510-1515.

21. Kaimal TNB (1999) Gamma-oryzanol from rice bran oil. J Oil Technol Assoc India 31: 83-93.

22. Tsushimoto G, Shibahara T, Awogi T, Kaneko E, Sutou S, et al. (1991) DNA-damaging, mutagenic, clastogenic and cell-cell communication inhibitory properties of gamma-oryzanol. J Toxicol Sci 16: 191-202.

23. Prasanth Kumar PK, SaiManohar R, Indiramma AR, Gopala Krishna AG (2012) Stability of oryzanol fortified biscuits on storage. J Food Sci Technol.

24. Kang MH, Naito M, Tsujihara N, Osawa T (1998) Sesamolin inhibits lipid peroxidation in rat liver and kidney. J Nutr 128: 1018-1022.

25. Hirose N, Doi F, Ueki T, Akazawa K, Chijiiwa K, et al. (1992) Suppressive effect of sesamin against 7, 12- Dimethylbenz [a] anthracene induced rat mammary carcinogenesis. Anticancer Res 12: 1259-1266.

26. Nakano D, Kwak CJ, Fujii K, Ikemura K, Satake A, et al. (2006) Sesamin metabolites induce an endothelial nitric oxide-dependent vasorelaxation through their antioxidative property-independent mechanisms: possible involvement of the metabolites in the antihypertensive effect of sesamin. J Pharmacol Exp Ther 318: 328-333.

27. Hirose N, Inoue T, Nishihara K, Sugano M, Akimoto K, et al. (1991) Inhibition of cholesterol absorption and synthesis in rats by sesamin. J Lipid Res 32: 629-638.

28. Swick RA (1999) Considerations in using protein meals for poultry and swine. ASA Technical Bulletin AN 21: 1-11.

29. Prakruthi Appaiah, Sunil L, Prasanth Kumar PK, Gopala Krishna AG (2014) Composition of coconut testa, coconut kernel, and its oil. J Am Oil Chem Soc 91: 917-924.

30. Prakruthi Appaiah, Sunil L, Prasanth Kumar PK, Gopala Krishna AG (2014) Phytochemicals and antioxidant activity of testa extracts of commercial wet and dry coconuts and cakes. Int Res J Pharma 7: 9-13.

31. Liu RH (2004) Potential synergy of phytochemicals in Cancer prevention: Mechanism of Action. ASNS 3479-3485.

32. Ayaz FA, Glew RH, Millson M, Huang H, Chuang L, et al. (2006) Nutrient contents of kale (Brassica oleraceae L. var. acephala DC.). Food Chem 96: 572-579.

33. IMNS (2010) Dietary reference intakes tables and application from Institute of Medicine of the National Academy of Sciences, IMNS, India.

34. NIN (2009) Nutrient requirements and recommended dietary allowances for Indians. A report of the expert group of the Indian Council of Medical research. National Institute of Nutrition, India.

14

Characteristics of Onion under Different Process Pretreatments and Different Drying Conditions

Alabi KP[1]*, Olaniyan AM[2] and Odewole MM[3]

[1]Department of Food, Agriculture and Bio-Engineering, College of Engineering and Technology, Kwara State University, Kwara State, Nigeria

[2]Department of Agricultural and Bioresources Engineering, Faculty of Engineering, Federal University Oye Ekiti, Ekiti State, Nigeria

[3]Department of Food and Bioprocess Engineering, Faculty of Engineering and Technology, University of Ilorin, KwaraState, Nigeria

Abstract

Introduction: Onion (*Allium cepa*) is an important spice crop often grown outdoors in temperate climates as an annual crop because of its adaptability to varying weather conditions. It is an underground vegetable which varies in size colour, firmness and strength of flavour. Onion is often called "poor man's orange" because it is a good source of vitamins, particularly Vitamin A and C. It is also a rich source of minerals such as iron, thiamine, niacin and manganese contents. Onion is said to be very useful against heart diseases and many bacterial species including bacillus subtilis, salmoneva, and *E. coli*. This vegetable crop is highly perishable in its natural state after harvest resulting in huge postharvest losses during storage, transportation and marketing in the production season and extreme scarcity in the off-season which can be checked by drying.

Material and methods: The main materials used were 192 samples of pre-treatment and 10 samples of untreated (control) fresh onion. Others equipments used were temperature controlled dryer, sensitive weighing balance, water baths (Shell Lab Model and HH-W420, XMTD-204 Model), thermo-hygrometer, desiccators, desiccants, stop watch, onion slicer, stainless tray, foil wrap, conical flask, measuring cylinder, NaCl and distil water. Agarry and AOAC methods were used for quantitative analysis and nutritional analysis respectively. Statistical analysis of all data obtained was done.

Results: Results showed that drying rate, water loss, solid gain, vitamin C, manganese and iron contents varied with different levels of OSC, OPD and OST at $p \leq 0.05$. However, drying rate, water loss, solid gain and all the quality parameters were influenced by all the process parameters.

Where;

OSC = Osmotic solution concentration

OST = Osmotic solution temperature

OPD = Osmotic process duration

Conclusion: Osmotic dehydration pretreatments had significant effect on process outputs (drying rate, water loss, solid gain, vitamin C, manganese and iron contents of onion.

Keywords: Onion (*Allium cepa*); Osmotic dehydration; Process pretreatments; Quality parameters

Introduction

Onion is an important underground vegetable compared with the other root and tuber crops (cassava, yam and potatoes) that has many nutritional and medicinal values that are beneficial to human beings.

Productivity of onion shows variable trends as the crop is susceptible to various weather variations. The optimum temperature required for its cultivation is 15 to 27°C and it can withstand drought fairly and do well in heavy rainfall areas [1]. Onion is among the most important vegetable grown in the tropic and a good source of calorie (50 gcal/100g). The growing and handling has received considerable attention in agricultural and food research and development and can be found in almost every fresh market in Africa.

Drying process is a thermo-physical and physio-chemical operation by which moisture is being removed from food materials. It involves the movement of water from the interior of the drying material to the surface from where it evaporates. Drying makes food materials suitable for safe storage and protects them against attack of insects, molds and other micro-organisms during storage [2]. In addition to preservation, dehydration helps to decrease the weight and bulk of material by significant amounts and improves the efficiency of product transportation and storage. Alam *et al.* [3] studied the effect of process parameters on the effectiveness of osmotic dehydration of summer onion. In the studied onion was treated in three sucrose (40, 50 and 60%), five salt (5, 10, 15, 20 and 25%) and five sucrose-salt (combine) solution (40:15, 45:15, 45:20, 50:15 and 55:15%) under

different osmo-drying systems and osmotic solution temperature of 5, 25, and 40°C at the same time. The result obtained showed that among different solution concentration and temperature for 6 hours contact time 50:15° brix at 40°C gave water loss (50.05%), solid gain (16.25%) and normalized solid content (2.34). The result also showed that the 60° brix sucrose solution gave 35.60%, 9.32%, 1.81 and 25° brix salt solution gave 33.50%, 12.21%, 2.25 water loss, solid gain and normalized solid content respectively (Figure 1).

Asiru *et al.* [4] studied the effects of drying temperature and drying duration on colour changes in cashew kernels during hot air drying. Cashew nuts were steamed in an autoclave at 121°C for 30 min at a pressure of 7.93×10^5 N/m³ and allowed to cool at 30°C ± 2°C for 24 hours. Result showed that both air temperature and drying time had significant effect on L-, a- and b- values of cashew kernel colour during over the temperature range studied.

---

*Corresponding author: Alabi KP, Lecturer II, Department of Food, Agriculture and Bio-Engineering, College of Engineering and Technology, Kwara State University, Kwara State, Nigeria

Email: kehinde.alabi@kwasu.edu.ng

Figure 1: Whole Onion Samples.

Afolabi and Adeniyi [5] mathematically modeled mass transfer during osmotic dehydration of Red Paprika. The drying kinetics was experimental determined using an explicit finite difference (EFD). Experimental condition such as osmotic solution concentration, osmotic process duration at constant osmotic solution temperature were investigated while measured parameters included predicted moisture content, solid contents, the rate of water loss and solid gain. The results showed that the difference equations from EFD predict moisture, solid contents, the rate of water loss and solid gain which were found to be pronounced at the first two hours of the process. Results also confirmed that EFD at the usage of 7×90 nodal points and $\frac{\Delta x}{\Delta x^2} \leq 0.5$ for all cases of osmotic conditions was efficient at predicting the process parameter in terms of water loss and solute impregnated.

Kalse *et al.* [6] studied microwave drying of onion slices. The effect of various power levels (0.25, 1.00, 1.50 and 2.25 kW) on mass reduction, water loss and diffusivity were studied. The results showed that the mass reduction and water loss increased with increase in power level. The results also showed that the moisture diffusivity varied in the range of 6.491×10^{-9} to 6.491×10^{-8} m²/s.

Rzepecka *et al.* [7] studied foam-mat dehydration of tomato paste using microwave energy. A comparative study of foam-mat dehydration of tomato paste by a conventional hot-air method and a microwave energy method were demonstrated. In the investigation, the thickness of the foam 3.17 mm (1/8 inch), 6.35 mm (1/4 inch) and 12.7 mm (1/2 inch) were tested. The drying tests were performed for five forward power setting between 150W and 350W with a 50W increment operated at a frequency of 2,450 MHz. The measured parameter includes drying rate and colour. The results showed that the moisture removal rate was faster at any of the examined levels of microwave power for each of the thickness, than for the hot air dehydration, especially for the samples of thickness 6.35 mm and 12.7 mm the difference was significant. The result also indicated that no lightness-darkness effects for samples dehydrated by microwave power where as maximum power setting at the maximum foam thickness gave slightly lower lightness-darkness effects with the hot air drying.

The demand for onion by the growing population has not been met despite the increase in the production of onion. This is as a result of wastes that come from biological and biochemical activities taking place in this fresh product. Their (coating) protection mechanisms are not well sealed making them more vulnerable to deterioration due

to rapid metabolic activities that take place within their cells. These activities causes sprout growth which reduces their quality when not kept under the best storage condition after harvest thereby lower their market value. Onion (moisture content 82% wb) is highly perishable after harvest and cannot be keep more up to 30 days.

Onion is available during the production season but very scarce and expensive during the off-season. This crop contains many nutritional and medical values that are beneficial to human beings to the extent that it should be made available for both industrial and domestic uses all the year round at it nutritional qualities. Drying is the most effective and reliable unit operations of postharvest preservation of onion and other highly perishable fruits and vegetables. Therefore, the main objective of this research work was to investigate the effects of some processing parameters on the drying rate, water loss, solid gain and post-drying quality attributes of onion. The specific objectives are: (i) to investigate the effects of osmotic solution concentration (OSC) as a pre-treatment factor on drying rate, water loss, solid gain and post- drying quality attributes of onion; (ii) to investigate the effects of osmotic solution temperature (OST) as a pre-treatment factor on drying rate, water loss, solid gain and post- drying quality attributes of onion and; (iii) to investigate the effects of osmotic process duration (OPD) as a pre-treatment factor on drying rate, water loss, solid gain and post- drying quality attributes of onion.

Materials and Methods

Experimental equipment

The major equipment used for this study is an experimental dryer which was designed and built prior to this study. Other and materials included matured and firm onions, sensitive weighing balance, water baths (Shell Lab Model and HH-W420, XMTD-204 Model), thermo-hygrometer, desiccators, desiccants, stop watch with alarm, onion slicer, stainless tray, foil wrap, conical flask, measuring cylinder, NaCl and distil water.

As shown in Figures 2 and 3, the dryer consists of heating chamber having three electrical heating coils of 1.8 kW each, connected directly to a centrifugal fan 0.5 hp and drying chamber. The heating coils are connected in series and the whole unit connected to the temperature regulator (0-400°C) which controls the temperature of the heaters. The drying cabinet measures 50 cm long, 50 cm wide and 80 cm high (with external dimension of 56 cm × 56 cm × 86 cm) consisting of three set of trays separated by 15 cm clearance. The drying chamber is double walled insulated with fibre glass with a thickness of 3 cm. The drying trays having an area of 50 cm × 50 cm are made from one inch square pipe with expanded metal having an aperture wide enough to allow free flow of heated air.

The heating chamber is trapezoidal in shape with the length of the side touching the drying chamber 60 cm while the opposite side touching the fan is 20 cm. The length of the chamber is 50 cm in order to accommodate the heating elements. To ensure that the hot air touches all the products simultaneously the heating chamber opened directly into the drying chamber. To avoid moisture condensation at the top of the dryer vents are provided with the aid of two galvanized pipes of four inch diameter. This was achieved by drilling holes of about 5 mm diameter for discharge of moisture laden air and for the placement of the thermo-hygrometer probe.

Experimental design

In order to investigate the effects of the processing parameters on

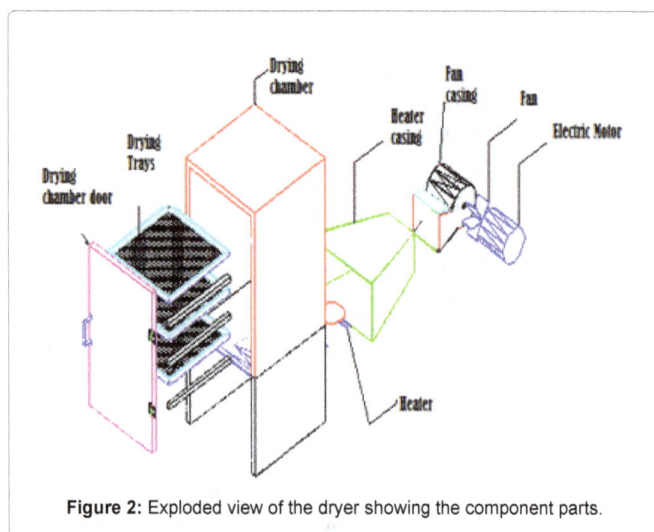

Figure 2: Exploded view of the dryer showing the component parts.

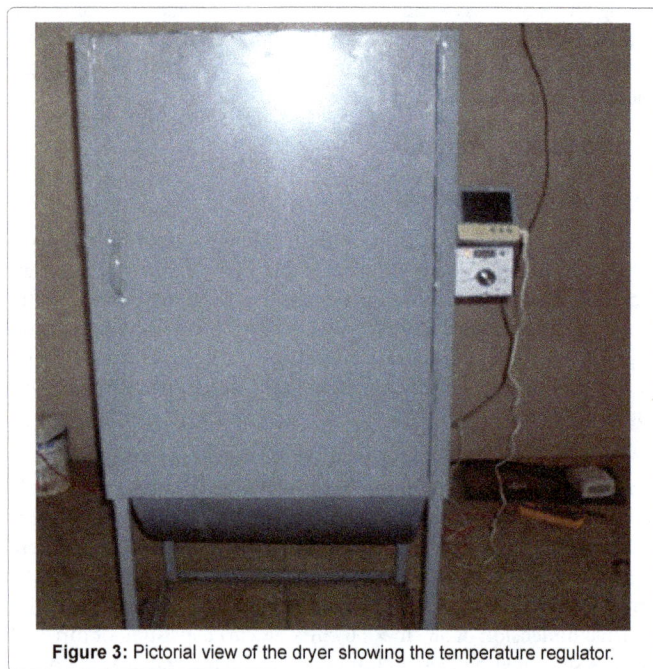

Figure 3: Pictorial view of the dryer showing the temperature regulator.

the drying rate, water loss, solid gain and post-drying quality attributes of onion, a 4 × 4 × 4 factorial experiment under randomized complete block design (RCBD) was used for the study. The design included four levels of osmotic solution concentration (5, 10, 15 and 20 g), four levels of osmotic solution temperature (35, 40, 45 and 50°C) and four levels of osmotic process duration (30, 60, 90 and 120 minutes). All tests were carried out in triplicates making a total of 192 experimental runs that were individually tested and measured.

Experimental procedure

Sample pretreatment: Fresh, mature onions that were healthy and free from mechanical injuries were purchased from Ipata market in Ilorin metropolis and kept in the Agricultural and Bio-systems Engineering Laboratory University of Ilorin to attain room temperature. Samples were weighed using a sensitive weighing balance with an accuracy of 1 g and range 0-5000 g (Figure 4).

Onions were rinsed in clean water at room temperature and cut with onion slicer to a thickness of 3 mm. 50 g of the sample were weighed and immersed in a hypertonic (NaCl) solution of four different concentrations (0.05, 0.10, 0.15 and 0.20 % w/w- mass ratio of fruit to solution was 1:2) for two simultaneous counter-current flows- an exist of water from the product to the solution and a migration of natural solids into the product. The samples were then immersed in a water bath continuously stirred to maintain a uniform temperature not higher than ± 1°C for the four temperature levels (35, 40, 45 and 50°C). Samples were removed from the osmotic solution after 30, 60, 90 and 120 minutes of immersion. All samples were drained, weighed and checked for sample weight, moisture content, and dry solids mass after osmotic dehydration pretreatment.

Drying procedure: The dryer was pre-heated to a temperature of 60°C by the means of temperature regulator while the samples were being prepared to ensure stability of the condition of the drying chamber. After arranging the trays in the dryer, the fan was switched on and set to a velocity of 0.5 m/s using the fan regulator with the speed measured with anemometer. The initial condition of the environment and the drying chamber was recorded immediately after loading. Reduction of masses of samples was noticed (less than 50 g that was introduced for osmotic dehydration). As a result of this, a uniform sample mass of 40 g was used for each sample combination, and properly re-arranged on the drying trays for drying with each experiment carried out in triplicates. The temperature of the exhaust air from the dryer was also measured and recorded. The drying samples were weighed at intervals of 1 h and drying continued until the desirable moisture contentment of 8% (db) was reached.

Output parameters

Drying rate: Drying rate is the quantity of moisture removed from products per unit time during the drying operation. In this study, drying rate was estimated by using equation 1 below:

$$\text{d} = \frac{d_m}{d_t} = \frac{m_i - m_f}{t} \qquad (1)$$

Where; d is the drying rate in g/h; d_m is change in mass of onion in g; d_t is change in time in h; t is the total time of drying in h; m_i and m_f are the initial and final mass of onion samples respectively in g.

Mass transfer: Two mass transfer parameters of onion were

Figure 4: Sample of sliced onion before dehydration.

determined during the osmotic dehydration process for each sample using Agarry et al. [8] method. The mass transfer parameters determined were water loss and solid gain.

$$WL = \frac{(M_O - m_O) - (M_t - m_t)}{M_0} \qquad (2)$$

$$SG = \frac{m_t - m_O}{M_O} \qquad (3)$$

Where; WL is the water loss in g/g; SG is the solid gain in g/g; M_0 is the initial weight of fresh onion in g; m_0 is the dry mass of fresh onion in g; M_t is the mass of onion after time t of osmotic treatment in g and m_t is the dry mass of onion after time t of osmotic dehydration pretreatment in g.

Post-drying qualities: The post-drying qualities of onion were determined at Chemistry Laboratory of the University of Ilorin using the AOAC [9] standards. The post-drying qualities determined included: Vitamin C, Manganese and Iron.

Statistical analysis: The data obtained from the experiments for drying rate, water loss, solid gain and post-drying qualities were subjected to the statistical Analysis of Variance (ANOVA) at 95% confidence level (p ≤ 0.05) using the SPSS computer software package. Further analysis by Duncan New Multiple Range Test (DNMRT) was used to compare the means among different levels of each experiment factors.

Results and Discussion

ANOVA of process variables on drying rate, water loss, solid gain and post-drying qualities

The result of the statistical analysis of variance (ANOVA) of the data obtained from the experiment is presented in table 1. From the analysis table, it is clear that all the process parameters examined had significant effect drying rate, solid gain and all the post-drying quality parameters except water loss and vitamin C. Osmotic solution concentration was significant on drying rate, solid gain and post drying quality except vitamin C and percent water loss; Osmotic solution temperature was significant on drying rate, solid gain and manganese content; osmotic process duration was not significant on drying rate, water loss but was significant on solid gain and post drying quality except vitamin C and iron; interaction of the process variables were significant on all of the output parameters except vitamin C on which all the process variables and their interactions were not significant, all at p ≤ 0.05. This implies that osmotic solution concentration, osmotic solution temperature and osmotic process duration had appreciable effects on the drying rate, water loss, solid gain and post drying attributes of onion. Therefore, while drying onion, these factors must be carefully controlled.

Effect of Osmotic solution concentration on drying rate, water loss, solid gain and post-drying qualities

The effect of osmotic solution concentration on drying rate is shown in Figure 5. The figure showed that, osmotic solution concentration had no appreciable effect on drying rate. However, the interaction between the first level of osmotic solution concentration (5% w/w) and second level of osmotic solution temperature (40°C) gave significantly higher mean drying rate. As shown in Figure 6, increase in osmotic solution concentration increased the water loss at first and second levels of osmotic solution temperature and decreased slightly with increasing osmotic solution temperature from 45 to 50°C. Soli gain slightly decreased with increasing osmotic solution temperature form 35 to 50°C (Figure 7) at all levels of osmotic solution concentration.

As indicated by Figure 8, osmotic solution concentration had no appreciable effect on vitamin C. However, the interactive effect between the third level of osmotic solution concentration (15% w/w), and second level of osmotic solution temperature (40°C) gave significantly higher mean vitamin C.

Manganese content increased with increasing osmotic solution concentration for all osmotic solution temperature (Figure 9). As seen from Figure 10, iron content at 20% w/w was higher when compared with that of 5% w/w. However, it is obvious that increase in osmotic solution concentration from 5% w/w to 20% w/w led to increase in iron content.

Effect of osmotic solution temperature on drying rate, water loss, solid gain and post-drying qualities

From table 2, it can be seen that drying rate increased as the osmotic solution temperature increased from 35°C to 40°C. This is in agreement with the finding of past researchers such as Alam et al. [3]. Duncan's new multiple range tests (Table 2) showed that all levels of osmotic solution temperature had effect on drying rate (at p ≤ 0.05). This implies that increase in osmotic solution temperature from 35°C to 40°C showed a progressive increase in drying rate. Water loss increased progressively (though not statistically significant) as the osmotic solution temperature increased from 35°C to 45°C at all level of osmotic solution concentration and osmotic process duration. However, all levels of osmotic solution temperature had effect on water loss (p ≤ 0.05). This implies that increase in osmotic solution temperature from 35 to 45°C resulted in progressive increase in water loss.

Solid gain decreased with increasing osmotic temperature for all level of osmotic solution concentration and osmotic process duration. Further analysis showed that solid gain between 45°C and 50°C were not significantly different. From the data, the highest solid gain was recorded at 35°C.

	Inputs			Interactions			
Outputs	**A**	**B**	**C**	**A × B**	**A × C**	**B × C**	**A × B × C**
Drying Rate (g/hr)	0.001*	0.001*	0.063	0.001*	0.028*	0.245	0.001*
Water Loss (g/g)	0.548	0.142	0.294	0.001*	0.017*	0.001*	0.001*
Solid Gain (g/g)	0.001*	0.001*	0.038*	0.001*	0.058	0.101	0.006*
Vitamin C (mg/100 g)	0.548	0.142	0.294	0.001*	0.000*	0.001*	0.000*
Manganese (mg/1000 g)	0.027*	0.000*	0.029*	0.005*	0.001*	0.000*	0.000*
Iron (mg/1000 g)	0.002*	0.987	0.057	0.050*	0.024*	0.021*	0.051*

*Significantly different at p ≤ 0.05

Table 1: Analysis of variance (ANOVA) of the effect of process variables on quantity and post-drying quality of onion: A-osmotic solution concentration; B-osmotic solution temperature; C-Osmotic process duration

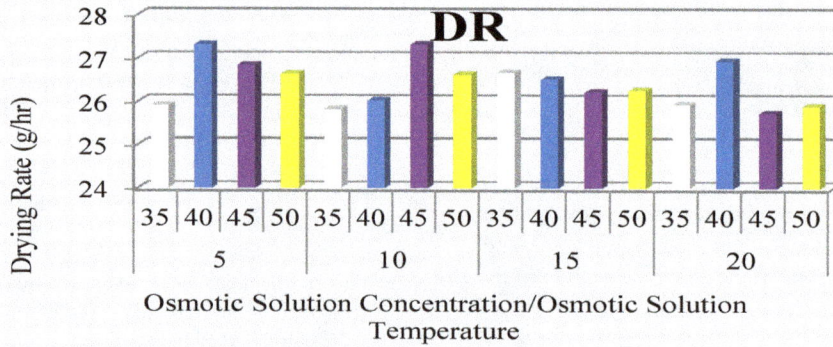

Figure 5: Effect of osmotic solution concentration on drying rate at different osmotic solution temperature.

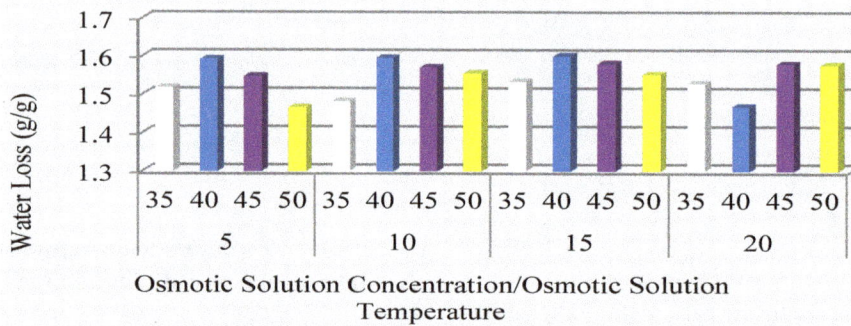

Figure 6: Effect of osmotic solution concentration on water loss at different osmotic solution temperature.

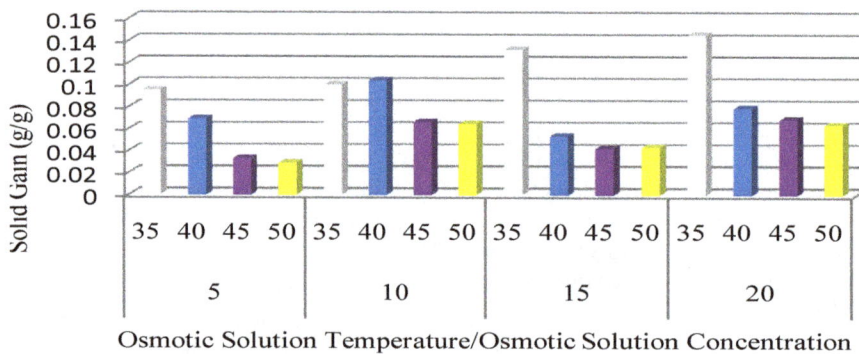

Figure 7: Effect of osmotic solution concentration on solid gain at different osmotic solution temperature.

Vitamin C between 35 and 50°C were not significantly different from each other but values at 50°C level of osmotic solution temperature was found to be the highest. Manganese increased with increasing osmotic solution temperature from 35 to 50°C but values from 40 to 50°C were not significantly different from each other. Iron content was almost the same for all osmotic solution temperature. Values of iron from 35 to 50°C were not significantly different.

Effect of Osmotic process duration on drying rate, water loss, solid gain and post drying qualities

Table 2 showed all the effect of osmotic process duration on drying rate, water loss, solid gain and post-drying qualities of onion. Drying rate increased as osmotic process duration increased from 30 to 60 minutes at different osmotic solution temperature and for all osmotic solution concentration.

Water losses were almost the same for all the process duration and were not significantly different. Values of solid gain were also almost the same for all process duration at different osmotic solution temperature for all osmotic solution concentration. Vitamin C content increased as osmotic process duration increasing from 30 to 90 minutes at different osmotic solution temperature and for all osmotic solution concentration but were not significant different. Manganese and iron

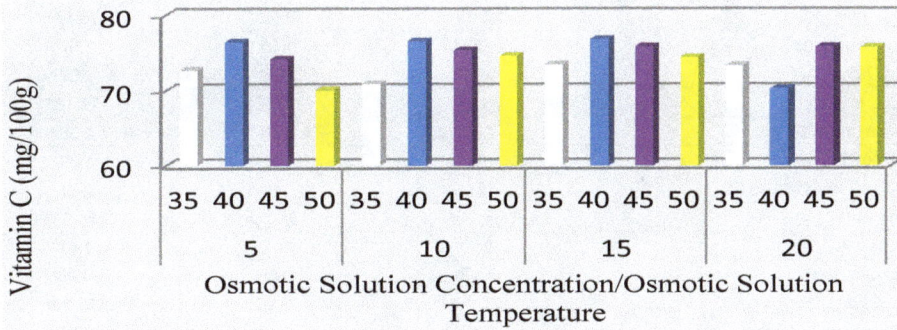

Figure 8: Effect of osmotic solution concentration on vitamin C at different osmotic solution temperature .

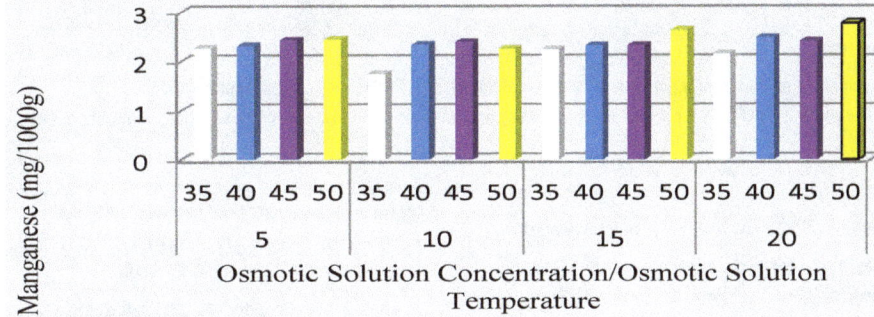

Figure 9: Effect of osmotic solution concentration on manganese at different osmotic solution temperature.

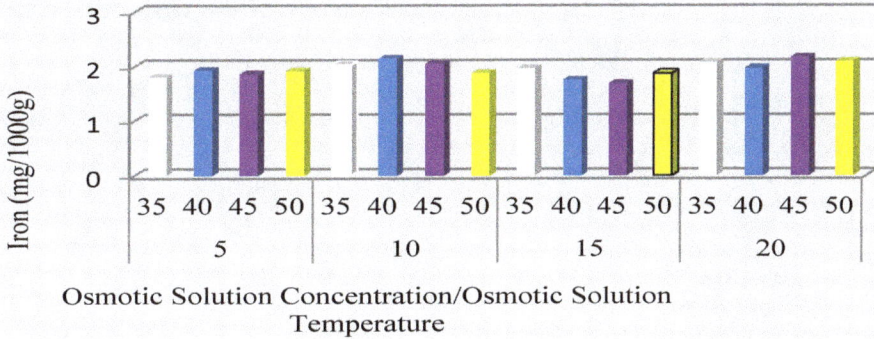

Figure 10: Effect of osmotic solution concentration on iron at different osmotic solution temperature.

content increased with increased in osmotic process duration from 30 to 60 minutes at different osmotic solution temperature and for all osmotic solution concentration.

Conclusion

Osmotic dehydration pretreatments had significant effect on process outputs (drying rate, water loss, solid gain, vitamin C, manganese and iron contents of onion.

Onion dried faster when treated 5% w/w at 50°C osmotic solution temperature for 30 minutes process duration. Vitamin C content can best be preserved in onion by drying at a temperature of 60°C, osmotic solution concentration of 5% w/w, osmotic solution temperature of 40°C and osmotic process duration of 30 minutes. Based on the result of this study, solid gain and iron content in dried onion can be maximized by drying at 60°C after pretreatment with osmotic solution concentration of 20% w/w and osmotic solution temperature of 50°C for 120 minutes process duration. Manganese was higher at drying temperature of 60°C, osmotic solution concentration of 20% w/w and osmotic solution temperature of 50°C for 30 minutes process duration. Further analysis should be carried out on shrinkage, microbial load and rehydration properties of dried onion as related to storage and preservation.

Osmotic Solution Concentration (g/g)	5	10	15	20
Drying Rate (g/h)	26.68 [a]	26.45 [b]	26.41 [b]	26.12 [c]
Water Loss (g/g)	1.53 [a]	1.55 [a]	1.57 [a]	1.54 [a]
Solid Gain (g/g)	0.06 [a]	0.08 [b]	0.07 [a]	0.09 [b]
Vitamin C (mg/100 g)	73.33 [a]	74.35 [a]	75.13 [a]	73.75 [a]
Manganese (mg/1000 g)	2.36 [a]	2.18 [b]	2.38 [a]	2.45 [a]
Iron (mg/1000 g)	1.88 [a]	2.03 [b]	1.83 [a]	2.08 [b]
Osmotic Solution Temperature (°C)	**35**	**40**	**45**	**50**
Drying Rate (g/h)	26.08 [a]	26.70 [b]	26.53 [c]	26.36 [c]
Water Loss (g/g)	1.52 [a]	1.56 [a]	1.57 [a]	1.54 [a]
Solid Gain (g/g)	0.12 [a]	0.08 [b]	0.05 [c]	0.05 [c]
Vitamin C (mg/100 g)	72.56 [a]	75.02 [a]	75.28 [a]	73.70 [a]
Manganese (mg/1000 g)	2.10 [a]	2.36 [b]	2.39 [b]	2.53 [b]
Iron (mg/1000 g)	1.97 [a]	1.96 [a]	1.95 [a]	1.95 [a]
Osmotic Process Duration (min)	**30**	**60**	**90**	**120**
Drying Rate (g/h)	26.58 [a]	26.31 [b]	26.46 [a]	26.31 [b]
Water Loss (g/g)	1.54 [a]	1.53 [a]	1.57 [a]	1.55 [a]
Solid Gain (g/g)	0.07 [a]	0.08 [b]	0.07 [a]	0.07 [a]
Vitamin C (mg/100 g)	73.61 [a]	73.06 [a]	75.43 [a]	74.46 [a]
Manganese (mg/1000 g)	2.22 [a]	2.36 [a]	2.32 [a]	2.49 [b]
Iron (mg/1000 g)	1.98 [a]	2.02 [a]	1.83 [b]	1.99 [a]

*Means with the same letter are not significantly different but means with different letters are significantly different (at $p \leq 0.05$).

Table 2: Duncan's Multiple Range Test (DNMRT) for the effect of process conditions on drying rate, water loss, solid gain and post drying qualities of onion.

References

1. FAO (2003) A global review of area and production of major vegetables crops.

2. Wack AL (1998) Soaking process in tenary liquids: experimental study of mass transport under natural and forced convection. Journal of food engineering 37: 451-469.

3. Alam MM, Islam MN (2013) Effect of process parameters on the effectiveness of osmotic dehydration of summer onion. International food research Journal 20: 391-396.

4. Asiru WB, Raji AO, Igbeka JC, Elemo GN (2011) Effect of drying temperature and drying duration on colour changes in cashew kernels during hot air drying. NIAE, Ilorin 32: 421-426.

5. Afolabi TJ, Adeniyi AG (2011) Numerical modeling of mass transfer during osmotic dehydration of red paprika. NIAE, Ilorin 32: 684-689.

6. Kalse SB, Patil MM, Jain SK (2012) Microwave drying of onion slices. Research Journal of chemical sciences 2: 57-60.

7. Rzepecka MA, Brygidyr AM, McConnell MB (1976) Foam-mat dehydration of tomato paste using microwave energy. Can Agric Eng 18: 36-40.

8. Agarry SE, Yusuf RO, Owabor CN (2008) Mass transfer in osmotic dehydration of potato. A mathematical model approach. Journal of Engineering and applied sciences 3: 190-198.

9. AOAC (1990) Official Methods of Analysis of the Association of Official Analytical Chemist. Arlinton, Virginia, USA.

15

Effect of Processing on the Characteristics Changes in Barnyard and Foxtail Millet

Nazni P* and Shobana Devi R

Department of Food Science and Nutrition, Periyar University, Salem, India

Abstract

Introduction: Millet grains, before consumption and for preparing food, are usually processed by commonly used traditional processing techniques to improve their edible, nutritional and sensory properties.

Background: The processing techniques aim to increase the physicochemical accessibility of micronutrients, decrease the content of anti-nutrients or increase the content of compounds that improve bioavailability.

Objectives: Thus, an attempt was made in the present study with the objectives to study the effects of boiling, pressure cooking, roasting and germination on functional, nutritional, anti-nutritional and pasting properties of barnyard millet and foxtail millet.

Materials and Methods: Physical properties of unprocessed millets, chemical, functional, anti-nutritional and pasting properties of both unprocessed and processed millets were analysed using standard techniques.

Results and Conclusion: Physical characteristics such as thousand grain weight, thousand grain volumes, hydration capacity and index, swelling capacity and index and cooking quality of the selected two unprocessed millets were considerably differed from each other. There was a significant variation with respect to functional, nutritional, anti-nutritional and pasting properties of selected two millets in response to different processing methods. Among them, germination reduces the anti-nutritional factors while roasting significantly increases the nutritional compounds. The improved functional and pasting properties of the selected two millets were observed in the germinated and roasted millet flours.

Keywords: Barnyard millet; Foxtail millet; Functional properties; Nutritional properties; Anti nutritional parameters and pasting properties

Introduction

Millet is a collective term referring to a number of small-seeded annual grasses that are cultivated as grain crops, primarily on marginal lands in dry areas in temperate, subtropical and tropical regions. The most important species are barnyard millet and foxtail millet. Foxtail millet is important in parts of Asia (mainly China) and Europe. The other species barnyard millet is locally important food grains restricted to smaller regions or individual countries. The various species differ in their physical characteristics, quality attributes, soil and climatic requirements and growth duration [1].

Cereal grains are the most important source of the world's food and have a significant role in the human diet throughout the world. Millets are also rich sources of phytochemicals and micronutrients [2]. As a food source it is non-glutinous and non- acid forming, so is soothing and easy to digest. It is considered to be one of the least allergenic grains. It is high in quality protein, contains high fibre, B-complex vitamins and also the vitamins A and E [3]. Millets may serve as a natural source of antioxidants in food applicants and as a nutraceutical and functional food ingredient in health promotion and disease risk reduction. Millets are also rich sources of non-nutritional components like phenols, tannins, phytate and flavonoids. These compounds serves as an antioxidant and millets could also be used as source of extremely beneficial phytochemicals in the pharmaceutical and food industry [4]. Therefore, millet grains are now receiving specific attention from the developing countries in terms of utilization as food as well as from some

developed countries in terms of its good potential in the manufacturing of bioethanol and biofilms [5].

Millet grains, before consumption and for preparing food, are usually processed by commonly used traditional processing techniques to improve their edible, nutritional and sensory properties [6]. The processing techniques aim to increase the physicochemical accessibility of micronutrients, decrease the content of antinutrients or increase the content of compounds that improve bioavailability [7]. Thus, an attempt was made in the present study with the objectives to study the effects of wet heat treatment (boiling and pressure cooking), dry heat treatment (roasting) and germination on physicochemical, functional, anti-nutritional and pasting properties of barnyard millet, and foxtail millet.

Materials and Methods

The selected two millets namely barnyard millet and foxtail millet were procured from local market of Vellore district, Tamil Nadu, India. The millets were cleaned properly and stored in sealed containers.

*Corresponding author: Nazni P, Department of Food Science and Nutrition, Periyar University, Salem, India, E-mail: naznip@gmail.com

Physical characterization of selected two unprocessed millet grains

Physical characteristics such as thousand grain weights, thousand grain volumes, hydration capacity and index, swelling capacity and index, cooking quantity/characteristics were analyzed for the selected two unprocessed millet grains using standard procedures in triplicates.

Processing of selected two millet grains in to flours

Food processing is the transformation of raw ingredients, by physical or chemical means into food, or of food into other forms. It is widely accepted that simple and inexpensive traditional processing techniques are effective methods of achieving desirable changes in the composition of grains. The processing methods such as boiling, germination, pressure cooking and roasting were implemented with raw sample to test the effect of processing on functional, chemical, anti-nutritional and pasting characteristics of millet grains.

Functional properties of raw and processed millet flours

The functional properties such as bulk density, swelling power, solubility, solid loss, water absorption capacity and oil absorption capacity were analyzed using standard procedure in triplicates. Bulk density of each sample was determined by the method described by Ige et al. [8]. Swelling power, solubility and solid loss of the raw and processed millet flours was determined using the modified method of Shin et al. [9]. The water and oil absorption capacity were determined using the method described by Lin et al. [10].

Nutrient analysis of raw and processed millet flours

Proximate composition such as pH, moisture, ash, total titrable acidity, crude protein, crude fibre, carbohydrate, fat, energy, amylose content, total starch and mineral composition like sodium, potassium, calcium, iron and phosphorus were determined for the selected two raw and processed millet flours using standard procedures.

Anti nutritional parameters of raw and processed millet flours

Anti-nutrients are natural or synthetic compounds that when present in foods reduce the availability of the nutrients. The anti-nutritional parameters such as tannin, total phenolics and trypsin inhibitor were determined using standard procedures as described by Sadasivam and Manickam [11].

Pasting properties of raw and processed millet flours

Pasting properties of the selected two raw and processed millet flours were analyzed using Rapid Visco Analyzer (RVA) as described by Tester and Morrison [12]. Parameters recorded were pasting temperature, peak viscosity, trough viscosity (minimum viscosity at 95°C), final viscosity (viscosity at 50°C), breakdown viscosity (peak-trough viscosity), and setback viscosity (final-trough viscosity). The pasting properties of each sample were determined from 3g of flour (dry basis) in 25ml of distilled water by using a Rapid Visco Analyzer (RVA) model 3D (Newport Scientific Pty. Ltd., Australia). The sample was heated from 50 to 95°C at the rate of 12°C/min with constant stirring at 160 rpm and held at 95°C for 2.5minutes (break down), then cooled at 50°C at the rate of 13°C/min (set back) and held for 2 minutes. The total cycle was 12.5 minutes. Pasting temperature was recorded as the temperature at which an increase in viscosity was first observed. The values reported included pasting temperature (°C), Peak viscosity (cP), final viscosity (cP), trough (lowest viscosity, cP), break down

(difference between peak viscosity and trough, cP), set back from peak (the difference between final viscosity and peak viscosity, cP) and set back from trough (the difference between final viscosity and trough, cP).

Statistical analysis

The final data was compiled and analyzed using suitable statistical methods. The results were represented as descriptive statistics such as mean, standard deviation and one way ANOVA. pvalues < 0.05 were considered significant. Analysis of variance (ANOVA) was used to test the differences among different processed millet flours. The data reported in tables are an average of triplicate observations subjected to one-way analysis of variance (ANOVA).

Results and Discussion

Physical characteristics of selected two unprocessed millet grains

The physical characteristics such as thousand grain weight, thousand grain volume, hydration capacity and index, swelling capacity and index and cooking quantity/characteristics of selected two unprocessed millets grains were presented in Table 1. The thousand grain weight of barnyard millet and foxtail millet were 3.69 g and 2.64 g respectively. The thousand grain volume of barnyard millet was 4.01 ml and 2.96 ml for foxtail millet. Grain volumes change significantly and most often, regularly at varying moisture contents.

The hydration capacity of the barnyard millet and foxtail millet were 2.93 g/1000 seeds and 2.01 g/1000 seeds with the hydration index of 78.39% and 76.9% respectively. The barnyard millet grain was found to have swelling capacity of 0.23 ml/1000 seeds with an index of 5.83% and foxtail millet grain was observed to have swelling capacity of 0.21 ml/1000 seeds with an index of 6.72%. The presence of high protein, lipid, fiber and larger amount of amylose-lipid complex in flour could inhibit the swelling of starch granules [13].

Cooking quality of the grain is an important criterion to assess the consumer acceptability [14]. The barnyard millet and foxtail millet grains required 12 minutes to get cooked. The increase in the weight of barnyard millet and foxtail millet grains after cooking were 175.6% and 197.6% respectively. The increase in the volume of barnyard millet and foxtail millet grains after cooking were 41.2% and 75.5% respectively.

Functional properties of selected two raw and processed millet flours

The functional properties such as bulk density, swelling power, solubility, solid loss, water absorption capacity and oil absorption capacity of the selected two raw and processed millet flours were discussed below.

S.no	Parameters	Barnyard millet	Foxtail millet
1.	Thousand grain weight (g)	3.69 ± 0.01	2.64 ± 0.005
2.	Thousand grain Volume (ml)	4.01 ± 0.01	2.96 ± 0.05
3.	Hydration Capacity (g/1000 seeds)	2.93 ± 0.05	2.01 ± 0.02
4.	Hydration Index (%)	78.39 ± 0.1	76.9 ± 0.05
5.	Swelling Capacity (ml/1000 seeds)	0.23 ± 0.05	0.21 ± 0.02
6.	Swelling Index (%)	5.83 ± 1.44	6.72 ± 0.02
7.	Gain in weight after cooking (%)	175.6 ± 10.6	197.6 ± 14.0
8.	Gain in volume after cooking (%)	41.2 ± 3.76	75.5 ± 2.07
9.	Cooking time (minutes)	12 minutes	12 minutes

Table 1: Physical characteristics of selected two unprocessed millet grains.

Bulk Density of the raw and processed millet flours

The bulk density of the selected two raw and processed millet flours was presented in Table 2. Bulk density is an important parameter that determines the packaging requirement of a product [15]. In comparison with raw millet flours, the bulk density of the boiled, pressure cooked and roasted samples were significantly increased (p < 0.05) whereas the bulk density of the germinated millet flours of selected two millets was decreased. The decreased bulk density of the germinated millet flour indicates low porosity or air spacing in the flour, therefore less auto-oxidation. This is an advantage in respect to spoilage, packing and transportation as goods in relation to weight [16].

Swelling power of the raw and processed millet flours

Swelling and water absorption capacities are important parameters in determining the sample consistency (solid, semi-solid and liquid) and are dependent on the compositional structure of the sample [17]. The swelling power of the selected two raw and processed millet flours was given in Table 3. The swelling power of the selected two processed millet flours at 90°C were consistently reduced when compared with raw samples and does not varied significantly with each other (p > 0.05). The swelling behaviour below 16 g/g is considered as highly restricted. This restricted swelling behaviour of the flour samples indicates its stability against shearing action when subjected to heat [18]. Therefore, in the present study, the selected two raw and processed millet flours showed restricted swelling behaviour indicate its resistant power towards heating.

Solubility behaviour of the raw and processed millet flours

Solubility is an indicator of the degree of starch granules dispersion after cooking [19]. Solubility behaviour of the selected two raw and processed millet flours at 90°C was presented in Table 4. Among all the processing techniques of selected two millets, germinated millet flour contained highest solubility behaviour. The solubility could imply to the amount of amylose leaching out from starch granule when swelling, therefore the higher the solubility the higher will be the amylose leaching [20]. Difference in solubility could also be attributed to different chain length distribution in the starch [21].

Solid loss of the raw and processed millet flours

Solid loss of the selected two raw and processed millet flours was shown in the Table 5. The processed barnyard millet flour exhibited loss of solids to an extent of 27.9% in boiled, 26.5% in germinated, 29.1% in pressure cooked, 30.1% in roasted and 30.7% in raw samples. The cooking of processed foxtail millet flour resulted in 27.5% of solid loss in boiled, 27% in germinated, 30.5% in pressure cooked, 28.4% in roasted and 29.3% in raw samples. A significant difference in solid loss was observed between different processed barnyard and foxtail millet flours (p < 0.05).

S.no	Processing techniques	Barnyard millet (g/ml)	Foxtail millet (g/ml)
1.	Boiling	0.62 ± 0.005[a]	0.70 ± 0.005[a]
2.	Germination	0.44 ± 0.005[b]	0.43 ± 0.005[b]
3.	Pressure cooking	0.52 ± 0.03[c]	0.53 ± 0.01[c]
4.	Roasting	0.55 ± 0.005[c]	0.53 ± 0.01[c]
5.	Raw	0.50 ± 0.005[c]	0.52 ± 0.005[c]

The values are expressed as the mean of three replicate samples ± SD. Values in a column with different superscripts differ significantly (p < 0.05).

Table 2: Bulk Density of the raw and processed millet flours.

S.no	Processing techniques	Barnyard millet (g/g)	Foxtail millet (g/g)
1.	Boiling	4.73 ± 0.3	4.83 ± 0.45
2.	Germination	4.53 ± 0.35	4.63 ± 0.4
3.	Pressure cooking	4.7 ± 0.17	5.16 ± 0.4
4.	Roasting	4.96 ± 0.15	4.9 ± 0.3
5.	Raw	5.2 ± 0.2	5.23 ± 0.4

The values are expressed as the mean of three replicate samples ± SD. Values in a column do not differ significantly (p > 0.05).

Table 3: Swelling power of the raw and processed millet flours.

S.no	Processing techniques	Barnyard millet (%)	Foxtail millet (%)
1.	Boiling	6.3 ± 0.41[a]	6.3 ± 0.55[a]
2.	Germination	6.8 ± 0.41[a]	7.1 ± 0.55[b]
3.	Pressure cooking	6.7 ± 0.60[a]	5.8 ± 0.60[c]
4.	Roasting	6.3 ± 0.55[a]	5.4 ± 0.45[d]
5.	Raw	6.2 ± 0.26[a]	6.2 ± 0.55[a]

The values are expressed as the mean of three replicate samples ± SD. Values with similar superscripts in a column do not differ significantly. Values with different superscripts in a column differ significantly.

Table 4: Solubility behaviour of the raw and processed millet flours.

S.no	Processing techniques	Barnyard millet (%)	Foxtail millet (%)
1.	Boiling	27.9 ± 1.45[a]	27.5 ± 0.87[a]
2.	Germination	26.5 ± 1.30[b]	27 ± 1.76[b]
3.	Pressure cooking	29.1 ± 1.55[c]	30.5 ± 1.21[c]
4.	Roasting	30.1 ± 1.34[d]	28.4 ± 0.96[d]
5.	Raw	30.7 ± 0.90[e]	29.3 ± 0.86[e]

The values are expressed as the mean of three replicate samples ± SD. Values with different superscripts in a column differ significantly.

Table 5: Solid loss of the raw and processed millet flours.

Water absorption capacity of the raw and processed millet flours

The water absorption capacity of the selected two raw and processed millet flours was presented in the Table 6. Water absorption capacity is important in the development of ready to eat foods, and high absorption capacity may assure product cohesiveness [22]. The variations in water absorption capacity of the selected two processed millets is highly attributed by several factors such as number of hydration positions, physical environment, pH, solvent, presence of lipids and carbohydrates [23]. In the present study, water absorption capacity was significantly increased (p < 0.05) in boiled, pressure cooked and roasted samples compared to raw samples, but germinated millet flours of selected two millets have least water absorption capacity. When the lipid content is high in the flour, the water absorption decreases because lipids block the polar sites of the proteins attenuating the absorption of water [24].

Oil absorption capacity of the raw and processed millet flours

Table 7 shows the oil absorption capacity of the selected two raw and processed millet flours. The high oil absorption capacity makes the flours suitable for facilitating enhancement in flavour and mouth feel when used in food preparations [25]. Oil absorption capacity was significantly increased (p < 0.05) in boiled, germinated; pressure cooked and roasted millet flours compared to raw samples. Variation in fat absorption may be due to the variation in protein concentration, degree of interaction with water and oil and conformational characteristics [26].

Nutrient analysis of selected two raw and processed millet flours

Proximate composition such as pH, moisture, ash, total titrable acidity, crude protein, crude fibre, carbohydrate, fat, energy, amylose content, total starch and mineral composition like sodium, potassium, calcium, iron and phosphorus analyzed for the selected two raw and processed millet flours were discussed below.

Proximate composition of selected two raw and processed millet flours

The proximate composition of raw and processed barnyard millet flour was tabulated in Table 8. The pH value of the boiled, germinated, pressure cooked, roasted and raw samples were 7, 6.5, 6.8, 6.7 and 7.1 respectively. The total ash content of the raw and processed barnyard millet flour ranged from 2 g to 2.5 g. The ash content of the samples is the reflection of minerals. Hence the barnyard millets are rich in minerals. The highest total titrable acidity was noticed in germinated barnyard millet flour 0.72 g and lowest value was seen in pressure cooked sample 0.32 g. The higher the moisture content the lower the dry matter yield on drying [27]. The raw barnyard millet flour contained high moisture content of 10 g, followed by germinated 9.5 g, boiled 8 g, pressure cooked 7.4 g and roasted 6.2 g [28-31], found that total protein

S.no	Processing techniques	Barnyard millet (g/g)	Foxtail millet (g/g)
1.	Boiling	2.40 ± 0.02ᵃ	1.86 ± 0.04ᵃ
2.	Germination	1.12 ± 0.01ᵇ	0.95 ± 0.02ᵇ
3.	Pressure cooking	2.31 ± 0.02ᶜ	2.06 ± 0.04ᶜ
4.	Roasting	1.59 ± 0.01ᵈ	1.39 ± 0.06ᵈ
5.	Raw	1.24 ± 0.04ᵉ	1.18 ± 0.05ᵉ

The values are expressed as the mean of three replicate samples ± SD. Values with different superscripts in a column differ significantly.

Table 6: Water absorption capacity of the raw and processed millet flours.

S.no	Processing techniques	Barnyard millet (g/g)	Foxtail millet (g/g)
1.	Boiling	1.14 ± 0.03ᵃ	1.17 ± 0.04ᵃ
2.	Germination	1.12 ± 0.03ᵇ	1.26 ± 0.02ᵇ
3.	Pressure cooking	1.39 ± 0.05ᶜ	1.15 ± 0.03ᶜ
4.	Roasting	1.16 ± 0.05ᵈ	1.12 ± 0.03ᵈ
5.	Raw	1.10 ± 0.01ᵉ	1.06 ± 0.02ᵉ

The values are expressed as the mean of three replicate samples ± SD. Values with different superscripts in a column differ significantly.

Table 7: Oil absorption capacity of the raw and processed millet flours.

S.no	Parameters	Boiling	Germination	Pressure cooking	Roasting	Raw
1.	pH	7.00	6.5	6.8	6.7	7.1
2.	Ash (g)	2.5	2.0	2.4	2.2	2.3
3.	Total titrable Acidity (g)	0.48	0.72	0.32	0.36	0.46
4.	Moisture (g)	8	9.5	7.4	6.2	10
5.	Crude Protein (g)	7.4	8.9	8.3	12.8	11.2
6.	Crude Fibre (g)	6.4	5.7	7.2	7.0	5.3
7.	Carbohydrates (g)	72	61	80	75	72
8.	Fat (g)	3.4	2.5	3.0	1.9	1.8
9.	Energy (Kcals)	348.2	302.1	380.2	368.3	349
10.	Total Starch (g)	17.2	7.8	13.59	10.32	12.6
11.	Amylose content (%)	15.22	10. 8	12	9.6	8

Table 8: Proximate composition of the raw and processed barnyard millet flour (per 100 gms).

increased after germination process. In this study, the results showed that roasted barnyard millet flour contained high protein content (12.8 g) when compared to other processed samples.

The fibre content of the raw and processed barnyard millet flour ranged from 5.3 g to 7.2 g, being the highest for pressure cooked flour 7.2 g and lowest for raw flour 5.3 g. It is generally accepted that the consumption of food naturally rich in dietary fibre is beneficial to the maintenance of health [32]. The carbohydrate content of the boiled barnyard millet flour was 72 g, 61 g for germinated, 80 g for pressure cooked, 75 g for roasted and 72 g for raw samples. The carbohydrate content of the germinated barnyard millet flour was comparatively lesser than the other processed samples. This results were in agreement with Mubarak [33], reported that germinated samples showed a significant reduction in the total carbohydrate. The decrease in the carbohydrate during the process of germination is due to the use of carbohydrate for metabolism by the sprouts [34].

Among the fat content of barnyard millet flour, the roasted and raw flour contained less fat content of 1.9 g and 1.8 g respectively. This results confirms with the report of Aremu, et al. [35] stated that kersting groundnut and cranberry bean upon roasting were found to greatly reduce crude fat content. The energy value of raw and processed flour varied from 302.1 kcals to 380.2 kcals. The lowest energy value was observed in germinated flour (302.1 kcals) and the highest energy value was observed in pressure cooked flour (380.2 kcals) due to its high carbohydrate and fat content. The total starch content of raw and processed barnyard millet flour ranged from 7.8 g to 17.2 g, being the highest in boiled 17.2 g and lowest in germinated flour 7.8 g. Germination of proso millet grains decreased the dry weight and the total starch content [36]. All the processing methods were exhibited increased amylose content of barnyard millet flour compared to raw flour. The boiled sample contained high amylose content of 15.22%, followed by pressure cooked 12%, germinated 10.8%, roasted 9.6% and raw sample of 8%.

Grains are the store houses of many chemical components including nutrients, phytochemicals, and non-nutritive plant protective functional constituents. The nutritional content of the raw and processed foxtail millet flour was tabulated in Table 9. The pH value of boiled, germinated, pressure cooked, roasted and raw foxtail millet flour was 6.7, 7.2, 6.5, 6.2 and 7.0 respectively. The ash content of the raw and processed foxtail millet flour ranged from 2.1 g to 4.2 g. This indicates the high mineral content of foxtail millet flour. Ash is the inorganic residue remaining after the water and organic matter have been removed by heating in the presence of oxidizing agents which provides a measure of the total amount of minerals within a food. Regarding the total titrable acidity, the highest value was found in germinated sample (0.55 g), followed by 0.54 g in raw, 0.45 g in roasted, 0.34 g in boiled and 0.32 g in pressure cooked samples. The moisture content of boiled (6.46 g), germinated (11.56 g), pressure cooked (7.3 g) and roasted foxtail millet flour (7.56 g) was comparatively lesser than the raw millet flour (15 g). This indicates the less susceptibility to microbial infection. The high protein content was seen in germinated flour (12.6 g) were in agreement with the report of Srichuwong et al. [37] stated that process of germination greatly attributed protein increase to protein synthesis due to inclusion of microbial cells in to the flour.

The crude fibre content ranged from 4.9 g to 6.1 g. processing of foxtail millet flour showed an increase in the fibre content compared to raw flour. Germination can increase dietary fibre content than other processing and increase mineral bioavailability [38,39]. The carbohydrate content of processed foxtail millet flour was significantly reduced when compared to raw millet flour except germinated sample which exhibits higher content of carbohydrates (72 g). Fat content was increased in roasted sample (3.2 g), followed by raw (2.9 g), pressure cooked (2.8 g), germinated (2.6 g) and boiled sample (1.9 g). The calorific value of foxtail millet flour exhibited 302.7 kcals in boiled, 361.8 kcals in germinated, 332 kcals in pressure cooked, 294.4 kcals in roasted and 333.3 kcals in raw samples. The high calorific value of germinated flour was due to its high protein and high carbohydrate content. The total starch content of the processed foxtail millet flour was drastically reduced compared to raw flour. It has been observed as 13.6 g of starch in boiled, 11.9 g in germinated, 8.62 g in pressure cooked, 7.85 g in roasted and 18.5 g in raw kodo millet flour. The pressure cooked sample contained high level of amylose (12.4%), followed by germinated (10.8%), boiled (6.8%), roasted (4.4%) and raw flour (4%).

Mineral composition of selected two raw and processed millet flours

The mineral composition of raw and processed barnyard millet flour were shown in the Table 10. The sodium content of the raw and processed barnyard millet flour ranged from 13 mg to 20 mg, being the highest for pressure cooked flour (20 mg) and lowest for boiled barnyard millet flour (13 mg). The potassium content of processed barnyard millet flour was significantly reduced compared to raw flour. Boiled sample has 210 mg of potassium, 215 mg in germinated, 225 mg in pressure cooked, 248 mg in roasted and 298 mg in raw samples. Iron content was decreased when grains were roasted [40]. This statement was in agreement with the current study findings revealed that roasted barnyard millet flour exhibited less iron content (3.28 mg) than other processing techniques. The calcium and phosphorus content of the processed barnyard millet flour were decreased when compared with raw millet flour. Processing techniques were considerably affected the calcium and phosphorus content of the barnyard millet. It has been

observed as 30 mg of calcium in boiled, 50 mg in germinated, 32 mg in pressure cooked, 40 mg in roasted and 60 mg in raw millet flour. Whereas phosphorus content of boiled barnyard millet flour was 252 mg, 210 mg in germinated, 227 mg in pressure cooked, 234 mg in roasted and 252 mg in raw flour.

Table 11 showed the mineral content of the raw and processed foxtail millet flour. When compared to raw flour, the sodium and potassium content of the processed foxtail millet flour were considerably increased. The sodium and potassium content was ranged from 8 mg to 25 mg and 252 mg to 356 mg respectively. Germinated foxtail millet flour found to have high iron content (5.8mg) followed by pressure cooked 3.8 mg, raw 3.7 mg, boiled 2.59 mg and roasted 2.4 mg. Germination or malting generally improves the nutrient content and digestibility of foods and it could be an appropriate food-based strategy to derive iron and other minerals maximally from food grains [41]. In this study, processing exhibited desirable changes in the calcium and phosphorus content. The calcium content of the raw and processed foxtail millet flour ranged from 31 mg to 60 mg, whereas the phosphorus content varied from 221 mg to 280 mg.

Anti-nutritional parameters of selected two raw and processed millet flours

The anti-nutritional parameters of the selected two raw and processed millet flours were analyzed to determine the effect of processing on anti-nutrients such as tannin, total phenolics and trypsin inhibitor.

Anti-nutritional parameters of raw and processed barnyard millet flour were shown in Figure 1. An increase in the tannin, total phenolics and trypsin inhibitor levels was noticed in processed barnyard millet flour when compared to raw flour. This was in agreement with the study conducted by Seifi et al. [42] also revealed that during processing, an increase in the content of tannin, total phenolics and trypsin inhibitor activity was noticed in little millets compared to the native millets.

The anti-nutritional parameters of raw and processed foxtail millet flour were shown in Figure 2. Among the different processing techniques, the pressure cooked foxtail millet flour showed high level of tannin content (0.41 mg/g). The total phenolics content was found to be high in roasted (68.1 mg/g) and pressure cooked (67.2 mg/g) foxtail millet flour. Longer the cooking time, the greater losses of the total phenolic compound measured. This could be due to the breakdown of phenolics or losses (leached out) during cooking as most of the bioactive compounds are relatively unstable to heat and easily solubilised. The trypsin inhibitor content of processed foxtail millet flour was considerably reduced with that of raw flour. Cooking by boiling, germinating and frying resulted in a significant reduction in the trypsin inhibitor content of tomatoes [43].

Pasting properties of the selected two raw and processed millet flours

Pasting properties are the important factors in determining the application values of flours and starches [44]. Tables 12 and 13 shows the pasting properties of the raw and processed barnyard millet flour and foxtail millet flour respectively. Pasting temperature is an indication of minimum temperature required for cooking the samples [45]. Pasting temperature of the raw and processed barnyard millet flour ranged from 48.25°C to 94.45°C, being the highest for roasted sample (94.45°C) and lowest for boiled sample (48.25°C). Pasting temperature of the raw and processed foxtail millet flour ranged from 48.5°C to 89.6°C being the highest for raw foxtail millet flour (89.6°C)

S.no	Parameters	Boiling	Germination	Pressure cooking	Roasting	Raw
1.	pH	6.7	7.2	6.5	6.2	7.0
2.	Ash (g)	2.8	4.2	3.4	2.6	2.1
3.	Total titrable Acidity (g)	0.34	0.55	0.32	0.45	0.54
4.	Moisture (g)	6.46	11.56	7.3	7.56	15
5.	Crude Protein (g)	8.4	12.6	7.7	10.4	6.8
6.	Crude Fibre (g)	6.1	5.1	5.9	5.1	4.9
7.	Carbohydrates (g)	63	72	69	56	70
8.	Fat (g)	1.9	2.6	2.8	3.2	2.9
9.	Energy (Kcals)	302.7	361.8	332	294.4	333.3
10.	Total Starch (g)	13.6	11.9	8.62	7.85	18.5
11.	Amylose content (%)	6.8	10.8	12.4	4.4	4

Table 9: Proximate composition of the raw and processed foxtail millet flour (per 100 gms).

S.no	Minerals	Boiling	Germination	Pressure cooking	Roasting	Raw
1.	Sodium (mg)	13	17	20	14	18
2.	Potassium (mg)	210	215	225	248	298
3.	Iron (mg)	3.3	7.59	8.9	3.28	9.0
4.	Calcium (mg)	30	50	32	40	60
5.	Phosphorus (mg)	252	210	227	234	252

Table 10: Mineral composition of the raw and processed barnyard millet flour (per 100 gms).

and lowest for both boiled and pressure cooked foxtail millet flour (48.5°C).

Peak viscosities attained during the heating portion of tests indicates the water binding capacity of starch mixture. This often correlates with the final product qualities [46]. High peak viscosity indicates the high swelling capacity of the starch granules. High peak viscosity of the raw barnyard millet flour (854cP) and roasted foxtail millet flour (1143cP) indicates its high water binding capacity resulting in more swelling of the starch granules. The ranges observed in trough viscosity of the raw and processed barnyard millet flour and foxtail millet flour were 21 to 741cP and 29 to 420cP respectively.

The breakdown is caused by the disintegration of gelatinized starch granules structure during continued stirring and heating, thus, indicating the shear thinning property of starch [47]. The low breakdown viscosity of both boiled and pressure cooked barnyard millet flour and foxtail millet flour (2cP) indicates its stability of the starches under hot conditions. Final viscosity indicates the ability of the starch to form a viscous paste. The high final viscosity of the raw barnyard millet flour (1235cP) and foxtail millet flour (1398cP) indicates its high resistance to shear.

The setback viscosity is an index of retrogradation. The setback viscosity is the increased in viscosity resulting from the rearrangement of amylose molecules that have leached out from the swollen starch granules during cooling and is generally used as a measure of gelling ability or retrogradation tendency of the starch [48]. The lower setback value of boiled barnyard millet flour (9cP) and roasted foxtail millet flour (-107cP) indicated its lowest rate of retrogradation and hence the product made of low set back viscosity flour will have prolonged shelf life period.

S.no	Minerals	Boiling	Germination	Pressure cooking	Roasting	Raw
1.	Sodium (mg)	22	25	19	16	8
2.	Potassium (mg)	294	320	275	356	252
3.	Iron (mg)	2.59	5.8	3.8	2.4	3.7
4.	Calcium (mg)	40	50	60	31	40
5.	Phosphorus (mg)	221	265	247	280	238

Table 11: Mineral composition of the raw and processed foxtail millet flour (per 100 gms).

Conclusion

There was a significant variation with respect to functional, nutritional, anti-nutritional and pasting properties of selected two millets in response to different processing methods. Among them, germination reduces the anti-nutritional factors while roasting significantly increases the nutritional compounds. The improved functional and pasting properties were observed in the germinated and roasted millet flours that make them good base ingredients in infant food formulation. Considering the above mentioned benefits, it is recommended that the germinated and roasted millet flours would be of use in food systems where these properties are required.

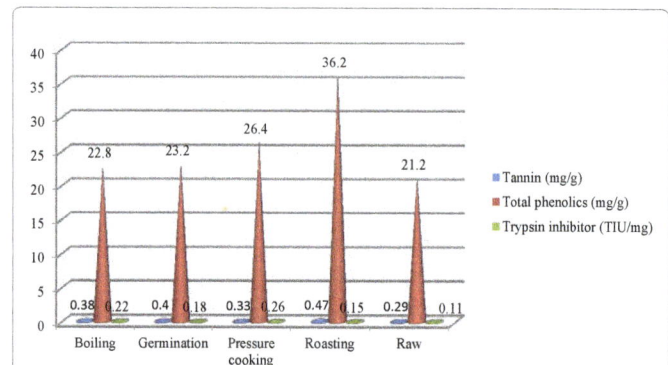

Figure 1: Anti-nutritional parameters of raw and processed Barnyard millet flour.

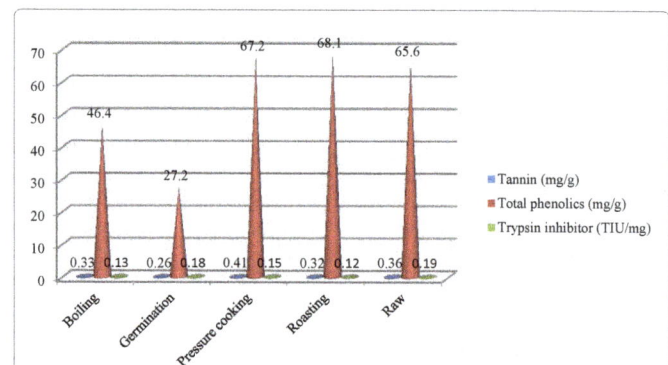

Figure 2: Anti-nutritional parameters of raw and processed Foxtail millet flour.

S.no	Parameters	Pasting temperature (°C)	Peak viscosity (cP)	Trough viscosity (cP)	Breakdown viscosity (cP)	Final viscosity (cP)	Setback viscosity (cP)
1.	Boiling	48.25	23	21	2	30	9
2.	Germination	49.55	70	21	49	32	11
3.	Pressure cooking	48.35	31	29	2	44	15
4.	Roasting	94.45	101	93	8	196	103
5.	Raw	90.4	854	741	113	1235	494

Table 12: Pasting properties of the raw and processed barnyard millet flour.

S:No:	Parameters	Pasting temperature (°C)	Peak viscosity (cP)	Trough viscosity (cP)	Breakdown viscosity (cP)	Final viscosity (cP)	Setback viscosity (cP)
1.	Boiling	48.50	41	39	2	59	20
2.	Germination	48.60	116	48	68	90	42
3.	Pressure cooking	48.50	31	29	2	37	8
4.	Roasting	50.2	1143	361	782	254	-107
5.	Raw	89.6	521	420	101	1398	978

Table 13: Pasting properties of the raw and processed foxtail millet flour.

Acknowledgement

The first author is grateful to DST-SERB for providing financial assistance to carry out the work.

References

1. Bavec F, Bavec M (2006) Millets in organic production and use of alternative crops. CRC Press, London.

2. Mal B, Padulosi S, Ravi SB (2010) Minor millets in South Asia: Learnings from IFAD-NUS project in India and Nepal. MS Swaminathan Research Foundation 1-185.

3. Devi PB, Vijayabharathi R, Sathyabama S, Malleshi NG, Priyadarisini VB, et al. (2014) Health benefits of finger millet (*Eleusine coracana* L.) polyphenols and dietary fiber: a review. J Food Sci Technol 51: 1021-1040.

4. Pradeep SR, Guha M (2011) Effect of processing methods on the nutraceutical and antioxidant properties of little millet (*Panicum sumatrense*) extracts. Food Chem 126: 1643-1647.

5. Li J, Chen Z, Guan X, Liu J, Zhang M, et al. (2008) Optimization of germination conditions to enhance hydroxyl radical inhibition by water soluble protein from stress millet. J Cereal Sci 48: 619-624.

6. Ahmed SM, Zhang Q, Chen J, Shen Q (2013) Millet Grains: Nutritional quality, processing and potential health benefits. Comprehensive Reviews in Food Science and Food Safety 12: 281-295.

7. Hotz C, Gibson RS (2007) Traditional food-processing and preparation practices to enhance the bioavailability of micronutrients in plant-based diets. J Nutr 137: 1097-1100.

8. Ige MM, Ogunsua AO, Oke OL (1984) Functional properties of the protein of some Nigerian oilseed, conophor seed and 3 varieties of melon seed. J Agri Food Chem 32: 822-825.

9. Shin M, Gang DO, Song JY (2010) Effects of protein and transglutaminase on the preparation of gluten-free rice bread. Food Sci Biotechnol 19: 951-956.

10. Lin MJY, Humbert ES, Sosulski FW (1974) Certain functional properties of sunflower meal products. J Food Sci 39: 368-370.

11. Sadasivam S, Manickam A (2005) Biochemical Methods. New Age International Publishers.

12. Tester RF, Morrison WR (1990) Swelling and gelatinization of cereal starches II. Waxy rice starches Cereal Chemistry 67: 551-557.

13. Phattanakulkaewmorie N, Paseephol T, Moongngarm A (2011) Chemical compositions and physico-chemical properties of malted sorghum flour and characteristics of gluten free bread. World Academy of Science, Engineering and Technology 57: 454-459.

14. Urbano G, Lopez-Jurado M, Frejnagel S, Gomez-Villalva E, Porres JM, et al. (2005) Nutritional assessment of raw and germinated pea (*Pisum Sativum* L.) protein and carbohydrate by in vitro and in vivo techniques. Nutrition 21: 230-239.

15. Parde SR, Johal A, Jayas DS, White NDG (2003) Physical properties of buckwheat cultivars. Canadian Bio-systems Engineering, Technical Note.

16. Merill AY, Walt BK (1973) Energy value of food basis and derivation. USA dept of agric 74: 2-4.

17. Omegie HNA, Ogunsakin R (2013) Assessment of chemical, rheological and sensory properties of fermented maize-cardaba banana complementary food. Food and Nutrition Sciences 4: 844-850.

18. Ugare R (2008) Health benefits, Storage quality and value addition of barnyard millet (Echinochloa frumentacaea Link). College of Rural Home Science. University of Agricultural Sciences, Dharwad.

19. Bhupender SK, Rajneesh B, Baljeet SY (2013) Physico chemical, functional, thermal and pasting properties of starches isolated form pearl millet cultivars. International Food Research Journal 20: 1555-1561.

20. Reungmaneepaitoon S, Sikkhamondhol C, Tiangpook C (2006) Nutritive improvement of instant fried noodles with oat bran. J Sci Technol 28: 89-97.

21. Bello-Pérez LA, Contreras-Ramos SM, Jìmenez-Aparicio A, Paredes-López O (2000) Acetylation and characterization of banana (*Musa paradisiaca*) starch. Acta Cient Venez 51: 143-149.

22. Housson P, Ayenor GS (2002) Appropriate processing and food functional

properties of maize flour. Afr J Sci Technol 3: 121-126.

23. Kinsella JE (1982) Relationship between structural and functional properties of food proteins. Food protein. Applied Science Publishers, London: 51-103.

24. Schoch TJ (1964) Swelling power and solubility of granular starches. Academic Press, New York 106-108.

25. Balogun IO, Olatidoye OP (2010) Functional properties of dehulled and undehulled velvet beans flour (mucuna utilis). Journal of Biological Sciences and Bioconservation.

26. Butt MS, Batool R (2010) Nutritional and functional properties of some promising legumes protein isolates. Pakistan J Nutr 9: 373-379.

27. Ajayi IA, Oderinde RA, Kajogbola DO, Uponi JI (2006) Oil content and fatty acid composition of some underutilized legumes from Nigeria. Food Chem 99: 15-120.

28. Ghadivel RA, Prakash J (2007) The impact of germination and dehulling on nutrients, anti-nutrients, in vitro iron and calcium bioavailability and in vitro starch and protein digestibility of some legume seeds. LWT 40: 1292-1299.

29. Kaushik G, Satya S, Naik SN (2010) Effect of domestic processing techniques on the nutritional quality of the soyabean. Mediterranean Journal of Nutrition and Metabolism 1: 39-46.

30. Khatoon N, Prakash J (2006) Nutrient retention in microwave cooked germinated legumes. Food Chem 97: 115-121.

31. Yadav BS, Yadav RB, Kumar M (2011) Suitability of pigeon pea and rice starches and their blends for noodle making. LWT Food Science and Technology 44: 1415-1421.

32. Champ M, Langkilde AM, Brouns F, Kettlitz B, Collet Y (2003) Advances in dietary fibre characterization. Definition of dietary fibre, physiological relevance, health benefits and analytical aspects. Nutrition Research Reviews 16: 71-82.

33. Mubarak AE (2005) Nutritional composition and anti-nutritional factors of mungbean seeds (Phaseolus aureus) as affected by some home traditional processes. J Food Chem 89: 489-495.

34. Akinijayeju O, Francis (2007) Effects of sprouting on the proximate composition of bambara nut flours. Processing of the annual conference/general meeting of NFST 25: 158-159.

35. Aremu MO, Olayioye YE, Ikokoh PP (2009) Effect of processing on nutritional quality of kersting groundnut (*Kerstingiella geocarpa* L.) seed flours. J Chem Soc Nig 34: 140-149.

36. Parameswaran K, Sadasivam S (1994) Changes in the carbohydrates and nitrogenous components during germination of proso millet (*Panicum miliaceum*). Plants Foods Hum Nutr 45: 97-102.

37. Srichuwong S, Suharti C, Mishima T, Isono M, Hisamatsu M (2005) Starches from different botanmical sources: Contribution of starch structure to swelling and pasting properties. Carbohydrate polymers 62: 25-34.

38. Sprouting (2008) Nutritional information. Wikipedia.

39. Deorthale GY, Rao U (2006) Polyphenoloxidase activity in germinated legume seeds. Journal of food science.

40. El-Adawy TA, Taha KM (2001) Characteristics and composition of watermelon, pumpkin, and paprika seed oils and flours. J Agric Food Chem 49: 1253-1259.

41. Platel K, Eipeson SW, Srinivasan R (2010) Bioaccessible mineral content of malted finger millet (*Eleusine coracana*), wheat (*Triticum aestivum*), and barley (*Hordeum vulgare*). J Agric Food Chem 58: 8100-8103.

42. Seifi MR, Alimardani R, Akram A, Asakereh A (2010) Moisture-depend physical properties of safflower (Goldasht). Advance Journal of Food Science and Technology 2: 340-345.

43. Sahlin E, Savage GP, Lister CE (2004) Investigation of the antioxidant properties of tomatoes after processing. Journal of Food Composition and Analysis 17: 635-647.

44. Sivashanthini K, Vivekshan Reval S, Thavaranjith AC (2012) Comparative study on organoleptic, microbiological and biochemical qualities of commercially and experimentally prepared salted and sun dried talang queen fish, scomberoides commersonianus. Asian Journal of Animal and Veterinary Advances 7: 1279-1289.

45. Kaur M, Singh N (2005) Studies of functional, thermal and pasting properties of flours from different chickpea (*Cicer arietinum* L.) cultivars. Food Chemistry 91: 403-411.

46. Taiwo OO, Jimoh D, Osundeyi E (2010) Functional and pasting properties of composite cassava-sorgum flour meals. Agriculture and Biology Journal of North America 1: 715-720.

47. Zhang D, Hamauzu Y (2004) Phenolics, ascorbic acid, carotenoids and antioxidant activity of broccoli and their changes during conventional and microwave cooking. Food Chemistry 88: 503-509.

48. Karim AA, Toon LC, Lee VP, Ong WY, Fazilah A, et al. (2007) Effects of phosphorus contents on the gelatinization and retrogradation of potato starch. J Food Sci 72: 132-138.

16

Control of *Listeria monocytogenes* on Alternatively Cured Ready-to-Eat Ham Using Natural Antimicrobial Ingredients in Combination with Post-Lethality Interventions

Lavieri NA[1], Sebranek JG[1*], Cordray JC[1], Dickson JS[1], Horsch AM[1], Jung S[2], Manu DK[3], Mendonça AF[3] and Brehm Stecher B[3]

[1]Department of Animal Science, Iowa State University, 215 Meat Laboratory, Ames, IA, USA
[2]Food Science and Human Nutrition Department, Iowa State University, 1436 Food Science Building, Ames, IA, USA
[3]Food Science and Human Nutrition Department, Iowa State University, 3399 Food Science Building, Ames, IA, USA

Abstract

Ready-to-Eat (RTE) meat and poultry products manufactured with natural or organic methods may be at greater risk for *Listeria monocytogenes* growth, if contaminated, than their conventional counterparts due to the required absence of preservatives and antimicrobials. Thus, the objective of this study was to investigate the use of commercially available natural antimicrobials in combination with post-lethality interventions for the control of *L. monocytogenes* growth and recovery on alternatively-cured RTE ham. Antimicrobials evaluated were cranberry powder (90 MX), vinegar (DV), and vinegar and lemon juice concentrate (LV1 X). Post-lethality interventions studied included high hydrostatic pressure at 400 MPa (HHP), lauricarginate (LAE), octanoic acid (OA), and post-packaging thermal treatment (PPTT). Viable *L. monocytogenes* on modified Oxford (MOX) and thin agar layer (TAL) media were monitored through 98 days of product storage at 4 ± 1°C. The post-lethality treatments of HHP, OA, and LAE significantly reduced initial viable *L. monocytogenes* numbers compared to the control, regardless of the antimicrobial ingredient used in the formulation while PPTT did not. Only when used in combination with DV and LV1 X did HHP, OA, and LAE exhibit sustained suppression, of *L. monocytogenes* recovery and growth throughout refrigerated storage. As a result, the use of natural antimicrobial ingredients such as DV and LV1 X in combination with post-lethality interventions such as HHP, LAE, and OA represents an effective multi-hurdle approach that could be instituted by manufacturers of organic and natural processed meat and poultry products for *L. monocytogenes* control.

Keywords: Ham; High-pressure; *Listeria monocytogenes*; Natural; Organic

Introduction

The popularity of natural and organic foods has been increasing for several years, and has led to noticeable market growth of these food categories [1,2]. In 2013, for example, organic foods in the United States experienced a 13% increase in sales compared to the previous year [3]. Similar increases are expected to continue in the future in spite of the price premiums typically associated with these products [4]. Natural and organic meat products, in particular, have accounted for a significant part of that growth. Stringent regulations that govern the production of natural and organic foods have prevented the use of certain traditional ingredients. For instance, in the manufacture of natural and organic processed meat products, such as boneless ham and frankfurters, the direct addition of nitrite or nitrate, curing ingredients used in the manufacture of such products, and that have strong antimicrobial properties, are not permitted. Additionally, lactate and diacetate, antimicrobials commonly found in ready-to-eat (RTE) meat and poultry products, and that is effective inhibitors of pathogens such as *Listeria monocytogenes*, are not permitted in the manufacture of natural or organic meat products. Thus, RTE meat and poultry products manufactured under uncured, natural, or organic methods are sometimes termed "alternatively cured" or "naturally cured". The requirements for these products suggest that they are likely to be at a greater risk than their conventional counterparts for growth of *L. monocytogenes* if contamination occurs, and previous reports have supported this concern as well [5-7].

The use of natural antimicrobials or post-lethality interventions in the manufacture of natural and organic meat products has beenstudied by several researchers and meat processors alike [8-11]. The United States Department of Agriculture Food Safety Inspection Services (USDA-FSIS) defines a post-lethality treatment as "…a lethality treatment that is applied or is effective after post-lethality exposure. It is applied to the final product or sealed package of product in order to reduce or eliminate the level of pathogens resulting from contamination from post-lethality exposure" [12]. High hydrostatic pressure processing (HHP), for example, is one such post-lethality intervention that takes place after the product has gone through the lethality or cooking step [12,13]. Other examples of post-lethality interventions include sprays or solutions such as lauric arginate (lauramide arginine ethyl ester or LAE) and octanoic acid (sometimes referred to as caprylic acid or OA) as well as post-packaging thermal treatment or pasteurization, all of which can be applied to the finished product. The USDA-FSIS lists lauric arginate as a safe and suitable ingredient for the production of meat and poultry products, and allows up to 44 mg/kg (ppm) (± a 20% tolerance) by weight of the product to be applied to the inside of a package as a processing aid [14]. When used at this level, lauric arginate is considered a processing aid, would not have to be declared on the label of the product, and could be used in the manufacture of uncured, no-nitrate-or-nitrite-added (alternatively-cured), RTE natural or organic meat and poultry products. Similarly, the USDA-FSIS also allows for octanoic acid to be used as a processing aid if applied to the surface of an RTE meat and poultry product at

---

*Corresonding author: Joseph G Sebranek, Department of Animal Science, Iowa State University, 215 Meat Laboratory, Ames, IA, USA 50011-3150
E-mail: sebranek@iastate.edu

a rate not to exceed 400 mg/kg octanoic acid by weight of the final product [14]. Octanoic acid is a saturated ($C_{8:0}$) fatty acid (pK_a 4.89) naturally found in coconut oil and bovine milk [15].

While natural sources of antimicrobials could potentially replace chemical preservatives as a means to address *L. monocytogenes* [10,16,17], it has also been shown that the anti listerial properties of antimicrobials can vary as a result of the fat content of the food [18] and other variables including protein content, pH, a_w, and other ingredients added.

Thus, there is significant concern for the potential recovery and growth of sub lethally injured and uninjured *L. monocytogenes* during the storage life of alternatively-cured RTE ham and frankfurters that do not include the antimicrobial agents normally used in conventional cured meats. Such concerns highlight the need for a combination of antimicrobial hurdles to be investigated and, eventually, implemented in order to fully address *L. monocytogenes* control in natural and organic RTE meat and poultry products.

Previous work in our laboratory [19] demonstrated that post-lethality interventions such as HHP, OA, and LAE can deliver an initial lethality for *L. monocytogenes*, but survivors will grow in processed meats following the treatment. Secondly, we have also observed that natural antimicrobials such as vinegar and vinegar and lemon juice concentrate can impart a bacteriostatic effect on this pathogen, thus suppressing subsequent growth, but without reducing the initial population.

Consequently, the objective of this study was to assess the commercially available natural antimicrobial ingredients that are currently allowed for natural and organic meat and poultry products when used in combination with post-lethality interventions to both reduce the initial contaminating population, and subsequently inhibit the recovery and growth of any *L. monocytogenes* survivors. We hypothesized that a combination of treatments that achieves both initial lethality and sustained suppression of growth of survivors would effectively improve the overall control of *L. monocytogenes* on alternatively-cured processed meat products.

Materials and Methods

Manufacture of hams

Thirteen ham formulations (twelve experimental and one control formulation) were manufactured at the Iowa State University Meat Laboratory using inside (*gracilis* and *semimembranosus*) ham muscles. The formulations consisted of 18.14 kg of ham insides, 3.66 kg water, 0.50 kg salt, 0.30 kg sugar and 74.84 g celery powder plus the selected antimicrobials. The ham muscles were obtained from a local processor and frozen prior to use to ensure uniformity of raw materials. The ham muscles were tempered to -2°C, then coarse ground through a grinder plate with 9.53-mm-diameter holes (Biro MFG Co., Marblehead, OH). Nonmeat ingredients (water, salt, sugar) were added and mixed with ground ham muscles at 26 rpm for 2 min using a double action, paddle-and-ribbon mixer (Leland Southwest, Fort Worth, TX). Pre-converted (nitrate converted to nitrite) celery powder (Veg Stable 504, Florida Food Products, Inc., Eustis, FL) containing 1.5% (wt/wt) nitrite was used as the natural, alternative source of nitrite. All products were formulated to contain 50 mg/kg (ppm) ingoing natural nitrite to represent the reduced ingoing nitrite concentration that is typical of many natural and organic processed meat products. Control hams were formulated without antimicrobials or post-lethality interventions to best represent the natural and organic hams currently produced. Three commercially available natural antimicrobial ingredients were evaluated in this study; cranberry powder (90 MX; Ocean Spray International, Middleboro, MA), buffered vinegar (DV; WTI Ingredients, Inc., Jefferson, GA), and buffered vinegar and lemon juice concentrate (LV1 X; WTI Ingredients, Inc.) (wt/wt). Each ingredient was added at a concentration (1.0%, 1.0%, 2.5%, respectively) recommended by the respective supplier. The pH of 10% solutions (w/v) of the 90 MX, DV, and LV1 X ingredients were 3.89, 5.87 and 5.57 respectively.

The hams and appropriate ingredients were mixed, then reground using a grinder plate with 6.35 mm diameter holes and stuffed into a 50 mm diameter impermeable plastic casing (Nalobar APM 45, Kalle USA, Gurnee, IL) using a rotary vane vacuum stuffer (RS 1040 C, Risco USA Corp., South Eaton, MA). All samples were then placed in a single-truck smokehouse (Maurer, AG, Reichenau, Germany) and heated to an internal temperature of 71.1°C. The hams were then placed in a 0°C cooler overnight to stabilize. The next day (day 0 of the experiment), the hams were sliced into approximately 12.0 mm thick slices using a hand slicer (SE 12 D, Bizerba, Piscataway, NJ), placed into barrier bags (B2470, Cryovac Sealed Air Corporation, Duncan, SC; oxygen transmission rate of 3-6 cc/m², 24 h at 4°C, 0% RH; water vapor transmission rate of 0.5-0.6 g/0.6 m² at 38°C (100% RH, 24 h), and vacuum-sealed (UV 2100, Multivac, Inc., Kansas City, MO). Hams for physicochemical analyses were placed in boxes, transferred to a holding cooler in the Iowa State University Meat Laboratory and stored at 4 ± 1°C until analyses were conducted. Hams for microbial analyses were placed in boxes with vacuum packaged ice, transferred to the Iowa State University Microbial Food Safety Laboratory in the Food Science and Human Nutrition Department for subsequent inoculation, and stored at 4 ± 1°C for the duration of the experiment. Two complete independent replications of the entire experiment were performed.

Product analyses

Proximate analysis was conducted for moisture, fat, and protein of homogenized control and treatment formulations on day 0 using AOAC methods 950.46, 960.63, and 992.15, respectively [20-22]. Samples were prepared in duplicate for each ham formulation.

Product pH was measured by placing a pH probe (FC20, Hanna Instruments, Woonsocket, RI) into homogenized (KFP715 food processor, Kitchenaid, St. Joseph, MI) samples from the control and treatments that were prepared by first blending the ground ham with distilled, de-ionized water in a 1:9 ratio, and then measuring the pH with a pH/ion meter (Accumet 925 pH/ion meter, Fisher Scientific, Waltham, MA). Calibration was conducted using phosphate buffers of pH 4.0, 7.0, and 10.0. Duplicate readings were taken for each product formulation on day 0.

Available moisture was determined using a water activity meter (AquaLab 4 TE, Decagon Devices Inc., Pullman, WA). Samples were cut into small pieces, placed in disposable sample cups, covered, and allowed to equilibrate to room temperature (5-10 min). Measurements were obtained on day 0 and were performed in duplicate for the control and all treatments. Calibration was performed using 1.00 and 0.76 sodium chloride water activity standards.

Residual nitrite concentration was determined utilizing AOAC method 973.31 [23]. Samples from each treatment were frozen at -20 ± 1°C on day 0 and evaluated in duplicate at a later date.

Inoculation of samples

L. monocytogenes strains Scott A NADC 2045 serotype 4b, H7969 serotype 4b, H7962 serotype 4b, H7596 serotype 4b, and H7762 serotype 4b were obtained from the Iowa State University Microbial Food Safety Laboratory in the Food Science and Human Nutrition Department. These strains were selected because each has been isolated from cases of food-borne disease outbreaks. Each strain was cultured separately in tryptic soy broth supplemented with 0.6% yeast extract (TSBYE) (Difco, Becton Dickinson, Sparks, MD) for 24 h at 35°C. A minimum of two consecutive 24 h transfers of each strain to fresh TSBYE (35°C) were performed prior to each experiment. The cells were harvested by centrifugation (10 min at 10,000 × g and 4°C) in a Sorvall Super T21 centrifuge (American Laboratory Trading, Inc., East Lyme, CT). The supernatant was discarded and the pelleted cells were re-suspended in 30.0 ml of sterile buffered peptone water (BPW) (Difco, Becton Dickinson). The total concentration of the five-strain mixed culture was approximately 10^9 CFU per ml based on the washed cell suspension. Two serial dilutions (100-fold each) of the cell suspension were prepared in BPW to give a final inoculum concentration of approximately 10^5 CFU per ml. This diluted five-strain mixed culture was used to inoculate the ham samples.

While in the Microbial Food Safety Laboratory, each packaged sample was reopened and the surface of the product was aseptically inoculated with 0.2 ml per package, using the diluted five-strain mixed culture of the pathogen. The viable cell concentration at inoculation was approximately 10^3 CFU per g of ham slice. The bags were then vacuum-sealed using a model A300/52 vacuum packaging machine (Multivac, Inc.) and stored at 4 ± 1°C for the duration of the experiment.

Post-lethality interventions

Four post-lethality interventions were evaluated in this study; high hydrostatic pressure (HHP), octanoic acid (OA), lauricarginate (LAE), and post-packaging thermal treatment (PPTT). Ham slices from each formulation were randomly assigned to these post-lethality interventions. All post-lethality interventions were applied to the product within two hours following inoculation on day 0 of the study.

The HHP parameters were 400 MPa, 4 min dwell time at 12 ± 2°C initial fluid temperature of the pressurization fluid. The 400 MPa HHP treatment was utilized for this study rather than the more common 600 MPa that is used for commercial products to allow a measurable number of the organisms to survive so that the effects of the antimicrobials in combination with HHP could be assessed. Inoculated samples were transported to the High Pressure Processing Laboratory at the Iowa State University Food Science and Human Nutrition Department and subjected to HHP treatment using a FOOD-LAB 900 Plunger Press system (Standsted Fluid Power Ltd., Standsted, UK). The pressurization fluid was a 50.0% propylene glycol (GWT Koilguard; GWT Global Water Technology, Inc., Indianapolis, IN) and 50.0% water solution (v/v). The average rate of pressurization was 350 MPa per min and depressurization occurred within 7 s. Adiabatic heating of the pressurization fluid was 4.6°C ± 0.8°C/100 MPa.

Octanoic acid (Octa-Gone; Eco Lab, Inc., Eagan, MN) was applied according to the manufacturer's recommendations. Octa-Gone contains approximately 3.6% octanoic acid (v/v). A 23.4% Octa-Gone solution (v/v) was prepared by mixing Octa-Gone with sterile de-ionized water at 4 ± 1°C. Based on average surface area measurements obtained as previously described, the OA solution was aseptically dispensed into the bag containing the ham slice (0.0186 ml per cm²) and vacuum-sealed.

Lauricarginate (Protect-M; Purac America, Lincolnshire, IL) was also applied according to the manufacturer's recommendations.

Protect-M contains approximately 10.0% lauricarginate (v/v). A 2.5% Protect-M solution (v/v) was prepared by mixing Protect-M with sterile de-ionized water at 4 ± 1°C. Based on the ham slice surface area measurements, the LAE solution was aseptically dispensed into the bag containing the ham slice ($7.19 × 10^{-3}$ ml per cm²) and vacuum-sealed.

PPTT was conducted by immersing packages of ham in water at 71.0 ± 1.0°C water for 30 s using a water bath (Isotemp-228, Fisher Scientific). Seven packages were immersed as a group so that water temperature would not change by more than 1.0°C. Water temperature was monitored throughout the process. Packages were held in heated water for the prescribed length of time and then placed on ice immediately to chill before placement in refrigerated storage.

Microbial analysis

Microbial analysis of ham samples for viable *L. monocytogenes* was conducted on days 1, 14, 28, 42, 56, 70, 84, and 98 of storage. On the appropriate day, two packages for each treatment were removed from the holding cooler, opened aseptically, and their contents placed inside a sterile Whirl-Pak stomacher bag (Nasco, Ft. Atkinson, WI). Fifty (50.0) ml of sterile BPW was added to each bag, and the bags shaken by hand for approximately 30 s. The rinse solution from each ham sample was then serially diluted (10-fold) in BPW to obtain pre-determined dilutions of the samples according to the sampling day. One ml (for undiluted rinsate, divided into three ~0.33-ml aliquots plated on three separate plates) or 0.1 ml of the appropriate dilution was surfaced plated on modified listeria selective agar (Oxford, MOX)(Difco, Becton Dickinson). The dry ingredients used to manufacture the MOX were 42.5 g of Columbia agar base (Difco, Becton Dickinson), 15.0 g of lithium chloride (Difco, Becton Dickinson), 1.0 g of esculin hydrate (Sigma-Aldrich, St. Louis, MO), and 0.5 g of ferric ammonium citrate (Difco, Becton Dickinson) per liter of de-ionized water. Additionally, an aliquot of 1.0 ml (for undiluted rinsate, divided into three ~0.33-ml aliquots plated on three separate plates) or 0.1 ml of the appropriate dilution was surface-platedon thin agar layer medium base (TAL) that was made according to Kang and Fung [24]. Within 48 h before use, MOX plates to be made into TAL were aseptically overlaid with 7.0 ml of sterile tryptic soy agar (Difco, Becton Dickinson) held at 55°C to facilitate the even distribution of the molten agar. Each sample was plated in duplicate. All inoculated plates were incubated in an inverted position at 35°C for 48 h, after which time they were removed from the incubator, and colonies typical of *L. monocytogenes* were enumerated. The populations (CFU per ml) were averaged and then converted to log_{10} CFU per g using the average weight of the sliced ham from the two replications of the experiment (n=40). The detection limit of our sampling protocols was ≥0.30 log_{10} CFU per g based on a sample weight of 25.0 g.

Statistical analysis

The overall design of the experiment was a factorial design. The generalized linear mixed models (GLIMMIX) procedure of Statistical Analysis System (version 9.3, SAS Institute Inc., Cary, NC) was used for statistical analysis. *L. monocytogenes* growth data were analyzed for treatment effects within day. Day and treatment x day interactions were also analyzed. The effects of each post-lethality intervention were analyzed separately for each natural antimicrobial ingredient studied. Likewise, the effects of each natural antimicrobial ingredient were analyzed separately for each post-lethality intervention studied. Where significant effects ($P<0.05$) were found, pair-wise comparisons between the least squares means were computed for each day using Tukey's honestly significant difference adjustment.

Results and Discussion

The mean weight of the ham slices was 24.57 ± 0.64 g, while the mean diameter, height, and surface area were 4.72 ± 0.06 cm, 1.31 ± 0.01 cm, and 54.51 ± 1.13 cm², respectively (data not shown; $n=40$ for all measurements). These ham slice dimensions were used to calculate ham slice surface area for LAE and OA treatment volumes of 0.39 and 1.01 ml per package, respectively. The dosages of each compound were calculated according to the respective manufacturer's recommendations as previously described. These dosages resulted in LAE and OA treatment concentrations of 39.82 and 343.03 mg/kg (ppm), respectively.

Physicochemical traits

Physicochemical characteristics of the hams can be found in Table 1. All treatments exhibited significantly lower a_w values than the control treatment ($P<0.05$). The DV and LV1 X treatments, in turn, resulted in significantly lower a_w values when compared to the 90 MX treatment ($P<0.05$). Final product pH was also affected by natural antimicrobial compound added. The pH of the control treatment was not significantly different from that of the DV treatment ($P>0.05$), but did significantly differ from both the LV1 X and the 90 MX treatments ($P<0.05$). These differences in pH most likely resulted from the presence of acidic compounds in the natural antimicrobial compounds utilized. Cranberry has been reported to contain phenolic acids and exhibit a high titratable acidity [25]. Xi et al. [16] obtained similar pH results when using different ingoing levels of cranberry powder in a cooked meat model system and in frankfurters [17]. Similarly, the vinegar and vinegar and lemon juice concentrates used in this study also contain acidic compounds, such as acetic and citric acid, and can be expected to result in the observed lower pH in products made with those ingredients. No significant differences in protein % and moisture % were found between the treatments ($P>0.05$). Fat %, however, was significantly lower in the 90 MX treatment compared to both the DV and LV1 X treatments ($P<0.05$). Although some of these differences were statistically significant, the differences were very small and were not expected to affect the results of this study.

The residual nitrite concentration found in the 90 MX treatment was lower ($P<0.05$) than that of the control and DV treatments. No significant differences between all other treatments were detected ($P>0.05$). Although all ham formulations were manufactured with 50 mg/kg (ppm) natural nitrite on an ingoing basis, the highest residual nitrite concentration observed in all of the treatments on day 0 of the study was 36.01 mg/kg (ppm) (control treatment). This indicates that part of the ingoing nitrite was depleted in curing and other reactions that took place, as expected, during product manufacture.

Treatment[b]	a_w	pH	Fat %	Moisture %	Protein %	Residual Nitrite (mg/kg)
Control	0.9819[a]	6.35[b]	1.96[ab]	75.84	18.09	36.01[bc]
90MX	0.9793[b]	6.05[a]	1.58[a]	75.82	17.95	31.32[a]
DV	0.9759[c]	6.24[bc]	2.26[b]	75.30	17.88	35.36[bc]
LV1X	0.9772[c]	6.18[ac]	2.32[b]	74.93	18.02	33.56[ac]
SE[c]	0.0005	0.04	0.19	0.29	0.20	1.11

[a]Values are least squares means. Within a column, means with different superscripts (a through c) are significantly different ($P<0.05$)

[b]Control, naturally-cured control; 90MX, cranberry powder; DV, vinegar; LV1X, vinegar and lemon juice concentrate

[c]Standard error of the differences of least squares means

Table 1: Effect of natural antimicrobial ingredients on physicochemical properties of naturally-cured RTE ham[a].

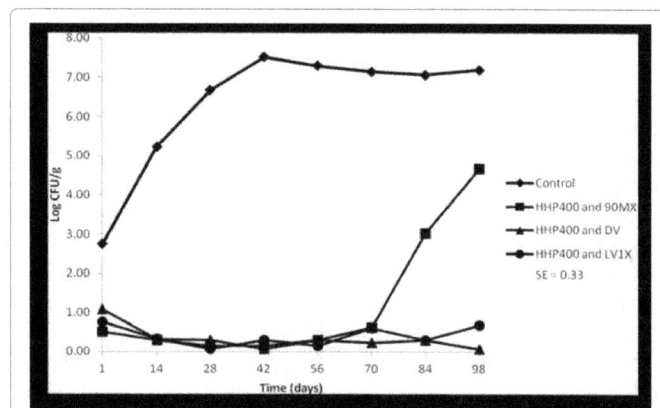

Figure 1: Effect of high hydrostatic pressure treatment in combination with natural 535 antimicrobials on viable *Listeria monocytogenes* (log10 CFU per gram) on alternatively-cured RTE ham stored at 4 ± 1°C, using modified Oxford medium.

Honikel [26] reported that as much as 65% of the ingoing nitrite can be depleted during product manufacture. Similarly, Xi et al. [17] reported that as much as 75% of the ingoing nitrite can be depleted during the manufacture of frankfurters. Factors such as product pH, cooking temperature, and addition of reducing agents have been long recognized as important factors affecting residual nitrite concentrations in meat systems [27]. Thus, the significant ($P<0.05$) decrease in pH brought about by the natural antimicrobial ingredients used in this study, especially cranberry powder, was expected to influence residual nitrite concentrations.

Viable *Listeria monocytogenes* populations

The growth mediums used, MOX and TAL, did not significantly differ ($P>0.05$) within treatment on any given day, indicating that, under the conditions of this study, the use of the TAL technique offered no significant advantage compared to using a traditional medium such as MOX. Thus, the discussion about viable *L. monocytogenes* populations as affected by treatment is limited to the results obtained using MOX.

The ham formulations included controls that were manufactured without antimicrobials or post-lethality treatments to provide comparison to the treatment combinations. The 400 MPa HHP treatment used in combination with all of the natural antimicrobial ingredients studied resulted in a significant ($P<0.05$) reduction in viable *L. monocytogenes* populations on day 1 when compared to the control treatment (Figure 1). More specifically, the HHP treatment resulted in populations that were 2.25, 1.99, and 1.67 $\log_{10}$ CFU per g lower ($P<0.05$) on day 1 when combined with 90 MX, LV1 X, and DV, respectively, and relative to the control treatment. The differences in $\log_{10}$ CFU per g reductions observed on day 1 in the different treatments subjected to HHP, however, were not significant ($P>0.05$) compared to each other, indicating that the three antimicrobial ingredients used did not influence the bactericidal properties of the HHP treatment applied. These results confirm the bactericidal properties of HHP at 400 MPa against *L. monocytogenes*. However, only when combined with DV or LV1 X was the initial reduction in viable *L. monocytogenes* achieved by 400 MPa HHP sustained throughout the duration of the study. The combination of 400 MPa and 90 MX resulted in an increase in the *L. monocytogenes* populations after day 70 that reached about 5 $\log_{10}$ CFU per g by the end of the study.

Damage to the cell membrane seems to be the likely mode of action for HHP, and it has been reported that damage to bacterial cell membranes can be extensive, often resulting in cell death [28,29]. Changes in membrane permeability, scarring around the cell wall, separation of the cell wall from the membrane and protein denaturation, as well as damage to transport systems have also been reported in HHP-treated microbial populations [30,31]. Thus, it is likely that the bacteriostatic effect observed in the HHP treatments combined with ingredients such as vinegar or vinegar and lemon juice concentrate was a result of the migration of growth inhibitory compounds present in these ingredients into the bacterial cells. As a result, the use of HHP at 400 MPa in combination with DV or LV1 X represents a promising multiple-hurdle approach for addressing the potential presence of *L. monocytogenes* in processed meats, andfor inhibiting the potential recovery and growth of those cells that remain viable over the refrigerated storage of the products. Further, it appears that the use of these antimicrobials may permit reduced HHP pressure of 400 MPa as an alternative to the higher 600 MPa that is currently used in commercial applications where HHP is used alone. Reduction of pressure used in the HHP process would increase product throughput for the process and result in lower maintenance cost, both of which are important in determining total cost of the treatment [32,33].

Combining OA with the natural antimicrobial ingredients evaluated in this study (Figure 2) yielded similar patterns to those obtained when combining HHP with the same ingredients in terms of viable *L. monocytogenes* populations observed. Significant (*P*<0.05) reductions in initial viable *L. monocytogenes* populations were observed when OA was combined with each of the natural antimicrobial ingredients evaluated after day 1 and compared to the control treatment. On day 1, compared to the control treatment, the *L. monocytogenes* populations were lower by 2.67, 2.52, and 2.33 $\log_{10}$ CFU per g when OA was combined with 90 MX, DV, and LV1 X, respectively. Burnett et al. [34] concluded that octanoic acid solutions acidified to pH 2.0 or 4.0, and applied to RTE meat and poultry, resulted in *L. monocytogenes* log reductions ranging from 0.85 to 2.89 $\log_{10}$ CFU per sample. The pH of the working solution of OA used in the current study was 3.01. It has been reported that the main mechanism by which medium and short chain fatty acids achieve microbial inactivation is through the diffusion of undisociated acids across the bacterial cells and the subsequent intracellular acidification [35]. Thus, it is likely that the bactericidal effects of OA on *L. monocytogenes* follow that mechanism.

Sustained inhibition of *L. monocytogenes* recovery and growth compared to the control was exhibited by treatments that combined OA with DV or LV1 X (*P*<0.05) but not with 90 MX (*P*>0.05), which

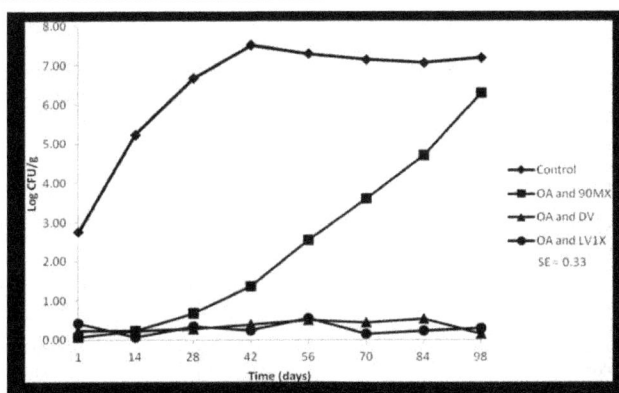

Figure 2: Effect of octanoic acid treatment in combination with natural antimicrobials on viable *Listeria monocytogenes* (log10 CFU per gram) on alternatively-cured RTE ham stored at 4 ± 1°C, using modified Oxford medium.

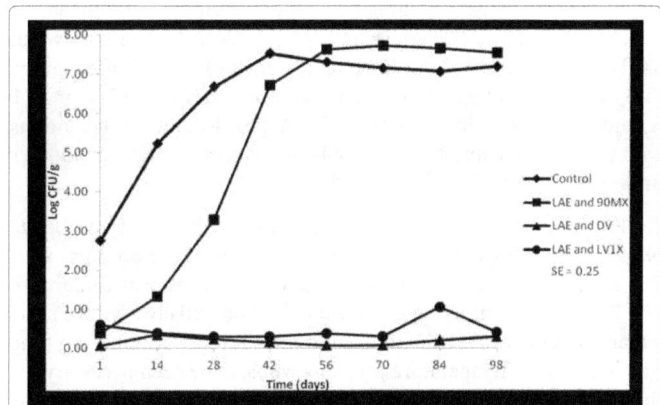

Figure 3: Effect of lauricarginate treatment in combination with natural antimicrobials on viable *Listeria monocytogenes* (log10 CFU per gram) on alternatively-cured RTE ham stored at 4 ± 1°C, using modified Oxford medium.

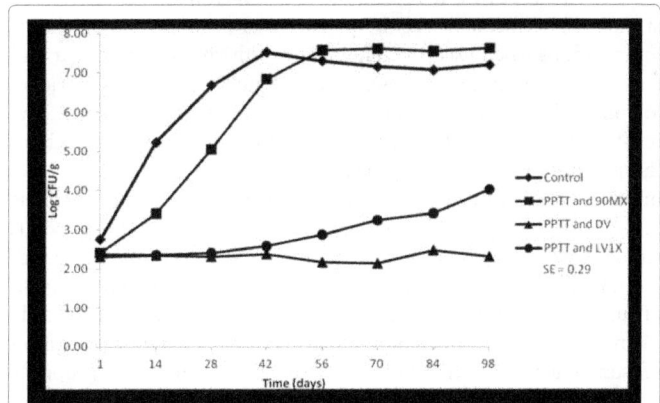

Figure 4: Effect of post-packaging thermal treatment in combination with natural antimicrobials on viable *Listeria monocytogenes* (log10 CFU per gram) on alternatively-cured RTE ham stored at 4 ± 1°C, using modified Oxford medium.

resulted in an increased population by over 6 $\log_{10}$ CFU per g after 98 days. Previous work in our laboratory [19] showed that OA, when applied alone to naturally-cured frankfurters and RTE ham using similar protocols, exerted an initial bactericidal effect on *L. monocytogenes* but failed to inhibit the organism's recovery and growth over the refrigerated life of the products. Thus, the use of OA in combination with DV or LV1 X, similar to the effect of HHP, represents a necessary multiple-hurdle approach for *L. monocytogenes* in alternatively-cured processed meats.

The effects of using lauricarginate in combination with natural antimicrobial ingredients on viable *L. monocytogenes* populations are shown in Figure 3. Again, on day 1 of the study, LAE in combination with DV, 90 MX, and LV1 X resulted in 2.67, 2.37, and 2.16 $\log_{10}$ CFU per g reductions, respectively, in viable *L. monocytogenes* populations (*P*<0.05) compared to the control but which were not different (*P*>0.05) from each other. Similar to patterns observed when combining HHP and OA with the specified antimicrobial ingredients, sustained inhibition of the recovery and growth of *L. monocytogenes* was only observed when LAE was combined with the DV or LV1 X ingredients. When LAE was used in combination with the 90 MX ingredients, on the other hand, significant (*P*<0.05) increases in viable *L. monocytogenes* populations were observed from day 0 to day 14 of the study, with the increase reaching more than 7 $\log_{10}$ CFU per g by 56 days and after. These findings are similar to other reports that found lauric arginate will exert a bacteriostatic effect on the pathogen only when used in combination with lactate or diacetate [36,37]. Consequently, the combination of a LAE post-lethality intervention with DV or LV1 X, much like combining HHP and OA post-lethality interventions with those same natural antimicrobial ingredients represents another promising multiple-hurdle approach.

For the PPTT treatment, no significant reduction in viable *L. monocytogenes* populations was observed in any of the products with PPTT ($P>0.05$) when compared to the control treatment (Figure 4). The PPTT treatment has been shown to be a potentially effective post-lethality treatment [38], but in the current study, a longer heating time or a higher final temperature probably would have been necessary for the products to achieve significant population reduction under the conditions used.

Conclusions

As evidenced by our results, the use of high hydrostatic pressure at 400 MPa, octanoic acid, or lauric arginate as post-lethality interventions when used in combination with vinegar or vinegar and lemon juice concentrate represent effective multiple-hurdle approaches to control *L. monocytogenes* if post-processing contamination occurs in alternatively-cured RTE ham. These combination treatments inhibit the potential recovery and growth of those cells that might survive initial lethality treatments and that might remain viable during the refrigerated storage of the products. It should be noted that previous studies have shown that these post-lethality interventions will reduce the initial bacterial population but will not affect subsequent growth of survivors. Further, the antimicrobial ingredients used in this study did not affect initial population numbers but provided for suppression of subsequent growth. Thus, the combination of the appropriate post-lethality treatment with an effective bacteriostatic ingredient is necessary to assure control of *L. monocytogenes* on natural and organic ready-to-eat processed meat products. While these treatments did not independently achieve and sustain reduction of *L. monocytogenes* populations during product storage, the combination of these hurdles provides a means for manufacturers of natural and organic processed meat and poultry products to achieve control of *L. monocytogenes*.

Acknowledgement

This project was supported by the American Meat Institute Foundation. The authors would like to recognize the companies that donated research materials: Florida Food Products, WTI Ingredients, Ocean Spray International, EcoLab, Purac America, and Kalle USA. Special thanks to Devin Maurer and Daniel Fortin for the invaluable contributions made throughout the conduction of these experiments.

References

1. Winter CK, Davis SF (2006) Organic foods. J Food Sci 71: R117-R124.

2. Sebranek JG, Bacus JN (2007) Cured meat products without direct addition of nitrate or nitrite: What are the issues? Meat Sci 77: 136-147.

3. 2014 Press Releases (2014) American appetite for organic products breaks through$35 billion mark.

4. Bacus JN (2006) Natural ingredients for cured and smoked meats. Proceedings Recipe Meat Conf, American Meat Science Association, Champaign, USA.

5. Schrader KD (2010) Investigating the control of *Listeria monocytogenes* on uncured, no-nitrate- or-nitrite-added meat products Iowa State University, Ames, Ia, USA.

6. Sullivan GA (2011) Naturally cured meats: quality, safety, andchemistry.Iowa State University, Ames, Ia, USA.

7. Valenzuela-Martinez C, Pena-Ramos A, Juneja VK, Korasapati NR, Burson DE, et al. (2010) Inhibition of *Clostridium perfringens* spore germination and outgrowth by buffered vinegar and lemon juice concentrate during chilling of ground turkey roast containing minimal ingredients. J Food Prot 73: 470-476.

8. Patterson MF, Mackle A, Linton M (2011) Effect of high pressure in combination with antilisterial agents on the growth of *Listeria monocytogenes* during extended storage of cooked chicken. Food Microbiol 28: 1505-1508.

9. Sebranek JG, Jackson-Davis AL, Myers KL, Lavieri NA (2012) Beyond celery and starter culture: advances in natural/organic curing processes in the United States. Meat Sci 92: 267-273.

10. Sullivan GA, Jackson-Davis AL, Niebuhr SE, Xi Y, Schrader KD, Sebranek JG, et al. (2012) Inhibition of *Listeria monocytogenes* using natural antimicrobials in no-nitrate-or-nitrite-added ham. J Food Protect 75: 1071-1076.

11. Sullivan GA, Jackson-Davis AL, Schrader KD, Xi Y, Kulchaiyawat C, Sebranek JG, et al. (2012) Survey of naturally and conventionally cured commercial frankfurters, ham, and bacon for physio-chemical characteristics that affect bacterial growth. Meat Sci 92: 808-815.

12. http://www.fsis.usda.gov/shared/PDF/Controlling_LM_RTE_guideline_0912.pdf

13. De Alba M, Bravo D, Medina M, Park SF, MacKey BM (2013) Combined effect of sodium nitrite with high pressure treatments on the inactivation of *Escherichia coli* BW25113 and *Listeria monocytogenes* NCTC 11994. Letters Appl Microbiol 56: 155-160.

14. http://www.fsis.usda.gov/wps/wcm/connect/7f981741-94f1-468c-b60d b428c971152d/7120_68.pdf?MOD=AJPERES

15. Jensen RG (2002) The composition of bovine milk lipids. J Dairy Sci 85: 295-350.

16. Xi Y, Sullivan GA, Jackson AL, Zhou GH, Sebranek JG (2011) Use of natural antimicrobials to improve the control of *Listeria monocytogenes* in a cured cooked meat model system. Meat Sci 88: 503-511.

17. Xi Y, Sullivan GA, Jackson AL, Zhou GH, Sebranek JG (2012) Effects of natural antimicrobials on inhibition of *Listeria monocytogenes* and on chemical, physical and sensory attributes of naturally-cured frankfurters. Meat Sci 90: 130-138.

18. Larson AE, Yu RR, Lee OA, Price S, Haas GJ, et al. (1996) Antimicrobial activity of hop extracts against *Listeria monocytogenes* in media and in food. Internet J Food Microbiol 33: 195-207.

19. Lavieri NA, Sebranek JG, Brehm-Stecher BF, Cordray JC, Dickson JS, et al. (2014) Investigating the control of *Listeria monocytogenes* on a ready-to-eat ham product using natural antimicrobial ingredients and post lethality interventions. Foodborne Path Disease 11: 462-467.

20. AOAC International (1990) Fat (crude) or ether extract in meat.Official method 960.39.Official Methods of Analysis, Arlington, Va, USA.

21. AOAC International (1990) Moisture in meat.Official method 950.46.Official Methods of Analysis. Arlington, Va, USA.

22. AOAC International (1993) Crude protein in meat and meat products.Official method 992.15.Official Methods of Analysis. Arlington,Va, USA.

23. AOAC International (1990) Nitrites in cured meat.Official Method 973.31. Official Methods of Analysis. Arlington, Va, USA.

24. Kang DH, Fung DY (1999) Thin agar layer method for recovery of heat-injured *Listeria monocytogenes*. J Food Protect 62: 1346-1349.

25. Lee C, Reed JD, Richards MP (2006) Ability of various polyphenolic classes from cranberry to inhibit lipid oxidation in mechanically separated turkey and cooked ground pork. J Muscle Foods 17: 248-266.

26. Honikel KO (2008) The use and control of nitrate and nitrite for the processing of meat products. Meat Sci 78: 68-76.

27. Cassens RG, Ito T, Lee M, Buege D (1978) The use of nitrite in meat. Bio Sci 28: 633-637.

28. Hugas M, Garriga M, Monfort JM (2002) New mild technologies in meat processing: high pressure as a model technology. Meat Sci 62: 359-371.

29. Myers K, Montoya D, Cannon J, DicksonJ, Sebranek J (2013) The effect of high hydrostatic pressure, sodium nitrite and salt concentration on the growth of *Listeria monocytogenes* on RTE ham and turkey. Meat Sci 93: 263-268.

30. Park SW, Sohn KH, Shin JH, Lee HJ (2001) High hydrostatic pressure inactivation of *Lactobacillus viridescens* and its effects on ultrastructure of cells. Internet J Food Sci Tech 36: 775-781.

31. Ritz M, Tholozan JL, Federighi M,Pilet MF (2001) Morphological and physiological characterization of *Listeria monocytogenes* subjected to high hydrostatic pressure. Appl Environ Microbiol 67: 2240-2247.

32. Jung S, Tonello Samson C, de Lamballerie Anton M (2011) Processing of foods with high hydrostatic pressure. Alternatives to Conventional Food Processing. A Proctor (Ed.) RSC Publishing, Cambridge, UK.

33. Doona CJ, Feeberry FF, Ross EW, Kustin KC (2012) Inactivation   kinetics of *Listeria monocytogenes* by high pressure processing: Pressure and temperature variation. J Food Sci 77: M458-M465.

34. Burnett SL, Chopskie JH, Podtburg TC, Gutzmann TA, Gilbreth SE, et al. (2007) Use of octanoic acid as a postlethality treatment to reduce *Listeria monocytogenes* on ready-to-eat meat and poultry products. J Food Protect 70: 392-398.

35. Sun CQ, O'Connor CJ, Turner SJ, Lewis GD, Stanley RA, et al. (1998) The effect of pH on the inhibition of bacterial growth by physiologicalconcentrations of butyric acid: implications for neonates fed on suckled milk. Chemico-Biol Interact 113: 117-131.

36. Porto-Fett AC, Campano SG, Smith JL, Oser A, Shoyer B, et al. (2010) Control of *Listeria monocytogenes* on commercially-produced frankfurters prepared with and without potassium lactate and sodium diacetate and surface treated with lauricarginate using the Sprayed Lethality in Container (SLIC®) delivery method. Meat Sci 85: 312-318.

37. Luchansky JB, Call JE, Hristova B, Rumery L, Yoder L, et al. (2005) Viability of *Listeria monocytogenes* on commercially-prepared hams surface treated with acidic calcium sulfate and lauricarginate and stored at 4°C. Meat Sci 71: 92-99.

38. CM, Sebranek JG, Dickson JS, Mendonca AF (2004) Combining pediocin (ALTA 2341) with postpackaging thermal pasteurization for control of *Listeria monocytogenes* on frankfurters. J Food Protect 67: 1855-1865.

17

Effects of Vital Gluten Enrichment on Qualities of Value Added Products

Mekuria B and Emire SA*

Food, Beverage and Pharmaceutical Industry Development Institute, Ministry of Industry, Private mailbag 33381, Addis Ababa, Ethiopia

Abstract

The aim of this research was to study the effects of vital gluten enrichment on qualities of pasta flour, bread and biscuit products. The proportions of vital gluten powder added to wheat flour and the blend of wheat flour and corn flour were 3% and 6%. The Faringraph values, wet gluten, protein, ash, color grade and falling number of wheat flour, corn flour, the composite flours of wheat and corn as well as vital gluten blended flours were determined and compared with the normal flours used for pasta, bread and biscuit production. The effects of vital gluten enhancement on product qualities were assessed by formulating pasta flour, bread and biscuit products; respectively. Consequently, the highest wet gluten content of flours for 3% and 6% vital gluten supplemented flours were 41.8% and 46.67%; respectively. Vital gluten supplemented flours for bread production exhibited better wet gluten content. The protein content was higher (13.10%) than the normal bread (10.53%). Enhanced protein and crispiness of biscuit was obtained by blending soft wheat and corn flours and enriching them with vital gluten. The sensory qualities of both bread and biscuit products were acceptable by panelists. It was also possible to obtain suitable and quality pasta wheat flour by enriching the locally available hard wheat varieties with vital gluten. It is consequently critical to enhance the local wheat varieties with vital gluten to enlarge their protein qualities which can be used for protein enriched value added products manufacturing industries.

Keywords: Enrichment; Sensory quality; Value added products; Vital gluten; Wheat varieties

Introduction

Wheat is inimitable among all the cereal grains grown in many parts of the World as a stable food, and forms the basis for numerous food products. Several thousand varieties of wheat are known where the three main types include *Triticum durum* (durum wheat) largely used for pasta production, *Triticum vulgare* or aestivum mainly used for bread production and *Triticum compactum* (soft wheat) used for biscuit, pastry and cakes production.

The inimitable properties of wheat gluten are given an intense attention by the food industries. Interest has extended to the commercial separation of gluten from the starch and soluble proteins of flour for food production applications. In actual fact, it is the cohesive properties of gluten that makes its commercial preparation a relatively simple process. Vital wheat gluten is now a significant ingredient in the food industry and an important item of world trade [1]. Food industries are using vital wheat gluten as an ingredient especially in those countries which produces wheat with low protein content.

The wheat quality is intricate and is complicated to articulate in terms of a single property. It may be defined in terms of its suitability for a particular purpose or use [2]. Variety is an imperative factor that manipulates grain quality. Generally, wheat is marketed according to the class and each class consists of a group of varieties with similar characteristics apt for similar purposes and end-use [3]. In Ethiopia hard and soft wheat varieties are widely cultivated in the highlands of South East, Central and North West parts. The Ethiopian Agricultural Research Institute, Kulumsa Agricultural Research Center is currently performing research activities on various wheat cultivars to boost up the small scale farms income and thereby support the milling industries [4].

The incorporation of vital gluten in a bread recipe increases the protein value of the final loaf. It will also enhance the loaf volume and crumb structure better than the one without gluten supplementation [5]. It also increases the yield of final product by increasing the water absorption of dough and also the finished products will have a longer shelf life due to the finer grain and softer texture. Toufeili showed that loaf volume increases when low protein flours are fortified with

vital gluten [6]. On the other hand, Spychaj and Gil reported that supplementation of common wheat flour used for pasta production with vital gluten will reduce the cooking loss yielding good cooking quality [7]. Losses that occur due to cracks and broken pasta can be minimal if the pasta flour contains adequate amount of protein. The rheological characteristics of all flours and the blends were measured and analyzed by farinograph. The farinograph measures (as torque) and records the resistance to mixing of dough as it is formed from flour and water [8]. The peculiar visco-elastic properties of wheat dough are the result of the presence of a three dimensional network of gluten proteins.

The visco-elastic properties enable dough to retain gas which is essential for production of baked products with a light texture. Rheological properties such as elasticity, viscosity and extensibility are important in the prediction of the processing parameters of dough and quality of end product.

The farinograph curve (Figure 1) provides useful information for bakers striving to produce consistent products with flour of variable quality. It is used to determine flour strength and to predict processing characteristics like water absorption (%), development time (min), stability (min), degree of softening (FU), and farinograph quality number [9].

The farinograph values in general provide a means to select quality flour for bakers. The faringram below as taken from Brabender measurement and control system shows the farinograph parameters (Figure 1).

---

*Coresonding author: Emire SA, Food, Beverage and Pharmaceutical Industry Development Institute, Ministry of Industry, Addis, Ababa, Ethiopia
E-mail: shimelisemire@yahoo.com

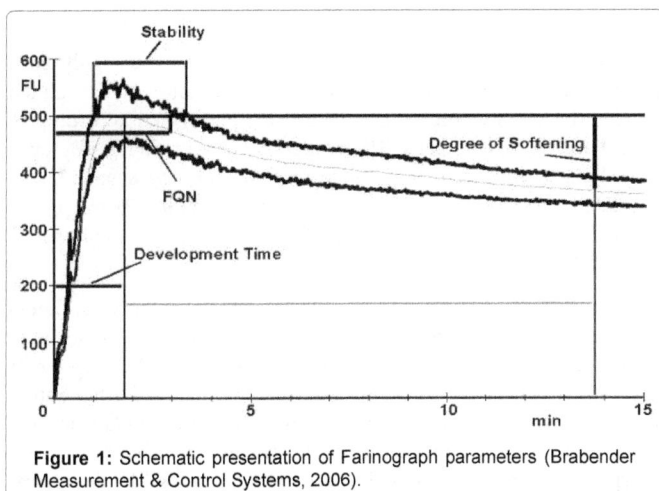

Figure 1: Schematic presentation of Farinograph parameters (Brabender Measurement & Control Systems, 2006).

The quality of wheat flour and flour products solely depends on the amount and quality of gluten contained by the flour used as an ingredient in producing such products. According to the wheat quality analysis report of Kaliti Food Share Company the protein content of the first grade hard wheat varieties that are used for pasta and bread making purposes are low [10]. As a result of this, most bread loaves made out of the local varieties have low water absorption, less volume, coarse grain and less soft crumb texture. The pasta products have high cooking loss, sticks after cooking (to each other or to the teeth or to the gum), are less tolerant to moderate over cooking, less firm and springy as well as have less smooth surface. Due to these quality attributes customers usually do not prefer pasta made out of local wheat. In order to satisfy the needs of their customers most of the pasta producers import high protein hard wheat. Import of wheat typically has high price, requires long purchasing processes and foreign currency. To alleviate these problems the production of vital gluten and supplementation of flours obtained from local hard wheat varieties is essential.

The inefficiency of utilizing other relatively cheaper cereal grains such as corn, sorghum, barely and rice for bread, pasta, biscuits and pastry production depends on the absence of gluten in these grains. The incorporation of these grains can be made possible through the supplementation of vital gluten to them. Gluten will provide them the necessary elasticity, gas retention capacity during fermentation process as well as increase their water absorption potential. These cereal grains are widely cultivated in our country and are readily available in the market but are less used for food. It is therefore, important to supplement flours of these cereals and utilize them for pasta, bread, biscuit, pastry and noodle production.

The other challenge that consumers face in their daily menu is that shortage of protein. Most of our foods are carbohydrate rich but less in protein. According to the wheat quality analysis report by Ethiopian Health and Nutrition Research Institute indicated that the protein contents of pasta and biscuit product samples collected from Kaliti Food Share Company, KFSC are 12% and 8%; respectively [10]. In order to produce protein affluent products, it is essential to supplement the flours of these products with vital gluten. This is because the prices of protein isolates such as soy protein isolate, whey protein, and caseinate are relatively high [1]. It is therefore essential to produce vital gluten from local wheat varieties and enrich the existing flour and flour products in order to obtain a good supply of protein.

The average protein content of locally released varieties is about 10% while the imported Russian milling wheat has a protein content of 14.5% [10]. Protein content and flour strength are the primary factors influencing pasta and bread quality. Wheat varieties with high protein content and strength usually yield bread loaves with good quality and pasta with firm and best cooking quality. It is therefore important to reconstitute local wheat flour with vital gluten. The need for the enrichment of the local hard wheat flours with vital gluten is a compulsory activity for bread and pasta producers in order to attain the best quality product in the Ethiopian context [11].

Taking into account the aforementioned facts, the production of vital gluten from hard wheat varieties that can be utilized in the enrichment of cereal based food products. In view of the importance of gluten production and the unparallel contribution of improved varieties; the present research was pursued with the rationale to evaluate the effects of vital gluten enrichment on quality of some value added food products.

Materials and Methods

Source of materials and sample preparation

Hard wheat varieties specifically Pavon 76, HAR 2501 (Hawi) and HAR 2536 (Simba) were released in 1982 and 2000 by the Ethiopian Agricultural Research Institute, Kulumsa Agricultural Research Center. All samples were collected from Arsi and Bale Agricultural Development Enterprises. All the wheat samples were collected from the two locations merely because the two areas are most known in high quality wheat production, and the released wheat varieties are available from small and private-owned large scale farmers. Moreover, wheat milling industries prefer the wheat varieties grown in these areas. Thus, the samples from the two locations represent the varieties. The samples were collected using random sampling system. They were then packed in polypropylene bags and brought to the laboratory. The bags were well tightened to avoid contamination and spillage. Each wheat variety was then thoroughly cleaned using 2 mm opening width laboratory sieve. Coarse impurities were handpicked. They were then packed in polypropylene bags again and stored at room temperature for analyses. Furthermore, graded wheat flour was collected from Kaliti Food Share Company (KFSC) while corn flour was purchased from the local market. All analyses were conducted by using analytical grade chemicals and reagents.

Blend formulation

Vital gluten powder was prepared by extracting gluten from wheat flour of the local variety HAR 2501 according to the method described by Kovacs [12]. The main process steps in gluten isolation include mixing, washing, pressing, drying, cooling, milling and packaging. Subsequently, a blend of wheat flour and vital gluten powder, corn flour and vital gluten powder and wheat flour, corn flour and vital gluten powder were formulated for subsequent applications in the preparation of pasta flour, bread and biscuit.

In the preparation of blended flour for pasta production, blends of local hard wheat flour with better gluten content and vital gluten powder, corn flour and vital gluten powder and hard wheat flour, corn flour and vital gluten powder were blended using mixer (Model: R100C, CAT). The 3% and 6% of vital gluten powder was added to both wheat and corn flours [7]. In addition to this, 13% of vital gluten powder was added to corn flour in order to see the potential of corn for pasta production. Wheat flour and corn flour were blended in 75 and 25% as well as 50 and 50% ratios; respectively.

In the preparation of blended flour for bread production similar ratio of vital gluten powder was used as that of blended flour for pasta production as indicated in Table 1. However, the wheat flour used for this purpose was low in gluten content than that used for pasta. 12% of vital gluten was added to corn flour this time to see how it will affect the characteristics of the flour to be used for bread production.

In the case of biscuit flour preparation wheat flour, corn flour and vital gluten powder were used in various ratios. Wheat flour and corn flour were mixed in 75:25, 50:50 and 25:75 ratios; respectively. All blended flours formulated for pasta, bread and biscuit products were characterized for moisture, protein, wet gluten, falling number, ash, color grade and particle size distribution. Based on the quality analyses results such as wet gluten, protein, falling number and faringraph values suitable blended flours were then selected for each product to be developed.

Preparation of pasta

Since there is no pilot pasta making machine in the country, the production of pasta with the vital gluten fortified flour was a limitation. However, the flour blends were characterized and compared with the normal pasta wheat flour. Nine blended flours were formulated and the quality characteristics of blended flours as well as pasta flour currently used by KFSC were analyzed. The quality characteristic of pasta flour was analyzed for comparison with the blended flours.

Bread processing

Bread was prepared by straight dough method bread production process (mixing and kneading, bulk fermentation, molding, rounding, intermediate proofing, molding, final proofing, baking, cooling and packaging). The normal flour, blended flours at ratios of 3% VGP + SWF, 3% VGP + (75% + 25 CF), 6% VGP + (75% SWF + 25 CF), 6% VGP + (50% SWF+ 50% CF) and 12% VGP + CF were employed in the production of bread.

The five blend formulations and the control flour were baked using the straight dough method [13]. The baking formula was 62% wheat flour or the blend, 0.3% yeast, 0.3% bread improver, 0.6% salt and 37.0% water. All ingredients were mixed in a dough mixer (Model: B15 mixer, England,1998) for 15 minutes. The dough was fermented in a bowl covered with polyethylene plastic for 30 minutes at room temperature. It was then knocked back and molded. The dough pieces were then allowed to ferment for 60 minutes in a proofing room of temperature 35°C and relative humidity of 80%. The fermented dough was baked at 250°C for 20 minutes [6,13-15].

Biscuit processing

Baking tests were carried out at the laboratory scale using the corn and vital gluten blend as well as wheat flour, corn flour and vital gluten blends. The biscuit processing method was performed according to Kaliti Food Share Company. The proportion of ingredients employed in the baking recipe were wheat flour or blends 77.82%, sugar 13.62%, salt 0.58%, shortening 3.89%, skimmed milk powder 0.19%, NH_4HCO_3 1.17%, $NaHCO_3$ 0.23%, $KHC_4H_4O_6$ 0.08%, $Na_2S_2O_5$ 0.01% and vanilla flavor 0.08%.

The production process in biscuit were creaming, mixing, dough relaxation, dumping and cutting, conveying, laminating (sheeting), shaping and cutting, baking, cooling and packaging. All ingredients other than wheat flour or blended flour were first creamed in laboratory mixer (model B15, China) for 5 minutes at high speed (294 rpm) and then mixed with flour and kneaded. The mixing time was at low speed (91 rpm) for 5 minutes and high speed for 10 minutes. The dough was allowed to relax, covered with polyethylene sheet for 30 minutes at room temperature (22°C). It was then laminated, stamped and cut in to pieces. The dough pieces were transferred to baking pans and baked at a temperature of 250°C for 20 minutes. The product was then cooled at ambient temperature and packed in polypropylene films.

Analysis methods

Proximate analysis: Proximate chemical composition analysis of the blended flours for pasta, biscuit and bread processing and their products including moisture, total ash, crude protein and crude fat were analyzed according to AOAC official methods 925.09, 923.03, 979.09 respectively of the official methods of analysis of AOAC (2005) International. Total carbohydrates including crude fiber were calculated by difference. Crude protein and crude fat were analyzed using Kjeldahl block digestion and system distillation (2200 Kjeltec Autodistillation, Foss Tecator, Sweden) and Sox Tec Service Unit 1046 (Foss Tecator, Sweden); respectively. The bread and biscuit products were analyzed for moisture, crude protein, crude fat, ash and total carbohydrate and water activity.

Moisture content: The moisture content of the wheat samples was determined by rapid moisture tester according to the method described by ICC No. 110/1. The samples were well mixed to attain uniformity and ground using laboratory grinder (Bühler). About 10 gram of the sample was placed in the moisture testing plate made of stainless steel. It was then placed in the oven for 10 minutes at a temperature of 130°C.

Blended flours	Parameters						
	Moisture (%)	Color grade	Wet gluten (%)	Ash (%)	Protein (%)	FN (sec.)	Particle size (< 180 µ)
SWF	14.0 ± 0.15a	-0.13 ± 0.07[a]	27.67 ± 0.51[d]	0.59 ± 0.01[a]	10.53 ± 0.16[c]	344 ± 7[a]	100 ± 0.00[a]
CF	10.57 ± 0.00[d]	8.69 ± 0.14[g]	ND	1.23 ± 0.05[f]	7.43 ± 0.39[e]	278 ± 17[b]	99.27 ± 0.49[a]
SWF + 3VGP	12.97 ± 0.06[b]	0.75 ± 0.36[b]	30.53 ± 0.61[c]	0.64 ± 0.03[b]	12.29 ± 0.32[b]	335 ± 8[a]	77.97 ± 12.1[d]
CF + 3VGP	11.13 ± 0.15[c]	9.01 ± 0.16[h]	ND	1.25 ± 0.05[f]	8.81 ± 0.51[d]	273 ± 8[b]	99.20 ± 0.10[a]
BFB1 +3VGP	11.7 ± 0.30[c]	2.52 ± 0.29[d]	29.8 ± 0.26[c]	0.83 ± 0.03[c]	13.10 ± 0.40[a]	277 ± 15[b]	97.63 ± 1.88[b]
SWF + 6VGP	12.77 ± 0.06[b]	1.50 ± 0.48[c]	32.50 ± 0.20 a	0.72 ± 0.02[c]	13.22 ± 0.21[a]	318 ± 10[c]	99.67 ± 0.49[a]
CF + 6VGP	10.23 ± 0.06[d]	7.09 ± 0.06[f]	ND	1.13 ± 0.01[e]	12.44 ± 0.28[b]	239 ± 15[e]	64.63 ± 6.43[e]
BFB1 + 6VGP	11.70 ± 0.00[c]	3.17 ± 0.42[d]	31.23 ± 0.38[b]	0.80 ± 0.01[c]	13.46 ± 0.25[a]	270 ± 4[c]	98.5 ± 1.57[b]
BFB2 + 6VGP	10.73 ± 0.15[d]	4.69 ± 0.24[e]	22.20 ± 0.53[e]	0.91 ± 0.01[d]	12.43 ± 0.38[b]	255+6[d]	98.17 ± 0.59[b]
CF+ 12 VGP	9.00 ± 0.15[e]	7.82 ± 0.43[f]	3.23 ± 0.59[f]	1.11 ± 0.02[e]	12.54 ± 0.36[b]	255 ± 16[d]	90.2 ± 2.75[c]

All [a-h] values are means of triplicate ± SD on dry matter basis means followed by different superscript within the same raw differ significantly (P ≤ 0.05)

Where: - ND- Not detected; SWF: Soft Wheat Flour; CF: Corn Flour; VGP: Vital Gluten Powder; BFB1 = 75% SWF + 25% CF; BFB 2= 50% SWF + 50% CF

Table 1: Effects of vital gluten enhancement on physico-chemical quality characteristics of blended flours for bread production (All values are in dry matter basis)

The moisture content of flour samples, vital gluten, bread and biscuit products as well as non gluten flour were determined using infrared moisture analyzer (MBA 310, England) where 10 gm of each sample was spread on clean aluminum foil plate and placed in the moisture analyzer. The temperature of the tester was adjusted at a temperature of 130°C and drying was carried out for 10 minutes.

Protein: The protein content of wheat samples was measured by grain analyzer (Minifra-2000T) where clean wheat samples were placed in a sample cell (cuvette) and protein content of the samples were read and recorded from the instrument. Protein content of the wheat flour, vital gluten powder and blended flours, bread and biscuit products were measured using kjeldahl apparatus according to the method described by ICC No105/2.

Fat: The fat content of vital gluten, bread and biscuit samples were determined using the soxhlet extraction method using petroleum ether as a solvent.

Ash: The ash content of wheat, flour and gluten samples were analyzed by burning the samples in a muffle furnace (Mr 170 E, Germany, 2004) at a temperature of 600°C according to the ICC standard method ICC No 104/1.

Total carbohydrates: The total carbohydrate content of both the bread and biscuit products was determined by difference.

Granularity (particle size): The granularity of both wheat flour and gluten powder and the blends were determined by sieving 50 g of the samples using 180 mesh size sieves (ES 1052:2005). The percentage through (granulation) of each sample was then calculated using equation 1:

$$\% \text{ Through} = \frac{W_{th}}{W_f} * 100 \tag{1}$$

Where: W_{th} = weight of flour through the sieve, W_f = initial weight of flour

Color grade: The color grade value of the flour samples was measured to make sure the brightness of the flours. About 30 g of flour sample was placed in a beaker containing 50 ml distilled water and made in to a paste by continuously stirring it with a glass rod for 45 seconds. The paste was then poured in to the sample cell and the sample cell containing the paste was inserted into the instrument and then the result was displayed within 90 seconds.

Falling number: Both the wheat and flour samples were evaluated for falling number using the Hagberg falling number apparatus (model 1500, Sweden, Perten Instruments, 2005) according to the ICC standard No 107/1 for the determination of the amylase activity of cereal and flour.

Farinograph measurement: The farinograph values such as water absorption, stability, dough development, degree of softening and farinograph quality number of the flours of the three varieties and blended flours were measured using the Brabender Farinograph according to the standard method of ICC No 115/1 for the determination of the farinograph values of wheat flour. The instrument automatically determines the amount of flour to be poured into the mixer of the farinograph based on the moisture content of the flour. The farinograph is equipped with a 300 g capacity mixer. Mixing was carried out for 20 minutes. The speed of the torque was adjusted to be 63 min^{-1}.

Water activity analysis: The water activity of vital gluten powder, bread and biscuit samples were determined using Aqua Lab Lite water activity measuring unit manufactured by Decagon a_w meter (Aqua Lab Lite, 2004). Each sample of the gluten powder, bread and biscuit samples were half filled in a small plastic cup supplied with the instrument and inserted in to the instrument. At last, the water activity of each sample was displayed automatically.

Sensory quality attributes of bread and biscuit products: The sensory attributes used to determine the sensory qualities of bread and biscuit samples include appearance, texture, flavor (taste + odor), color, crispiness and overall acceptability. The samples were evaluated by 25 trained consumer panelists to assess acceptability of the bread and biscuit samples using (a five-point hedonic scale) with 5 representing the highest score (excellent) and 1 the lowest score (poor) for every quality parameter. All samples for sensory evaluation were conducted at room temperature.

Experimental design: Data were analyzed using analysis of variance (ANOVA) followed by Least Significance Difference (LSD) at 5% level of significance. Statistical analysis was performed using SPSS/15 software for windows.

Results and Discussion

Effects of vital gluten enhancement on physico-chemical quality characteristics of blended flours for pasta production

The normal pasta flour prepared from foreign hard wheat currently used for pasta production by KFSC has a wet gluten content of 41.33% and a crude protein content of 13.30%. Local hard wheat flour blended with 3% VGP (3% VGP + HWF/ Hard Wheat Flour/) has a wet gluten content of 41.80% and a protein content of 14.20% (Table 2). The protein content of blended flours (3% VGP + HWF) had a higher protein than the normal pasta flour (14.20>13.3). The blend of wheat flour, corn flour and vital gluten powder (BFP1 + 6% VGP) can also be utilized for pasta production because this blended ratio of flour has produced 43.9% wet gluten and 15.85% protein. The results revealed that by blending (75% HWF + 25% CF) + 6% VGP it was possible to get higher wet gluten and protein than the normal pasta flour. The blend of 13% VGP and corn flour has given better protein content 19.07% flour than the normal flour but produced low wet gluten content. This may be due to the hindrance of corn husk particles present in corn flour caused by improper milling during the development of dough. The blend of 6% VGP and HWF has the highest wet gluten and protein contents which are 46, 67% and 16.12%; respectively. This blend of flour can produce the best pasta. Thus, from the blend ratios it can be deduced that the production of cooked pasta aldente quality to meet customer expectations and criteria for instance bright color, clean and smooth surface, firmness, springiness, lack of stickiness and cooking loss, tolerance to moderate overcooking and to have a high protein pasta is possible by incorporation of vital gluten powder in local hard wheat varieties.

The color grade value and ash content of blended flours from wheat flour, corn flour and vital gluten powder is slightly higher than the normal flour. This is due to the fact that corn flour whose outer cover (husk) was not completely removed in this experiment. The color of corn flour can be adjusted by using proper corn milling process.

Farinograph values of blended flours for pasta production

The rheological properties of flour dough were measured by farinograph instrument and the results of farinograph values of blended flours for pasta production are presented in Figures 2a-2e. The water

Blended flours	Parameters						
	Moisture (% w/w)	Color grade	Wet gluten (% w/w)	Ash (% w/w)	Protein (% w/w)	FN (sec)	Particle size (< 180 µm)
PF	14.43 ± 0.42[f]	-.77 ± 0.57[a]	41.33 ± 0.45[c]	0.61 ± 0.01[b]	13.30 ± 0.51[d]	400 ± 0.0[0]	98.6 ± 0.37
HWF	14.03 ± 0.06[f]	-2.33 ± 0.14[a]	35.03 ± 0.91[e]	0.55 ± 0.03[a]	11.27 ± 0.51[e]	387 ± 16	99.27 ± 0.35
CF	10.57 ± 0.38[c]	8.69 ± 0.14[f]	ND	1.23 ± 0.05[d]	7.43 ± 0.39f	272 ± 8	99.27 ± 0.49
HWF +3VGP	13.63 ± 0.06[e]	-0.56 ± 0.04[a]	41.8 ± 0.10[c]	0.64 ± 0.00[b]	14.20 ± 0.23[c]	359 ± 6	78.22 ± 0.11
CF +VGP	11.13 ± 0.15[d]	9.01 ± 0.16[f]	ND	1.25 ± 0.06[d]	8.81 ± 0.51[f]	273 ± 8	99.20 ± 0.07
BFP1 +3VGP	10.37 ± 0.15[c]	1.81 ± 0.08[c]	37.17 ± 0.35[d]	0.82 ± 0.03[b]	11.79 ± 0.25[e]	296 ± 15	99.55 ± 0.40
BFP2 +3VGP	9.10 ± 0.10[b]	4.27 ± 0.04[d]	27.90 ± 0.56[f]	1.10 ± 0.02[c]	11.10 ± 0.09[e]	261 ± 17	77.49 ± 0.94
HWF +6VGP	13.23 ± 0.12[a]	0.23 ± 0.09[b]	46.67 ± 0.87[a]	0.75 ± 0.01[b]	16.12 ± 0.16[b]	331 ± 2	97.90 ± 1.45
CF +6VGP	10.23 ± 0.06[c]	7.09 ± 0.06[e]	3.43 ± 0.70[h]	1.13 ± 0.01[c]	12.44 ± 0.28[d]	239 ± 15	64.63 ± 6.42
BFP1 +6VGP	10.50 ± 0.36[c]	1.95 ± 0.17[c]	44.03 ± 0.60[b]	0.90 ± 0.02[c]	15.85 ± 0.33[b]	291 ± 5	96.22 ± 5.12
BFP2 +6VGP	9.40 ± 0.10[b]	4.01 ± 0.04[d]	34.77 ± 2.08[e]	1.09 ± 0.04[c]	14.04 ± 0.17[c]	277 ± 5	99.53 ± 0.25
CF + 13VGP	8.26 ± 0.38[a]	9.11 ± 0.04[f]	16.60 ± 1.32[g]	1.30 ± 0.06[d]	19.07 ± 0.10[a]	232 ± 4	83.68 ± 13

All [a-h] values are means of triplicate ± SD on dry matter basis means followed by different superscript within the same raw differ significantly (P ≤ 0.05)

Where: ND- not detected; PF: Pasta Flour; HWF: Hard Wheat Flour; CF: Corn Flour; VGP: Vital Gluten Powder; BFP1 = 75% HWF + 25% CF; BFP2 = 50% HWF + 50% CF

Table 2: Effects of vital gluten enhancement on physico-chemical quality characteristics of blended flours for pasta production (All values are in dry matter basis).

absorption of the normal flours and those of the blended ones corrected to 14% moisture basis ranged between 58.5 and 65.0. The highest water absorption was recorded for BFP1 + 6VGP. The faringraph quality number ranged between 61 and 122 in that the highest value was recorded for PF. The highest values of stability and dough development were obtained for PF. The HWF + 3VGP and HWF+ 6VGP have exhibited the characteristics of strong flour with respect to faringraph quality number, stability and dough development time. Hence they are suitable for pasta production.

Effects of vital gluten enhancement on quality characteristics of blended flours for bread production

The moisture content of all blended flours ranged between 9.00 and 14.00 which are within the standard requirement as stated by Ethiopian standard (ES, 2005). The blend of CF + 12% VGP has the lowest moisture content while the normal soft wheat flour has showed the highest value. The blended flour of CF + 3% VGP has the highest color value 9.01 while the normal soft wheat flour has the lowest color grade value. Low values of color grade are the preferred value in producing white bread. Proper milling of corn must be carried out using corn mill to maintain the color value of flour.

The highest wet gluten value was obtained for the flour blend SWF + 6VGP which is presented in Table 2. The highest values can be preferred by bread bakers for they absorb much water, increase the protein content of bread, impart better gas retention and increase the volume of loaf [1,5]. The highest protein content of blended flours was recorded for BFB1 + 6% VGP (75% SWF + 25% CF + 6% VGP) while the lowest value was recorded for that of corn flour.

Farinograph values of blended flours for bread production

The Farinograph values of blended flours for bread production are presented in Figures 3a-3f. The flour blends SWF + 3% VGP has the highest stability (mixing tolerance) while the flour blend CF + 12% VGP has the lowest value. With regard to dough development, the highest value was recorded for BFB2 + 6% VGP. The flour blend SWF + 3% VGP has also the highest farinograph value indicating that the flour is strong flour. The highest values of stability, dough development and farinograph values of the flour indicate that this flour blend is strong flour and suitable for bread production [6]. The highest water absorption was recorded for the flour blend BFB2 + 6% VGP. This may

be due to inclusion of corn flour in the blend. The lowest value was achieved for the flour blend SWF + 3% VGP. Thus the flour blends SWF + 3% VGP, BFB1+ 3% VGP, SWF + 6% VGP and BFB1 + 6% VGP were suitable for bread production. It was also possible to incorporate 25% of corn flour in the bread production recipe.

The farinograms for CF + 12 VGP and BFB2 + 6% VGP blended flours indicates that they are abnormal curves as depicted in Figures 3e and 3f. This may be due to the fact that the inclusion of corn husk in the corn flour hindered the development of gluten.

Proximate composition and water activity of bread products

The proximate composition and water activity of the normal and vital gluten supplemented flours were used in the production of bread. The proximate composition and a_w of bread loaves are summarized in Table 3. The highest crude protein 11.33 was recorded for soft wheat flour supplemented with 3% VGP while the lowest was for 50% SWF+ 50% CF + 6 VGP. Accordingly, the protein content of bread increased as a result of the addition of vital gluten. The crude fat content of all samples ranged between 0.99 and 2.30 where the highest fat content was recorded for BFB1 + 3% VGP. This may be due to the high oil content of corn flour which was prepared without de-germination of corn. The moisture content of bread ranged between 33.70 and 39.93 with the bread made with BFB1 + 3% VGP having the highest value. The values for all samples are optimal and may not have adverse effect on the quality attributes of bread (ES, 2005). The highest ash content of bread (060) was obtained for CF + 9% VGP blended flour, and this might be due to the high ash content of corn flour as shown in Table 2.

The loaf volume of bread produced from the normal and blended flours with codes 010,020,030,040, 050 and 060 were 1385, 1327.5, 1565, 1092.5, 1501 and 394; respectively (Figures 4a-4f). The highest volume was observed for bread with code 030 and the lowest to be for that of 06. In general terms the volume of bread loaves made with codes 030 and 050 yielded the highest volumes. The increase in loaf volume upon the incorporation of vital gluten is the expected result as indicated by investigators [5]. From this, it can be concluded that the addition of vital gluten to low protein wheat can increase the volume of bread and hence better quality of bread can be produced. The loaves of bread obtained from the normal and blended flours are shown in

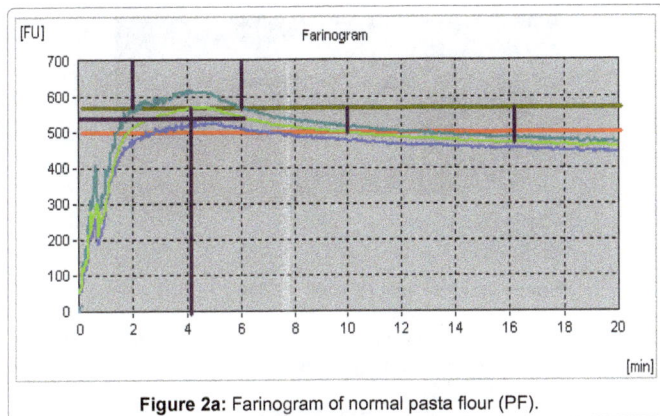

Figure 2a: Farinogram of normal pasta flour (PF).

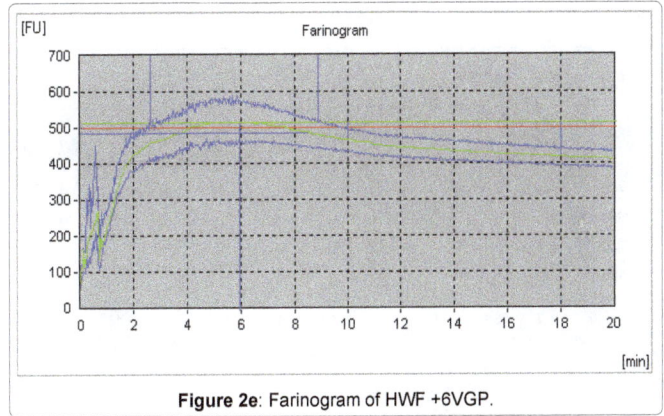

Figure 2e: Farinogram of HWF +6VGP.

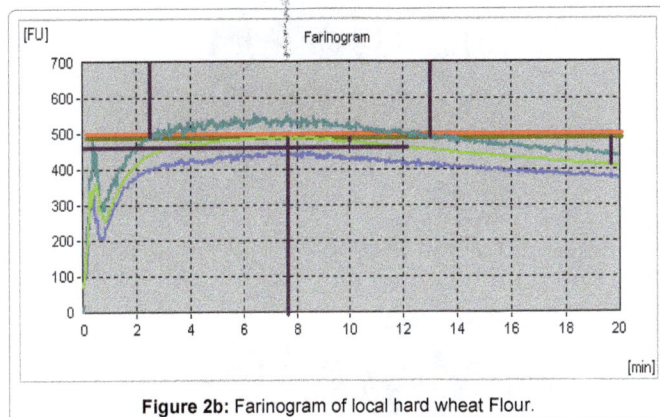

Figure 2b: Farinogram of local hard wheat Flour.

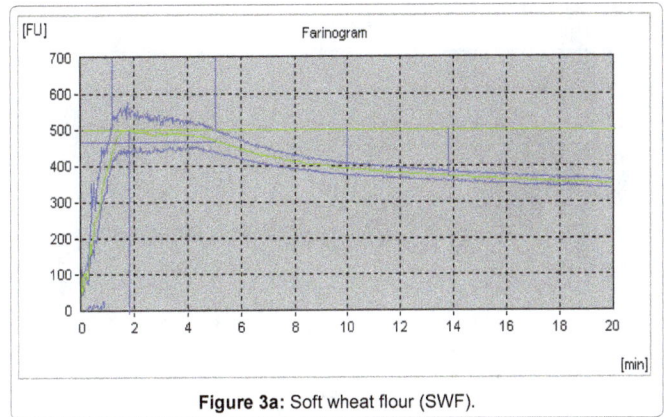

Figure 3a: Soft wheat flour (SWF).

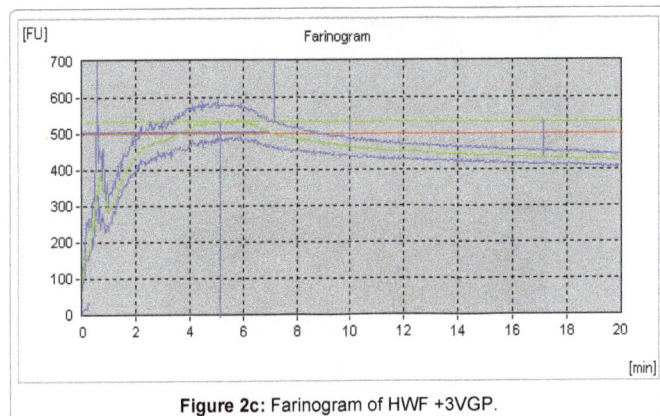

Figure 2c: Farinogram of HWF +3VGP.

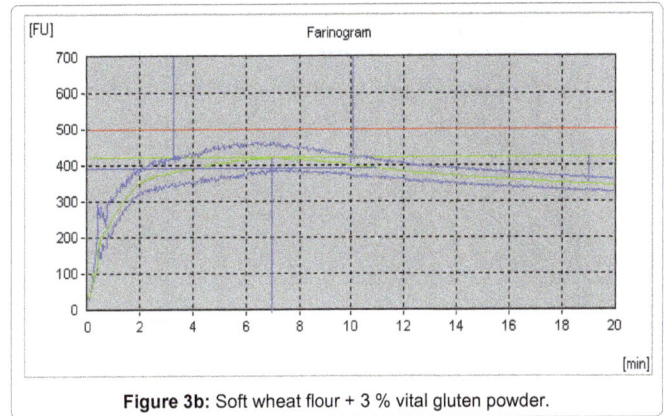

Figure 3b: Soft wheat flour + 3 % vital gluten powder.

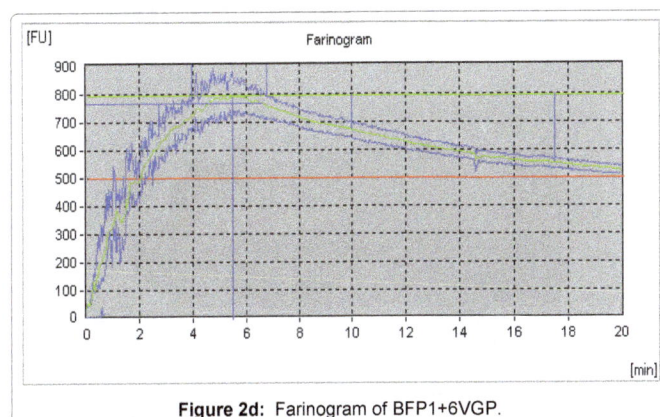

Figure 2d: Farinogram of BFP1+6VGP.

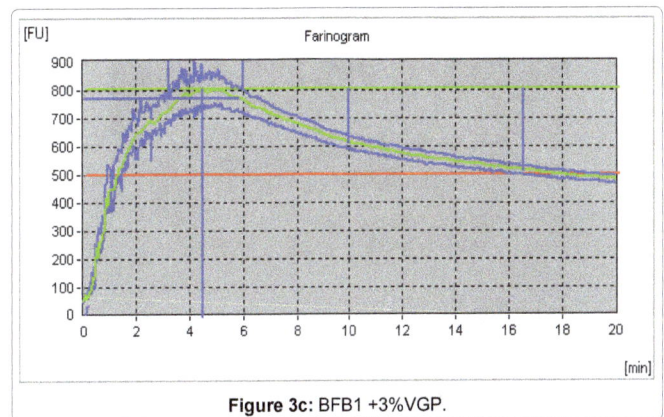

Figure 3c: BFB1 +3%VGP.

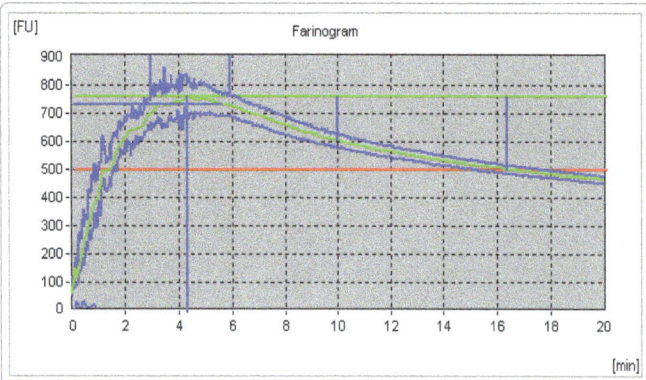

Figure 3d: BFB1 +6%VGP).

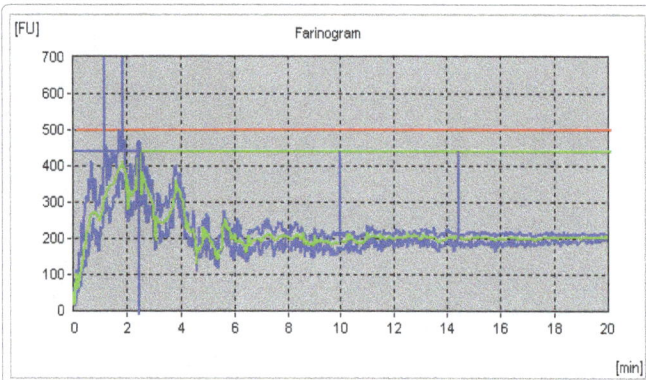

Figure 4b: Bread from 75%SWF + 25 % CF + 3 % VGP

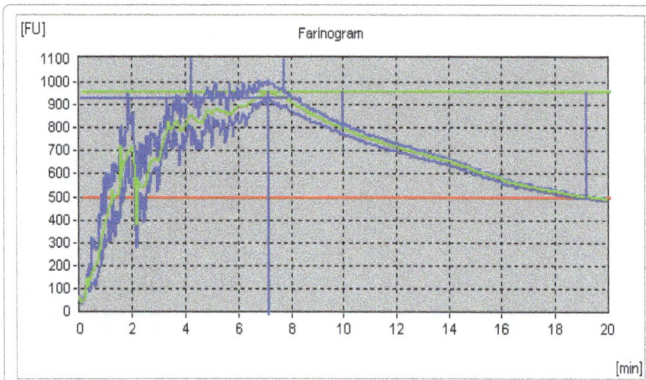

Figure 3e: CF+12%VGP.

Figure 4c: Bread from SWF + 3 % VGP.

Figure 3f: BFB2 +6%VGP.

Figure 4d: Bread from (50%WF + 50 % CF) + 6 % VGP.

Figure 4e: Bread from 75%WF + 25 % CF + 6% VGP.

Figure 4a: Bread from normal wheat.

Figure 4f: Bread from CF + 12 % VGP.

Figures 4a-4f.

Effects of vital gluten enhancement on quality characteristics of blended flours for biscuit production

The moisture content of all flour samples ranged from 8.60% to 14.0% while the protein content ranged from 7.43 to 13.46%. Blended flours BFB1 + 3VGP, BFB2 + 3VGP< BFB3 + 3 VGP, BFB1 + 6 VGP, BFB2 + 6 VGP, BFB3 + 6VGP had high protein values (Table 4) which in turn were implicated in the final biscuit product. The color grade values were obtained to be between -0.13 to 9.01. Negative color grade values are quality indicator of brightness of flour [16].

The more negative color grade value, the brighter is the flour (Color Grader Series 4 PCGA ,1997). The visco-elasticity of flour is less demanded by biscuit producers as a result of which the farinograph values of blended flours for biscuit were not determined.

Proximate composition and water activity of biscuit products

The protein content of biscuit products ranged from 8.83 to 11.76% (Table 5) and this result has similarity with the findings obtained by Olaoye [17]. The highest value was recorded for the product with code 060 (CF+ 9% VGP) while the lowest value was for 010(normal soft wheat flour). Thus, incorporation of vital gluten in to soft wheat flour or corn flour and the blend of both would improve the nutritional value of biscuit. The highest fat content of biscuit was obtained for product with code 060 and the same was true for the ash content. The highest total carbohydrate value was obtained for the biscuit with code 050(25% SWF + 75% CF + 6%VGP). From the proximate composition analysis it was observed that biscuit product with better protein was obtained by enhancing soft wheat and corn flours with vital gluten. It

was also possible to incorporate corn flour up to 75% in the biscuit production recipe.

Sensory quality evaluation of bread and biscuit

The sensory hedonic mean scores of the bread and biscuit samples are presented in Tables 6 and 7. The mean scores for sensory attributes decreased in increasing the proportion of corn flour in the blend. The highest value was recorded for bread made with 3%VGP and SWF while the lowest value was for 12% VGP and CF. Based on the sensory quality evaluation, bread made with 3% VGP and soft wheat flour is the best quality bread which fulfilled most of the quality attributes. The samples with code 010, 020, 030 and 050 are generally acceptable by consumers while the samples with code 040 and 060 were less acceptable by the panelists due to their appearance, texture and flavor or overall acceptability.

The sensory hedonic mean scores of the biscuit samples are shown in Table 7. The mean scores decreased in increasing the proportion of corn flour in the blend. The highest value 4.33 was recorded for biscuit made with the normal soft wheat flour while the lowest value 2.33 was for 9% VGP and CF mixed flour.

From the results obtained it is possible to produce biscuits with good overall acceptability by most blending ratios except for the blend CF+ 9%VGP. The blend CF + 9% VGP have low overall acceptability.

Conclusions

The enrichment of vital gluten powder to local wheat varieties flours which have low protein (10.53%) or the blends of wheat flour and corn flour were satisfactory in producing better protein flours (14.20%) for pasta and bread manufacturing. The enrichment of vital gluten with corn flour produced acceptable protein blended flour. However, it is

Code	Moisture(%)	Crude protein (%)	Crude fat (%)	Ash (%)	Total carbohydrates (%)	a_w
010	38.10 ± 2. 66[b]	9.10 ± 0.50[c]	0.99 ± 0.54[d]	1.04 ± 0.13[c]	50.77 ± 3.01[c]	0.93 ± 0.004[b]
020	39.93 ± 3.59[a]	10.62 ± 0.66[b]	2.30 ± 1.07[a]	0.75 ± 0.68[a]	46.39 ± 3.22[e]	0.93 ± 0.007[b]
030	38.67 ± 1.10[b]	11.33 ± 0.16[a]	1.19 ± 0.63[c]	0.97 ± 0.07[b]	47.85 ± 1.79[d]	0.93 ± 0.006[b]
040	36.60 ± 3.84[c]	8.62 ± 0.64[d]	0.85 ± 0.39[e]	1.53 ± 0.07[e]	52.40 ± 3.08[a]	0.94 ± 0.004[a]
050	33.70 ± 2.85[d]	10.74 ± 0.92[c]	1.24 ± 0.86[b]	1.42 ± 0.19[d]	52.90 ± 2.99[a]	0.91 ± 0.028[c]
060	36.53 ± 4.36[c]	9.20 ± 1.13[d]	1.27 ± 0.55[b]	1.65 ± 0.57[f]	51.34 ± 3.33[b]	0.90 ± 0.013[d]

All [a-f] values are means of triplicate ± SD on dry matter basis;   Means followed by different superscript within the same raw differ significantly (P ≤ 0.05)

Where:- Bread samples, 010= normal soft wheat; **020**= 50% SWF + 50% CF + 3% VGP; **030** = 25% SWF + 75% CF + 3% VGP; **040**= 75% WF + 25% CF +3% VGP; **050** = 25% SWF + 75% CF + 6% VGP and **060** = 06=CF + 9% VGP.

Table 3: Proximate composition and water activity of bread.

Blended flours	Parameters						
	Moisture (%)	Color grade	Wet gluten (%)	Ash (%)	Protein (%)	FN (sec.)	Particle size (< 180 µ)
SWF	14.0 ± 0.15[a]	-0.13 ± 0.07[a]	27.67 ± 0.51[c]	0.59 ± 0.01[a]	10.53 ± 0.16[d]	344 ± 7[a]	100 ± 0.00[a]
CF	10.57 ± 0.00[c]	8.69 ± 0.14[g]	ND	1.23 ± 0.05[f]	7.43 ± 0.39[f]	278 ± 17[b]	99.27 ± 0.49[a]
CF + 3VGP	11.13 ± 0.15[b]	9.01 ± 0.16[h]	ND	1.25 ± 0.05[f]	8.81 ± 0.51[e]	273 ± 8[b]	99.20 ± 0.10[a]
BFB1 +3VGP	11.70 ± 0.30[b]	2.52 ± 0.29[b]	29.8 ± 0.26[b]	0.83 ± 0.03[b]	13.10 ± 0.40[a]	277 ± 15[b]	97.63 ± 1.88[c]
BFB2+3VGP	10.80 ± .20[c]	2.52 ± 0.29[b]	21.1 ± 0.10[e]	0.94 ± 0.02[c]	13.20 ± 0.09[a]	258 ± 5[c]	77.9 ± 0.90[e]
BFB3+3VGP	10.13 ± 0.12[c]	5.46 ± 0.34[e]	8.43 ± 2.54[f]	1.06 ± 0.05[d]	11.75 ± 0.20[c]	252 ± 1[c]	69.43 ± 7.47[f]
CF + 6VGP	10.23 ± 0.06[c]	7.09 ± 0.06[f]	ND	1.13 ± 0.01[e]	12.44 ± 0.28[b]	239 ± 15[e]	64.63 ± 6.43[f]
BFB1 + 6VGP	11.70 ± 0.00[b]	3.17 ± 0.42[c]	31.23 ± 0.38[a]	0.80 ± 0.01[b]	13.46 ± 0.25[a]	270 ± 4[b]	98.5 ± 1.57[b]
BFB2 + 6VGP	10.73 ± 0.15[c]	4.69 ± 0.24[d]	22.20 ± 0.53[d]	0.91 ± 0.01[c]	12.43 ± 0.38[b]	255 ± 6[c]	98.17 ± 0.59[b]
BFB3+6VGP	9.63 ± 0.25[d]	6.23 ± 0.46[f]	6.60 ± 2.10[g]	1.03 ± 0.03[d]	12.23 ± 0.64[b]	247 ± 7[d]	87.13 ± 6.03[d]
CF + 9VGP	8.60 ± 0.20[e]	7.67 ± 0.09[f]	ND	1.19 ± 0.06[e]	11.44 ± 0.36[c]	242 ± 1[d]	88.46 ± 5.75[d]

All [a-g] values are means of triplicates ± SD; Means followed by different superscript within the same raw differ significantly (P ≤ 0.05).

Table 4: Effects of vital gluten enhancement on quality characteristics of blended flours for biscuit production (All values are in dry matter basis).

Code	Moisture (%)	Crude protein (%)	Crude fat (%)	Ash (%)	Total carbohydrates (%)	a_w
010	9.35 ± 0.21[a]	8.83 ± 0.51[d]	4.28 ± 0.32[d]	1.20 ± 0.18[d]	76.34 ± 0.89[b]	0.55 ± 0.02[d]
020	5.87 ± 0.57[d]	9.45 ± 0.12[c]	4.27 ± 0.30[d]	1.33 ± 0.34[e]	79.09 ± 0.27[a]	0.45 ± 0.01[b]
030	7.68 ± 0.59[c]	9.31 ± 0.40[c]	5.85 ± 0.28[b]	1.21 ± 0.19[d]	75.96 ± 0.60[c]	0.52 ± 0.01[c]
040	8.57 ± 0.59[b]	10.74 ± 0.33[b]	4.57 ± 0.48[c]	0.88 ± 0.18[a]	75.23 ± 0.95[c]	0.54 ± 0.01[d]
050	4.47 ± 0.12[e]	10.27 ± 0.33[b]	4.46 ± 0.23[c]	1.00 ± 0.33[b]	79.80 ± 0.69[a]	0.37 ± 0.03[a]
060	5.89 ± 0.62[d]	11.76 ± 0.24[a]	6.46 ± 0.89[a]	2.11 ± 0.51[c]	73.78 ± 0.96[d]	0.44 ± 0.01[b]

All[a-e] values are means+SD of triplicates;-Means followed by different superscript within the same raw differ significantly (P ≤ 0.05)

Where:- Biscuit samples, 010= normal soft wheat; **020**= 50%SWF + 50 % CF + 3 % VGP; **030** = 25% SWF + 75% CF + 3% VGP; **040**= 75%WF + 25 % CF + 3% VGP; **050** = 25% SWF + 75% CF + 6 % VGP and **060** = 06= CF + 9% VGP

Table 5: Proximate composition and water activity of biscuit (All values are in dry matter basis).

Types of bread	Sensory attributes			
	Appearance	Texture	Flavor	Overall acceptability
010	4.07 ± 0.76[a]	4.27 ± 0.76[a]	4.07 ± 0.76[b]	3.93 ± 0.81[b]
020	4.13 ± 0.46[a]	3.97 ± 0.40[b]	4.00 ± 0.35[b]	4.10 ± 0.52[a]
030	4.20 ± 0.20[a]	4.20 ± 0.20[a]	4.40 ± 0.53[a]	4.13 ± 0.12[a]
040	1.87 ± 0.12[c]	1.47 ± 0.12[d]	2.53 ± 0.23[d]	1.47 ± 0.12[d]
050	3.07 ± 0.12[b]	3.33 ± 0.12[c]	3.67 ± 0.23[c]	3.67 ± 0.12[c]
060	1.33 ± 0.12[d]	1.27 ± 0.12[e]	1.80 ± 0.20[e]	1.20 ± 0.00[e]

All[a-e] values are means of triplicates+SD; Means followed by different superscript within the same raw differ significantly (P ≤ 0.05)

Table 6: Sensory quality evaluation of bread.

Types of biscuit	Sensory attributes					
	Color	Appearance	Texture	Flavor	Taste	Overall acceptability
010	3.80 ± 0.20[a]	4.33 ± 0.23[a]	4.27 ± 0.12[a]	4.00 ± 0.20[a]	4.33 ± 0.31[a]	4.27 ± 0.12[a]
020	3.13 ± 0.70[c]	3.00 ± 0.20[b]	3.07 ± 0.31[c]	3.13 ± 0.12[c]	3.27 ± 0.31[c]	3.00 ± 0.53[c]
030	3.27 ± 0.42[b]	2.87 ± 0.31[c]	2.87 ± 0.31[d]	3.00 ± 0.35[d]	2.87 ± 0.12[e]	2.87 ± 0.50[d]
040	2.80 ± 0.53[d]	2.87 ± 0.12[c]	3.00 ± 0.20[c]	3.20 ± 0.20[b]	3.13 ± 0.31[d]	3.07 ± 0.31[c]
050	3.73 ± 0.12	3.33 ± 0.12	3.33 ± 0.42[b]	3.13 ± 0.23[c]	3.67 ± 0.70[b]	3.40 ± 0.40[b]
060	2.87 ± 0.31	2.33 ± 0.64	2.67 ± 0.64[d]	2.47 ± 0.81[d]	2.80 ± 0.40[e]	2.33 ± 0.23[e]

All[a-e] values are means of triplicates+SD; Means followed by different superscript within the same raw differ significantly (P ≤ 0.05)

Where:- Biscuit samples: 010= normal soft wheat; **020**= 50% SWF + 50% CF + 3% VGP; **030** = 25% SWF + 75% CF + 3 % VGP; **040**= 75%WF + 25% CF +3% VGP; **050** = 25% SWF + 75% CF + 6% VGP and **060** = 06= CF + 9% VGP

Table 7: Sensory quality attributes of biscuit.

hardly possible to get a good yield of wet gluten from the blend of corn flour and vital gluten due to the oily nature and husk inclusion in the corn flour. This problem can perhaps be avoided using typical corn milling machines.

In the formulation of value added food products (pasta, bread and biscuit) the enhancement of vital gluten revealed appreciable results. An equivalent flour quality as compared with the imported pasta wheat flour was obtained from the blends. It is therefore possible to incorporate vital gluten to local hard wheat varieties for the production of pasta without importing foreign wheat for pasta production. The utilization of corn in the production of pasta was acceptable up to a percentage of 25% together with wheat flour and vital gluten powder.

Increased volume (1565 cm³) and nutritional quality bread products were obtained from the blends of vital gluten and soft wheat flour, and corn flour without alteration of the normal bread production process. The research work has indicated that the utilization of locally available wheat varieties via enhancement with vital gluten powder for the production of value added products. The sensory evaluation of bread and biscuit samples showed acceptable results by panelists.

Furthermore, the present research encourages food technologists, breeders and agronomists in various Agricultural Research Centers and Higher Learning Institutes, who are currently striving hard to release high quality wheat varieties which in turn boost up wheat growers to produce large quantities of wheat from potential areas and win over import substitution.

References

1. Day L, Augustine MA, Batey IL, Wrigley CW (2006) Wheat gluten uses and Industry needs. Trends in Food Science and Technology 17: 82-90.

2. Pasha I, Anjum FM, Morris CF (2010) Grain hardness: A major determinant of wheat quality. Food Science and Technology International 16: 511-522.

3. Halverson J, Zeleny L (1988) Criteria of wheat quality. In: Pomeranz Y (ed). Wheat Chemistry and Technology. St Paul. MN.

4. Crop Development Department (2005) Crop Variety Register, Addis Ababa, Ethiopia.

5. Czuchajowska Z, Paszczynska B (1996) Is wet gluten good for baking? Journal Cereal Chemistry 73: 483-489.

6. Toufeili I, Ismail B, Shadarevian S, Baalbaki R, Khatkar B, et al. (1999) The role of gluten proteins in the baking of Arabic Bead. Journal of Cereal Science 30: 255-265.

7. Spychaj R, Gil Z (2005) Effects of adding dry gluten powder to common wheat flour on the quality of pasta. Electronic Journal of Polish Agricultural Universities 8.

8. ICC (2000) Association of International Cereal science and Technology. Vienna, Austria.

9. Nordson BKG GmbH (2006) Instruction manual of farinograph-E with USB port. Duisburg, Germany.

10. KFSC (2006) Kaliti food share company (KFSC) quality analysis report of wheat collected from Arsi and Bale Agricultural Development Enterprises. Addis Ababa, Ethiopia.

11. Mekuria B, Admassu S (2012) Grain quality evaluation and characterization of vital gluten powder from bread wheat varieties grown in Arsi and Bale, Ethiopia. East African Journal of Sciences 5: 35-41.

12. Kovacs MIP, Fu BX, Woods SM, Khan K (2004) Thermal stability of wheat gluten protein: its effect on dough properties and noodle texture. Journal of Cereal Science 39: 9-19.

13. Federation of bakers (2002) The Federation of Bakers of the UK's largest baking companies manufacture sliced and wrapped bread, bakery snacks and other products.

14. Edema MO, Sanni LO, Abiodun I (2005) Evaluation of Maize-soybean flour blends for sour maize bread production in Nigeria. African Journal of biotechnology 4: 911-918.

15. Olaoye OA, Onilude AA, Idowu OA (2006) Quality characteristics of bread produced from composite flours of wheat, plantain and soy beans. African Journal of Biotechnology 5: 1102-1106.

16. Technical manual of Color Grader (Series 4) PCGA (1997). Satake Corporation UK Division, England.

17. Olaoye OA, Onilude AA, Oladoye CO (2008) Breadfruit flour in biscuit making: effects on product quality. African Journal of Food Science 1: 20-23.

18

Physical, Textural and Sensory Characteristics of Gluten Free Muffins Prepared with Teff Flour (Eragrostistef (ZUCC) Trotter)

Tess M, Bhaduri S, Ghatak R and Navder KP*

CUNY School of Public Health at Hunter College, 2180 Third Avenue, 10035, New York, USA

Abstract

Since the enrichment of gluten-free cereal products is not mandatory there is a need for improving nutritional content of gluten-free diets by incorporating alternative gluten-free grains that are naturally abundant in nutrients. This study examined the effects of substitution of rice flour (control) with teff flour at 25%, 50%, 75% and 100% on the physical, textural, and sensory characteristics of gluten free muffins. A decrease in height of baked muffins was observed with an increase in the percentage of teff flour. Muffins with 75% and 100% teff flour had very viscous batters with significantly lower line spread tests compared to control rice muffins. Specific gravity was not significantly different between teff muffins, but all teff variations were significantly lower than the control. Textural measurements made using TA.XT Plus Texture Analyzer (Texture Technologies Corp., Scarsdale, NY) showed no significant difference between the control, 25% and 50% teff muffins but the 75% and 100% teff muffins were significantly harder. Springiness was significantly lower when the teff muffins were compared to the control, but no differences were found between teff variations. Substitutions up to 50% with teff flour were acceptable to the panelists. Friedman's rank test showed no significant difference in the overall liking between control, 25% and 50% teff muffins. This study demonstrates that substituting 50% rice flour with teff not only produces acceptable gluten free muffins, but these are more nutritious because of their higher protein (27%), iron (2095%), calcium (25%) and fiber (221%) contents.

Keywords: Teff flour; Gluten-free muffin; Texture; Sensory

Introduction

Celiac disease, also known as gluten sensitive enteropathy, is an autoimmune chronic disease causing inflammation of the upper small intestine in genetically predisposed individuals. It is triggered by ingestion of wheat gliadin and prolamins of rye and barley [1] which release these peptides during digestion and cause flattening of the intestinal mucosa [2] due to loss of normal villi and inflammation [3] leading to mal-absorption of nutrients like iron, folic acid, calcium and fat-soluble vitamins [4,5]. The cornerstone treatment of celiac disease is a lifelong strict withdrawal of wheat, rye, barley and an adherence to a gluten free diet [3]. Many of the gluten-free products are not enriched, and therefore do not provide the same levels of thiamin, riboflavin, niacin [6] iron and folate found in enriched and fortified wheat products [7]. Recent studies have consequently shown nutritional inadequacy of these nutrients associated with the gluten-free diet [8,9]. Since the fortification and enrichment of gluten-free cereal products is not mandatory in the US, there is a need for improving the nutritional content of gluten-free diet by incorporating alternative gluten-free grains that are naturally abundant in these nutrients.

Teff [Eragrostistef (ZUCC) Trotter] is a grain commonly used in Ethiopia. Its small size (1-1.5 mm) prevents the separation of the germ from the endosperm in teff flour [10]. It is reported to have a higher content of iron, calcium, phosphorus, copper, and thiamine compared to other grains like, wheat, barley, and sorghum [11]. It is also reported to be free of gliadin [12,13] and could be suitable for use in the diet of patients with celiac disease [14,15].

The goal of this study was to test the acceptability of gluten free muffins made with teff flour, and to compare the physical, textural and sensory properties of teff muffins with control muffins made with rice flour.

Materials and Methods

Muffin preparation

The muffin formulation [16] is presented in Table 1. Muffins were prepared with 0%, 25%, 50%, 75%, and 100% teff flour as a replacement for rice flour (both flours were provided by Bob's Red Mill, Milwaukie, Oregon). Milk, oil, and egg were mixed together for 1 min at speed 5 with an electric hand mixer (Kitchen Aid Ultra Power 5). Flour, sugar, baking powder, and salt were mixed together in a separate bowl, and then were sifted into with the wet ingredients at speed 4 for 10 seconds. Muffin pans were filled with the batter (65-66 g each) and were baked for 21 minutes or until done at 204°C in a preheated oven. Following a five-minute setting period, muffins were removed from the pans and allowed to cool on wire racks for one hour after which analyses were performed.

Physical measurements

A Vernier caliper (Monostat Corp, Merenschwand, Switzerland) was used to measure height and percent increase/decrease in height was determined from initial and final heights. Initial and final weights were obtained using a top loading electronic balance (OHAUS-Explorer, Pinebrook, NJ). Percent moisture loss upon baking was determined from weight of muffin batter and weight of muffin after baking. Moisture was determined by moisture analyzer (OHAUS-Explorer, Pinebrook, NJ). Specific gravity was measured using a pycnometer (Fisher Scientific, Pittsburg, PA). Line spread test was performed using a line spread chart. All tests were performed in triplicate.

Texture analysis

Texture profile analysis (TPA) of muffins was performed using

***Corresponding author:** Navder KP, CUNY School of Public Health at Hunter College, 2180 Third Avenue, 10035, New York, USA
E-mail: knavder@hunter.cuny.edu

TA.XT Plus Texture Analyzer (Texture Technologies Corp, Scarsdale, NY). The instrument was equipped with a 5 kg load cell and calibrated to a force sensitivity of 1 g. The test was performed on cubes (2.5 cm side) taken from the center of the muffin. The test speed was 5 mm s^{-1} at 75% of the original height; the post test speed was 5 mm s^{-1} and there was a 5 s interval between the two compression cycles. A trigger force of 5 g was selected. The compression of 75% was performed with a 36 mm diameter acrylic cylinder probe, and the cubes were compressed twice. The TPA primary parameters hardness, springiness, and cohesiveness and the secondary texture parameter chewiness were calculated from the curves [17].

Sensory evaluation

The study was approved by the Hunter College Institutional Review Board. Muffins were evaluated by 89 untrained consumers. The panelists had to be at least 18 years of age and not allergic to any food in order to participate in this study. Panelists were informed that they would be evaluating gluten free muffins, and they were presented with control muffin first (coded "000") and then with the other four samples. The order of presentation of the teff muffins was also random. The panelists were asked to evaluate the samples in relation to the control muffin. Appearance, flavor, taste, and overall liking was evaluated using a 5-point hedonic scale, with 1 for "dislike extremely" and 5 for "like extremely". They were also instructed to rank the products in the order in which they liked them, with 1 for "least liked" and 5 for "most liked". They were also asked how often they ate muffins, if they had tried gluten free products before, and if anyone in their family had celiac disease. Water and unsalted crackers were provided to panelists to cleanse their palates between samples. Data acquisition was done using FIZZ software (Biosystems, France).

Nutritional analysis

Nutrient content of muffins was analyzed using Nutritionist Pro software (Axxya Systems, 2007, Stafford, TX). The Nutritionist Pro food database was expanded by adding analysis of teff and rice flour provided by Bob's Red Mill. The calories, fat, fiber, protein, iron, and calcium content of each muffin was analyzed and the changes in nutrients were calculated.

Statistical analysis

Sensory and instrumental data were analyzed using one-way analysis of variance with post-hoc testing by Least Significant Difference (LSD) multiple comparisons using SPSS (version 18 for Windows 2008, SPSS Inc, Chicago, IL). $P < 0.05$ was considered significant. Friedman two-way analysis of variance based on ranks was conducted using FIZZ software (Biosystemes, France).

Results and Discussion

Physical measurements

The objective measurements are shown in Table 2. As the percentage of teff flour increased, the muffin batter became more viscous and dense, had low line spreads because of less fluidity, and had low specific gravity. The teff flour muffins also were more compact and had reduced heights and volume. Similar decrease in bread volume was also found when resistant starch (corn, tapioca or a combination) was used in gluten-free bread [18]. Moisture content of the final products revealed no difference between the control, 25%, 50% and 75% teff muffins. Moisture loss was not significantly different between the control, 25% and 50% teff muffins, while the 75% and 100% teff muffins had significantly greater moisture loss.

Texture profile analysis

A typical texture profile analysis curve obtained for muffins is shown in Figure 1 and the primary TPA parameters-hardness, springiness and cohesiveness, and secondary parameter of chewiness are shown in Table 2.

Hardness, defined as the maximum peak force during the first compression cycle (first bite) (Figure 1), was not increased in the 25% and 50% teff muffins when compared to the control, but the 75% and 100% variations were significantly harder. Similar results with an increase in hardness were found in breads prepared with 10%, 20%, and 30% buckwheat flour [19]. However, the substitution of corn and potato starch in gluten-free bread with varying proportions of resistant starch did not significantly change the hardness of bread crumb [20].

Springiness is related to the height that the food recovers during the time that elapses between the end of the first bite and the start of the second bite (Figure 1). Springiness is associated with freshness in a product with a high quality muffin having higher springiness values. Control rice muffin was significantly springier than any variation of teff muffins. There was no difference in the springiness of the various teff muffin variations when compared to each other. Similar results were also found when 10%, 20% and 30% of corn or tapioca starch was incorporated in gluten-free bread with no significant effect on springiness [20].

Cohesiveness is defined as the ratio of the positive force during the second compression to that during the first compression (Figure 1), this parameter is the strength of internal bonds which make up the body of the product. All teff muffins had significantly decreased cohesiveness compared to the control rice muffin, and lower compression energy was required. Typically, a more cohesive product retains more gas, and has a higher volume. No difference was seen in both, cohesiveness and percentage increase in height in control, 25% and 50% teff muffins,

Ingredients (g)	100% Rice (Control)	Teff 25%	Teff 50%	Teff 75%	Teff 100%
Rice flour	200.0	50.0	100.0	150.0	200.0
Teff flour	-	150.0	100.0	50.0	-
Sucrose	47.6	47.6	47.6	47.6	47.6
Baking powder	5.6	5.6	5.6	5.6	5.6
Salt	4.0	4.0	4.0	4.0	4.0
Milk (2% milk fat)	174.2	174.2	174.2	174.2	174.2
Egg	76.0	76.0	76.0	76.0	76.0
Oil	53.4	53.4	53.4	53.4	53.4

Table 1: Formulas for muffins.

but the 75% and 100% teff muffins had decreased cohesiveness and consequently, lower volume. This was most clearly evident in 100% teff muffin, where the decrease in cohesiveness (12.5%) was even visually apparent, as this muffin crumbled easily with simple manipulation.

Chewiness is measured as the product of hardness, cohesiveness and springiness (Figure 1) and is defined as the energy needed to masticate a solid food to a state ready for swallowing. Changes in chewiness parameter varied, a decrease in chewiness was observed with up to 50% substitution, while an increase was seen with 75% and 100% teff muffins.

Sensory evaluation

The mean sensory scores for appearance, flavor, taste and overall liking are presented in Table 2, and the Friedman sum of rank scores are shown in Figure 2.

Appearance of control rice muffins was significantly different from any of the teff muffins, which was excepted since teff flour has a dark brown color. No significant differences were found in flavor, taste and overall liking of control, 25% and 50% teff muffins. Similar results were found when corn flour was substituted with amaranth flour in gluten-free sponge cake [21]. Comparable scores were also obtained when

control carrot cake (corn flour) was compared with gluten-free carrot cake, where corn flour was substituted with 64% milled linseed flour [21]. The flavor, taste and overall liking of 75% and 100% teff muffins were rated significantly lower than control, 25% and 50% teff muffins. Mean scores for overall liking for control, 25% and 50% teff muffins were between 4 (like slightly) and 3 (neither like nor dislike), while the mean scores for 75% and 100% teff muffins were between 3 (neither like nor dislike) and 2 (dislike slightly). This data indicates that 50% rice flour can be substituted with teff flour without any significant changes in overall liking of muffins. Increasing beyond 50% substitution with teff flour resulted in unacceptable muffins from sensory point of view. Another study showed that the overall acceptability of cookies supplemented with flax seed flour decreased with increase in flax seed flour from 20% to 30% [22].

Friedman's rank test showed no significant difference in ranking based on liking among control, 25% teff, and 50% teff muffins (Table 2). Muffins prepared with 75% and 100% teff flour received lower rankings. Figure 2 shows a comparison between sum of ranks from panelists who had not (n=50) versus panelists who had consumed (n=39) gluten free products. The gluten-free product consumers found

Characteristic	100% Rice (Control)	25% Teff	50% Teff	75% Teff	100% Teff
Physical measurements of batters and baked muffins					
Line spread (mm)	39.7 ± 5.39	35.4 ± 5.24	26.5 ± 5.37	21.5 ± 7.39	21.6 ± 8.91
Specific gravity	1.14 ± 0.010	1.11 ± 0.049	1.10 ± 0.297	1.09 ± 0.279	1.10 ± 0.310
Moisture (%)	41.9 ± 0.593	41.7 ± 0.568	42.0 ± 0.637	42.1 ± 0.841	42.4 ± 0.728
% increase in height	113.0 ± 4.93	113.2 ± 3.89	113.0 ± 4.48	109.9 ± 4.95	107.6 ± 6.19
% moisture loss	13.8 ± 0.818	13.9 ± 1.46	13.9 ± 0.745	14.4 ± 0 .996	14.7 ± 0.771
Texture profile analysis parameters					
Hardness (N)	562.1 ± 94.5	597.0 ± 95.2	593.3 ± 67.8	687.0 ± 91.72	812.9 ± 128.2
Springiness (mm)	1.38 ± 0.640	1.10 ± 0.392	1.08 ± 0.347	1.06 ± 0.322	0.99 ± 0.307
Cohesiveness	0.830 ± 0.030	0.793 ± 0.025	0.781 ± 0.037	0.750 ± 0.029	0.732 ± 0.016
Chewiness (N.mm)	639.7 ± 324.8	524.6 ±185.1	508.3 ± 174.4	549.6 ±149.5	590.8± 93.66
Sensory evaluation					
Appearance	4.20 ± 0.7	3.83 ± 0.87	3.73 ± 0.85	3.65 ± 1.08	3.62 ± 1.11
Flavor	3.37 ± 0.92	3.49 ± 0.94	3.36 ± 0.96	2.72 ± 1.15	2.55 ± 1.09
Taste	3.33 ± 0.9	3.51 ± 0.98	3.33 ± 1.02	2.65 ± 1.16	2.51 ± 1.10
Overall liking	3.36 ± 0.91	3.47 ± 0.97	3.34 ± 1.04	2.72 ± 1.12	2.61 ± 1.11
Friedman ranking test (sum of ranks)	299.0	295.0	284.0	236.0	221.0

Table 2: Physical measurements, texture profile analysis, and palatability ratings, overall liking, and ranking ratings of muffins.

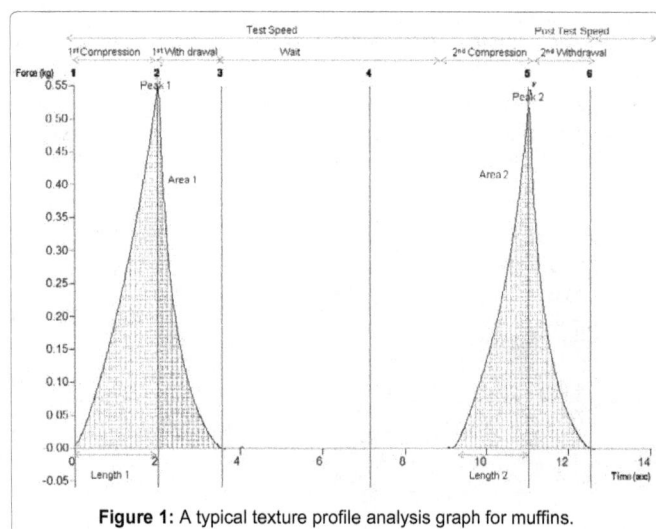

Figure 1: A typical texture profile analysis graph for muffins.

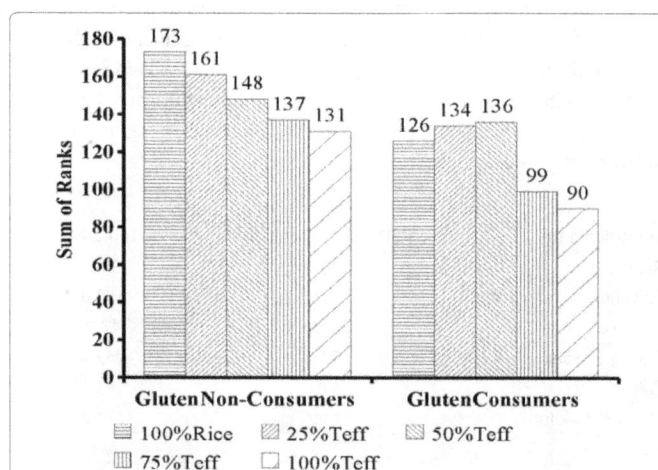

Figure 2: Friedman's sum of ranks test for control rice muffin and teff muffins for non-consumers.

Nutrient/ Serving (1 muffin, 55 g)	100% Rice (Control)	25% Teff	50% Teff	75% Teff	100% Teff
Calories (kcal)	184	184.6	185.3	185.9	186.5
Increase in calories		0.34%	0.69%	1.03%	1.38%
Fat (g)	7.796	7.899	8.001	8.104	8.207
% Increase in Fat		1.32%	2.63%	3.95%	5.27%
Fiber (g)	0.571	1.204	1.838	2.471	3.105
% Increase in Fiber		110.8%	221.9%	332.7%	443.8%
Protein (g)	3.154	3.582	4.01	4.438	4.866
% Increase in protein		13.57%	27.14%	40.71%	54.28%
Iron (mg)	0.304	3.488	6.673	9.857	13.042
% Increase in Iron		1047%	2095%	3142%	4190%
Calcium (mg)	58.894	66.312	73.73	81.148	88.566
% Increase in Calcium		12.59%	25.19%	37.78%	50.38%

[1]Values are means ± standard deviations.
[2]Means in same column with same letters are not significantly different at $P<0.05$ determined by ANOVA
[3] Values were determined with TA.XT Plus Texture Analyzer (5-kg capacity) (Texture Technologies Corp., Scarsdale, NY) equipped with 36 mm diameter acrylic cylinder probe.
[4] Values are means ± standard deviations or sum of ranks
[5]Palatability ratings, overall liking, and ranking ratings of muffins as evaluated by 89 untrained panelists
[6]Muffins were evaluated on a 5-point hedonic scale: 1=dislike extremely, 2=dislike slightly, 3=neither like nor dislike, 4=like slightly, 5=like extremely.
[7]Muffins were ranked using the Ranking Test scale: 1=least liked and 5=most liked.

Table 3: Nutrition information for gluten free muffins.

the 50% teff muffin acceptable whereas consumers who had not tried gluten free products liked the variation only up to 25% level. Panelists who had never consumed gluten free products found no difference between the control and 25% teff but had significantly lower ratings for the 50, 75 and 100% teff muffins. However, panelists who had previously consumed gluten-free products ranked the 50% teff muffins highest and no significant difference was found between the control, 25 and 50% teff muffins.

Nutrition analysis

Results of nutrition analysis are presented in Table 3. Substitution of rice flour with teff flour at all levels increased fiber (111-444%), iron (1047-4190%), and calcium (30-50%) content of the muffins. This is important since adequate amounts of these nutrients are usually lacking in the gluten-free diet and recent studies have shown nutritional inadequacy associated with these diets [7,8]. In addition, since many gluten free products are neither enriched, fortified nor naturally rich sources of fiber and micronutrients [7], celiac disease patients do not meet recommended nutrient intakes [23,24]. Another study demonstrated that nutrient profile of the gluten-free diet was improved by incorporating quinoa grain in the diet grains [25]. This present study showed that incorporation of teff flour at 50% level produced gluten free muffins that were an excellent source of iron and a good source of dietary fiber.

Conclusions

Since the overall acceptability was unchanged and the substitution of rice flour with teff flour improved nutritional content, our results show that replacing rice flour up to 50% with teff produces acceptable muffins. Nutrition analysis showed that substitution of rice flour up to 50% with teff produces muffins that are excellent source of iron and good source of dietary fiber. Since many gluten free products are neither naturally rich nor enriched or fortified this study confirms that teff flour substitution improves the nutritional value of gluten-free muffins.

This study could be useful to dietetics professionals who can recommend teff flour as an alternative for patients with celiac disease.

The apparent improved nutritional profile could benefit the nutritional adequacy in celiac patients.

Acknowledgements

The authors would like to thank Bob's Red Mill for generously providing the flours, Benhilda Wekwete for helping with the instrumental analysis, and Biosystemes, FIZZ sensory software for providing trial license for sensory data acquisition.

References

1. Murray JA (1999) The widening spectrum of celiac disease. Am J ClinNutr 69: 354-365.

2. McGough N, Cummings JH (2005) Coeliac disease: A diverse clinical syndrome caused by intolerance of wheat, barley and rye. Proc Nutr Soc 64: 434-450.

3. Wieser H, Koehler P (2008) The biochemical basis of celiac disease. Cereal Chem 85: 1-13.

4. Feighery C (1999) Fortnightly review - coeliac disease. Br Med J 319: 236-239.

5. Niewinski MM (2008) Advances in celiac disease and gluten-free diet. J Am Diet Assoc 108: 661-672.

6. Thompson T (1999) Thiamin, riboflavin, and niacin contents of the gluten-free diet: Is there cause for concern? J Am Diet Assoc 99: 858-862.

7. Thompson T (2000) Folate, iron, and dietary fiber contents of the gluten-free diet. J Am Diet Assoc 100: 1389-1396.

8. Hallert C, Grant C, Grehn S, Grännö C, Hultén S, et al. (2002) Evidence of poor vitamin status in coeliac patients on a gluten-free diet for 10 years. Aliment Pharmacol Ther 16: 1333-1339.

9. Dickey W, Kearney N (2006) Overweight in celiac disease: Prevalence, clinical characteristics, and effect of a gluten-free diet. Am J Gastroenterol 101: 2356-2359.

10. Mohammed MIO, Mustafa AI, Osman GAM (2009) Evaluation of wheat breads supplemented with teff (eragrostistef (ZUCC.) trotter) grain flour. Australian Journal of Crop Science 3: 207-212.

11. Mengesha MH (1966) Chemical composition of teff (eragrostistef) compared with that of wheat barley and grain sorghum. Econ Bot 20: 268-273.

12. Spaenij Dekking L, Kooy Winkelaar Y, Koning F (2005) The Ethiopian cereal tef in celiac disease. N Engl J Med 353: 1748-1749.

13. Hopman E, Dekking L, Blokland M, Wuisman M, Zuijderduin W, et al. (2008) Tef in the diet of celiac patients in the Netherlands. Scand J Gastroenterol 43: 277-282.

14. Mamo T, Parsons JW (1987) Iron nutrition of eragrostis-tef (teff). Trop Agric 64: 313-317.

15. Hall A, Kassa T, Demissie T, Degefie T, Lee S (2008) National survey of the health and nutrition of school children in ethiopia. Tropical Medicine and International Health 13: 1518-1526.

16. Johnson FCS (1988) Utilization of american-produced rice in muffins for gluten-sensitive individuals. Family and Consumer Sciences Research Journal 17: 175-183.

17. Bourne MC (1978) Texture profile analysis.

18. Korus J, Witczak M, Ziobro R, Juszczak L (2009) The impact of resistant starch on characteristics of gluten-free dough and bread. Food Hydrocoll 23: 988-995.

19. Torbica A, Hadnadev M, Dapcevic T (2010) Rheological, textural and sensory properties of gluten-free bread formulations based on rice and buckwheat flour. Food Hydrocoll 24: 626-632.

20. Krupa U, Rosell CM, Sadowska J, Soral Smietana M (2010) Bean starch as ingredient for gluten-free bread. J Food Process Preserv 34: 501-518.

21. Gambus H, Gambus F, Pastuszka D, Wrona P, Ziobro R, et al. (2009) Quality of gluten-free supplemented cakes and biscuits. Int J Food Sci Nutr 60: 31-50.

22. Hussain S, Anjum FM, Butt MS, Khan MI, Asghar A (2006) Physical and sensoric attributes of flaxseed flour supplemented cookies. Turk J Biol 30: 87-92.

23. Thompson T, Dennis M, Higgins LA, Lee AR, Sharrett MK(2005) Gluten-free diet survey: Are americans with coeliac disease consuming recommended amounts of fibre, iron, calcium and grain foods? J Hum Nutr Diet 18: 163-169.

24. Ohlund K, Olsson C, Hernell O, Ohlund I (2010) Dietary shortcomings in children on a gluten-free diet. Journal of Human Nutrition and Dietetics 23: 294-300.

25. Lee AR, Ng DL, Dave E, Ciaccio EJ, Green PH (2009) The effect of substituting alternative grains in the diet on the nutritional profile of the gluten-free diet. J Hum Nutr Diet 22: 359-363.

19

Influence of Meat Processing on the Content of Organochlorine Pesticides

Muresan C[1]*, Covaci A[2], Socaci S[1], Suharoschi R[1], Tofana M[1], Muste S[1] and Pop A[1]

[1]Faculty of Food Science and Technology, University of Agricultural Sciences and Veterinary Medicine, Cluj-Napoca, Romania
[2]Toxicological Center, University of Antwerp, Universitetsplein, Wilrijk B-2610, Belgium

Abstract

The purpose of this study was to identify pathways to reduce Organochlorine pesticides (OCP) contamination, through various thermal methods used for meat processing and to establish mathematic models predictive for the influence of thermal treatments on the OCP residues content in meat. By cold smoking, a reduction of less than 1% in the OCP content was observed, while for frying, the reduction was up to 48%. Warm smoking and pasteurization as combined treatments, determined as well a reduction in the OCP content of maximum 15 and 16%, respectively. Baking reduced also the level of OCPs with a maximum of 56%. Stewing under pressure caused the most dramatic reduction in the OCP levels (up to 92%). Using the Mc Donald's polynomial regression, predictive mathematical models for the variation of OCP levels with the applied heat treatments were computed. These models allow a good selection of the appropriate industrial food processing with the ultimate goal to reduce OCP residues.

Keywords: OCPs; Meat; Thermal processing; Predictive mathematical models

Introduction

Organochlorine pesticides (OCPs) were widely used worldwide until restrictions were introduced in the late 1970s. However, despite these measures, OCPs are still among the most prevalent environmental pollutants and are present in food for human consumption [1]. They are mainly found in lipid-rich food of animal origin, such as meat, fish and dairy products, which constitute an important part of our daily diet. It has been suggested that food of animal origin is responsible for more than 90% of the average human intake of polychlorinated biphenyls PCBs and OCPs [2]. One common feature of OCPs is that they bio-accumulate in the food chain and in the human body, and, as a consequence, represent a potential risk for the human health [3,4]. Therefore, it is important to reduce levels of OCPs in food of animal origin. Studies on the influence of thermal processing of foods contaminated with OCP residues have showed this treatment as being the determining factor for their significant decrease through food processing [5-11]. Thus, thermal treatments, such as frying, baking, grilled, boiling and microwave cooking were the best options to reduce the pesticide load in raw meat [12-17].

Recent research on this topic has mainly focused on several pollutants, such as arsenic, polychlorinated dibenzo-p-dioxins and dibenzofurans (PCDD/Fs), polychlorinated biphenyls (PCBs), hexachlorobenzene (HCB), polycyclic aromatic hydrocarbons (PAHs), polychlorinated diphenyl ethers (PCDEs), and polybrominated diphenyl ethers (PBDEs), in several fish and meat food products [7,12,15,18,19]. The reduction in the content of pollutant residues through processing of food products is due to the variable lipid content of the analyzed foodstuffs, as well as to the variation in the lipid content in samples of the same food item. However, taking into account that PCDD/Fs, PCBs, PBDEs, PCDEs, and other organic pollutants are associated with the fat portion of foods, cooking methods that release or remove fat from the product will tend to reduce the total amount in the cooked food. Results showed that the influence of cooking on the levels of these contaminants depends not only on the particular cooking process, but even more on the specific food item [7]. Dietary exposure to these environmental pollutants can be thus reduced by discarding the fat, which is released from foods during cooking. The purpose of meat thermal treatment is to improve its hygienic and flavour qualities. Therefore, the present study aimed to build a prediction model on the

influence of thermal treatments in order to reduce the meat content of OCPs. Based on the obtained experimental results and using Mc. Donald's polynomial regression, predictive mathematical models were computed for the dependency of OCP levels in meat with the applied heat treatments.

Materials and Methods

Samples collection, storage and processing

The material for analysis consisted of pork neck meat and was sampled from one commercial unit in Cluj-Napoca (Romania). A number of 18 pork neck meat samples were injected in several places with a mixture of OCPs using a 15% saline solution containing the contaminants in a concentration of 200 ng/g fat. The fortified samples were kept for six days at a temperature of 0-4°C. The control (6 meat samples) and fortified samples were subjected to six different thermal treatments. The working parameters for the thermal treatments are presented in Table 1.

Chemicals and materials

An OCP standard mixture containing: aldrin, α-chlordane,

Processing operation	Code	Temperature(°C)	Time
Cold smoking	AFR	45	3 h
Warm smoking	AFC	85	1 h
Pasteurization	PST	74	2 h
Frying	FR	180	10 min
Baking	CO	110	1 h
Stewing under pressure (cooking)	IN	130	1 h

Table 1: Processing operations codes and working parameters for thermal treatments.

*Corresponding author: Muresan C, Faculty of Food Science and Technology, University of Agricultural Sciences and Veterinary Medicine, Cluj -Napoca, Romania, E-mail: crina.muresan@usamvcluj.ro

γ-chlordane, 4,4'-DDD, 4,4'-DDE, 4,4'-DDT, dieldrin, endosulfan I, endosulfan II, endosulfan sulfate, endrin, endrin aldehyde, endrin ketone, heptachlor, heptachlor epoxide and methoxychlor (Restek Corp, SUA) was used to construct the calibration curves. A stock mixture of OCP standards in hexane was prepared and stored at 4°C. The calibration curve of each OCP wa s constructed using standard samples with six different concentrations (5, 10, 50, 100, 200 and 500 ng/g) of the standard mixture solution.

The areas of the obtained peaks were plotted as function of the concentration. Each sample and standard solution was analyzed in duplicate using GC-ECD. The method was validated by determining recoveries, correlation coeficients, relative standard deviations (RSD) and detection limits (Table 2). The precision of the method was satisfactory with RSDs < 20% for all OCPs.

Fat extraction by automatic extractor

A Soxhlet apparatus (VELP Scientifica) was used for the determination of the fat content in the meat samples. Ten gm of meat were accurately weighed and homogenized with 10 g of Celite. The mixture was quantitatively transferred into an extraction thimble. Before use, the extraction thimble was washed with 15 mL of acetone and 15 mL of hexane. Fat extraction was performed by Soxhlet extraction with 90 mL of petroleum ether at 70°C by placing the extractor slider into the "immersion" position for 10 min and, successive ly, into the "washing" position for 180 min.

Cryogenic extraction of pesticides

In order to measure the fat content, 0.5 g of fat was weighted and mixed with 3 mL of methylene chloride-acetonitrile mixture (25:75, v/v). The mixture was centrifuged at 3000 rpm at -15°C for 20 min. The supernatant layer was transferred into a different tube and slowly heated in a water bath at 40°C to melt the fat. The extraction was re peated with another 3 mL of acetonitrile- methylene chloride mixture and centrifugation was repeated. Second supernatant layer was added to the first one. The organic phase was evaporated at 35°C under a nitrogen stream in order to reach a final volume of 2 to 3 ml (solution A).

Partitioning on C18 SPE cartridge

A SPE vacuum manifold, SPE Florisil cartriges (6 mL, 2000 mg) and SPE Chromabond C18 cartridges (6 mL, 500 mg) were used for the

purification of the obtained extracts. The C18 cartridge was activated by eluting twice with 5 ml of petroleum ether, then 5 ml of acetone and finally 5 ml of methanol. Eluted solutions were discarded. The solution A was added in cartridge and left for 3 min, then eluted with 10 mL acetonitrile at a flow rate of one drop per every three seconds. The eluted solution was collected in a suitable container. The collected eluant was concentrated on a rotary evaporator and dissolved in 5 ml n-hexane (solution B).

Clean-up by Florisil SPE cartridge

Florisil SPE cartridge was conditioned by elution with 10 ml of n-hexane. Solution B was added in the cartridge and left for 3 min, and then was eluted with 10 mL of ethyl ether-petroleum ether (98:2) with a flow rate of one drop per second, for all OCPs, except endrin and dieldrin. The eluted fraction was collected in an evaporation flask. For endrin and dieldrin, the Florisil SPE cartridge was further eluted with 12 ml of petroleum ether-diethyl ether (85:15) with a flow rate of one drop per every three seconds. The second fraction was collected in the same flask as the first part. Finally, the elution mixture was brought to dryness and redissolved in 5 ml hexane.

Analysis

The purified extracts were analyzed using a Shimadzu 2010 Gas Chromatograph (GC) equipped with an electron capture detector (ECD). The column temperature started at 150°C for 2 min, The raised to 200°C at 4°C/min, kept at 200°C for 5 min, raised to 230°C at 5°C/ min, kept at 230°C for 5 min, raised to 300°C at 2°C/min, and then kept at 300°C for 5 min. The temperatures of injector and detector were 250°C and 315°C, re spectively. The injection volume was 1 µL. The flow rates of the carrier gas (He) and make up gas (N2) were maintained at 19.2 and 3.0 mL/min, respectively. A Restek capillary column (RTx-5, 20 m length and internal diameter of 0.18 mm, with 0.4 µm film thickness) was used for the separation of OCPs.

Statistical analysis

The statistical analysis was achieved with GraphPad Prism v5.00 (Graph Pad, San Diego, CA, USA). Analysis of variance (one-way Anova) was used to compare the level of each OCP between the thermal treatments. The predictive mathematical model used to assess the influence of the various thermal treatments on the reduction of OCP contents in the fortified samples was established using the Mc Donald's polynominal regression.

Results and Discussion

Organochlorine pesticide residues determination in the control sample

For each thermal treatment (AFR, AFC, PST, FR, CO, IN), a control sample was used. Table 3 presents the results obtained for the OCP residues (ng/g fat) in the control sample. It can be observed that the control pork meat sample contained DDT, DDE and DDD residues at 33.8 ng/g fat, a value which falls under the maximum residual limit (MRL, Σ DDTs 1000 ng/g). Covaci reported residues of OCPs in organs and fat tissue from Romanian pigs over the MRL, while the pork samples analyzed by Covaci had OCP concentrations under the MRL.

Head treatments efficiency applied to reduce Organochlorine pesticides level in fortified pork neck samples

After fortification at a concentration of 200 ng/g fat, the pork meat samples were subjected to the thermal treatments presented in Table 1.

Compounds	Mean recovery (± SD)	Correlation Coefficient (R2)	RSD %	Detection limits (ng/g)
heptachlor	97.4 ± 1.0	0.9998	10.6	1
aldrin	70.6 ± 8.4	0.9999	13.1	1
heptachlor epoxide	83.5 ± 1.1	0.9997	7.0	5
γ chlordane	77.7 ± 9.3	0.9998	12.9	5
α chlordane	73.6 ± 1.1	0.9996	10.4	5
4.4 DDE	70.5 ± 9.3	0.9996	12.3	5
endosulfan I	77.2 ± 9.2	0.9999	10.3	1
dieldrin	67.9 ± 1.2	0.9993	10.4	1
endrin	100.6 ± 1.6	0.9999	10.5	1
4.4' DDD	74.2 ± 1.2	0.9996	6.9	1
endosulfan II	67.9 ±1.5	0.9997	11.9	1
4.4 DDT	83.2 ±1.2	0.9991	10.6	1
endrin aldehyde	77.7 ± 1.5	0.9993	15.3	1
methoxychlor	102.5 ± 1.1	0.9997	15.9	1
endosulfan sulfate	111.0± 1.2	0.9992	11.8	1
endrin ketone	77.2 ± 2.7	0.9994	10.9	1

Table 2: Mean recoveries, correlation coefficients, relative standard deviation (RSD%) and detection limits.

As a consequence of applied heat treatments to pork neck samples, variations from control samples in the levels of OCPs were observed (Table 4). A reduction of only 1% in the OCP content from control sample was observed after cold smoking (AFR). Warm smoking (AFC) and pasteurization (PST) as combined treatments determined a higher reduction in the OCP contents from control sample of up to 15 and 16%, respectively. Both frying (FR) and baking (CO) led to an even higher reduction in the OCP content of up to 46 and 56%, respectively. Stewing under pressure and cooking caused the most dramatic decrease in the OCP levels up to 92%. The above mentioned tendencies refer to the reduction of the degree of contamination from one treatment to another, making possible a hierarchy for the decontamination processes, from less efficient to most efficient. Thus, the contamination reduction ratios vary in the following order:

AFR < AFC+PST < PST+AFC < FR < CO < IN

Organochlorine pesticide	Measured (ng/g fat)	MRL (ng/g fat)
aldrin+dieldrin	ND	200
endrin	ND	50
endrin ketone	ND	n.s.
Σ „drins"	ND	
α chlordane	ND	50
γchlordane	ND	n.s.
Σ chlordanes	ND	
4,4'-DDT	7.9	
4,4'-DDE	23	
4,4'-DDD	2.9	
Σ DDTs	33.8	1000
heptachlor	ND	200
heptachlor epoxide	ND	200
Σ heptachlors	ND	
endosulfan I	ND	50
endosulfan II	ND	50
Σ endosulfans	ND	
methoxychlor	ND	10

ND-not detected; n.s-not specified:

Table 3: Organochlorine pesticide levels (ng/g fat) in the control sample, a commercially available pork neck, lipid content 23.5%. Levels are compared against the Maximum Residue Levels (MRLs) set in the European legislation [27].

The main cause in the reduction of the OCP content was the loss of fat during treatments and, only in to a smaller extent, the reduction was due to chemical transformations that are taking place. Heat treatments were effective, since they reduced the OCP content of the meat, as noted in other studies [5,6,8,20,21].

Zabic et al., determined the levels of HCB in raw skinless Chinook salmon and carp fillet sampled from the Great Lakes. HCB concentrations were also measured after baking, charbroiling, and canning salmon, as well as after pan and deep fat frying carp. The cooking procedures significantly reduced HCB content (with more than 40%) and only few significant differences were found among the various cooking methods.

In another study, cooking processes were further tested to assess their potential to reduce the levels of HCB and other pesticides in fish. Lean skinless lake trout and siscowets (fat laketrout) from the Great Lakes were cooked by baking and charbroiling. In all cases, cooking reduced the levels of HCB in comparison to the raw samples (17.5% in baked and 19.5% in charbroiled samples, respectively). Bayen examined the effects of various cooking processes on the potential loss of various POPs, including PBDEs, from salmon. Losses of PBDEs as the result of the cooking processes, and without skin removed, were the following: 42%, 25%, 32% and 44% for pan- frying, microwave cooking, boiling and baking, respectively. The authors concluded that the reduction of POP burden (including PBDEs) in cooked fish would be a result of efficient lipid removal during the cooking process rather than by the type of cooking method used.

The statistical analysis of obtained data showed significant correlations between pairs extremely comparable for all treatments except stewing under pressure (cooking), where the correlation is insignificant (Table 4).

Organochlorine pesticides weight variability during applied heat treatments

The evolution of OCP contribution in the fortified meat samples after the thermal treatments is presented in Figure 1. For heptachlor, its share increased after frying (5.22%) and baking (5.36%) compared

Organochlorine pesticides	% from M						
	M	AFR	AFC+PST	PST+AFC	FR	CO	IN
heptachlor	100.0	99.8	89.8	88.1	68.6	55.9	11.8
aldrin	100.0	99.3	86.9	84.8	64.3	52.4	10.6
Heptachlor epoxide	100.0	99.0	88.9	88.0	67.6	51.8	18
γ chlordane	100.0	99.2	85.4	84.6	63.9	45.3	18.6
4.4'DDE	100.0	99.0	88.0	85.4	54.6	47.8	7.6
endosulfan I	100.0	99.2	86.0	84.6	63.7	43.4	15.9
dieldrin	100.0	99.4	89.3	84.7	64.5	53.7	10.7
4.4'DDD	100.0	99.1	88.3	85.8	54.7	48.2	9.2
endosulfan II	100.0	99.2	87.5	85.4	64.9	44.5	15.3
4.4'DDT	100.0	99.3	88.5	86.1	55.1	48.9	10.1
endrin aldehyde	100.0	99.2	88.5	84.6	67.6	49.5	17.1
methoxychlor	100.0	99.0	85.0	84.2	68.6	49.6	9.9
endrin sulfate	100.0	99.4	86.9	84.4	66.5	50.9	8.9
endrin ketone	100.0	99.3	87.0	84.3	54.6	46.3	15.3
% fat	23.5	23.08	21.37	20.08	17.03	15.21	9.29

M-Control; AFR-Cold smoking; AFC+PST-Warm smoking and pasteurization; PST+AFC-Pasteurization and warm smoking; FR-Frying; CO-Backing; IN-Stewing under pressure (cooking) M-AFR ***extremely significant P≤0.001; M-(AFC+PST) ***extremely significant P≤0.001; M-(PST+AFC) ***extremely significant P≤0.001; M-FR***extremely significant P≤0.001; M-CO***extremely significant P≤0.001; M-IN not significant P>0.05

Table 4: Head treatments efficiency applied to reduce organochlorine pesticides level in fortified pork neck samples.

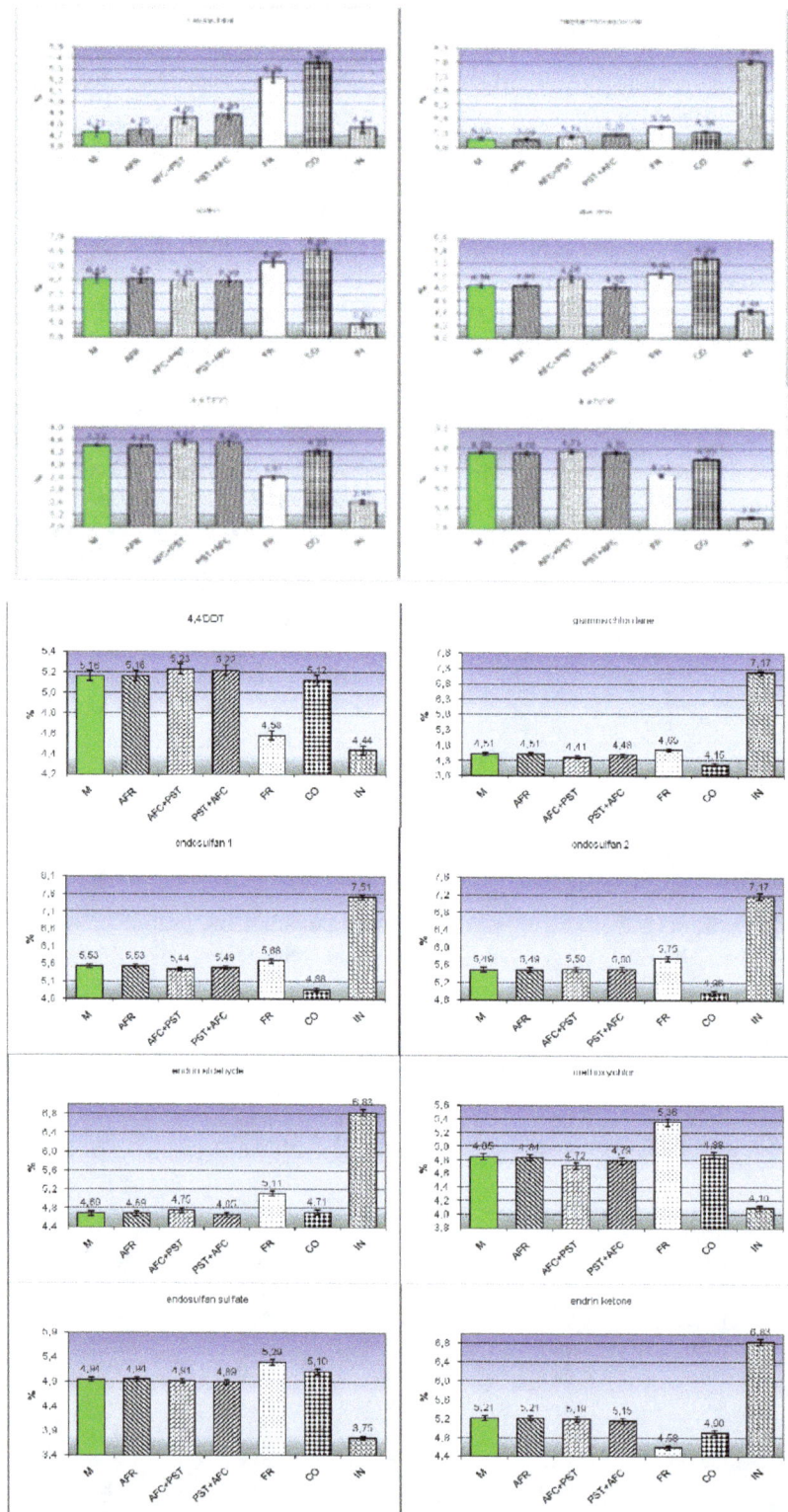

Figure 1: Evolution of OCPs weight in heat treatments applied (average fat content 23.5%). M-Control: AFR- Cold smoking; AFC+PST-Warm smoking and pasteurization; PST+ AFC-Pasteurization and warm smoking; FR-Frying; CO-Backing; IN-Stewing under pressure (cooking).

to the control sample (4.72%). For this compound, the same trend was also noticed for cooking (5.36%). One explanation could be the high resistance of heptachlor to thermal treatment as suggested in the literature (22).

Heptachlor epoxide has the same increase tendency compared to the control (5.09%) regardless of the applied treatment, reaching a maximum level in the cooked sample (7.85%). In the cold smoking treatment (AFR), the heptachlor epoxide concentration showed a slight increase (5.08%) compared to the control sample. Perellò showed that the influence of cooking on organic pollutants depends not only on the cooking process, but also on the type of food. Aldrin reached a maximum share for cooking (6.82%) and a minimum for baking (5.8%) compared to a share of 6.41% in the control sample (M). Compared with to aldrin, the concentration of dieldrin was lower in the control sample (4.84%), as well as in the cooked and baked samples (4.43 and 5.28%, respectively). One explanation could be the possible oxidation of aldrin to dieldrin, as already reported in the literature [10,22,23]. The lowest levels of endrin aldehyde were determined for the pasteurization combined with the warm smoking treatment (4.65%). Endosulfan I and II had the same trend compared to the control for the thermal treatments applied in this study. It may be noted that endosulfan I is more stable than endosulfan II during the studied thermal treatments. Clearly, the applied thermal treatments led to a partial elimination or degradation of OCPs. In particular, heptachlor and heptachlor epoxide appeared to be very resistant to the applied thermal treatments compared with dieldrin and DDT. Changes (degradation) can be chemically explained; for example, 4,4'-DDT is transformed to 4,4'-DDD through reductive dechlorination and to 4,4'-DDE through dehydrochlorination. Also, heptachlor can be oxidized to heptachlorepoxide [24,25].

In food products processed through frying, baking and stewing (home cooking), the reduction in the OCP content is higher compared to industrial processing of meat products (pasteurization, warm smoking, or cold smoking) [7,13]. The duration and temperature of the meat processing have both an important role in the fat degradation. In the baking and cooking processes, lipid hydrolysis can occur at temperatures above 100°C [15].

The advanced thermodegradation of fats is possible at high temperatures (110, 130, and 180°C) and time periods (0.5-1 h) in processes, such as stewing, baking and frying [5]. During the thermal treatment, a break in the cells of the fat tissues takes place; the grease is removed and further dispersed into the meat mass. Depending on the meat processing, a part of the fat can be eliminated in the boiling mass (cooking or stewing) or removed through melting as it occurs in pasteurization, warm smoking and frying processes [26,27].

Designing mathematical models to predict the effect of heat treatments on OCP levels in meat

Table 5 presents the predictive mathematical models for all studied treatments. Statistical techniques are used in this study to model and predict, on a laboratory scale, the degree of reduction in the OCP

content by controlling the limiting factors of the food processing treatments. This modelling aims to estimate the impact of an thermical process (cold smoking, smoking, warm smoking, pasteurization, frying, baking, and cooking) on the OCP content. The study used one control and 6 working variants, thus one variable set and their influence was studied. A prediction for the reduction degree in the contamination of meat samples subjected to 6 thermal treatments could be achieved. The coefficient of determination r^2 indicates the response variation within the mathematical model, with values ranging between 0 and 1. The r^2 value was found to be close to 1.0, which denotes a high correlation between the observed and predicted values. A higher value indicates a stronger relation between the studied variables (x, y). The final value, y, of OCP residues is function of the OCP residues, x, before cooking treatments.

Cold smoking, warm smoking, pasteurization, and baking had r^2 values close to 1. For the stewing under pressure treatment, a predictive mathematical model could be developed only for some OCPs, namely for aldrin, 4,4'-DDD, 4,4'-DDT, methoxychlor and endosulfan sulfate. Statistical and mathematical techniques for designing experiments are useful tools to estimate the behavior of various compounds. Predictive mathematical models obtained from the experimental results depend on the behavior of the compounds analyzed. They can be thus used to choose the most appropriate thermal treatment method to reduce contamination. After applying various thermal treatments, the initial variation of OCPs from the original conditions is continuously modifying between treatments, being influenced by "intrinsic factors" linked to the physico-chemical properties of individual OCPs, such as: different partition coefficients between the lipidic and aqueous phase driven by differences in solubility; different volatility governed by different boiling points and by different stability of OCPs at high temperatures.

Our results show that the influence of cooking on the levels of OCPs in meat does not depend only on the particular cooking process, but even more on the specific food item. Lipophilic OCPs are associated with the fat portion of foods, thus cooking methods that release or remove fat from the product, will tend to reduce the total amount of OCPs in the cooked food. Dietary exposure to these environmental pollutants can be reduced by discarding the fat, which is released from foods during cooking [14,16,17]. Fortunately, this practice is already common among consumers who wish to reduce their fat intake.

Acknowledgement

Prof. Dr. Laslo Cornel, for the valuable help provided in the elaboration of the PhD thesis: Researches Concerning the Influence of Processing on organochlorine pesticide residues from meat and meat products.

References

1. Hura C, Leanca M, Rusu L, Hura BA (1999) Risk assessment of pollution with pesticides in food in the Eastern Romania area. Toxicol Lett 107: 103-107.

2. Witczak A (2012) Influence of smoking on indicatory PCB congeners residues levels in fish sliced. Pol. J Food Nutr Sci 62: 31-39.

3. Covaci A, Hura C, Schepens P (2001) Selected persistent organochlorine pollutants in Romania. Science of the Total Environment 280: 143-152.

4. Covaci A, Gheorghe A, Schepens P (2004) Distribution of organochlorine pesticides, polychlorinated biphenyls and a-HCH enantiomers in pork tissues. Chemosphere 56: 757-766.

5. Kubacki SJ, Lipowska T (1980) The role of food processing in decreasing pesticide contamination of foods. Food and Health: Science and Technology, Applied Science Publishers, London.

6. Gonzalez SM, Visweswariah K (1984) Efecto de la coccion sobre el

Processing operation	1st degree equation	RI
Cold smoking (AFR)	y = 0.992x + 0.053	0.9999
Warm smoking (AFC)	y = 0.836x + 5.102	0.9959
Pasteurization (PST)	y = 0.828x + 3.277	0.9975
Frying (FR)	y = 0.464x + 23.838	0.8919
Baking (CO)	y = 0.468x + 2.586	0.9532
Stewing under pressure (IN)	y = 0.712x + 3.539	0.8875

Table 5: Mathematical models of first degree equations for thermal treatments.

contenidoresidual de hexaclorociclohexano (BHC) en carrie de pollo broiler. Agricultura Tecnica 44: 39-43

7. Ariño A, Herrera MP, Conchello A, Perez C (1992) Hexachlorobenzene 282 residues in Spanish meat products after cooking, curing, and long term ripening. J Food Prot 55: 920-923

8. Conchello P, Herrera A, Ariño A, Lázaro R, Pérez-Arquillué C, et al. (1993) Effect of grilling, roasting, and cooking on the natural hexachlorobenzene content of ovine meat. Bull of Environmental Contamination and Toxicology 50: 828-833.

9. Rose M, Thorpe S, Kelly M, Harrison N, Startin J, et al. (2001) Changes in concentration of five PCDD/F congeners after cooking beef from treated cattle. Chemosphere 43: 861-868.

10. Sengupta D, Aktar W, Alam S, Chowdhury A (2010) Impact assessment and decontamination of pesticides from meat under different culinary processes. Environ Monit Assess 169: 37-43.

11. Rawn DFK, Breakell K, VSerigin V, Tittlemeir SA, Gobbo LD, et al. (2013) Impacts of Cooking Technique on Polychlorinated Biphenyl and Polychlorinated Dioxins/Furan Concentrations in Fish and Fish Products with Intake Estimates. J Agric Food Chem 61: 989-997.

12. Bayen S, Barlow P, Lee HK, Obbard JP (2005) Effect of cooking on the loss of persistent organic pollutants from salmon. Toxicology and environmental health 68: 253-265.

13. Domingo JL (2011) Influence of Cooking Processes on the Concentrations of Toxic Metals and Various Organic Environmental Pollutants in Food: A Review of the Published Literature. Critical reviews in Food Science and Nutrition 51: 29-37.

14. Hori T, Nakagawa R, Tobiishi K, Iida T, Tsutsumi T, et al. (2005) Effects of cooking on concentrations of polychlorinated dibenzo-p-dioxins and related compounds in fish and meat. J Agric Food Chem 58: 8820-8828.

15. Mureşan C, Tofană M, Laslo C (2010) Effect of several technological and kitchen treatments on hexachlorocyclohexan residues in pork meat. Bulletin of the University of Agricultural Sciences and Veterinary Medicine Cluj-Napoca Agriculture 67: 312-318.

16. Perelló G, Martí-Cid R, Castell V, Llobet JM, Domingo JL, et al. (2009) Influence of various cooking processes on the concentrations of PCDD/Fs, PCBs and PCDEs in food. J Food Prot 21: 178-185.

17. Perelló G, Martí-Cid R, Llobet JM, Castell V, Domingo JL, et al. (2009) Concentrations of polybrominated diphenyl ethers, hexachlorobenzene and polycyclic aromatic hydrocarbons in various foodstuffs before and after cooking. Food and Chemical Toxicology 47: 709-715.

18. Liem AKD, Fürst P, Rappe C (2000) Exposure of populations to dioxins and related compounds. Food Addit Contam 17: 241-259.

19. Zabik ME, Booren AL, Zabik MJ, Welch R, Humphrey H, et al. (1996) Pesticide residues, PCBs and PAHs in baked, charbroiled, salt boiled and smoked Great Lakes lake trout. Food Chem 55: 231-239.

20. Zabik ME, Zabik MJ, Booren AM, Nettles M, Song JH, et al. (1995) Pesticides and total polychlorinated biphenyls in Chinook salmon and carp harvested from the Great Lakes: Effects of skin-on and skin-off processing and selected coking methods. J Agric Food Chem 43: 993-1001.

21. Katayama A, Matsumura F (1993) Degradation of organochlorine pesticides, particulary endosulfan, by Trichoderma harzianum. Environ Toxicol Chem 12: 1059-1065.

22. Kipčić D, Vukušić J, Šebečić B (2001) Monitoring of chlorinated hydrocarbon pollution of meat and fish in Croatia. Food Technol Biotechnol 40: 39-47.

23. Morgan KJ, Zabik ME, Funk K (1972) Lindane, dieldrin and DDT residues 328 in raw and cooked chicken and chicken broth. Poultry Science 51: 470-475

24. Bai Y, Zhou L, Li J (2006) Organochlorine pesticide (HCH and DDT) residues in dietary products from Shaanxi province, People's Republic of China. Bull Environmental Contamination and Toxicology 76: 422-438.

25. Wilson ND, Shear NM, Paustenbach DJ, Price PS (1998) The effect of cooking practices on the concentration of DDT and PCB compounds in the edible tissue of fish. J Exp Anal Environ Epidemiol 8: 423-440.

26. Perello G, Marti-Cid R, Castell V, Llobetc JM, Domingoa JL, et al. (2010) Influence of various cooking processes on the concentrations of PCDD/PCDFs, PCBs and PCDEs in foods. Food Control 21: 178-185.

27. EC Regulation (2008) Regulation (EC) no. 299/2008 the European Parliament and of the Council amending Regulation (EC) no. 396/2005 on maximum residue levels of pesticides in or on food and feed.

20

Nutritional Evaluation and Sensory Characteristics of Biscuits Flour Supplemented with Difference Levels of Whey Protein Concentrates

Mohammed AA[1], Babiker EM[2], Khalid AG[3], Mohammed NA[4]*, Khadir EK[2] and Eldirani[5]

[1]Department of Quality Control, Seen Milles, Khartoum North, Sudan

[2]Department of Food Science and Technology, Faculty of Agriculture, University of Khartoum, Khartoum North, Shambat, Sudan

[3]Department of Nutrition and Food Technology, Faculty of Science and Technology, Omdurman Islamic University, Omdurman, Sudan

[4]Faculty of Applied Medical Science, Department of Community Health, Albaha University, Albah City, Kingdom of Saudi Arabia

[5]Department of Food Science and Technology, Faculty of Agriculture, Omdurman Islamic University, Omdurman, Sudan

Abstract

Proximate analysis and amino acid profile were carried for biscuit flour and whey protein. The gluten quantity and quality was tested for biscuit flour and biscuit flour-whey mixture with different concentration 0, 5, 10 and 15% whey. The results of the proximate analysis showed that there was no difference between protein 11.3% and carbohydrate (74.87%) for biscuit flour and protein (11.7%), carbohydrate (74.47%) for whey. The moisture content of biscuit flour was 10.97% which was higher than whey (5.47%) with highly significant difference at level of (p>0.05). The fat and ash contents of Biscuit flour were significantly (p>0.05) lower than the other one. The biscuit flour had lower content in essential amino acids especially limiting amino acid (Lysine) compared to whey protein. The gluten quantity and quality was affected by supplementation with whey and decreased with increased the concentration of whey. The overall quality of biscuits made from mixture showed high acceptability, Biscuit flour blended with 10% spray-dried whey showed best biscuit.

Keywords: Whey; Proximate analysis; Amino acid; Wheat flour; Biscuits; Gluten

Introduction

Wheat is a type of grass grown all over the world for its highly nutritive and useful grain. It is one of the top three most produced crops in the world, along with corn and rice. Wheat has been cultivated for over 10,000 years and probably originates in the Fertile Crescent, along with other staple crops. A wide range of wheat products are made by humans, including most famously flour, which is made from the grain itself. There are six different classes of wheat: Hard Red Winter, Hard Red Spring, Soft Red Winter, Hard White, Soft White and Durum. The end products are determined by the wheat's characteristics, especially protein and gluten content. The harder the wheat, the higher the protein content in the flour. Soft, low protein wheat is used in cakes, pastries, cookies, crackers and oriental noodles. Hard, high protein wheat is used in breads and quick breads. Durum is used in pasta and egg noodles [1]. Whey protein is a mixture of some of the proteins naturally found in milk. The major proteins found in whey include beta-lacto globulin and alpha-lactalbumin. Whey protein has one of the highest protein digestibility-corrected amino acid scores (PDCAAS; a measure of protein bioavailability) and is more rapidly digested than other proteins, such as casein (another milk protein), [2]. Increasing knowledge of the nutritional and health benefits of dairy proteins (casein and whey and their bioactive peptides and amino acids) is leading to recognition of their potential as value-added ingredients in many functional foods and beverages, not only for weight management, but also for other health benefits [3]. Dairy proteins, specifically casein and whey, are high quality protein sources that provide all the essential amino acids, and in particular the branched chain amino acid leucine, which has been shown to specifically stimulate the synthesis of new muscle protein [4]. Whey proteins is of high biological value compared to most other protein; has high content of sulfur amino acid important for the biosynthesis is of glutatione, atripeptide with antioxidant, anti-carcinogenic and immune simulating properties and the highest natural source of branched chain amino acid which may stimulate muscle

protein synthesis [5]. Biscuits are widely accepted and consumed in many developing countries, so its need supplementation with another sources of proteins such as legumes and milk proteins to improve their nutritional values, the objective of this work was to determine the proximate composition and amino acids profiles of biscuit wheat flour and spry-dried whey (whey powder), to investigate the effect of whey protein supplementation on quantity and quality of biscuit wheat flour gluten and to show the effect of supplementation with different levels of whey protein on sensory characteristics of Biscuits quality.

Materials

Biscuit wheat flour (Australian wheat, 72% extraction rate) samples were collected from the local market (Omdurman market). Liquid whey was collected from cheese makers in Khartoum north (helat koko). Alaseel (hydrogenated vegetable oils), skimmed milk and sugar which were used in process of biscuit were purchased from local market. All chemicals and reagents used in this study were of analytical grade.

Methods

Spray drying of whey

Whey protein, which is drained off the coagulated cheese cured during the cheese making process, was collected from cheese making

---

*Corresponding author: Mohammed NA, Faculty of Applied Medical Science, Department of Community Health, Albaha University, Albah City, Kingdom of Saudi Arabia, E-mail: azhari1933@gmail.com

in Khartoum north. The liquid was evaporated to make 30% whey concentration. Because whey contains 93%water and only 0.6 proteins, it must be concentrated to produce the various whey ingredients [6]. The concentrated whey was then spry dried into whey powder with 67% yield. It can be observed that the dried whey ingredients are manufactured after pasteurized and clarified of liquid whey. The main advantages of spray drying are rapid draying, large scale continuous production, low labor costs and simple operation and maintenance [7].

Whey protein preparation

The whey protein, a bye product of cheese making, was collected, concentrated to 30% by the evaporator (70-75°C), and dried by spry drier [6]. The powder yield was calculated by the following equation:

$$Yield\% = \frac{Actual\ weight}{Theoretical\ weight} \times 100$$

Where:

Actual weight = weight obtained from spry drier

$$Theoretical\ weight = \frac{Concentration\ of\ whey\ (30\%)}{Volume\ used\ in\ spry\ drier}$$

The resultant powder was divided into two portions; one portion was kept in plastic container and stored in a deep freezer at -18°C for further analysis. The other portion was mixed with wheat flour to make concentrations of 5, 10 and 15%, kept in plastic containers and stored in a deep freezer at -18°C until required.

Determination proximate analysis

The proximate chemical composition of each of the biscuit wheat flour and the whey powder was performed according to AOAC [8] method. Carbohydrates were determined by difference.

Determination of amino acids

Amino acid composition of samples was measured on hydrolysates using amino acids analyzer (Sykam-S7130/Germany) based on high performance liquid chromatography technique. Sample hydrolysates were prepared following the method of Moore and Stein. Two hundred milligrams of sample were taken in hydrolysis tube. Then 5 ml 6N HCl was added to sample into tube tightly closed and incubated at 110°C for 24 hours. After incubation period, the solution was filtered (Whatman No. 1) and 200 ml of the filtrate were evaporated to dryness at 140°C for an hour. Each hydrolysate after dryness was diluted with one milliliter of 0.12 N, pH 2.2 citrate buffer (11.8 g trisodium citrate+ 6 g citric acid + 14 ml biobiglycol + 12 ml 32% HCl + 2.0 g phenolinone liter), the same as the amino acids standards (amino acid standards H, Pierce. Inc., Bockford). An amount of 150 μgL of sample hydrolyzate was injected in the cation separation column at 130°C. Ninhydrin solution (reaction reagent) and an eluent buffer (The buffer system contained solvent A, pH3. 45, and solvent B, pH 10. 85) were delivered simultaneously into a high temperature reactor coil (16 m length) with a flow rate of 0.7 mL/min. The buffer/ninhydrin mixture was heated in the reactor at 130°C for two minutes to accelerate chemical reaction of amino acid

with ninhydrin. The products of the reaction mixture were detected at wavelength of 570 nm and 440 nm on a dual channel photometer. The amino acid composition was calculated from the areas of standards obtained from the integrator and expressed as percentages.

Gluten quantity and quality attributes of base and substituted flours

Gluten quantity and quality were carried on the base wheat flour and wheat flour- spray-dried whey mixture by the Glutomatic system (Perten Instrument) according to ICC standard method No. 155 (2000). The wet gluten content, gluten index and dry gluten content were determined as follows: Ten gram of samples was mixed with 4.8 ml of 2% sodium chloride solution for 20 seconds in test chamber. The dough was washed with 2% NaCl for 10 min. When the glutomatic was stopped, the gluten ball was carefully centrifuged through special sieve. The percentage of wet gluten remaining on the sieve after centrifugation was defined as gluten index. The part of the gluten remaining on the sieve and the part, which passes through it, were collected and weighed and were defined as wet gluten content.

$$Wet\ gluten\ content\% = \frac{Weight\ of\ wet\ gluten \times 100}{Weight\ of\ sample}$$

Then the total wet gluten was dried in Glutork heater to give the dry gluten.

$$Dry\ gluten\ content\% = \frac{Weight\ of\ dry\ gluten \times 100}{Weight\ of\ sample}$$

$$Gluten\ index\% = \frac{Total\ wet\ gluten\ wt - passed\ gluten\ wt}{Total\ wet\ gluten\ wt}$$

Statistical analysis

The statistical analysis of the sample was done according to the method described by Mead and Gurnow [9], data generated was subjected to Statistical Package for Social Sciences (SPSS). Means (± SD) were tested using one-factor analysis of variance, and then separated using Duncan's Multiple Range Test (DMRT).

Results and Discussion

Proximate chemical composition

The proximate chemical composition of each of the biscuit wheat flour and the whey powder was determined. The results expressed on dry weight basis, are presented in table 1.

Moisture content: Wheat flour gave relatively higher moisture content (10.97%), compared to whey powder which gave (5.47%) with high significant difference (p ≤ 0.5). The moisture content of the wheat flour was lower than the range of 13.0-15.5% [10]. The result obtained was in an agreement with the value of 10.5% for wheat flour (all purpose) and 10.1% for special flour [11]. The moisture content of the whey powder was found to be higher than the value of 3.8% moisture reported.

Ash content: Wheat flour gave lower ash content (0.87%) compared

Samples	Moisture	Ash	Protein	Fat	Fiber	Carbohydrate
Biscuit wheat flour	10.97 ± 0.09[a]	0.87 ± 0.09[b]	11.30 ± 0.06[a]	1.10 ± 0.12[b]	0.87 ± 0.15[a]	74.87 ± 0.09[a]
Whey powder	5.47 ± 0.18[b]	5.60 ± 0.12[a]	11.73 ± 0.18[a]	2.63 ± 0.15[a]	0.00 ± 0.00[b]	74.47 ± 0.29[a]

*Values are means of three triplicates.

**Mean values (± SD) having different superscript letters in columns differ significantly (p ≤ 0.05)

Table 1: Proximate chemical composition of biscuit wheat flour and whey powder.

to the whey powder (5.6%) with significant difference at the level of (p ≤ 0.05). The value of ash content of the biscuit flour was slightly higher than the upper limit of the range 0.38-0.84% [12], for Sudanese wheat flour. It is also higher than the level of 0.73% [11] for special flour. The ash content of the whey powder was lower than the value of 7.0%.

Protein content: The protein level of the biscuit flour and the whey powder are 11.3% and 11.73% respectively. Insignificant difference at (p ≤ 0.05) was observed between protein levels of biscuit flour and whey powder. Earlier findings showed protein levels of 11% in white flour, 10.97% in Australian wheat and 10.17% in special flour [14,11]. The protein content of the whey powder was lower than the level of 13.0%. Variation in whey protein content may be attributed to differences in the procedure adopted for whey powder production.

Fat content: The biscuit flour gave a lower fat content of 1.1%, compared the whey powder which gave 2.63% with highly significant difference at (p ≤ 0.5). The value of fat content of biscuit flour was within the range of 1-2% [15,16]. The special flour was reported to contain 1.67% fat. The level of fat in the biscuit flour, under study, was lower than the range of 2.15-2.35% fat in four Sudanese wheat cultivars (Deberia, Elneilain, Condor and Sasaraib). The fat content of the whey powder was higher than the value of 0.9% fat.

Fiber content: The biscuit flour gave a value of 0.87% fiber which was lower than the range of 2.10-2.85% for Deberia, Elneilain, Condor and Sasaraib cultivars. A lower fiber level 0.28% was reported by Bashir [11]. Obviously the whey powder was not found to contain crude fiber.

Carbohydrate: The carbohydrate level in the biscuit flour was found to be 74.87% and that of the whey powder were 74.47% with no difference (p ≤ 0.05). The carbohydrate of wheat flour was higher than the range of 65-70% and lower than the value of 75.9% [13]. The carbohydrate of the whey powder was slightly lower than the whey carbohydrate level (75.3%). The proximate chemical composition indicated that the biscuit wheat flour and the whey powder contained similar amounts of carbohydrates and proteins. The base wheat flour was found to possess lower levels of fat and ash and higher moisture content. Furthermore, the base wheat flour contain crude fiber which was completely absent in the whey powder.

Amino acid profile

Table 2 shows the biscuit wheat flour and whey powder composition

Amino acids	Biscuit wheat flour (g/100 g protein)	Whey powder (g/100 g protein)
Essential		
Histidine.	2.2	2.1
Isoleucine.	7.4	9.93
Leucine.	12.4	13.6
Lysine.	0.83	2.7
Methionine.	2.1	2.2
Phenylalanine.	7.9	3.4
Valine.	10.13	8.7
Non-essential		
Aspartic acid.	5.4	7.4
Glutamatic acid.	18.7	5.9
Serine.	2.6	1.7
Cystine.	1.04	1.6
Alanine.	11.4	-
Tyrosine.	-	1.9

Table 2: Amino acid profile for biscuit wheat flour and whey powder.

Treatments	Dry Gluten	Wet Gluten	Gluten index
control	27.50 ± 0.06a	9.20 ± 0.06[a]	91.13 ± 2.32[c]
A	26.53 ± 0.03[a]	9.30 ± 0.30[a]	96.27 ± 1.40[ab]
B	22.10 ± 1.20[b]	6.83 ± 0.07[b]	93.07 ± 0.12[bc]
C	14.30 ± 0.15[c]	5.23 ± 0.19[c]	99.13 ± 0.09[a]

*Each value in mean of three measurements.

**Mean values (± SD) having different superscript letters in columns differ significantly (p ≤ 0.05)

Where:

Control: Biscuit wheat flour (base).

A: 95% biscuit wheat flour + 5% whey powder

B: 90% biscuit wheat flour + 10%whey powder

C: 85% biscuit wheat flour + 15% whey powder

Table 3: Effect of whey protein substitution on quantity and quality of biscuit wheat flour gluten.

in g/100 g protein. The non-essential amino acids are those the body can synthesize and therefore non-essential in the diet. The essential amino acids, on the other hand are very important from nutritional point of view since the body cannot make and should therefore be supplemented in the diet. The level of the essential amino acids varies considerably in both biscuit wheat flour and whey powder. The amino acid profile of biscuit wheat flour showed that most essential amino acids were present, in varying amounts. Leucine (12.4 g/100 g protein) and Valine (10.13 g/100 g protein) were the predominant essential amino acids; While Lysine was present in a relatively small quantity (0.83 g/100 g protein). On the other hand the amino acid profile of the whey powder also revealed the presence of most essential amino acids. While Leucine, Isoleucine, and Valine were the predominant, Histidine was present in the smallest amount (1.9 g/100 g protein). Whey powder protein can complement biscuit wheat flour by contributing the essential amino acid Lysine (2.7 g/100 g protein). Whey and some of its derivatives were claimed to be used as a low-cost source of protein, carbohydrates and claimed, efficiently fortifying a wide variety of food produces. Whey proteins are considered to be high quality protein that contains all of the amino acids required by human. Milk proteins were found to contain relatively high amounts of lysine, a dietary essential amino acid, that in sometimes limiting in the diets for human, particularly those high in cereals.

Quality attributes

Gluten quantity and quality: The gluten quantity (wet and dry gluten) and quality (gluten index) of the 100% biscuit wheat flour (control) and the biscuit flour- whey powder mixtures made by mixing the base biscuit flour with three levels of whey powder (5, 10, and 15%) were determined. The wet gluten, dry gluten and gluten index were therefore estimated and the results are presented in table 3.

Wet gluten: The results presented in table 3 show that the wet gluten percent was affected by mixing biscuit flour with the various levels of whey powder. While the 100% biscuit wheat flour (control) gave the highest wet gluten content (27.5%), the addition of the three concentrations of whey powder (5, 10, and 15%) reduced the obtained level to 26.53, 22.10, and 14.30% wet gluten respectively. A significant difference (p ≤ 0.05) was observed between the various flour mixtures made by the addition of 5%, 10% and 15% whey powder. Insignificant difference was observed between the control and the biscuit wheat flour to which 5% whey powder was added. The reduction of the wet gluten percent could be due to the fact that part of the biscuit flour was replaced by whey powder. The wet gluten of dough's prepared from biscuit wheat flour with 0, 5, 10, 15, 20 and 25% of decorticated pigeon pea were found to be 31.2, 30.2, 30.5, 28.7, 27.3 and 23.1% respectively [17].

Dry gluten content: The biscuit wheat flour (control) gave 9.20% dry gluten; while biscuit wheat flour substituted by 5%, 10% and 15% whey powder gave 9.30%, 6.83% and 5.23% dry gluten respectively. The highest dry gluten level was found in the control biscuit flour and the flour substituted by the 5% whey powder. Which were none significantly different (P≤0.05) in their dry gluten level. The lowest dry gluten level was seen in the biscuit wheat flour and substituted by 15% whey powder.

Gluten index: The biscuit wheat flour (control) gave 91.13 gluten index, while wheat flour with 5, 10, 15% whey powder gave 96.27, 93.07 and 99.13% gluten indices respectively. The highest gluten index was obtained by the addition of 15% whey powder, while the lowest value was found in the control base flour. All flour mixtures were significantly different with respect to their gluten index. A lower value of gluten index (80.29) for biscuit wheat flour [17]. The results indicated that substitution by the varying levels of whey powder reduced the gluten quantity, the lowest being in the flour substitution by 15% whey. On the other hand, substitution has resulted in increased gluten quantity; the highest was noticed in the biscuit wheat flour substituted by 15% whey. The increased gluten index lowered the quality of the biscuit flour (Table 3).

Sensory evaluation of biscuits

Table 4 shows the score of sensory evaluation of biscuit made from 100% biscuit wheat flour (control) and biscuits made from wheat flour replaced by three different levels of whey powder. Biscuits were evaluated for color, aroma, texture, taste and overall acceptability. Increasing levels of whey powder resulted in increased scores of color. The reaction of amino acids with carbohydrate (millered reaction) could be responsible for the color formation in the final product [18-21].

Conclusion

The values obtained increased from 3.7 (control) to the highest score of 7.46 (15% replacement). A significant difference (p ≤ 0.05) was observed in score of color between biscuits made from the control and those from whey powder-supplemented flours. Increasing the levels of supplementation with whey powder also resulted in a significant increase in the score of aroma of biscuits. The values obtained were 4.01 (control), 5.31 (5% substitution), 6.53 (10% substitution) and 6.53 (15% substitution). The score of texture was increased with increasing level of whey powder supplementation with significant difference at a level of (p ≤ 0.05). The values obtained were 1.99 (control), 4.5 (5%

Treatment	Color	Aroma	Texture	Taste	Overall Acceptability
Control	3.70 ± 0.79[b]	4.01 ± 0.67[b]	1.99 ± 0.35[b]	4.09 ± 0.77[ab]	4.09 ± 0.77[ab]
A	5.78 ± 0.63[a]	5.31 ± 0.53[ab]	4.97 ± 0.39[a]	5.02 ± 0.58[bc]	5.21 ± 0.45[a]
B	7.07 ± 0.58[a]	6.54 ± 0.47[a]	6.16 ± 0.77[a]	6.93 ± 0.46[a]	6.87 ± 0.43[a]
C	7.46 ± 0.47[a]	6.53 ± 0.59[a]	6.35 ± 0.62[a]	6.65 ± 0.46[ab]	6.49 ± 0.80[a]

*Each value is a measure of three determinations.
**Mean values (± SD) having different superscript letters in columns differ significantly (p ≤ 0.05).
Control: Biscuit wheat flour (base).
A: 95% biscuit wheat flour + 5% whey powder.
B: 90% biscuit wheat flour + 10%whey powder.
C: 85% biscuit wheat flour + 15% whey powder.
Table 4: Sensory attributes of biscuits.

substitution) 6.16 (10% substitution), and 6.35 (15% substitution). The score of taste was increased with increasing whey powder levels in the substituted flours with significant difference at a level of (p ≤ 0.05). The values obtained were increased from 4.09 (control) to the higher value of 6.53 (10% replacement). The sensory evaluation of biscuits showed that there was a significant difference in the overall acceptability of biscuits made by different levels of whey powder supplementation. Biscuits made from wheat flour supplemented by 10% whey powder showed the best panelist's scores for overall acceptability of the biscuits produced. The panelists scores indicated that all sensory attributes evaluated were improved by inclusion of increasing levels whey powder in the base flour (100% biscuit wheat flour) used for the production of biscuits. Earlier reports found that milk and milk derivatives were used for color improvement, water absorbing, and spread control properties and flavor in baked goods. Adding of milk was reported to help the product to brown during baking and add to its nutritive value. Whey protein is widely used in various food applications due to their interesting functional properties such as thickening and fat and flavor binding capacity.

References

1. Vieira ER (1996) Elementary food science fourth edition Chapman and Hall. Fifth Avenue, New York.

2. Dalgleish DG, Senaratne V, Francois S (1997) Interactions between α-lactalbumin and β-lactoglobulin in the early stages of heat denaturation. J Agric Food Chem 45: 3459-3464.

3. Huth PJ, Layman DK, Brown PH (2006) Emerging health benefits of dairy proteins. Dairy council J Nutr 134: 961.

4. Ha E, Zemel MB (2006) Emerging health benefits of dairy proteins. Dairy council (Digest) 77: 19-21.

5. Mc Bean LD (2003) Emerging health benefits of whey. National Dairy Council. Digest 74: 15-44.

6. Resch JJ (2004) Processing parameter effects on the functional properties of derivatized whey protein ingredients. Department of food science. North Carolina state university.

7. Singh RP, Heldman DR (1993) Introduction to food engineering. Food dehydration, Academic Press, San Diego, CA.

8. AOAC (1984) Association of Official Agriculture Chemists. Washington Dc, USA.

9. Mead R, Gurnow RN, Harted AM (1993) Statistical Methods in Agriculture and Experimental Biology. Chapman & Hall, London.

10. Kent-Jones, Amos Aj (1967) Modern Cereal Chemistry. Northern Pub Co Ltd. Liverpool.

11. Bashir RE (2005) Evaluation of four wheat flours for bread, biscuits and cake making. Faculty of agricultura, University of Khartoum, Sudan.

12. Badi SM, Hosny RC, Casady AJ (1976) Pearl millet 1-Characterization by SEM. Amino acid analysis, Lipid composition and prolamin solubility. Cereal Chem 53: 478-542.

13. Giami SY, Achinewhu SC, Ibaakee C (2005) The quality and sensory attributes of cookies supplemented with fluted pumpkin (Telfairia occidentalis Hook) seed flour. Intr J Food Sci Technol 40: 613-620.

14. Mahmoud IA (2003) Biscuits from composite flour of wheat and sorghum. University of Khartoum, Sudan.

15. Pyler EJ (1973) Baking Science and technology. The Siebel Publishing Company, Chicago.

16. Kent-Jones, Amos AJ (1967) Modern Cereal Chemistry. Northern Pub Co Ltd, Liverpool.

17. Abd Elatief HA (2007) Fortification of wheat flour with decorticated pigeon pea flour and protein isolate for bakery products. University of Khartoum, Sudan.

18. Matz SA (1968) Cookie and Cracker Technology. Avi Publishing Co Inc,Wesport, Connecticut.

19. Bryant CM, Mc Clements DJ (1998). Molecular basis of protein functionality with special consideration of cold-set gels derived from heat denatured whey. Trends in Food Science and Technology 9: 143-151.

20. Burrington KJ (1998) The Whey Applications program coordinator at the University of Wisconsin-Madison Center for Dairy Research.

21. Mohamed EA (2000) Evaluation of four local wheat cultivars with special emphasis on protein fraction. University of Khartoum, Sudan.

21

Performance of Pickle Production Processing and Marketing in Sindh Pakistan

Sanaullah Noonari[1]*, Irfana Noor Memon MS[1] and Abdul Sami Kourejo[2]

[1]*Assistant Professor, Department of Agricultural Economics, Faculty of Agricultural Social Sciences, Sindh Agriculture University, Tandojam, Pakistan
[2]Research assistant, Department of Agricultural Economics, Faculty of Agricultural Social Sciences, Sindh Agriculture University, Tandojam, Pakistan

Abstract

Pickle products are packed in polythene bags then are placed in the Glass bottles or bucket. The pickle producers in Shikarpur area on average spent a total cost of production of Rs.1575700.00 in the Rs.607500.00 raw material cost, Rs.320900.00 packaging material cost, Rs.536000.00 human resource cost and Rs.130000.00 marketing cost respectively on capital inputs. The selected pickle producers in Shikarpur area on average revenue generate of Rs.2350000.00 and pickle net income per unit Rs.1340700.00, Rs.2350000.00 gross income and Rs.1575700.00 total expenditure respectively in the study area. The pickle production unit gross income Rs. 2350000.00 and total expenditure is Rs.1575700.00 in the study area therefore they availed input output ratio of 1:1.49 and pickle production unit net income Rs.774300.00 and total expenditure Rs.1575700.00 in the study area therefore they availed input output ratio of 1:0.49 respectively in the study area.

Keywords: Pickle; Raw material; Polythene bags; Bucket; Input output ratio; Shikarpur

Introduction

Pickle is one of the oldest and most successful methods of food preservation known to human. The optimization of pickle quality depends on maintenance of proper acidity, salt concentration, temperature and sanitary conditions. Pickle products add spice to meals and snacks. The skillful blending of spice, sugar and oil with fruit and vegetable gives crisp, firm texture and pungent, sweet-sour flavor. Pickles serve as appetizers and help in digestion by aiding flow of gastric juices. Fermented pickles also have beneficial bacteria that can control harmful intestinal microbes. Pickle in Shikarpur is manufactured in twenty-five different agriculture products like mango (carry), chillies, lemon onion etc. and is manufactured in three categories oil, vinegar, and water. The export quality of pickle is manufactured in soybean oil. There are twelve pickle manufacturing factories in Shikarpur and fifteen pickle selling shops in the city. Different manufacturers have opened their outlets in different cities of the country like in Quetta to export the product to Afghanistan and Iran. Pickle manufacturing process is consisting on six steps. When raw material are brought from different parts of the province/country (like mango) is being cut into different parts according to the requirement of buyer after then it is washed with fresh water, then it is mixed in mixing machine at this stage different spices are mixed with raw material. The raw material is stored in brine tank where they can be preserved for twelve months. After that according to the requirement of buyer mixed raw material is taken from the brine tank and are washed. After washing raw material are mixed in mixing tub(container) where different spices are mixed after then manufactured pickle is packaged in different sizes of jars and are sold. Pickle manufactured in vinegar takes fifteen days to be ready for eating purpose while pickle manufactured in soybean oil takes one month to be eaten [1]. Pickles are generally of three types namely pickles in vinegar, citrus juice, brine and oil. Besides the basic fruit/vegetable, the substances that are generally added to pickles are vinegar, sugar, salt, oil and spices. The presence of these ingredients makes the product highly acidic in nature. Due to the acid nature and/or the presence of oil in the pickle, the packaging material to be used should be oil and acid resistant. Spoilage of pickles could be due to microbial contamination or oxidation/ rancidity of the oil used. A good packaging material for pickle can prevent spoilage. A good package for pickles should have the attributes such as aroma retention, excellent protection against light, moisture and oxygen, excellent seal integrity for containment, grease, oil and acid resistance, good aesthetics and appearance. When food is available in ample it is preserved for further consumption [2]. Pickle has a large variety of pickles (known as Achar in Punjabi, Hindi, Bengali, Uppinakaayi in Kannada, Lonacha in Marathi, orukai in Tamil, oragaya in Telugu), which are mainly made from varieties of mango, lime, tamarind and Indian gooseberry (amla), chilli. Vegetables such as brinjal, carrots, cauliflower, tomato, bitter gourd, green tamarind, ginger, garlic, onion, and citron are also occasionally used. These fruits and vegetables are generally mixed with ingredients like salt, spices, and vegetable oils and are set to mature in moisture less medium. In Pakistan, pickles are known locally as Achaar (in Urdu) and come in a variety of flavors. A popular item is the traditional mixed Hyderabadi pickle, a common delicacy prepared from an assortment of fruits (most notably mangos) and vegetables blended with selected spices [3].

Pickling process in India differs from other regions mainly due to additional spice mixture added to them post anaerobic fermentation. Pickles are main side dishes and many varieties of vegetables are used. However, raw mango or tender mango is the most popular variety of fruit used for pickling. There are multiple varieties of mango pickles prepared depending on the region and the spices used but, broadly there are two types of - whole baby mango pickle or cut mango pickle. Whole baby mango pickle is a traditional variety very popular in

*Corresonding author: Sanaullah Noonari, Assistant Professor, Department of Agricultural Economics, Faculty of Agricultural Social Sciences, Sindh Agriculture University, Tandojam, Pakistan
E-mail: sanaullahnoonari@gmail.com

Southern India and uses baby mangoes that are few weeks old. There are special varieties of mangoes specifically used just for pickling and they are never consumed as ripe fruit. Baby mangoes are pickled and added with spice mixture in a very careful process which ensures pickles are preserved for years [4-6].

Keeping in the view competitiveness of Shikarpur district Sindh province of Pakistan pickle production processing and marketing, the study was conducted with the following specific objectives.

Objectives

ı. To describe socio-economic status of pickle producers in the study area.

ıı. To study the economics of production and marketing in Shikarpur Sindh.

ııı. To identify issues and suggest policy measures for promoting on pickle production in the study area.

Methodology

The Shikarpur district of Sindh province was selected purposively for the present study because it is largest pickle producers it is one of the district within Sindh province in terms of pickle production. The district Shikarpur is situated in the northern part of Sindh Pakistan [7-9]. The summers are very hot and the winters are mild to cold.

Data collection

This study was used the primary data in present study. Primary data was collected from sample of 60 respondents. A list of pickle producers of each selected producers were prepare selection was made on proportional random sampling method [10-14]. They were purposively selected from Shikarpur district Sindh.

Secondary data

Secondary data collected from various Government departments used to determine the overall growth rate of pickle production [15,16]. The Secondary data was collected from literature and publication including report, research papers,   etc.

Economic analysis estimation methods

Data were analyzed by developing equations for estimating fixed costs, variable costs, total cost of production, total revenue, net revenue Input-Output ratio and benefit cost ratio [11,17-19]. A brief description of each term is given as follows:

Estimation of factory inputs

For estimation of factory inputs for pickle on the sample unit, the following formula was used.

$$Fip = (Fm \times Fss) + Ap + Rfe) / Af$$

Where:

Fip = Factory input per unit of pickle

Ap = Area pickle production

Fm= Factory Machine

Fss= Factory Support Structure

Rfe= Rate of factory expenditures.

Af= Area factory

Estimation of human resource cost

The extent of labour inputs for various cultural operations involved in pickle production was estimated by applying the following formula:

$$Lit = (Mn\ Hc) + (Lwd \times Wr) + (Swd \times Hc) / Af$$

Where:

Lit =Labour input per unit of pickle production

Hc =Hiring charges.

Mn =Machine work hour.

Lwd=Labour work day.

Wr =Wage rate

Swd=Supervisor work day

Af=Area factory

Estimation of capital inputs

The following formula was used to compute per unit (factory) cost of the capital inputs.

$$Cipu = (Qs \times Pr) + (Qe \times Pr) + Qr \times Pr) / Af$$

Where:

Cipu =Capital inputs per unit/factory of pickle

Qs = Quantity of used.

Pr = Price per unit of input.

Qe = Quantity of electricity

Qr = Quantity of Raw Material Inventory

Af=Area factory

Marketing cost

The marketing cost was estimated by using the following formula:

$$Mc = Qm (Rl+Tr / Af)$$

Where;

Mc = Marketing cost.

Qm = Quantity of produce marketed.

Rl = Rate of loading.

Tr = Transportation rate.

Af=Area factory

Estimation of returns

The estimation of returns was developed by using the following formula:

$$VP = (Qs \times Pr) / Af$$

Where:

VP = Value of Product.

QS = Quantity Sold.

Pr = Price per unit.

As = Area factory

Total cost of production

Total cost of production was estimated by using the following formula:

TC=TFC+TVC

Where:

TC =Total Costs of Production

Net returns

Net returns were estimated by using the following formula:

NR = TI - TC

Where;

NR = Net Returns

TI = Total Income

TC = Total Cost

Results

This chapter provides results of the study including current status of pickle production practices and issues of pickle producers. Pickle has always been in consistent demand in the subcontinent as a compulsory add-on to be served with food. This indicates that there is a substantial potential for this business.

Age

Table 1 shows that majority 46.66 percent of the respondents were middle aged followed by young 36.66 percent and old 16.66 percent. It means that half of the respondents belonged to middle age category.

Literacy level

Table 2 shows that 6.66 percent of the respondents were primary education while 13.33 percent had literacy level middle standard, 20.00 percent were metric and college level education and 60.00 percent of the respondents had literacy level graduation.

Experience

Table 3 shows that pickle 30.00 percent respondents were Up to 10 years' experience, 45.00 percent respondents were 11-20 years and 25.00 percent respondents were above 26 years in the study area.

Type of business

Table 4 shows that the majority 61.66 percent of the respondents was owner and only 38.33 percent respondents were renter/tenant.

Age	No. of respondents	Percentage
Young (< to 30)	22	36.66
Middle aged (30-50)	28	46.66
Old (> 50)	10	16.66
Total	60	100.00

Table 1: Distribution of the respondents according to their age.

Literacy level	No. of respondents	Percentage
Primary	4	6.66
Middle	8	13.33
Matric and Collage	12	20.00
Graduation	36	60.00
Total	60	100.00

Table 2: Distribution of the respondents according to their literacy level.

Varieties pickle

Table 5 shows that 86.00 percent of the respondents were make mango oil pickle, 46.00 percent of the respondents were make lemon oil pickle, 73.00 percent of the respondents were make mixed oil pickle, 76.60 percent of the respondents were make mango vinegar pickle, 33.30 percent of the respondents were make lemon vinegar pickle 16.60 percent of the respondents were make chilli vinegar pickle and 25.00 percent of the respondents were make other pickle in the study area.

Product pack/weight

Table 6 shows that pickle product will first be packed in polythene bags which will then be placed in the Glass bottles or bucket. Therefore, polythene bags requirement would be equal to the number of bottles and buckets produced.

Machinery and equipment

Table 7 shows that pickle production machinery like mixing, filling, labeling packaging and others production capacity of per hour above 100 Kg and different costs/prices of machinery in the study area.

Office equipment

Table 8 shows provision of Rs.80000.00 for procurement of office furniture and equipment. This would include computer, printer, fax, table, desk, chairs etc.

Experience	No. of respondents	Percentage
Up to 10 years	18	30.00
11-25 years	27	45.00
Above 26years	15	25.00
Total	60	100.00

Table 3: Distribution of the respondents according to their type of tenure.

Type of business	No. of respondents	Percentage
Owner	23	38.33
Rental	37	61.66
Total	60	100.00

Table 4: Distribution of the respondents according to their type of tenure.

Varieties pickle	Number	Percentage
Mango oil	52	86.00
Lemon oil	28	46.00
Mixed oil	44	73.30
Mango vinegar	46	76.60
Lemon vinegar	20	33.30
Chili vinegar	10	16.60
Others pickle	15	25.00

Table 5: Different varieties of pickle production in the study area.

Product pack	Units	Pack weight
Glass bottles/ polythene bags	Grams	330
Glass bottles/ polythene bags	Grams	450
Glass bottles/ polythene bags	Kg	1
Glass bottles/ polythene bags	Kg	1.8
Polythene bags	Kg	5
Bucket	Kg	16
Bucket	Kg	32

Table 6: Different product pack/ weight of pickle in the study area.

Raw material

Table 9 shows that overall total raw material cost incurred towards various product was estimated and observed to be Rs.607500.00 followed by Rs.307500.00 in Mango/ Chilli/Garlic/Lime, Rs.4000.00, Rs.98000.00, Rs.51000.00, Rs.15000.00, Rs.48000.00, Rs.24000.00, Rs.60000.00 in salt, Chilli (Red), Saunf, Methi Seeds, Kalongi, Rai and Cooking oil for the production in the study area.

Packaging materials

Table 10 shows that overall total packaging material cost incurred towards various product was estimated and observed to be Rs.320900.00 followed by Rs.150000.00, Rs.120000.00, Rs.15000.00, Rs.15000.00, Rs.14400.00, Rs.6000.00, Rs.7500.00, Rs.8000.00 in glass bottles/ polythene bags 330 gram, glass bottles/ polythene bags 450 grams, Glass bottles/ polythene bags 1 kg, Glass bottles/ polythene bags

1.8 kg, Polythene bags 5 kg, Bucket 16 kg and Bucket 32 kg for the production in the study area.

Human resource

Table 11 shows that overall total human resource cost incurred towards various products was estimated and observed to be Rs.134000.00 followed by Rs.25,000.00, Rs.15,000.00, Rs.24000.00, Rs.35000.00, Rs.30000.00, Rs.5000.00 in owner /manager, supervisor, machine operators, packaging staff, workers and watchman respectively.

Marketing costs

Table 12 shows that marketing cost spent a sum of Rs.130000.00, this included Rs.22500.00 for loading, Rs.85000.00 for transportation and Rs.22500.00 of unloading respectively on marketing cost in the study area.

Total cost of production

Table 13 shows that the selected pickle producers in Shikarpur area on average spent a total cost of production of Rs.1575700.00 in the Rs.607500.00 raw material cost, Rs.320900.00 packaging material cost, Rs.536000.00 human resource cost and Rs.130000.00 marketing cost respectively on capital inputs.

Revenue generation

Table 14 shows that the selected pickle producers in Shikarpur area on average revenue generate of Rs. 2350000.00 respectively in the study area.

Net Income

Table 15 shows that pickle net income per unit Rs.774300.00, Rs.2350000.00 gross income and Rs.1575700.00 total expenditure respectively in the study area.

Description	Capacity Unit/Hr	Cot Rs/unit	Total Rs.
Mixing Machine	100 kg / hr	50,000.00	50,000.00
Utensils, sieve, pans	100 kg / hr	50,000.00	50,000.00
Filling Machine	100 kg / hr	100000.00	100000.00
Packing Machine	100 kg / hr	40000.00	40000.00
Labeling Machine	100 kg / hr	60000.00	60000.00
Laser print/coding Machine	100 kg / hr	80000.00	80000.00
Generator 10 KVA	–	200000.00	200000.00

Table 7: Different machinery and equipment used in the pickle production.

Description	Quantity	Cost	Amount
Computers	1	25000.00	25000.00
Printer	1	12000.00	12000.00
Telephones	1	5000.00	5000.00
Fax machines	1	8000.00	8000.00
Furniture	—	80000.00	80000.00

Table 8: Different machinery and equipment used in the pickle production.

Ingredients	Quantity (Kg)	Cost (kg)	Amount(Rs.)
Mango/ Chilli/Garlic/Lime	20500	15.00	307500.00
Salt	800	05.00	4000.00
Chilli (Red)	700	140.00	98000.00
Saunf	300	170.00	51000.00
Methi Seeds	300	50.00	15000.00
Kalongi	300	160.00	48000.00
Rai	300	80.00	24000.00
Cooking oil	500	120.00	60000.00
Total			607500.00

Table 9: Different raw materials and costs used in the pickle production.

Description	Quantity (Unit)	Cost(Unit)	Amount(Rs.)
Glass bottles/ polythene bags 330 gram	15000	10.00	150000.00
Glass bottles/ polythene bags 450 grams	10000	12.00	120000.00
Glass bottles/ polythene bags 1 kg	1000	15.00	15000.00
Glass bottles/ polythene bags 1.8 kg	800	18.00	14400.00
Polythene bags 5 kg	300	20.00	6000.00
Bucket 16 kg	250	50.00	7500.00
Bucket 32 kg	100	80.00	8000.00
Total			320900.00

Table 10: Different packaging materials and cost used in the pickle production.

Description	No. of Employees	Salary per month (Rs.)	Amount(Rs.)
Owner /Manager	1	25,000.00	25,000.00
Supervisor	1	15000.00	15000.00
Machine operators	2	12000.00	24000.00
Packaging Staff	5	7000.00	35000.00
Workers/labours	6	5000.00	30000.00
Watchman	1	5000.00	5000.00
Total			134000.00
Grand total Rs. 134000.00 x 4 months season of production =			536000.00

Table 11: Different human resource cost used in the pickle production.

Particulars	Mean
Loading	22500.00
Transportation	85000.00
Unloading	22500.00
Total	130000.00

Table 12: Per unit expenditure incurred on marketing cost in the study area.

Particulars	Mean
Raw material cost	607500.00
Packaging material Cost	320900.00
Human resource cost	536000.00
Marketing cost	130000.00
Total	1575700.00

Table 13: Per unit total cost of production in the study area.

Description	Production (Unit)	Sale price	Amount(Rs.)
Glass bottles/ polythene bags 330 gram	15000	40.00	600000.00
Glass bottles/ polythene bags 450 grams	10000	70.00	700000.00
Glass bottles/ polythene bags 1 kg	1000	130.00	130000.00
Glass bottles/ polythene bags 1.8 kg	800	200.00	160000.00
Glass bottles/ polythene bags 5 kg	300	500.00	150000.00
Bucket 16 kg	250	1500.00	375000.00
Bucket 32 kg	100	3000.00	300000.00
Total			2350000.00

Table 14: Per unit revenue generate in the study area.

Particulars	Mean
Gross Income (Rs) A	2350000.00
Total Expenditure (Rs) B	1575700.00
Net Income (Rs) A-B=C	774300.00

Table 15: Per unit net income in the study area.

Input-output ratio

Table 16 shows that pickle production unit gross income Rs.2350000.00 and total expenditure is Rs.1575700.00 in the study area therefore they availed input output ratio of 1:1.49 respectively in the study area.

Conclusion and Suggestions

The research study on an analysis of pickle production processing and marketing in Shikarpur district Sindh was concluded for the findings during study were the most efficient to produce the pickle at profitable level. The industrial infrastructure is the web of personal, economic, social and legal relationships that support the production of pickle commodities. It includes, most visibly, input suppliers and output processors. However, it also includes the formal and informal business relationships between individual farms.

Shikarpur district is a main pickle production processing and marketing area in Sindh Pakistan. Thus, the district can have a potential to produce more pickle for demand of growing population, there is also need for study the efficient pickle production and issues in the production process for policy making.

In our survey we found that pickle consumption is very popular in Sindh Pakistan and with the exception of one respondent rest all the respondents consume some or the other type of pickle. We also found that in spite of availability of readymade pickles the popularity of homemade pickles has not decreased. Thus there is large market for pickle producing units which goes untapped because people go for homemade pickles. We also found that that popularity of mango pickles is fairly consistent among the respondents with almost one-third preferring mango pickles. Communication and sharing of knowledge on interventions in production, processing and utilization of pickle and pickle by products is recommended

a) Pickle companies should more aggressively tap the Mango pickle and Lemon pickle segment.

b) Packaging of 250 gms should be produced by the companies as it contains optimum quantity which is sufficient for a nuclear family of 3-4 people.

Production Area	Gross Income(Rs.)	Total Expenditure(Rs.)	Input-output ratio
Unit	(A)	(B)	A/B=C
1	2350000.00	1575700.00	1:1.49

Table 16: Per unit input-output in the study area.

c) As consumers are educated and well aware now a days, they are very health conscious and hence nutritional value, manufacturing and expiry date should be mentioned on the packaging very clearly.

d) Companies should aggressively advertise coupled with effective sales promotions for improving customer recall and brand image.

e) Good packaging, proper oil content, certification from food agencies etc. should be maintained to give customers a home -made taste and push him to buy from market rather than making at home.

f) Attempt to manage human resource cost should be focused through performance measurement and performance based compensation.

g) Encouraging training and skill of self & employees through experts and exposure of best practices is route to success.

h) Small business loans should be provided on soft and simple terms and conditions to producers.

References

1. Akbudak B, Ozer MH, Uylaser V, Karaman B (2007) The effect of low oxygen and high carbon dioxide on storage and pickle production of pickling cucumbers cv. Octobus. Journal of Food Engineering 78: 1034-1046.

2. Khan SH, Muhammad F, Idrees M, Shafique M, Hussain I, et al. (2005) Some studies on spoilage fungi of pickles. J Agri Soc Sci 1: 4-15.

3. Hassan A, Raghuram P (2001) Pickle processing and marketing Agricultural Marketing. Iran Journal of Agricultural Economics 35: 104-108.

4. Verma RA, Tripathi VK, Tripathi MP, Singh S (2012) Studies on the effect of packaging materials on mango pickle during storage. Indian Food Packer 40: 22-25.

5. Amr AS, Jabay OA (2004) Effect of salt iodization on the quality of pickled vegetables. Food, Agriculture and Environment 2: 151-156.

6. Shinde AK, Wagh RG, Joshi GD, Waghmare GM, Kshirsagar PJ, et al. (2004) Pickle purpose mango variety- hybrid-4 (Konkan ruche). Indian Food Packer 58: 54-58.

7. Tomar MC, Singh UB, Bisht NS (2001) Dehydrated mixed vegetable pickle-A study on its preparation and storage. Indian Food Packer 5: 53-60.

8. Bhagwati S, Deka BC (2004) Screening of bamboo species for pickle preparation. Indian Food Packer 58: 49-53.

9. Jha SN, Narsaiah K, Sharma AD, Singh M, Bansa S, et al. (2010) Quality parameters of mango and potential of non- destructive techniques for their measurement-a review. J Fd Sci Technol 47: 1-14.

10. Naik VG (2005) An Economic analysis of mango Production, Processing and Export in South Konkan region of the Maharashtra. University of Agricultural Sciences, Dapoli.

11. GOP (2013) Agriculture: Economic Survey of Pakistan. Ministry of Food and Agriculture, Government of Pakistan, Islamabad.

12. Gupta GK (2008) Standardization of concentrations of additives for development and processing of oil less mango pickle. Indian Food Packer 27: 15-17.

13. Yalim S, Ozdemir Y (2003) Effects of preparation procedures on ascorbic acid retention in pickled hot peppers. International Journal of Food Sciences and Nutrition 54: 291-296.

14. Wadkar SS, Talathi JM, Dalvi MB, Pawar MB, Naik KV, et al. (2001) Economic analysis of kokum fruit processing units. Proceedings of First National Seminar on Kokum.

15. Narayana CK, Maini SB (1996) Changes in chemical composition of sweet pickle of turnip as affected by different types of containers. Indian Food Packer 7: 23-24.

16. Premi BR, Sethi V, Bisaria G (2002) Preparation of instant oilless pickle from aonla (Emblica officinalis gaertn.). Indian Food Packer 26: 72- 74.

17. Sheth M, Nandwana V (2004) Storage stability of commercial mango pickles in oil. Indian Food Packer 35: 73-77.

18. Geeta T, Prakash Jamun (2006) Quality characteristics of lime pickles prepared using different salts. The Ind J Nutr Dietet 43: 103-111.

19. Usha RM, Rama Rao SN, Girija BR, Nagaraja KV (1992) Studies on quality standards of Indian commercial pickle. Indian Food Packer 28: 27-33.

22

Exploring the Nutritional Quality Improvement in Cereal Bars Incorporated with Pulp of Guava Cultivars

Jahanzeb M[1], Atif RM[1], Ahmed A[2]*, Shehzad A[1], Sidrah[1] and Nadeem M[1]

[1]National Institute of Food Science and Technology, University of Agriculture, Faisalabad

[2]Wheat Research Institute, Ayyub Agricultural Research Institute, Faisalabad

Abstract

Fruits play an important role in human diet because they are dynamic bases of minerals, vitamins and dietary fiber. They are rich sources of iron, phosphorous, calcium, and magnesium and contribute 90% of dietary vitamin C. Yellow and green fruits are rich in vitamin A (beta carotene) folic acid, niacin and thiamine which are vital for normal functioning of the human body. Due to the high perishability, Guava fruit ripen quickly in a few days after harvesting at room temperature. Due to its delicate nature it cannot be stored for a longer period of time. The surplus quantity of the fruit remains unsold and goes to waste during peak harvest season. Extension in post-harvest shelf life and preservation of guava fruit is the pre-requisite for the economical and efficient utilization of this important fruit commodity in Pakistan. Guava pulp is considered as a rich source of fiber, ash, polyphenols, and sugars. This study was aimed to develop cereal based bars using different varieties of guava pulp from two cultivars (Gola and Surahi) (GP) at different concentrations (10% and 15%). Prepared bars were subjected to the proximate analysis. The results demonstrated high moisture, protein and lower level of fat contents in the bars. Bars with guava pulp of Gola cultivar (15%) showed the high level of moisture (6.34 ± 0.03 to 6.47 ± 0.02), protein (4.69 ± 0.02 to 4.61 ± 0.01), fiber (3.85 ± 0.05 to 3.84 ± 0.06) and fat (3.42 ± 0.05 to 3.06 ± 0.03) contents. Then the bars were evaluated for sensory acceptance by trained personal. The sensory analysis showed satisfactory acceptance of the bars containing 10% guava pulp (GP) regarding the aroma, flavor, and textural attributes. The bars containing 15% GP received satisfactory acceptance regarding to aroma only but their texture was not too good. The bars were stored for 14 days and the effect of storage were studied on the bars.

Keywords: GP (Guava pulp); Polyphenols; Sensory analysis; Proximate composition

Introduction

Guava (*Psidium guajava* L.) being a climacteric and highly perishable fruit has rapid post-harvest ripening [1]. It is called common man's fruit and also referred as the apple of the tropics and "*apple of poor man*". Guava has the priority over other fruits due to its high nutritional and commercial values.

Numerous studies have been carried out in order to replace wheat flour with flour made from fruit pulp (dry) in the preparation of bakery products such as biscuits due to economic constraints, business requirements, new consumption trends, and specific eating habits [2,3]. Residues of different fruits can be important source of nutrients to satisfy consumer demand for healthier products. Many food industries are finding ways to add functional ingredients to their products [4]. According to Aquino et al., [2] when fruits and their pulp added to foods. They were able to enhance the taste, texture, aroma, color and nutritional value of the products produced. Santucci et al. [5] stated that the flour mixed with different fruit fractions or using pulp improves not only the nutritional quality of bars but improve their palatability, making them more accepted by consumers.

Currently, the food sector has to deal with a high rate of food waste produced by fruit processing for various products such as jams, juices, ice cream, bakery products, sweets and others. The use of fresh pulp fruits is an important new step for the food industry. Use guava processing waste, such as guava peel, could increase the raw material yield minimizing the problems caused by the disposal of large amount of industrial waste and also expand alternative food production [6]. The Guava as a fruit has not much higher shelf life but when converted to different products the shelf life of the fruit as a new product is increased and thus post-harvest losses are lessened. An economically and technologically feasible alternative would be to produce flour from guava peels in order to either make new products, such as cookies, or partially replace wheat flour to improve the product's nutritional quality since guava has good antioxidant potential and high vitamin C and phenolic compound levels and pigments such as β-carotene and lycopene [7].

Guava is used in many products to enhance the shelf life and to increase the nutritional value of the product. Guava is being processed into different products e.g. jam, jelly, marmalade, beverages and many other different products. Being a climacteric fruit guava exhibit a rapid rise in rate of respiration and production of ethylene during ripening [8]. Guava fruit shelf life ranges from 2 to 4 days at ambient temperature. Numerous postharvest handling methods including controlled/ modified atmosphere and cold storages have been recommended to extend the storage life and maintain quality of guava fruit. Its delicate nature, short post-harvest life, and susceptibility to chilling injury and diseases, limit the potential for export of guava fruit. Thus, the current research was done to show that the shelf life can be increased by converting guava pulp in cereal based bars. During current study innovative bars filled with Guava pulp was made. It can increase the nutritional value of the cereal bars and introduce a new flavor to the customer.

---

*Corresponding author: Ahmed A, Wheat Research Institute, Ayyub Agricultural Research Institute, Faisalabad, Pakistan, E-mail: mjahanzeb2296@yahoo.com

Materials and Method

Procurement of raw material

Guava (*Psidium guajava* L.) fruit was purchased from the Samundari, Faisalabad Pakistan. The fruits were washed under running clean tap water, then the fruits were cut into small pieces and blanched to inactivate the enzyme of guava, then after it, the fruits were put into the fine pulper for pulp extraction. Material preparation and the physicochemical analysis were performed at the Laboratory of Food Safety at Faculty of Food, Nutrition and Home Sciences, University of Agriculture, Faisalabad, Pakistan.

Preparation of bars

Weighed pulp was filled in plastic bottles then stored at freezing temperature. The pulp was stored for about one week to get the desired hard texture for filling in the bars. The pulp was again weighed to calculate the yield, packed in sealed glass containers stored at room temperature for 12h prior to bar preparation. Then the pulp was added into the dough sheets to make Guava bars (Table 1).

Proximate analysis

Moisture: Moisture content of the bars was determined with slight modifications in methods used by Florina. Weighed samples were placed in the hot air oven for 5-6 hours at 95°C.

$$\text{Moisture (\%)} = \frac{\text{Wt. of original sample} - \text{Wt. of dried sample} \times 100}{\text{Wt. of original sample}}$$

Ash: The samples were analyzed for total ash with some difference in Florina.

$$\text{Ash} = \frac{\text{Wt. of ash} \times 100}{\text{Wt. of sample}}$$

Fat: The crude fat contents of samples was determined by following procedure described in AOAC [9] method No. 30-10.01.

$$\text{Crude Fat (\%)} = \frac{\text{Wt. of fat} \times 100}{\text{Wt. of sample}}$$

Crude protein: The crude protein contents in sample was estimated according to the Kjeldahl's method as described in Sackey.

$$\text{Nitrogen (\%)} = \frac{\text{Vol. of } H_2SO_4 \times 250 \times 0.0014 \times 100}{\text{Wt. of sample} \times \text{Vol. of sample}}$$

Total protein (%) = % Nitrogen × 5.57

Fiber: Fiber percentage was calculated according to Sackey Augustina Sackle and Kwaw Emmanuel method with minor modification. The percentage of crude fiber was calculated after igniting the samples according to the expression given below:

$$\text{Crude fiber (\%)} = \frac{\text{Wt. of residue left} - \text{Wt. of ash} \times 100}{\text{Wt. of original sample}}$$

Total sugars (Reducing and non-reducing): Total sugars was found by Lane and Eynon described by Ranganna with some modification.

$$\text{\%Total Sugars} = \frac{\text{Fehling solution factor} \times \text{dilution factor} \times 100}{\text{Titration value}}$$

Ascorbic acid: Ascorbic acid content was determined by indophenol's titration method described by AOAC [9] with some modifications.

$$\text{Ascorbic acid mg / 100 ml} = \frac{1 \times R \times V \times 100}{R_1 \times W \times V_1}$$

1 = Constant

R_1 = ml of dye used for standard solution

R = Dye used during sample

V = Volume of sample diluted with oxalic acid solution

W = Weight of sample

Sensory analysis of guava bars: Sensory analysis of product was performed by a panel of trained judges. Evaluation was done on 9 hedonic scale.

Results and Discussion

The graphical depiction of compositional result shows parameters one by one. There are few studies in the literature on the use of fruits pulps for making bars. No studies were identified in the literature reporting data of proximate composition of guava pulp; only a few studies on fresh fruit, guava powder and guava pomace residues were found. The lack of data on guava pulp may explain the differences observed between the data analyzed and those available in the literature because the flour undergoes prior physical processes, such as heating, which may change its physical and chemical attributes.

Moisture

It is obvious from the results that there was a gradual increase in moisture during the storage. The pulp concentration had a significant effect on bars, while the storage had the non-significant effect. Owing to high moisture content, increase in pulp concentration increases moisture value. This is because the pulp has high concentrations of the moisture that's why the moisture content is increasing by increasing the pulp concentration.

The moisture content of the guava bars was 6.28% to 6.40%, which was increased during the storage period because the plastic containers tends to impermeable to the moisture and thus the moisture of the bars were retained.

In a study by Munhoz et al. [10], the moisture content obtained from guava pulp was 12.55%, and the moisture content of flour obtained from the pulp with guava peel was 13.24%; these results are higher than those obtained in the present study. It was found the average moisture content of 9.72% in guava (*Psidium guajava* L) flour; a value that is somewhat comparable to the present study. In graph series 1 shows storage days (Figures 1 and 2).

Fiber

Storage impact was also had non-significant effect on crude fiber. The content of crude fiber remained almost same during whole storage period. As the concentration of guava pulp is different, the percentage of crude fiber contents depends upon the concentration of pulp used in different treatments. It is depicted from the results that guava pulp is the richest source of dietary fiber.

Treatments	Concentration
T_0	Controlled
T_1	Guava Pulp (Gola) (10%)
T_1'	Guava Pulp (Gola) (15%)
T_2	Guava Pulp (Surahi) (10%)
T_2'	Guava Pulp (Surahi) (15%)

Table 1: Treatment plan.

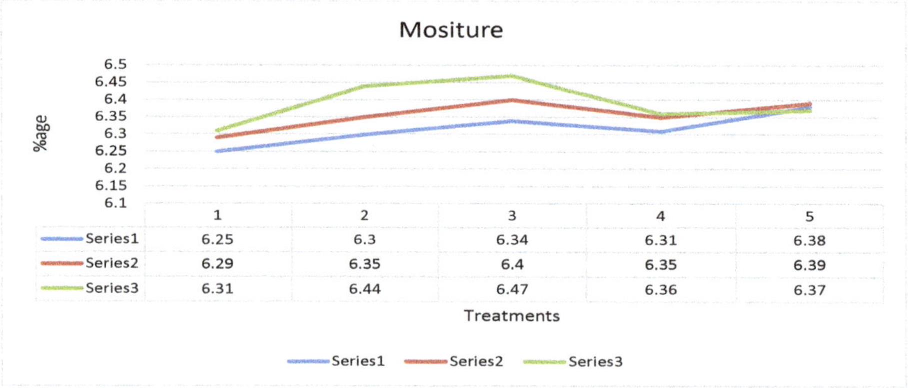

Series 1 = at 0 day of storage
Series 2 = at 7 days of storage
Series 3 = at 14 days of storage

1 = T_0
2 = T_1
3 = T_1'

4 = T_2
5 = T_2'

Figure 1: Graph for moisture content.

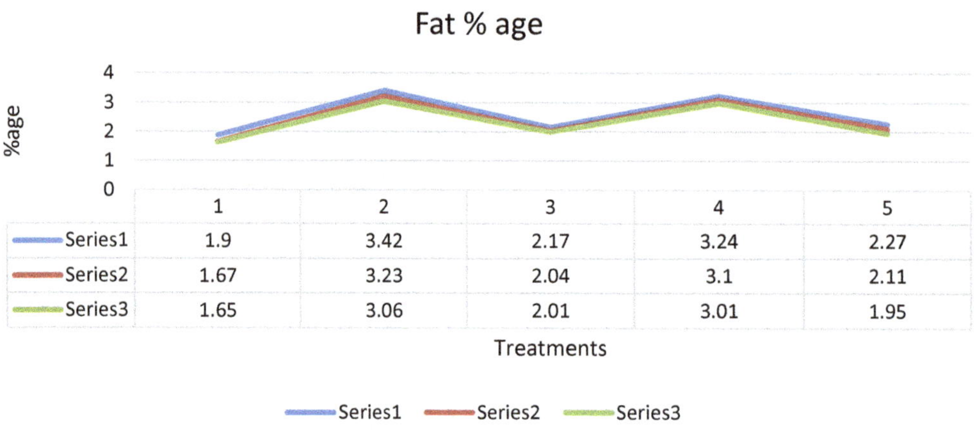

Figure 2: Graph for fat content.

It is vivid from the results that there is significant difference in crude fiber contents for different treatments of bars. Means for crude fiber of different samples (Graph) demonstrated that T_0 showed the lowest value 0.52% of fiber content. The highest value for crude fiber contents 3.52% were found in T_1'. It is depicted from the results that guava pulp is the richest source of dietary fiber so bars have more fiber content than control. Leelavanthi and Tzia [11] prepared high fiber bars and found that dietary fiber content that was about three times higher than the control bars (Figure 3).

Protein

Analysis of variance for protein contents of different treatments during storage has been given in graph. It is obvious from the results that there is significant difference in protein contents for different treatments of bars. Means for protein contents of different treatments demonstrated that T_2' showed the highest value (4.65%) of protein content. The lowest protein contents (2.90%) were found in T_0. Protein contents increased as we added guava pulp at different concentrations [12].

Means for protein contents of different treatments demonstrated that T_2' showed the highest value (4.65%) of protein content. The lowest protein contents (2.90%) were found in T_0. Protein contents increased as we added guava pulp at different concentrations (Figure 4).

Ash

The effect of pulp showed that the ash content was increasing in the bars. As the guava pulp has more fiber percentage thus, the ash content is increasing by addition of the pulp. Gola has the more ash percentage as compare to Surahi. Mean values (graph) for ash contents of different treatments showed the lowest value (0.69%) of ash content by T_0 while the highest value of ash contents 3.23% was found in treatment T_1'. While T_1, T_2 and T_2' had 3.16%, 2.32% and 4.24% ash respectively [13] (Figure 5).

Total sugars

It is evident from the results that there was a gradual increase in total sugars of guava bars with increasing in storage periods. The mean

Fiber Percentage

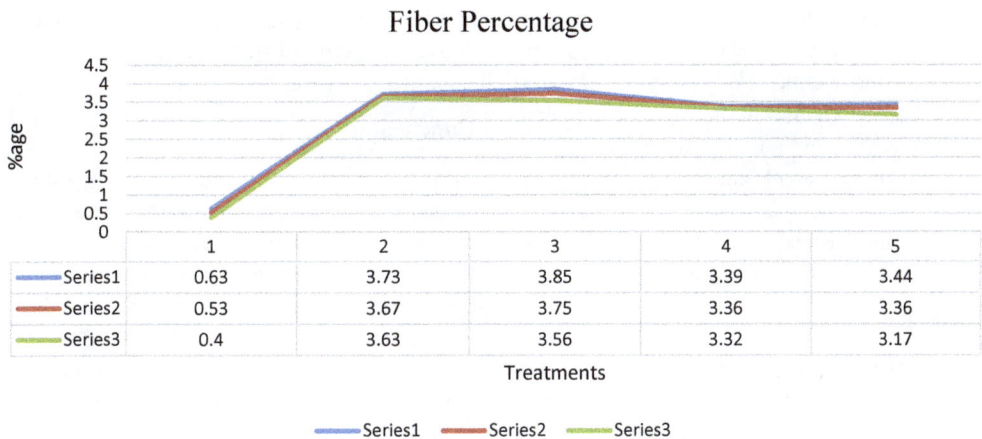

Treatments	1	2	3	4	5
Series1	0.63	3.73	3.85	3.39	3.44
Series2	0.53	3.67	3.75	3.36	3.36
Series3	0.4	3.63	3.56	3.32	3.17

Figure 3: Graph for fiber content.

Crude Protein

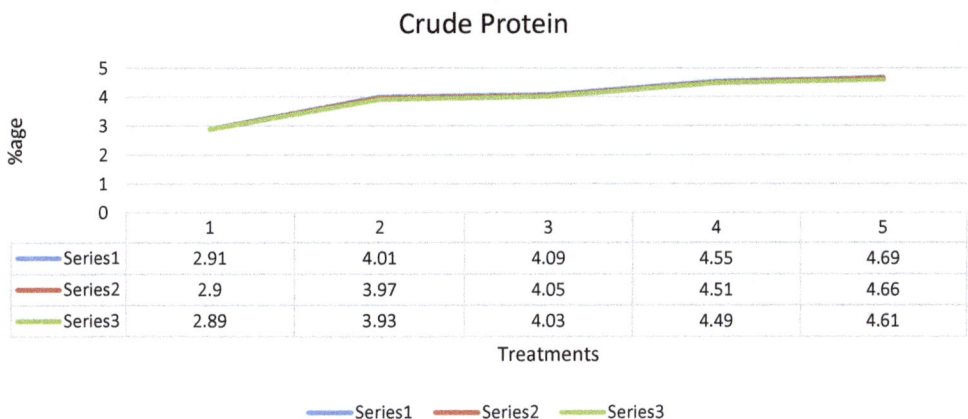

Treatments	1	2	3	4	5
Series1	2.91	4.01	4.09	4.55	4.69
Series2	2.9	3.97	4.05	4.51	4.66
Series3	2.89	3.93	4.03	4.49	4.61

Figure 4: Graph for crude protein content.

Ash

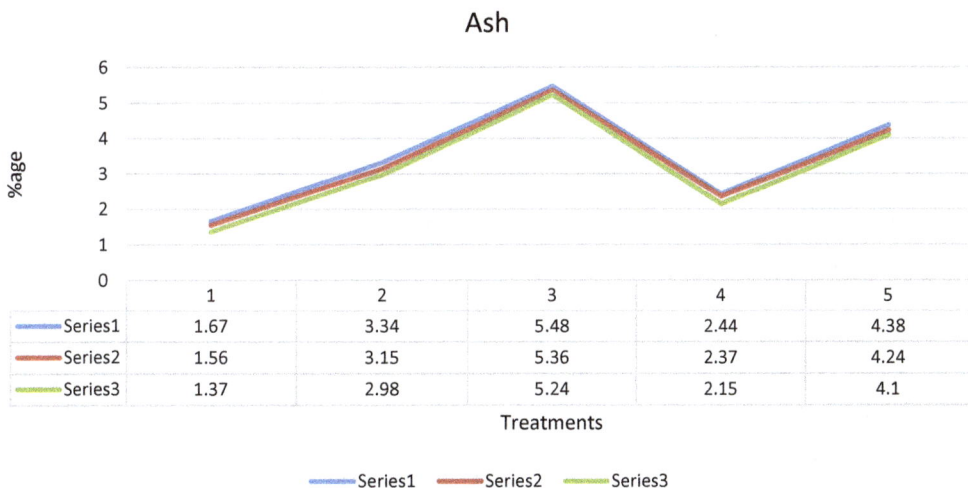

Treatments	1	2	3	4	5
Series1	1.67	3.34	5.48	2.44	4.38
Series2	1.56	3.15	5.36	2.37	4.24
Series3	1.37	2.98	5.24	2.15	4.1

Figure 5: Graph for ash content.

values of total sugars for storage periods increased from 46.62 to 49.15 at 0 to 14 days. The increase in TSS and total sugars would be attributed to the conversion of starch and other insoluble carbohydrates into sugars. Mean values regarding the treatments of total sugars of guava bars were 43.49, 45.63, 46.22, 51.13 and 52.48 for T_0, T_1, T_1', T_2 and T_2' respectively [13-15]. The lowest score was observed for T_0 and the highest score was in case of T_2'. It is apparent from the results that there was a gradual increase in total sugars of guava bars with increasing in storage periods. The mean values of total sugars for storage periods increased from 46.62 to 49.15 at 0 to 14 days (Figure 6).

Total phenolic content (TPC)

TPC was decreasing as the storage days increasing because it is a phenolic content and thus, sensitive to storage. TPC was increasing by increasing the concentration of the pulp. Amongst treatments, a similar behavior was shown by all the treatments indicating a steady decrease in total phenolic content during the course of storage. The maximum decrease in the total phenolic content value was noted for T_0 as it varied from 131.67 to 107.67 and 94.67 at 0 to 7th and 14th day, respectively (Figures 7-9).

Conclusion

It is apparent from mean squares regarding the ascorbic acid of treated guava that significant variations were recorded for the effect of treatments and storage period. Moreover, their interaction was also found to be momentous. As the vitamin-C (ascorbic acid) is a sensitive compound thus the ascorbic acid activity is decreasing as the storage days increasing. From means, it is inferred that the maximum value for ascorbic acid in the treated guava sample was recorded in T_2 and T_1' as 141.92 mg/100 g and 142 mg/100 g followed by 138.83 mg/100g in T_2', respectively.

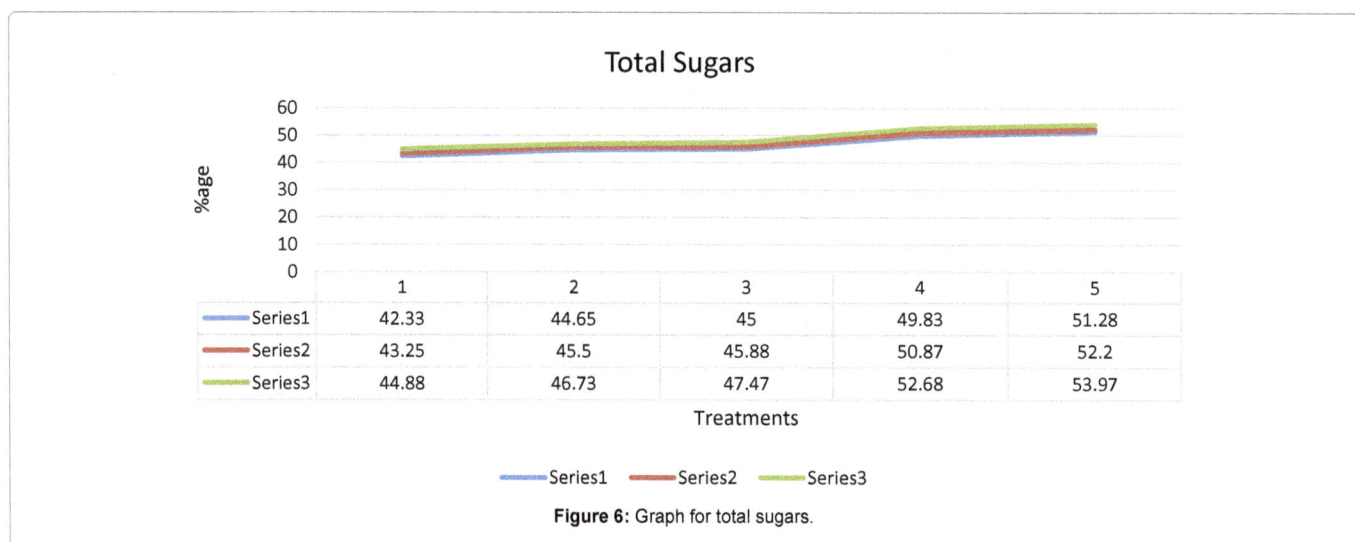

	1	2	3	4	5
Series1	42.33	44.65	45	49.83	51.28
Series2	43.25	45.5	45.88	50.87	52.2
Series3	44.88	46.73	47.47	52.68	53.97

Figure 6: Graph for total sugars.

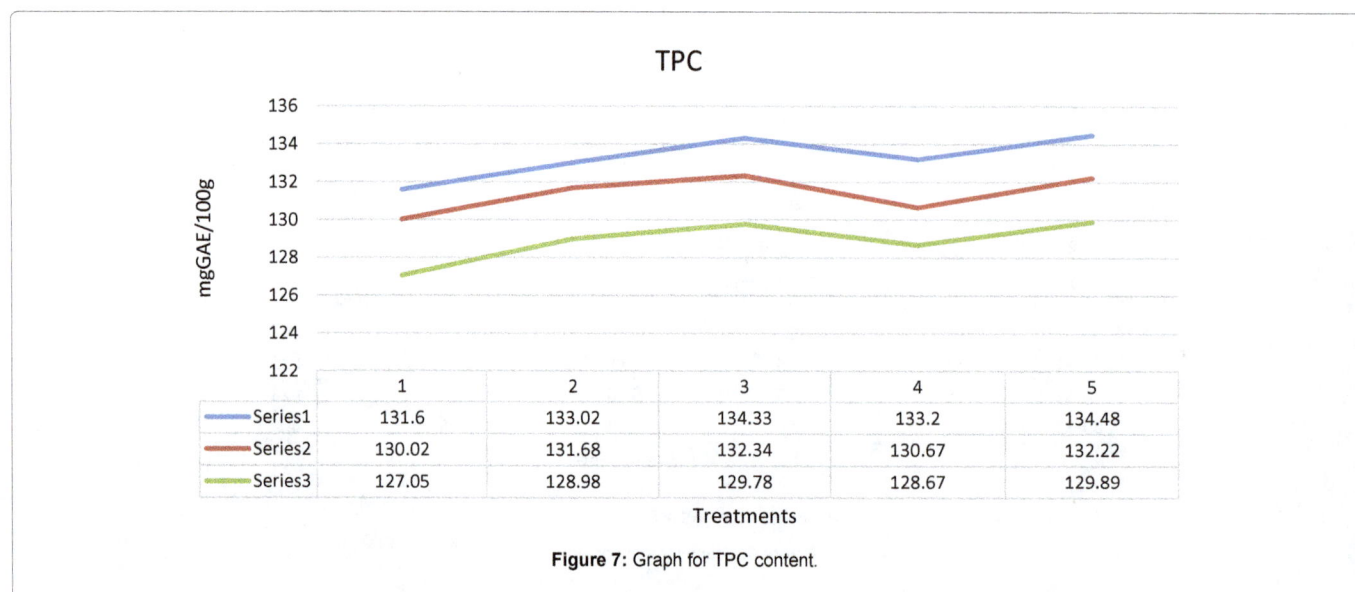

	1	2	3	4	5
Series1	131.6	133.02	134.33	133.2	134.48
Series2	130.02	131.68	132.34	130.67	132.22
Series3	127.05	128.98	129.78	128.67	129.89

Figure 7: Graph for TPC content.

Overall Acceptability

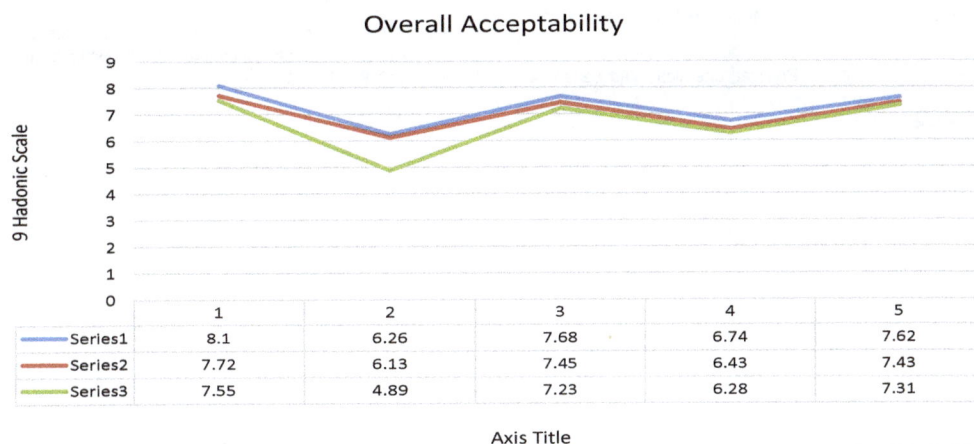

	1	2	3	4	5
Series1	8.1	6.26	7.68	6.74	7.62
Series2	7.72	6.13	7.45	6.43	7.43
Series3	7.55	4.89	7.23	6.28	7.31

Axis Title

Figure 8: Graph for overall acceptability.

Ascorbic Acid

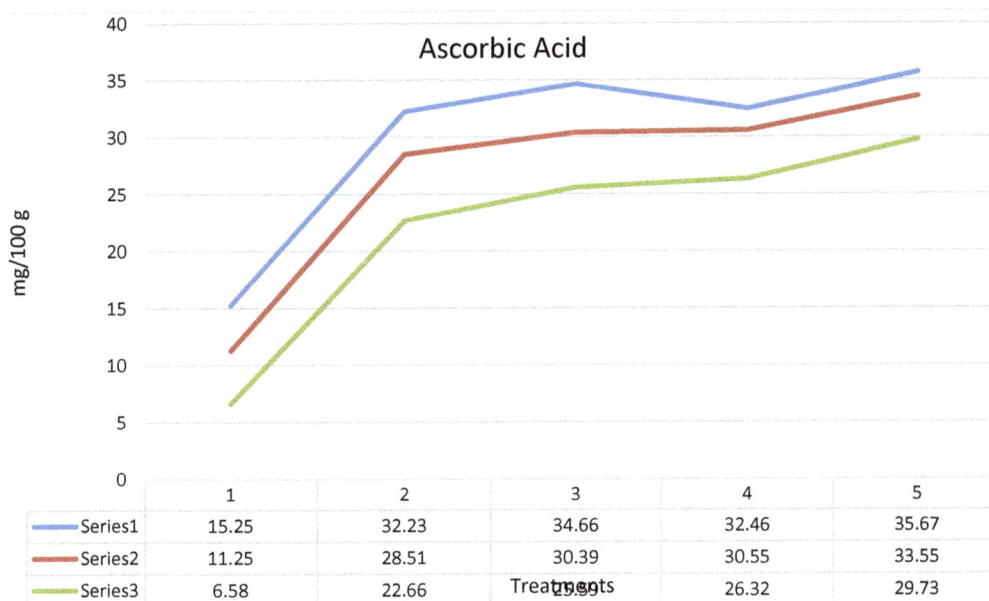

	1	2	3	4	5
Series1	15.25	32.23	34.66	32.46	35.67
Series2	11.25	28.51	30.39	30.55	33.55
Series3	6.58	22.66	Treatments	26.32	29.73

Figure 9: Graph for ascorbic acid.

References

1. Bashir HA, Abu-Goukh AA (2003) Compositional changes during guava fruit ripening. Food chem 80: 557-563.

2. Aquino AC, Leo KMM, Figueiredo AVD, Castro AA (2010) Physical-chemical and sensory acceptance of cookies like cookies made with waste acerola meal. J Inst Adolfo Lutz 69: 379-386.

3. Perez PMP, Germani R (2007) Elaboration of crackers, high in dietary fiber, using eggplant flour (Solanum melongena L). Food Sci and Tech 27: 186-192.

4. Assisi LM, Zavareze ER, Raünz AL, Dias ARG, Gutkoski LC (2009) Nutritional, technological and sensory properties of cookies with wheat flour substitute for oatmeal or parboiled rice flour. Food and Nutrition 20: 15-24.

5. Santucci MCC, Alvim ID, Faria EV, Sgarbieri VC (2003) Effect of enrichment biscuits type water and salt with yeast extract (Saccharomyces sp.). Sci and Tech Food 23: 441-446.

6. Kobori CN, Jorge N (2005) Characterization of some seed oils of fruits for utilization of industrial waste. Sci and Agrotecnologia 29: 1008-1014.

7. Oliveira DS, Aquino PP, Ribeiro SM, Proenca RP, Pinheiro-Phillips HMP (2011) Vitamin C, carotenoids, total phenolic and antioxidant activity of guava, mango and papaya coming from Ceasa of Minas Gerais. Acta Scientiarum: Health Sci 33: 89-98.

8. Mercado-Silva E, Bautista PB, Velasco MG (1998) Fruit development, harvest index and ripening changes of guavas produced in central Mexico. Postharvest Biology and Tech 13: 143-150.

9. AOAC (2006) Association of Official Analytical Chemists. Official methods of analysis of AOAC. AOAC press, Arlington, VA, USA.

10. Munhoz CL, Sanjinez-Argandoña EJ, Smith Junior MS (2010) Guava pectin extraction dehydrated. Food Sci and Tech 30: 119-125.

11. Leelavathi K, Tzia C (2011) Use of endoxylanase treated cereal brans for development of dietary fibre enriched cakes. Innovative Food Sci and Emerging Tech 13: 207-214.

12. Fernandes AF, Pereira J, Germani R, Oiano-Neto J (2008) Efeito da substituição parcial da farinha de trigo por farinha de casca de batata (Solanum tuberosum Lineu). Food Sci and Tech 28: 56-65.

13. Ishimoto FY, Harada AI, Branco IG, Conceição WAS, Coutinho MR (2007) Alternative use of the bark of yellow passion fruit (*Passiflora edulis var. FlavicarpaDeg.*) for the production of biscuits. Magazine Exact and Natural Sciences.

14. Mauro AK, Silva VLM, Freitas MCJ (2010) Phisical, chemical, and sensorial characterization of cookies made with kale stalk flour (KSF) and spinach stalk flour (SSF) rich in nourishing fiber. Food Science and Techonology 30: 719-728.

15. Ghafar MFA, Prasad KN, Weng KK, Al.Ismai (2010) Flavonoid, hesperidine, total phenolic Contents and antioxidant activities from citrus species. African J of Biotech 9: 326-330.

23

Preparation and Storing Constancy Assessment of Orange Lemonade Drink

Tariq Kamal[1]*, Matilda Gill[2], Ismail Jan[3] and Taimur Naseem[4]

[1]Department of Agricultural Extension Education and Communication, the University of Agriculture, Peshawar, Pakistan
[2]Department of National Institute of Food Science and Technology, University of Agriculture, Faisalabad, Pakistan
[3]Department of Statistics and Mathematics, the University of Agriculture Peshawar, Pakistan
[4]Department of Soil Sciences, the University of Agriculture, Peshawar, Pakistan

Abstract

The orange lemonade drink was prepared in which different concentrations of preservatives (sodium benzoate and potassium metabisulphite) were added along with 10% sugar while some samples were sugar free in order to obtain the best combination. These samples were studied for physicochemical (pH, % acidity, TSS, ascorbic acid, reducing sugar and none reducing sugar) and organoleptic evaluation (colour, flavour, taste and overall acceptability). The results were studied and compared after interval of 15 days for total of 90 days storage period at room temperature. Ascorbic acid content decreased in all the samples during storage. The minimum loss in ascorbic acid content was observed in T8 (27.01%) and maximum in T0 (50.28%). Increase in titratable acidity was observed during storage. Maximum increase was observed in T6 (30.34%) while the minimum increase was observed in T1 (10.45%). pH was slightly decreased during storage. Maximum decrease was observed in sample T5 (3.82%) while minimum decrease was observed in sample T3 (1.25%). TSS was increased and maximum increase was observed in sample T4 (13.3%) and minimum in T3 (4.04%). Reducing sugar increased during storage. Maximum increase was observed in T7 (22.36%) while minimum in T4 (9.30%). Non reducing sugar considerably decreased. Maximum decrease was noticed in To (73.3%) while minimum in T4 (22.2%). Sugar acid ratio decreased during storage. Maximum decrease was observed in T5 (15.97%) while minimum in T0 (2.98%). Organoleptically for colour factor, T7 obtained maximum score (7.20) while minimum was obtained by T0 (6.57). Results for statistical analysis for 90 days storage and internal comparison were found significant (P<0.05).

Keywords: Preparation; Concentration of preservatives; Organoleptic assessments

Introduction

Citrus fruit is botanically a hesperidium a particular kind of berry with leathery rind and divided internally into segments. The orange is a hybrid of ancient cultivated origin, possibly between pomelo (*Citrus maxima*) and tangerine (*Citrus reticulata*). It is a small flowering tree growing to about 10 m tall with evergreen leaves, which are arranged alternately, of ovate shape with crenulate margins and 4–10 cm long. The orange fruit is a hesperidium, a type of berry. Oranges originated in Southeast Asia. The fruit of *Citrus sinensis* is called *sweet orange* to distinguish it from *Citrus aurantium*, the bitter orange. In a number of languages, it is known as a "Chinese apple" (e.g. Dutch *Sinaasappel*, "China's apple", or "Apfelsine" in German). The name is thought to ultimately derive from the Dravidian word for the orange tree, with its final form developing after passing through numerous intermediate languages. (Wikipedia, the encyclopedia) Citrus ranks second to apple in world trade. Citrus is grown throughout the world in tropical and subtropical climates. The soil and climatic conditions of Pakistan especially N.W.F.P are congenial for the production of citrus fruits [1]. The important varieties produced are Oranges *(citrus sinensis osbeck)* and Lemon *(citrus lemon burman)*. The reproductive tissue surrounds the seed of the angiosperm lemon tree. The lemon is used for culinary and nonculinary purposes throughout the world. The fruit is used primarily for its juice, though the pulp and rind (zest) are also used, primarily in cooking and baking. Lemon juice is about 5% (approximately 0.3 moles per liter) citric acid, which gives lemons a tart taste, and a pH of 2 to 3. This makes lemon juice an inexpensive, readily available acid for use in educational science experiments. Lemons are also known for their sourness. Because of the tart flavor, many lemon-flavored drinks and candies are available on the market, including lemonade (Wikipedia, the encyclopedia). Large numbers of industries have started processing these fruits into pure citrus juice and

ready to drink juices etc. strong national and international demand for citrus products will provide stimulus to maintain increasing levels of production. Citrus fruits are the rich source of vitamin C. about 80% of the vitamin C in our diet comes from citrus fruits [2]. Horticultural crops not only provide human beings with nutritional and healthy foods, but also generate a considerable cash income for growers in many countries. However, horticultural crops typically have a high moisture content, tender texture, and high perishability. If not handled properly, a high-value nutritious product can deteriorate and rot in a matter of days or even hours. Therefore, a series of sophisticated technologies have been developed and applied in post-harvest handling of horticultural crops in the last few decades. Unfortunately, many Asian countries have not been able to use this advanced equipment, owing to cost or adaptability problems. Post-harvest losses, therefore, remain high [3].

Beverages are one of the important food items in our diet providing vitamin C and other nutrients to our body. Citrus juices are the emerging beverages which can be prepared by using the appropriate combinations of sucrose. The first artificial sweetener introduced for commercial use was saccharin [4]. Other sweeteners have been used commercially and many synthetic sweeteners have been introduced.

*Corresonding author: Tariq Kamal, Department of Agricultural Extension Education and Communication, The University of Agriculture, Peshawar, Pakistan
E-mail: tariqkamal10@gmail.com

However, at the present time saccharin is the only nutritive sweetener approved by the Food and Drug Administration for use in foods and beverages in the United States [5]. Global fruit and vegetable juices market sets sights on 53 billion liters by 2010, according to new report by Global Industry Analysts, Inc. Fruit and vegetable juices market is witnessing excellent growth primarily due to the increasing target audience focusing on health and nutritional issues. Juices represent one of the most competitive segments in the beverages industry vying intensely with alternative beverages such as bottled water, RTD drinks, sports and energy drinks and various other herbal concoctions promising a unique taste and flavor. World market for fruit and vegetable juices is forecast to reach 53 billion liters by 2010. Per capita consumption of fruit juices has been witnessing rapid growth, primarily driven by rising awareness over the importance of maintaining healthy and nutritious eating habits. Declining consumption of alcohol and the switch towards nonalcoholic beverages, advent of new products such as chilled and vitamin-fortified juice blends, and rising popularity of juices fortified with fiber, calcium and vitamins are expected to push up sales in the market. Worldwide fruit and vegetable juices market is portended to reach 53 billion liters by 2010, as stated in a recent report published by Global Industry Analysts, Inc. North America and Europe represent leading markets, accounting for about 60%. However, maximum growth is anticipated from Asia-Pacific, which is set to be the fastest growing market [6]. Global fruit juices market dominated the worldwide market for fruit and vegetable juices, capturing about 94% share. In chilled ready to serve juices market, European market is forecast to reach 11.3 billion liters in 2010. In vegetable juices market, the US is forecast to generate about 492 million liters in 2010. Flavor, price, and brand image are critical factors influencing purchasing decisions of consumers. Fruit and Vegetable Juices Market: A Global Strategic Business Report Major players in the marketplace include Del Monte Foods, Cadbury Schweppes, Minute Maid Company, Odwalla Inc, Nestle SA, Ocean Spray Cranberries, Tropicana Products Inc, and Welch Foods. Fruit and Vegetable Juices Market: A Global Strategic Business Report`, published by Global Industry Analysts, Inc., provides a comprehensive review of market trends, product profile, recent developments, mergers, acquisitions, profiles of major players and other strategic industry activities. Analysis is presented for major geographic markets such as US, Canada, Japan, France, Germany, Italy, the UK, Spain, Russia, Asia-Pacific, Latin America and Middle East. Analytics for the period 2000 through 2015 are provided in terms of product segments including fruit juices (frozen concentrates, chilled ready to serve juices and shelf stable juices) and vegetable juice [6]. Sodium benzoate may be used as a preservative (if declared on the label). Benzoic acid and sodium benzoate are generally regarded as safe up to a maximum permitted level of 0.1%. In most countries, the maximum permissible quantities generally range between 0.15-0.25 percent. Sorbic acid and its salt are some of the most widely used food preservatives in the world. As food preservatives, sorbates have found wide application in various foods, especially as yeast and mold inhibitors. Effective antimicrobial concentrations of sorbates in most foods are in the range of 0.05%-0.03%. In high sugar products (e.g. jams, jellies) smaller quantities of Sorbic acid are adequate for preservation, because of synergistic action of sorbet with sugar [7]. The use of benzoic acid as a food preservative has been limited to those products which are acid in nature. It is used as antimycotic agent, and most yeasts and fungi are inhibited by 0.05-0.1% of the undissociated acid. Food poisoning and spore forming bacteria are generally inhibited by 0.01-0.02% of undissociated acid, but many spoilage bacteria are much more resistant. Benzoic acid has been widely used to preserve beverages, fruit products, bakery products and other food products [8].

In the last decade there has been considerable increase in demand for ready to serve orange juice. Due to lack of preservation facilities in our country, this research will contribute.

The main objectives of the research

I.   To study the effect of chemical preservatives on the preservation quality of ready to serve orange juice.

II.   To determine the physicochemical changes taking place in the juice stored at room temperature.

III.   The findings will help the beverage industry and consumers will have a juice with increase shelf.

Materials and Methods

Preparation of sample

After thoroughly washing the oranges and lemons, juice was extracted; this was used for the research. The juice was then passed through muslin cloth to remove any undesirable materials. Orange juice and lemon juice was mixed together in the ratio of 9:1 and then this drink was named "orange lemonade." Sugar was added according to the likings of different people i.e. 25 gm in each sweetened bottled.

Proposed plan of study

Orange lemonade was than treated with sodium benzoate and potassium metabisulphite according to the following procedures:

T_o = orange lemonade (unpasteurized) + no-preservatives + no-sugar   (control)

T_1 = orange lemonade + sucrose

T_2 = orange lemonade unsweetened + 0.1% sodium benzoate

T_3 = orange lemonade + sucrose + 0.1% sodium benzoate

T_4 = orange lemonade unsweetened + 0.1% potassium metabisulphite

T_5 = orange lemonade + sucrose + 0.1% potassium metabisulphite

T_6 = orange lemonade unsweetened + 0.05% sodium benzoate + 0.05% potassium metabisulphite

T_7 = orange lemonade + sucrose + 0.05% sodium benzoate + 0.05% potassium metabisulphite

T_8 = orange lemonade unsweetened + 0.05% sodium benzoate + 0.05 % potassium metabisulphite (unpasteurized)

T_9 = orange lemonade + sucrose + 0.05% sodium benzoate + 0.05% potassium metabisulphite (unpasteurized)

Storage

Preserved orange juice was stored for a period of three months at room temperature. This product was studied for physicochemical and organoleptic evaluation at interval of 15 days for a total period of 90 days.

Physicochemical analysis

Ascorbic acid: The ascorbic acid was determined by the titramitric method as described in AOAC [9].

Preparation and standardization of the dye solution: Fifty mg of 2, 6 dichlorophenol indophenols dye and 42 mg of sodium bicarbonate were weighed, dissolved in distilled water and volume was made up to 250 ml. 50 mg of standard ascorbic acid was taken in 50 ml of volumetric flask and the volume was made up with 0.4% oxalic acid. 2 ml of this ascorbic acid solution was titrated against dye solution until light pink color was obtained which persisted for 15 seconds.

Titration of the sample: Ten ml of the sample was taken in 100 ml of volumetric flask and volume was made up to the mark by adding 0.4% oxalic acid. 10 ml of prepared sample was taken in the flask and was titrated against dye until light pink color appeared, which persisted for 15 seconds. Three consecutive readings were taken for each sample.

Calculation: The ascorbic acid was calculated by using the following formula;

$$Ascorbic\ acid\ (mg/100g) = \frac{F * T * 100}{S * D}$$

$F =$ Factor from standardization $= \dfrac{ml\ of\ ascorbic\ acid}{ml\ of\ dye\ used}$

$T =$ ml of dye used for sample

$S =$ ml of diluted sample taken for titration

$D =$ ml of sample taken for dilution

Titratable acidity %: Titratable acidity was determined by the standard method as reported in AOAC [9].

Standardization of the NaOH solution: About 6.3 g of oxalic acid was weighed dissolved in distilled water and the volume was made to 1000 ml by adding more distilled water. This is stock solution. About 4.5 g of NaOH pellets were taken and dissolved in distilled water and volume was made up to 1000 ml. The burette was then filled with roughly prepared 0.1 N NaOH. 10 ml of 0.1 N oxalic acid was taken in a conical flask in triplicate. Two or three drops of phenolphthalein as indicator were added to each conical flask. The 0.1 N NaOH oxalic acid was titrated against 0.1 N NaOH solutions until pink light color was appeared, which persist for 15 seconds. Three consecutive readings were taken and the normality of NaOH was calculated using the formula:

$$N_1 V_1 = N_2 V_2$$

Where,

$N_1 =$ Normality of oxalic acid solution

$V_1 =$ Volume of oxalic acid solution

$N_2 =$ Normality of NaOH solution

$V_2 =$ Volume of NaOH solution

Titration of samples: Ten ml of the sample was taken in 100 ml volumetric flask and diluted up to the mark. 10 ml of these samples were taken in a titration flask and add two or three drops of the phenolphthalein as indicator, then titrated against exact 0.1 N NaOH solution, until light pink color appeared, which persisted for 15 seconds. Three consecutive readings were taken and acidity was calculated by using the formula.

$$Acidity(\%) = \frac{0.067 * ml\ of\ NaOH\ used * 100 * 100}{A * B}$$

Where,

A= Sample taken for dilution

B= Sample taken for titration

pH: pH was determined by standard method of AOAC [9]. For the determination of pH of samples, the pH meter was used. First it was standardized by using buffer solutions of known pH (4 and 9) then 10 ml of sample was taken in a clean beaker and probe was directly dipped into the sample to record the pH value.

Total soluble solids: The Total Soluble Solids (TSS) was determined at room temperature by the recommended method of AOAC [9] using refractometer. The drop of representative sample was placed on the dry refractometer prism and readings were taken in "brix" while directing the prism towards light source, added the correction factor according to temperature.

Total sugars: Reducing sugar was determined by Lane and Eynon method as described in AOAC [9].

Reducing sugars:

i. Reagents: Fehling A: Dissolved 34.65 g of $CuSO_4.5\ H_2O$ in 500 ml of distilled water. Fehling B: 173 g of potassium tartarate and 50 g of NaOH were taken in beaker. About 100 ml of water was added and dissolved the chemicals by stirring. The solution was transferred to 500 ml flask and volume was made up to the mark with distilled water. Methylene blue was used as indicator.

ii. Procedure: Ten ml of sample was taken in 100 ml volumetric flask and made up to the mark with distilled water. The burette was filled with this solution. Then 5 ml of Fehling A and 5 ml of Fehling B solution along with 10 ml distilled water was taken in a conical flask. The flask was heated until boiling without disturbing the flask. Sample solution was added from the burette drop by drop while boiling until the color became brick red in flask. A drop of methylene blue was added as indicator in the boiling solution of without shaking the flask. If color changes from red to blue for a moment, reduction isn't complete and added more pulp solution till red color persisted.

iii. Calculations: The orange juice and lemon juice was mixed 5 ml of Fehling A+5 ml of Fehling B will reduce, 0.05 g of reducing sugar.

5 ml of Fehling A+5 ml of Fehling B = X ml of 10 % sample solution = 0.05 g of reducing sugar 100 ml of 10 % sample solution will contain

$$\frac{0.05 * 100}{X\ ml\ of\ 10\ \%\ sample\ solution} = Yg\ of\ reducing\ sugar$$

$$\%\ reducing\ sugar\ in\ sample = \frac{Y * 100}{10}$$

Non-reducing sugar:

i. Procedure: Ten ml of the sample was taken in a volumetric flask and made the volume up to the mark with distilled water. 20 ml of this solution was taken in a flask and 10 ml of 1 N HCl was added, and then heated this solution for 5-10 minutes. After cooling 10 ml of 1 N NaOH was added and made this solution up to 250 ml. This sample solution was taken in a burette. 5 ml Fehling A and 5 ml Fehling B solution along with 10 ml of distilled water was taken in a conical flask and boiled. When boiling started, it was titrated against the sample solution from the burette till changed to red-bricked color. It is tested with methylene blue as indicator till brick red color persisted.

ii. Calculations

X ml of sample solution contains = 0.05 g if reducing sugar.

$$250\ ml\ of\ sample\ sol.\ Contains = \frac{250 * 0.05}{X\ ml} = Y\ g\ of\ reducing\ sugars$$

This 250 ml of sample solution was prepared from 20 ml of 10% solution.

So 20 ml of 10% solution contain Y g of reducing sugar.

100 ml of 10% solution contain $= \dfrac{Y * 100}{10} = P$ g of reducing sugar

This 100 ml was prepared from 10 ml sample.

10 ml sample contain P g of reducing sugar.

100 ml solution contain $= \dfrac{P * 100}{10} = Q$ g of reducing sugar

Q g of reducing sugar = inverted sugar + Free reducing sugar.

Non-reducing sugar = Total reducing sugar + Free reducing sugar.

Sugar/Acid ratio: TSS/acid ratio was determined by standard method as described in AOAC [9]. The TSS/Acid ratio was calculated by the following formula:

$$\text{Sugar / Acid ratio} = \frac{\text{Total soluble solids}}{\text{Total acidity}}$$

Organoleptic evaluation: Selected samples of juice were evaluated organoleptically for color, flavor, and overall acceptability by using 9-point Hedonic scale method as described by Larmond. Samples were presented to trained judges to compare them and assign them score between 1-9, where 1 represents extremely disliked and 9 represent extremely liked. Tap water was provided for oral rinsing.

Statistical analysis: All the data regarding different parameters were statistically analyzed by Randomized Complete Block Design (RCBD) as recommended by Steel and Torrie, 1980 and the means were separated by least significant difference (LSD) test [10].

Results and Discussions

Ascorbic acid

Initially the ascorbic acid content of samples (T_0 to T_9) was 35.0, 35.1, 34.7, 36.1, 34.8, 37.2, 34.7, 37.1, 34.8 and 37.0 mg/100 g, which was gradually decreased to 317.4, 19.5, 20.1, 21.0, 22.6, 24.4, 25.0, 26.0, 25.4 and 25.7 mg/100 g respectively during 90 days of storage period. The mean values of ascorbic acid content significantly (P<0.05) decrease from 35.65 to 22.71 mg/100 g during storage. For treatments maximum mean values were recorded in sample T_7 (31.94) followed by T_5 (31.81 mg/100 g), while minimum mean values were recorded in

sample T_0 (26.64) followed by T_2 (27.75 mg/100 g). Maximum decrease was observed in sample T_0 (50.28%) followed by T_1 (44.44%), while minimum decrease was recorded in sample T_8 (27.01%) followed by T_6 (27.95%) (Table 1). The statistical analysis showed that all treatments and storage intervals had a significant effect (P<0.05) on ascorbic acid content of orange lemonade drink during storage. Similar results have been observed by Mehmood et al. [11] who found that ascorbic acid decreased in apple juice during storage. Zeb et al. [12] also found that ascorbic acid decreased in the grape juice during storage under room temperature. These results are in agreement with the findings of Kinh et al. [13] who recorded a decrease in ascorbic acid content in apple pulp (Table 2).

pH

Initially the pH values of the samples (T_0 to T_9) were 3.19, 3.17, 3.18, 3.19, 3.10, 3.14, 3.15, 3.18, 3.17 and 3.16 which gradually decreased to 3.13, 3.09, 3.15, 3.07, 3.02, 3.09, 3.12, 3.10 and 3.07 respectively during 90 days of storage. The mean pH value significantly (P<0.05) decreased from 3.16 to 3.09 during storage. For treatment maximum mean values were observed in sample T_3 (3.16) followed by T_0 (3.14) while minimum mean value was recorded in sample T4 and T5 (3.09) followed by sample T8 and T9 (3.11). During storage maximum decrease was observed in sample T_5 (3.82%) followed by T_9 (2.84%), while minimum decrease was observed in sample T_3 (1.25%) followed by T_7 (1.88%) (Table 3). The statistical analysis revealed that storage intervals and treatments had a significant (P<0.05) effect on pH. Similar results were obtained by Zeb et al. [12] who reported pH decreases during processing and storage. As the pH decreased there was a proportional increase in acidity during storage of grape juice. The decrease in pH is due to increase in acidity during storage period. Our results are in agreement with the finding of Cecilia and Maia [14], who observed a decrease in pH of high pulp content apple juice during storage. This decrease may be due to the formation of free acids and pectin hydrolysis [15]. These results are in agreement with the findings of Saini and Pal [16], who observed a decrease in pH of kinnow juice. The increase in acidity might be due to acidic compound formed by the degradation of reducing sugar and pectin. Similar trend was also found during storage of canned orange juice by El Warraki et al. [17].

Total soluble solids (TSS)

The TSS values of samples (T_0 to T_9) on day first was 11.0, 19.0,

Treatments	Storage intervals (Days)							% Decrease	Means
	Fresh	15	30	45	60	75	90		
T0	35.0	32.8	29.4	26.7	23.9	20.0	17.4	50.28	26.46e
T1	35.1	33.8	30.1	28.3	24.9	22.2	19.5	44.44	27.70cde
T2	34.7	32.6	30.5	27.4	25.2	22.0	20.1	42.07	27.50de
T3	36.1	34.9	31.3	29.2	26.4	23.6	21.0	41.82	28.93bc
T4	34.8	32.7	31.5	28.3	26.2	24.1	22.6	35.05	28.60cd
T5	37.2	35.5	34.2	32.0	30.8	28.6	24.4	34.40	31.81a
T6	34.7	32.5	31.5	30.3	29.0	27.2	25.0	27..95	30.03b
T7	37.1	35.7	34.5	32.3	30.8	27.2	26.0	29..91	31.94a
T8	34.8	32.7	30.0	25.8	25.7	25.6	25.4	27.01	28.57cd
T9	37.0	35.7	32.5	32.5	31.2	28.5	25.7	31.87	31.87a
Means	35.65a	33.89b	31.55c	29.28d	27.41e	24.9f	22.71g		

Mean values followed by different letters are significantly (P<0.05) different from each other.

LSD value for storage interval   = 1.164

LSD value for treatments   =1.391

Table 1: Effect of storage intervals and treatments on ascorbic acid content of orange lemonade.

Treatments	Storage intervals (Days)							% Decrease	Means
	Fresh	15	30	45	60	75	90		
T0	3.19	3.15	3.13	3.12	3.15	3.14	3.13	1.91	3.14bc
T1	3.17	3.17	3.13	3.12	3.12	3.10	3.09	2.52	3.13de
T2	3.18	3.18	3.16	3.15	3.14	3.12	3.10	2.51	3.15ab
T3	3.19	3.17	3.16	3.14	3.14	3.17	3.15	1.25	3.16a
T4	3.10	3.10	3.07	3.09	3.10	3.09	3.07	0.96	3.09f
T5	3.14	3.12	3.10	3.08	3.07	3.07	3.02	3.82	3.09f
T6	3.15	3.15	3.12	3.11	3.11	3.13	3.09	1.90	3.12de
T7	3.18	3.13	3.12	3.11	3.14	3.12	3.12	1.88	3.13cd
T8	3.17	3.15	3.10	3.10	3.09	3.09	3.10	2.20	3.11e
T9	3.16	3.15	3.11	3.10	3.11	3.10	3.07	2.84	3.11e
Means	3.163a	3.147b	3.12c	3.112c	3.117c	3.113c	3.094d		

Mean values followed by different letters are significantly (P<0.05) different from each other.

LSD value for storage interval   = 0.01268

LSD value for treatments   = 0.01516

Table 2: Effect of storage intervals and treatments on pH of orange lemonade.

Treatments	Storage intervals (Days)							% Decrease	Means
	Fresh	15	30	45	60	75	90		
T0	11.0	11.2	12.5	12.5	12.5	12.6	12.6	12.6	12.13d
T1	19.0	19.0	19.5	19.6	19.8	19.8	19.9	4.52	19.51b
T2	10.5	11.0	11.1	11.3	11.5	11.7	11.7	10.25	11.26e
T3	19.0	19.2	19.2	19.4	19.4	19.6	19.8	4.04	19.37b
T4	11.0	11.5	12.0	12.1	12.2	12.5	12.7	13.3	12.00d
T5	18.5	18.6	19.0	19.2	19.2	19.5	19.7	6.09	19.10c
T6	11.0	11.0	11.0	11.2	11.5	11.7	11.9	7.56	11.33e
T7	19.0	19.1	19.5	19.5	19.7	19.7	20.0	5.00	19.50b
T8	11.0	11.0	11.2	11.2	11.5	11.7	11.8	6.77	11.34e
T9	19.0	19.3	19.5	19.6	20.0	20.2	20.5	7.31	19.73a
Means	14.9f	15.09e	15.45d	15.56cd	15.73bc	15.9ab	16.06a		

Mean values followed by different letters are significantly (P<0.05) different from each other.

LSD value for storage interval   = 0.1701

LSD value for treatments   = 0.2033

Table 3: Effect of storage intervals and treatments on TSS of orange lemonade.

10.5, 19.0, 11.0, 18.5, 11.0, 19.0, 11.0 and 19.0 °brix, which were gradually increased to 12.6, 19.9, 11.7, 19.8, 12.7, 19.7, 11.9, 20.0, 11.8 and 20.5 °brix respectively during 90 days storage. The mean TSS values significantly (P<0.05) increased from 14.9 °brix to 16.06 °brix during storage. For treatments maximum mean values were recorded in sample T_9 (19.73) followed by T_1 (19.51) °brix while minimum mean value were observed in T_2 (11.26) followed by T_6 (11.33). During storage maximum increase was observed in sample T_4 (13.3%) followed by T_0 (12.6%), while minimum increase was recorded in sample T_3 (4.04%) followed by T_1 (4.52%) (Table 4).

The statistical analysis showed that storage intervals and treatments had a significant (P<0.05) effect on TSS of Orange Lemonade drink. These results are in confirmation with the work Zeb et al. [12] who reported that significant increase occur in TSS in grape juice stored at room temperature. These results are in agreement with the findings of Rodrique [18] reported that total soluble solids of mixed orange and carrot juice increased during storage. Gilani [19] also agreed that there was increase in TSS of mango squash prepared from different mango cultivars. Also Kinh et al. [13] reported an increase in TSS of apple pulp preserved with chemical preservative. Shah et al. [20] mentioned that increase in soluble content of the product may be due to the

solubilization of fruit constituents during storage.

Overall acceptability

Initially the mean score of judges for overall acceptability of samples (T_0 to T_9) was 8.4, 8.2, 8.1, 8.0, 7.9, 8.1, 8.3, 8.5, 8.0 and 8.0, which were gradually decreased to 1.0, 1.5, 2.0, 2.1, 3.5, 3.0, 4.5, 4.9, 2.9 and 3.0 respectively during 90 days storage. The overall mean scores of judges for overall acceptability significantly (P<0.05) decreased from 8.15 to 2.84 during storage. For treatments maximum mean values were recorded in sample T_7 (6.61) followed by T_6 (6.29), while minimum mean score was recorded in T_1 (4.04), followed by T_0 (4.41). During storage maximum decrease was observed in sample T_0 (88.09%) followed by T_1 (81.70%), while minimum decrease was observed in sample T_7 (42.35%) followed by T_6 (45.78%). The statistical analysis showed that storage intervals and treatments had a significant (p<0.05) effect on the flavor of carrot and kinnow juice during storage. Rosario [21] observed that increasing storage time and temperature cause progressive degradation, which leads to decrease in overall acceptability. These results in agreement with the findings of Martin [22], who observed the decrease in sensory qualities of pasteurized orange juice bottled in clear glass bottles. The loss of overall acceptability is attributed to the degradation of ascorbic acid and furfural production as described by Shimoda and Osajima

Treatments	Storage intervals (Days)							% Decrease	Means
	Fresh	15	30	45	60	75	90		
T0	8.4	6.5	5.5	4.5	3.0	2.0	1.0	88.09	4.41e
T1	8.2	5.3	4.3	3.5	3.0	2.5	1.5	81.70	4.04e
T2	8.1	7.4	6.4	5.4	3.6	3.6	2.0	75.30	5.21d
T3	8.0	7.6	6.6	5.6	3.5	2.9	2.1	73.75	5.19d
T4	7.9	7.3	6.2	5.2	4.2	3.9	3.5	55.69	5.46cd
T5	8.1	7.1	6.6	5.8	5.1	4.9	3.0	62.96	5.80bc
T6	8.3	7.9	6.5	6.0	5.7	5.1	4.5	45.78	6.29ab
T7	8.5	8.0	7.0	6.5	5.9	5.5	4.9	42.35	6.61a
T8	8.0	6.7	6.2	5.2	4.0	3.1	2.9	63.75	5.16d
T9	8.0	6.9	6.1	5.1	4.2	3.5	3.0	62.5	5.26cd
Means	8.15a	7.07b	6.14c	5.28d	4.22e	3.7f	2.84g		

Mean values followed by different letters are significantly (P<0.05) different from each other.

LSD value for storage interval   = 0.4803

LSD value for treatments   = 0.5741

Table 4: Mean score of judges for overall acceptability of orange lemonade.

[23]. Furfural level accumulated during storage was useful indicator of the overall acceptability in orange juice.

Conclusion and Recommendations

This research work was conducted in order to make a new flavored acceptable drink by mixing two fruit juice i.e. Orange and Lemon; this drink was thus called "Orange Lemonade Drink". To make the shelf life better, different combinations of chemical preservatives in different quantities was used in order to obtain best possible results. The overall result showed that samples T_7 (orange lemonade + sucrose + 0.05 % sodium benzoate + 0.05% potassium metabisulphite) retained maximum nutrients stability and overall acceptability followed by T_6 (orange lemonade unsweetened + 0.05% sodium benzoate + 0.05% potassium metabisulphite during storage at room temperature.

a)   It is obvious from the findings of this research work that certainly it can improve the nutritional status of the population and similar research work should be carried out with different preservatives individually as well as with combination.

b)   This research work was carried out at ambient temperature, so the research should also be carried out at refrigeration. In this research nutritive sweetener (sucrose) was used to make the acceptable blend of juice for the consumers, so it is also recommended that non-nutritive sweeteners like (saccharin, aspartame etc.) should also be used to carry out the research.

c)   High intense light and room temperature may affect Vit. C content, so the research work should also be conducted in plastic, plastic colored bottles and as well as in tin canes in order to check the effect of packaging material on the quality of the juice during storage. Glass bottles were used in this research.

d)   Same research work should be done on other fruit juices and squashes like strawberry, lemon and litchi etc. and should also be carried out over a long period of time i.e. 9-12 months.

References

1.   Agric Stat Pakistan (2007-2008) Agricultural Statistic of Pakistan. Govt. of Pakistan. Ministry of Food, Agri. and Livestock (Economic Wing). Islamabad, Pakistan 45-46.

2.   Livingston GE, Chang CM (1983) Nutrients in Processed Foods. Fats, carbohydrates. Publishing Sci. group. Inc. Actan, Massachusetts 179.

3.   Fu W (1999) Postharvest Handling in Asia 2. Horti. Crops. Dept. of Horti, Nat. Taiwan Univ. FFTC Pub01-02.

4.   Beck K (1969) Synthetic sweeteners: Past, present, and future. Avi Pub. Co.

5.   Dopty R, Vanminen S (1975) Hand Book of sugars (2ndedn), The Avi Pub Co Inc 41-42.

6.   Pak. Horti. Dev. Exp. Bd. (2008). Int Food & Agribusiness Mngmnt Rev 11: 01-02.

7.   Lueck E (1980) Antimicrobial Food Additives. Spring Verlag, New York. Sensory Studies. 18: 163-176.

8.   Chichester DF, Tanner WF (1981) Antimicrobial Food Additives. In: "Handbook of Food Additives". (2ndedn), (Edited by Furia. T. E.) 115.

9.   AOAC (2000) Official methods of analysis. Association of Official Analytical Chemists (13thedn), Washington, DC. U.S.A.

10. Chochron WG, Cox GM (1965) Experimental design. John Willey & Sons, Inc. New York.

11. Mehmood MH, Oveisi MR, Sadeghi N, Jannat B, Hadjibabaie M, et al. (2008) Antioxidant properties of peel and pulp hydro extract in ten Persian pomegranate cultivars. J Bio Sci 11: 1600-1604.

12. Zeb A, Ullah I, Ahmad A, Ali K, Ayub M (2008) Grape juice preservation with benzoate and sorbate. J Advances in Food Sci 31: 17-21.

13. Kinh Shearer AEH, Dunne CP, Hoover DG (2001) Preparation and preservation of apple pulp with chemical preservatives and mild heat. J Food Prot 28: 111-114.

14. Cecilia E, Maia GA (2002) Storage stability of cashew apple juice preserved by hot fill and aseptic process. Dept. of Food Tech. Univ. of Ceara, Brazil CEP 60: 511-110.

15. Imran A, Rafiullah K, Ayub M (2000) Effect of added sugar at various concentrations on the storage stability of guava pulp. Sarhad J Agric 16: 89-94.

16. Saini SPS, Pal D (1996) Concentrational behaviour of Kinnow juice. J Scientific & Ind. Research. 55: 890-896.

17. El-Warraki AG, Abdel Rehman NR, Abdallah MA, Abdel Fattah TA (1977) Physical and chemical properties of locally canned orange juice. Annual of Agric Sci Moshtohor 6: 195-209.

18. Rodrique D, Arranz JI, Koch S, Frigola A, Rodriqo MC (2003) Physicochemical characteristics and quality of refrigerated Spanish orange-carrot juices and influence of storage conditions. J of Food Sci 68: 2111-2116.

19. Gillani SSN (2002) Development of mango squash from four different cultivars of mango. M.Sc Thesis. Dept. of Food Sci. Tech. NWFP. Agri. Univ. Peshawar.

20. Shah WH, Sufi NA, Zafar SI (1975) Studies on the storage stability of guava fruit juice. Pak J Sci Ind Res 18: 179-183.

21. Rosario MJG (1996) Formulation of ready to drink blends from fruits and vegetable juices. J Philippines 9: 201-209.

24

Optimization of Iron Rich Extruded *Moringa oleifera* Snack Product for Anaemic People Using Response Surface Methodology (RSM)

Vivek K[1], Pratibha Singh[2] and Sasikumar R[3]*

[1]*Department of Food Process Engineering, National Institute of Technology, Rourkela, Odisha, India*
[2]*Indian Institute of Crop Processing Technology, Thanjavur, Tamil Nadu, India*
[3]*Department of Agri-Business Management and Food Technology, North Eastern Hill University, Tura Campus, Tura, Meghalaya, India*

Abstract

Moringa oleifera, a multiuse tree has numerous medicinal properties. Most parts of the *Moringa* tree are edible among these leaves are the most nutritious part of the tree and are having good quantity of iron content. Incorporation of nutritional properties via addition of natural components present in fruits and vegetables is a relatively novel concept. One of the most effective ways of achieving this is via extrusion technology. The present study deals with the development of extruded snacks by incorporation of *Moringa oleifera* leaf powder in finger millet using a lab-scale twin screw extruder. The main aim of this study is to optimize the extrusion process using Response Surface Methodology (RSM). The effect of feed moisture, blend ratio and barrel temperature on product responses viz. mass flow rate (MFR), expansion ratio (ER), bulk density (BD), water absorption index (WAI) and sectional expansion index (SEI) of the extruded product were studied. The blend of *Moringa* and finger millet was extruded at different moisture content (19% to 25%), barrel temperature (120°C to 140°C) and blend ratio (0% to 15%). Increase blend ratio had showed increase in WAI, MFR but decrease in ER, SEI and BD. The optimized sample was obtained at 25% M.C, 5% blend ratio, and 140°C barrel temperature and it has iron content of 5 ± 0.10 mg/100 g.

Keywords: *Moringa oleifera*; Physico-chemical properties; Extrusion technology; Optimization; RSM

Introduction

Extrusion is one of the important processing techniques in food processing industries i.e. snack food industries [1]. Indian snack market has been increasing rapidly with an annual growth rate of about 15% to 20% [2]. Extruded snacks were generally made from the grain based materials. Grains have an excellent swelling and binding power with the *Moringa* leaves powder because of the starch content in grains. The high temperature and short time extrusion (HTST) cooking has many benefits over low temperature and long-time (LTLT) cooking, because of inactivation of anti-nutritional factors [3]. HTST provides better quality extruded products and also found to be efficient in terms of energy. It was widely used to produce expanded snacks, modified starches, Ready to Eat (RTE) cereals, baby foods, pasta and pet foods [4]. The use of extrusion cooking has unique advantages includes versatility, high productivity, low cost, consumer acceptable product shapes, high product quality and product of new food with no or negligible effluents. Expansion of starchy products by extrusion was widely studied by Colonna et al. [5]. Extrusion cooking used in this study was accomplished through the application of heat directly by means of steam injection to the blend. Various changes have observed in ingredients during the extrusion cooking process were the gelatinization of starch, destruction of natural toxic substances, denaturation of proteins and the diminishing of microbial counts in the final product [6].

Moringa tree (*Moringa oleifera*) is a multipurpose tree native to the foot hills of Himalayas in north western India and is cultivated throughout the tropic [7]. *Moringa* leaves are considered to be an important source of several nutrients [8]. It consists of 17 g (28% RDA/Day) protein, 24.65 mg iron (230% RDA/Day) and 48.23 g carbohydrates (34% RDA/Day) because of the high nutritional value these leaves were used in blend with finger millet for preparing extruded snack [3,9,10].

Finger millet (*Eleusine coracana*) is considered as an annual plant widely grown as a cereal in the arid areas of Asia and Africa. It is an important minor millet crop of India. It is a rich source of many major nutrients such as protein, fibre, iron and calcium. It contains of significant number of micronutrients [11]. The state of Karnataka is the leading producer of ragi, accounting for 58% of India's ragi production Ministry of Agriculture Report-November [2]. Since finger millet is abundantly available and is also a rich source of iron (3.2 mg/100 g) it was used as a base material of extrudate. The proximate analysis value of *Moringa oleifera* leaf powder, finger millet power and extruded product were mentioned in the (Table 1). The present study deals with the development and dissemination of expanded snacks based on *Moringa* leaf powder using a lab-scale twin screw-extruder (SYSLG-IV, China) for industrial and nutraceutical applications in order to fully exploit the nutritional potential of *Moringa oleifera* leaves for the benefit of mankind and also enhance the consumption of iron in diet thereby eliminating iron deficiency and providing the right nutritional supplement to anaemic people and to find out the best suitable proposition for consumer consumption. Therefore, objective of this study is to optimize the extrusion process of *Moringa oleifera* leaves-finger millet powder in terms of feed moisture, blend ratio and barrel temperature for evaluating the physico-chemical properties (MFR, ER, BD, WAI and SEI) under optimum conditions.

***Corresponding author:** Sasikumar R, Department of Agri-Business Management and Food Technology, North Eastern Hill University, Tura Campus, Tura-794002, West Garo Hills, Meghalaya, India
E-mail: sashibiofoodster@gmail.com

Parameters	*Moringa* leaf powder (/100 g)	*Ragi* powder (/100 g)	Extruded product (/100 g)
Carbohydrate	48.23 ± 0.18 g	71.94 ± 0.21 g	68.10 ± 1.45 g
Protein	17 ± 0.08 g	7.1 ± 0.21 g	10.03 ± 0.17 g
Fiber	15.33 ± 0.06 g	3.992 ± 0.09 g	6.410 ± 0.22 g
Fat	7.8 ± 0.5 g	3.6 ± 0.21 g	4.60 ± 0.10 g
Iron	24.65 ± 0.12 mg	3.71 ± 0.05 mg	6.91 ± 0.18 mg
Moisture	9.5 ± 0.01 g	8.5 ± 0.02 g	7.65 ± 0.01 g

Table 1: Proximate analysis of extrudate.

Materials and Methods

Sample preparation

Preparation of *Moringa* leaf powder and feed for extrusion: *Moringa* leaves were collected from Indian Institute of Crop Processing Technology (IICPT) campus and sorted for infected and damaged leaves. The leaves were washed with tap water and surface dried. Then they were dried in hot air oven at 55°C for 6 hours and ground using a food mixer grinder (Philips, HL1632). The *Moringa* leaf Powder was packed in polyethylene (PE) bags and stored at 30°C for further studies.

Preparation of ragi powder: Finger millet was obtained from the local market and impurities were removed using sieves manually. It was then powdered using hammer mill in IICPT. 0.1 mm sieve was used for obtaining fine powder. Ground ragi was packed and stored at 30°C in polyethylene (PE) bags for further studies. Pure ragi extrudate is shown in Figures 1 and 2.

Preparation of blend: Both the finger millet and *Moringa* powder were mixed in the batch flour mixer (Jas -PM-1945) to various proportion i.e. 5%, 10% and 15% of *Moringa* powder in 95%, 90% and 85% of finger millet powder were made. *Moringa oleifera* leaf based snack product (10:95%) is shown in Figure 3.

Chemical analysis

Proteins, Carbohydrates, fibre, fat, iron and Moisture analysis of raw materials and extruded snack was carried out using standard procedures of AOAC. All analysis is expressed as the mean (± SD) of triplicate analysis and shown in Table 1.

Extruder conditions

The extruder used for the study was co-rotating and intermeshing twin screw extruder (SYSLG-IV, China) as shown in Figure 2. The barrel diameter of the extruder was 5 mm and its length to diameter (L/D) was 18:1. The extruder had four barrel zones. Temperature of the 1st, 2nd and 3rd were maintained at 40°C, 70°C and 100°C respectively throughout the experiment while the 4th zone was adjusted according to the experimental design. The die with 2.25 mm was used for the extrusion. Raw material fed into the extruder with a volumetric single screw feeder (Modena, Italy) which was attached on top of the extruder. The feed rate and screw speed was kept constant throughout the experiment at 150 rpm (Table 2).

Processing conditions

Moringa leaves and finger millet were used to make iron rich extruded product. In the preparation of this expanded product, moisture content in the blends was 19%, 22% and 25%. *Moringa* leaves powder to Finger millet blend ratio (BR) were 0: 100, 5: 95, 10: 90 and 15: 85 and temperature levels of 120°C, 140°C and 160°C were taken for each set of moisture content and blend ratio shown in Table 3. Once cooked, the product is forced through a die at the extruder discharge

end where it expands rapidly with some loss in moisture. After expansion, cooking and drying the extrudate develops a rigid structure and maintains a porous structure. The Overall effects of independent variables on mass flow rate, sectional expansion index (SEI), expansion ratio was studied. The effects of independent variable on responses were analyzed with RSM–Three level factorial models to check the significance of responses at p<0.05. The different Independent variables used in this study are shown in Table 2.

Figure 1: Extruded snack product based on (a*)* *Ragi* extrudate; (b) *Moringa oleifera* leaf.

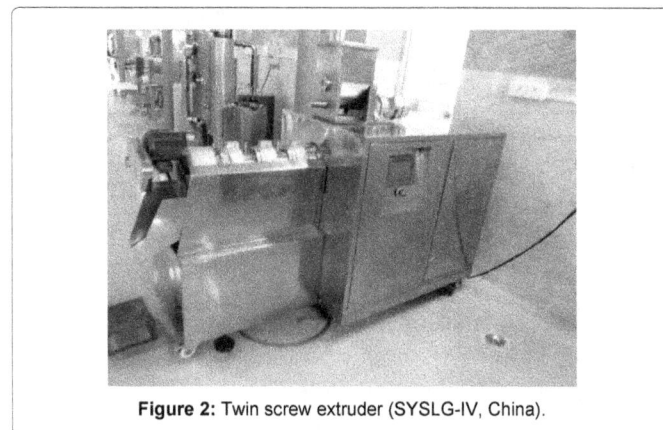

Figure 2: Twin screw extruder (SYSLG-IV, China).

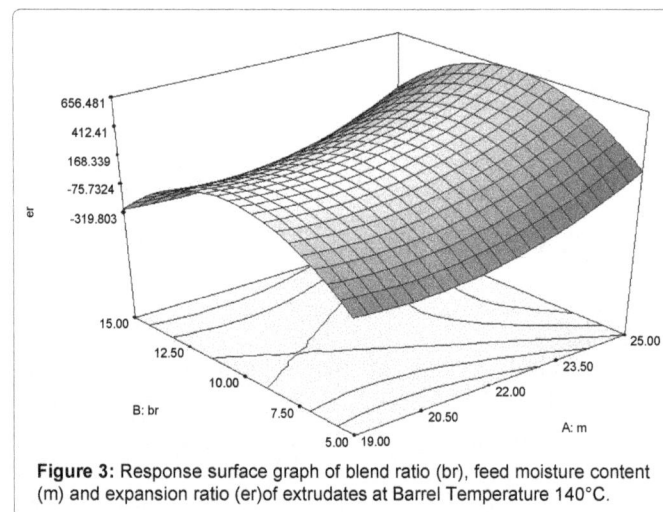

Figure 3: Response surface graph of blend ratio (br), feed moisture content (m) and expansion ratio (er)of extrudates at Barrel Temperature 140°C.

S. no.	Parameters
1	Moisture content in blend, (% wb)
2	Blend ratio of MLP in finger millet (%)
3	Barrel temperature (°C)

Table 2: Independent variables.

Variables	Level		
	-1	0	1
Feed moisture (%)	19	22	25
Blend ratio	5	10	15
Temperature (°C)	120	140	160

Table 3: Coded levels for independent variables.

Determination of product responses

Expansion ratio (ER): The ratio of diameter of extrudate and the diameter of die was used to express the expansion of extrudate [12]. The diameter of extrudate was determined as the mean of 10 random measurements made with Vernier callipers. The extrudate expansion ratio was calculated using the following equation.

$$ER = \frac{Extrudate\ diameter}{Die\ diameter}$$

Sectional expansion index (SEI): Sectional expansion index was measured by taking the ratio of the square of diameter of extrudate to the square of diameter of the die [13]. Random samples were selected from the extruded mass and their diameter was measured using screw gauge. For SEI was calculated using the following equation.

$$SEI = \left\{ \frac{Diameter\ of\ the\ extrudate}{Diameter\ of\ the\ dye} \right\}^2$$

Mass flow rate (MFR): It is defined as the ratio of the weight of the sample collected to the time taken to collect the sample [9]. It was measured by collecting the extrudate in polyethylene bags for a specific period as soon as it came out of the extruder and its weight was taken instantly and calculated using the following equation.

$$MFR\ (gm\ /\ \sec) = \frac{Weight\ of\ the\ sample\ collected}{Time\ taken\ to\ collect\ the\ sample}$$

Bulk density (BD): The bulk density was determined by the volumetric displacement method. Volume was measured using the 100 ml measuring cylinder. It is the ratio of the weight of the sample to the volume of the replaced sample was measured using the following equation [14].

$$BD\ (kg\ /\ cc) = \frac{Weight\ of\ the\ sample}{Volume\ replaced}$$

Water absorption index (WAI): It measures the volume occupied by the starch after swelling in excess water, which maintains the integrity of the starch in aqueous dispersion. It can also be used as an index of starch gelatinization. Water Absorption Index was calculated using the following method [15].

$$WAI = \frac{W_2 - W_1}{W_1}$$

Where,

W_1 is the weight of ground extrudate sample.

W_2 is the weight of ground extrudate sample after keeping in water.

Experimental design

Three level factorial designs were performed for three independent variables using design of experts 6.0.8. The independent variables considered were (1) Moisture content (2) Blend ratio (BR) and (3) Barrel temperature and the dependent variables were (1) ER (2) SEI (3) MFR (4) BD and (5) WAI. The coded and the responses for the three-level factorial model between levels and variables were given in the Table 4. Coded levels of each independent variables taken for Moisture

content (19% to 25%), BR (5% to 15%) and Barrel temperature (120°C to 160°C) are shown in Table 3.

Results and Discussion

Physical properties of extrudates were calculated with different percentage of *Moringa* and finger millet in blend at different temperature (120°C, 140°C, 160°C) and moisture content (10%, 15% and 20%) and found operating condition for maximum and minimum value of physical properties and the effects of independent variables on each parameter was analysed using response graph at 140°C as shown in Figures 3-7.

Expansion ratio

The multiple regression analysis of the expansion ratio versus feed moisture content (A), blend ratio (B), and barrel temperature (C) yielded following polynomial model.

$$ER = 2.84 + 0.74\ A - 0.077\ B - 0.022\ C + 1.64\ A^2 + 0.12\ B^2 - 0.044\ C^2 - 0.20\ AB - 0.021\ AC + 0.056\ BC$$

In this case moisture content of feed, blending ratio, and barrel temperature were found highly influencing variables on the ER of extrudates. The results showed that the expansion ratio increases

Exp no.	x_1^a	x_2^a	x_3^a	BD	ER	SEI	WAI	MFR
1	1	-1	1	346.97	5.42	23.079	71.82	52
2	0	-1	0	178.9	2.67	15.2013	45.14	93.2
3	0	1	1	172.03	3.28	14.2024	67.61	87.04
4	-1	1	0	149.058	3.98	17.9734	60.92	95.09
5	0	0	0	159.25	2.8	14.2991	59.57	91.62
6	0	0	0	159.25	2.8	14.2991	59.57	91.62
7	0	-1	-1	178.9	2.63	14.2071	54.1	87
8	1	-1	-1	342.85	5.75	23.088	71.66	55
9	0	1	-1	171	3.28	14.2298	68.55	89
10	0	0	1	160.02	2.81	14.2753	60.91	86.02
11	0	0	0	159.25	2.8	14.2991	59.57	91.62
12	-1	-1	-1	150.352	3.96	15.4711	63.65	98.7
13	1	1	1	339.14	5.27	23.318	72.76	114
14	-1	0	0	138.702	4.12	16.7601	64.73	97.15
15	-1	1	-1	152.368	3.35	17.1285	58.25	87.5
16	1	1	0	312.56	4.51	23.501	76.1	105.99
17	1	0	0	324.12	4.87	23.026	71.91	102.87
18	1	0	1	325.1	4.87	23.169	72.14	112
19	0	1	0	171	3.3	14.8102	62.91	89.78
20	1	-1	0	368.58	6.64	23.901	74.71	52.98
21	-1	-1	0	157.006	3.41	15.5612	65.74	93.65
22	0	0	0	159.25	2.8	14.2991	59.57	91.62
23	-1	-1	1	156.672	3.8	15.8152	62.59	98.5
24	1	0	-1	358.21	4.98	23.153	72.1	106
25	0	0	0	159.25	2.8	14.2991	59.57	91.62
26	-1	0	-1	138.915	4	16.5322	60	102.15
27	0	0	0	159.25	2.8	14.2991	59.57	91.62
28	1	1	-1	340.95	5.15	23.365	72.8	108
29	-1	0	1	141.561	4.03	16.5935	61.08	102.52
30	-1	1	1	153.81	3.41	17.418	57.52	87.53
31	0	-1	1	179.46	2.63	14.2001	54.39	84.82
32	0	0	-1	159.25	2.81	14.2359	61.75	87

x_1^a: Feed moisture (%); x_2^a: Blend ratio (*Moringa* leaf powder: finger millet powder)
x_3^a: Temperature (°C); BR: Blend Ratio; ER: Expansion Ratio; SEI: Sectional Expansion Index; WAI: Water Absorption Index; MFR: Mass Flow Rate.

Table 4: Experimental design for extrusion experiment with coded and actual variable levels - Three level factorial designs.

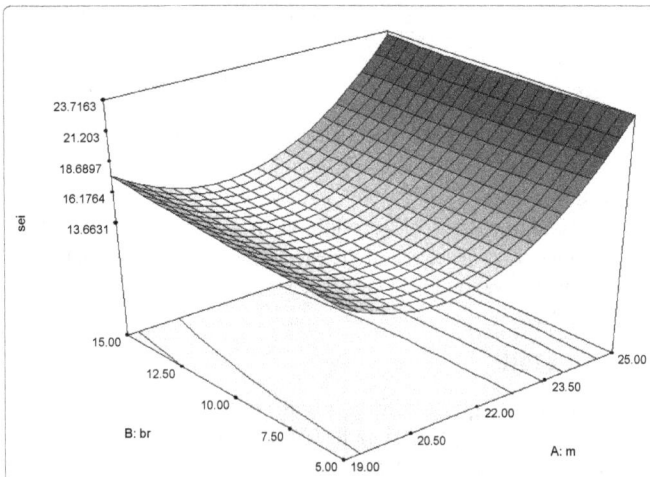

Figure 4: Response surface graph of blend ratio (br), feed moisture content (m) and sectional expansion index (SEI) of extrudates at Barrel Temperature 140°C.

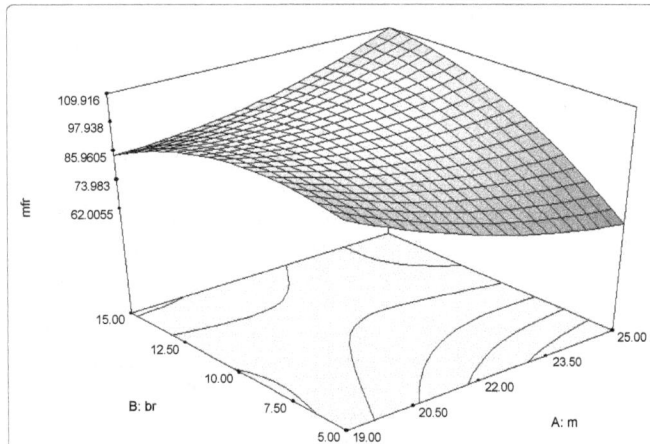

Figure 5: Response surface graph of blend ratio (br), feed moisture content (m) and mass flow rate (mfr) of extrudates at Barrel Temperature 140°C.

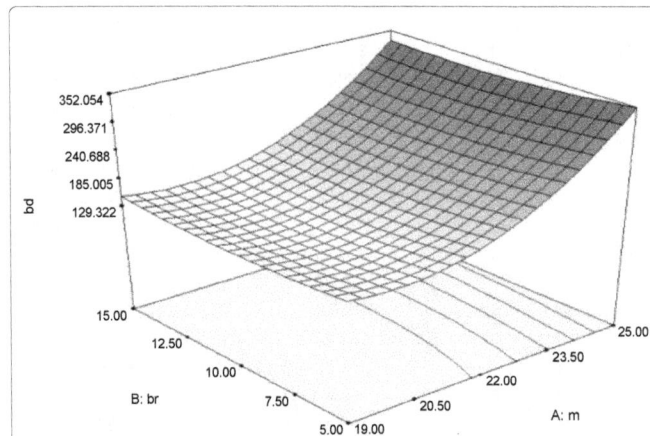

Figure 6: Response surface graph of blend ratio (br), feed moisture content (m) and Bulk Density (bd) of extrudates at Barrel Temperature 140°C.

with feed moisture content but decreases with the blend ratio it was observed that the increase in barrel temperature increases expansion ratio however for the increase in barrel temperature showed a

negative effect on ER shown in Figure 3. Expansion ratio followed almost increasing trend for moisture content with maximum value at 25% moisture content and minimum at 22% moisture content. Yagc and Gogus [16] mentioned increase in expansion ratio with the gradual increase in feed moisture of rice based extruded product. Earlier studies had also reported higher bulk density for extrudated products having lower expansion ratio [17,18]. The maximum value of Expansion ratio was achieved at 0% blend ratio for 25% moisture content of extrudate whereas the minimum value was obtained for 10% blend ratio for 22% moisture content for extrudate. With respect to Barrel temperature the maximum value (4.47) obtained at 140°C and minimum (4.39) at 160°C. With respect to Blend ratio maximum value (5.78) at 0% and minimum (3.92) at 10%. With respect to feed moisture content, maximum value (6.88) at 25% and minimum (2.81) at 22%. The design gives the "Model f-value" of 20.68 and R^2-value of 89.43% which implies the model is significant (p<0.05). Bhattacharya [18] had reported that the expansion of extruded product is dependent on the degree of gelatinization. Balasubramanian et al. [19] reported positive effect of moisture content on expansion because starch gelatinization dependent on the available feed moisture. Expansion ratio with respect to variations in barrel temperature followed almost increasing trend with maximum values for 140°C barrel temperature and minimum values at 120°C barrel temperature.

Sectional expansion index

The multiple regression analysis of the sectional expansion index versus feed moisture content (A), blend ratio (B), and barrel temperature (C) yielded following polynomial model.

$$SEI= 14.41 + 3.35 A + 0.30 B + 0.037 C + 5.59 A^2 + 0.20 B^2 - 0.30 C^2 - 0.46 AB - 0.061 AC - 9.417E - 003 BC$$

In this case, all the independent variables gave positive effect on the SEI. The combined effect of barrel temperature with high feed moisture content decreased the SEI and increase with the blend ratio [12]. Sectional Expansion Index of extrudate was studied at different feed moisture contents, barrel temperature and blending ratios shown in Figure 4. It was observed that the SEI increases with feed moisture and increase with decrease in blend ratio but concluded that the SEI decreases with increase in feed moisture content and blend ratio [20]. The maximum value of SEI was achieved at 15% blend ratio for 25%

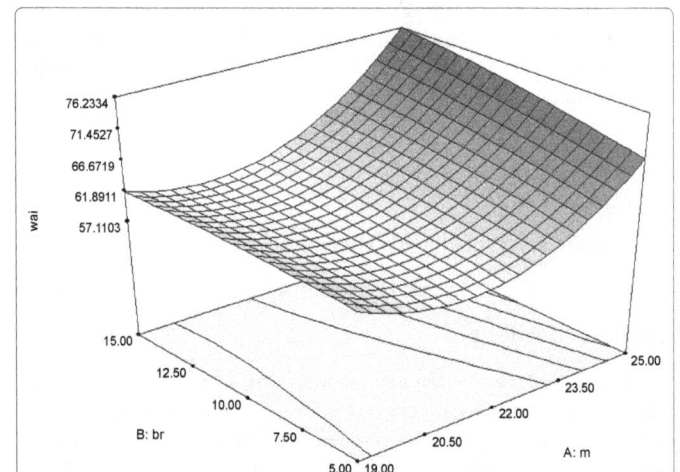

Figure 7: Response surface graph of blend ratio (br), feed moisture content (m) and water absorption Index (WAI) of extrudates at Barrel Temperature 140°C.

moisture content whereas the minimum value was obtained for 0% blend ratio for 19% moisture content shown in Figure 4. SEI followed almost increasing trend for moisture content with maximum value at 25% moisture content and minimum at 19% moisture content but SEI decreased with the increase in moisture content as reported by Badrie and Mellowes [21]. SEI with respect to variations in barrel temperature followed almost increasing trend. Maximum values for 140°C barrel temperature and minimum values at 120°C barrel temperature. Borah et al. [15] had also reported positive effect of barrel temperature on the SEI of rice extrudates. With respect to Barrel temperature, it was observed that the maximum value (16.86) obtained at 140°C and minimum (16.53) at 120°C. With respect to Blend ratio maximum value (18.43) at 15% and minimum (12.4) at 0%. With respect to feed moisture content, maximum value (20.86) at 25% and minimum (14.34) at 22%. The design gives the Model f-value of 456.32 and R^2-value of 99.47% which implies the model is significant (p<0.05).

Mass flow rate

The multiple regression analysis of the mass flow rate versus feed moisture content (A), blend ratio (B), and barrel temperature (C) yielded following polynomial model.

$$MFR = 93.35 - 3.00\,A + 8.23\,B + 0.23\,C + 5.44\,A^2 - 9.86\,B^2 + 0.97\,C^2 + 15.73\,AB + 0.73\,AC + 0.79\,BC$$

It was observed that MFR increases with the increase in blend ratio and barrel temperature but decreases with the decrease in feed moisture but Medeni [22] showed that the MFR decreases with the increase in feed moisture content. Mass Flow Rate of extrudate was studied at different feed moisture contents, barrel temperature and blending ratios shown in Figure 5. Model equation showed that moisture content gave negative effect on MFR. MFR quantifies the processing performance of the extruder [23]. The maximum value of mass flow rate was achieved at 15% blend ratio for 25% moisture content whereas the minimum value was obtained for 5% blend ratio for 25% moisture content shown in Figure 5. Mass flow rate with respect to variations in barrel temperature followed almost increasing trend with maximum values for barrel temperature 140°C and minimum values at barrel temperature 120°C. The design gives the Model f-value of 8.20 and R^2-value of 77.04% which implies the model is significant (p<0.05). With respect to Barrel temperature maximum value (108) at 120°C and minimum (55) at 140°C with respect to Blend ratio maximum value (102.15) at 15% and minimum (80) at 5% with respect to feed moisture content maximum value (98.10) at 19% and minimum (78.20) at 25%.

Bulk density

The multiple regression analysis of the bulk density versus feed moisture content (A), blend ratio (B), and barrel temperature (C) yielded following polynomial model.

$$BD = +159.72 + 95.56\,A - 5.43\,B - 1.00\,C + 74.56\,A^2 + 12.00\,B^2 + 2.99\,C^2 - 4.75\,AB - 3.43\,AC - 0.86\,BC$$

Model equation showed that moisture content gave positive effect on BD, blend ratio and barrel temperature showed the negative effect over the bulk density shown in Figure 6. Bulk Density of extrudate was studied at different feed moisture contents, barrel temperature (140°C) and blend ratios. It was observed that the trend was increasing for 25% moisture content and gradual decreasing trend has been seen for 19% and 22% moisture content for all blend ratios shown in Figure 6. As the Screw Speed increases bulk density decreases [24]. Extrudates fortified with the cabbage powder shows no significant increase in the bulk density with increase in feed moisture content [25]. The maximum

value of BD was achieved at 15% blend ratio for 25% moisture content whereas the minimum value was obtained for 0% blend ratio for 25% moisture content (Figure 6). BD followed decreasing trend for moisture content with maximum value at 19% moisture content and minimum at 25% moisture content. Bulk Density with respect to variations in barrel temperature followed a decreasing trend with maximum values for 120°C barrel temperature and minimum values at 160°C barrel temperature. The design gives the Model f-value of 327.79 and R^2-value of 99.26% which implies the model is significant at p<0.05. Values of "Prob> F" less than 0.0500 indicate model terms are significant. Values greater than 0.1000 indicate the model terms are not significant. By comparing mean values it was noted that maximum values were obtained at 15%, followed by 10%, 5% and 0%, with respect to blend ratio. With respect to moisture content the mean values indicated that maximum values were obtained for 19%, followed by 22%, then 25%. Similarly, with respect to barrel temperature, maximum values were achieved at 120°C, then 140°C, and finally 160°C. With respect to Barrel temperature, maximum value (202.77 kg/m³) at 120°C and minimum (187.25 kg/m³) at 160°C. With respect to Blend ratio maximum value (217.99 kg/m³) at 15% and minimum (148.84 kg/m³) at 0%. With respect to feed moisture content, maximum value (278.67 kg/m³) at 25% and minimum at 19% (153.37 kg/m³).

Water absorption index

The multiple regression analysis of the water absorption index versus feed moisture content (A), blend ratio (B), and barrel temperature (C) yielded following polynomial model.

$$WAI = +59.62 + 5.64\,A + 1.87\,B - 0.11\,C + 7.84\,A^2 - 0.37\,B^2 + 0.042\,C^2 + 1.56\,AB + 0.073\,AC - 0.092\,BC$$

WAI found to be increased with the increase in feed moisture content and blend ratio while WAI decreases with the increase in barrel temperature shown in Figure 7. Water Absorption Index of extrudate was studied at different feed moisture contents, barrel temperature (140°C) and blend ratios. It was observed that the trend was increasing for all blend ratios and moisture content of extrudates. The maximum value of WAI was achieved at 15% blend ratio for 25% moisture content whereas the minimum value was obtained for 0% blend ratio for 25% moisture content shown in Figure 7. WAI decreases due to degradation of starch [26]. Water absorption index increases with the feed moisture content [27]. Water Absorption Index with respect to variations in barrel temperature followed an almost decreasing trend with maximum values for 140°C barrel temperature and minimum values at 160°C barrel temperature. The design gives the Model f-value of 6.70 and R^2-value of 73.27% which implies the model is significant (p<0.05). By comparing mean values, it was noted that maximum values were obtained at 15%, followed by 10%, 5% and 0%, with respect to blend ratio. With respect to moisture content the mean values indicated that maximum values were obtained for 25%, followed by 19%, then 22%. Similarly, with respect to barrel temperature, maximum values were achieved at 140°C, then 120°C and finally 160°C. With respect to Barrel temperature, maximum value (59%) at 140°C and minimum (58.34%) at 160°C. With respect to Blend ratio maximum value (66.39%) at 15% and minimum (40.52%) at 0%. With respect to feed moisture content, maximum value (62.69%) at 25% and minimum (55.45%) at 22%.

Optimization of the extrusion cooking condition

The optimized sample was obtained by numerical optimization from design of expert 6.0.8, which gives a desirability function of 0.64. Equal importance of 3 was given to all the 3 independent variables (Feed moisture, Blend ratio and Barrel temperature). Based on their

Response	Goal	Lower limit	Upper limit	Lower weight	Upper weight	Importance
BD	minimize	138.702	368.58	1	1	3
ER	maximize	2.63	6.64	1	1	5
SEI	maximize	14.2001	23.901	1	1	5
WAI	maximize	45.14	76.1	1	1	5
MFR	minimize	52	114	1	1	3

BR: Blend Ratio; ER: Expansion Ratio; SEI: Sectional Expansion Index; WAI: Water Absorption Index; MFR: Mass Flow Rate

Table 5: Optimized parameter in the response optimizer.

Optimal solution			Predicted responses				
Feed moisture (%)	Blend ratio (%)	Temperature (°C)	ER	SEI	BD (g/cc)	WAI	MFR (g/s)
25	5:95	140	5.31	22.65	333.584	67.99	63.25

BR: Blend Ratio; ER: Expansion Ratio; SEI: Sectional Expansion Index; WAI: Water Absorption Index; MFR: Mass Flow Rate

Table 6: Optimized solution obtained using the response optimizer.

relative contribution to quality of final product importance of 5 was given to ER, SEI, WAI and importance of 3 was given to BD and MFR shown in Table 5. As shown in Table 6 at the optimal condition feed moisture, blend ratio and barrel temperature was found to be 25%, 5:95 and 140°C, respectively. Balasubramanian et al. [19] had reported better physico chemical properties for millet based extrudates developed at 120°C having 14% moisture content.

Conclusion

It was concluded that the best quality extrudate was obtained at 25% M.C, 5: 95% blend ratio, and 140°C barrel temperature. RSM – Three level factorial designs were used for optimizing the extrusion process for simplified understanding of individual operating variables. There is a huge variation in the effect of feed moisture content and blend ratio on the various physico-chemical parameters like Sectional Expansion Index, Mass Flow Rate, and Expansion Ratio of extruded *Moringa oleifera* snack product. It was observed that higher the bulk density of extrudate lower was the expansion ratio. Expansion ratio increases with increase in feed moisture and decreases with increase in blend ratio and barrel temperature. Bulk density increases with the increase in moisture and decreases with increase in blend ratio and barrel temperature. The maximum value of Expansion ratio was achieved at 0% blend ratio for 25% moisture content. As the moisture content increases expansion ratio increases. Sectional expansion index increases with the feed moisture content, blend ratio and barrel temperature. It was also concluded that the mass flow rate of the extrudate was also decreases with increase in feed moisture content and increases with increase in blend ratio and barrel temperature. Water absorption index increases with the increase in feed moisture content, blend ratio and barrel temperature.

References

1. Delgado-Nieblas CE, Aguilar- Palazuelos A, Gallegos-Infant E, RochoGuzman N, Zazueta-Morales J, et al. (2012) Characterization and optimization and extrusion cooking for the manufacture of third-generation snacks with winter squash (*Cucurbita moschata* D) flour. Cereal Chem 89: 65-72.

2. MOFPI (2005) Vision 2015, strategy and action plan for food processing industries in India. Ministry of Food Processing Industries 21: 152-153.

3. Walberg JL, Leidy MK, Sturgill DJ, Hinkle DE, Ritchey SJ, et al. (1988) Macronutrient content of a hypo-energy diet affects nitrogen retention and muscle function in weight lifters. Int J Sport Medicin 9: 261-266.

4. Deshpande HW, Poshadri A (2011) Physical and sensory characteristics of extruded snacks prepared from Foxtail millet based composite flours. Int Food Res J 18: 751-756.

5. Colonna P, Tayeb J, Mercier C (1989) Extrusion-cooking of starch and starchy products. AACC, St Paul, MN, USA.

6. Balfour D, Sonkar C, Sharma S (2014) Development and quality evaluation of extruded fortified corn snack. Int J Food Nutri Sci.

7. Lockett CT, Calvert CC, Grivett LE (2000) Energy and micronutrient composition of dietary and medical wild plants consumed during drought, study of rural Fulani, Northeastern Nigeria. Int J Food Sci Nutri 51: 195-208.

8. Makkar HPS, Becker K (1997) Nutrients and antiquality factors in different morphological parts of the Moringa oleifera tree. J Agri Sci 128: 311-322.

9. Okeng-Ogwal, Kasolo JN, Gabriel S, Bimenya O, Joseph JW (2010) Phytochemicals and uses of Moringa oleifera leaves in Ugandan rural communities. J Medicin Plant Res 4: 753-757.

10. Rajangan J, Azahakia Manavalan RS, Thangaraj T, Vijayakumar A, Muthukrishan N (2001) Status of production and utilisation of Moringa in Southern India. The miracle tree/ The multiple attributes of moringa, CTA, USA.

11. Deosthale YG, Mohan VS, Visweswara Rao K (1970) Varietal differences in protein, lysine and leucine content of grain sorghum. J Agri Food Chem 18: 644.

12. Fan J, Mitchell JR, Blanchard JMV (1996) The effect of sugars on the extrusion of maize grits: The role of the glass transition in determining product density and shape. Int J Food Sci Technol 31: 55-65.

13. Alvarez-Martinez L, Kondury KP, Harper JM (1988) A general model for expansion of extruded products. J Food Sci 53: 609-615.

14. Anderson RA (1982) Water absorption and solubility and amylograph characteristics of roll-cooked small grain products. Cereal Chem 59: 265-269.

15. Borah A, Mahanta CL, Kalita D (2015) Quality attributes of extruded breakfast cereal from low amylose rice and seeded banana (*Musa balbisiana*, abb). J Food Res Technol 3: 23-33.

16. Yagc S, Gogus F (2008) Response surface methodology for evaluation of physical and functional properties of extruded snack foods developed from food-by-products. J Food Eng 86: 122-132.

17. Jin Z, Hsieh F, Huff HE (1995) Effects of soy fibre, salt, sugar and screw speed on physical properties and microstructure of corn meal extrudates. J Cereal Sci 22: 185-194.

18. Bhattacharya S (1997) Twin-screw extrusion of rice-green gram blend: extrusion and extrudate characteristics. J Food Eng 32: 13-99.

19. Balasubramanian S, Singh KK, Patil RT, Onkar KK (2011) Quality evaluation of millet-soy blended extrudates formulated through linear programming. J Food Sci Technol 49: 450-458.

20. Basediya AL, Sheela Pandey SP, Shrivastava D, Khursheed AK, Anura N (2011) Effect of process and machine parameters on physical properties of extrudate during extrusion cooking of sorghum, horse gram and defatted soy flour blends. J Food Sci Technol 50: 44-52.

21. Badrie N, Mellowes WA (1991) Effect of extrusion variables on cassava extrudates. J Food Sci 56: 1334-1337.

22. Maskan M, Altan A (2011) Advances in food extrusion technology. CRC publisher and distributor.

23. Ayadi FY, Muthukumarappan K, Rosentrater KA, Brown M (2011) Single-screw extrusion processing of distillers dried grains with so lubles (DDGS)-based yellow perch (*Perca flavescens*) feeds. Cereal Chem 88: 179-188.

24. Chevanan N, Muthukumarappan K, Rosentrater KA (2007) Extrusion studies of aquaculture feed using distillers dried grains with solubles and whey. Food Bioprocess Technol 2: 177-185.

25. Stojceska V, Ainsworth P, Plunkett A, Ibanoglu S (2008) The recycling of brewer's processing by-product into ready-to-eat snacks using extrusion technology. J Cereal Sci 47: 469-479.

26. Pelembe LA, Erasmus C, Taylor JRN (2002) Development of a protein rich composite sorghum-cowpea instant porridge by extrusion cooking process. LWT-Food Sci Technol 35: 120-127.

27. Chang YK, Silva MR, Gutkoski LC, Sebio L, Da Silva AP (1998) Development of extruded snacks using jatoba (*Hymenaea stigonocarpa* mart) flour and cassava starch blends. J Sci Food Agri 78: 59-66.

25

Quality Evaluation of Coffee-Like Beverage from Date Seeds (*Phoenix dactylifera*, L.)

Sami Ghnimi*, Raisa Almansoori, Baboucarr Jobe, Hassan MH and Kamal-Eldin A

Department of Food Science, College of Food and Agriculture, United Arab Emirates University

Abstract

Quality characteristics of coffee-like beverage from roasted date seeds (*Phoenix dactylifera*, L.) were determined and compared with those of traditional Arabic coffee. The date seed beverage was found to have lower amount of total phenolic compounds and to be a less powerful antioxidant than Arabic coffee. Phytochemical screening showed that date seed extracts contained steroids, tannins, and coumarins, while caffeine, terponoids, saponins, alkaloids, anthraquinones, and anthocyanins were not detected. The levels of trace elements in the date seed extracts were within the ranges reported for Arabic coffee with the exception of cadmium (0.16-0.42 compared to <0.1 in Arabic coffee). The sensory evaluation revealed that the date seed extracts were acceptable but were slightly lower in quality compared to Arabic coffee. The date seed extracts were lighter in color, less cloudy, less bitter, and have less coffee flavor compared to the reference Arabic coffee. Before roasted date seed extracts will be approved for human consumption, their possible estrogenic effects need to be evaluated.

Keywords: Date seeds; *Phoenix dactylifra, L.;* Coffee-like beverage; Caffeine-free; Arabic coffee

Introduction

Fruits of date palm tree (*Phoenix dactylifera, L.*) are widely consumed in the Middle East and North Africa (MENA) where they play an important role in the nutrition and socioeconomic activities of the people. Date production in the MENA region amounts to 4,878,449 metric tons and in UAE to 900,000 metric tons [1]. The fruit of the date palm, considered a staple food in some countries, is composed of a fleshy pericarp (85-90%) and a seed (10%-15%) [2]. The utilization of date fruits and its waste is very important to date cultivation and to increase income to the sector [3]. The date seed is considered a waste product of many date processing plants producing pitted dates, date syrup and date confectioneries. At present, seeds are used mainly as animal feeds by the cattle, sheep, camel, and poultry industries. Roasted and powdered date seeds are used by some rural communities as coffee substitutes and coffee-like preparations made from date seeds are available in some Arabian markets in the Kingdom of Saudi Arabia (KSA) and United Arab Emirates (UAE). Since coffee drinking is a popular practice in the Arabian region where dates are widely grown, it was postulated that roasting date seeds into coffee-like beverage could represent an alternative product for those who want to enjoy the characteristic flavor and aroma of coffee without raising caffeine intake level. Cohen *et al.* [4] described and patented a roasting process of three different types of interventions resulting in the making of products that look like coffee in taste and texture. In these interventions, the date kernels were subjected to different types of roasting regime ranging from 150-300ºC. However, information regarding the safety, physical and chemical characteristics of roasted date seed extracts are not available.

Although the chemical and physical changes occurring during date seed roasting are not fully described, it is logical to postulate that some of these changes may be similar to what is happening in coffee roasting. Coffee roasting is an interesting thermal process in which final characteristic flavor and aroma are developed due to complex chemical reactions occurring during roasting. Franca *et al.* [5] divided roasting process of coffee beans into three successive stages: (i) drying, (ii) roasting or pyrolysis, and (iii) cooling. The drying phase is characterized by slow release of water and volatiles substances and bean color changes from green to yellow. Pyrolysis occurs in the second stage, resulting in complex physical and chemical changes in the beans. This stage is associated with release of CO_2, water and volatile substances and the color of beans become brown as a result of caramelization and Maillard reactions. If the roasting process is not halted, beans will burn and therefore cooling is necessary to terminate these reactions [6].

The evaluation of the competitiveness of Arabian coffee alternatives/substitutes depends on the economy of their production, their sensory acceptability and safety. Sensory evaluation, a critical component to this process, is important in generating new product ideas and improving existing products [7]. Safety is the other most important aspect that should be evaluated especially for *hitherto* unexploited food raw materials. In the case of date seeds, phytochemical screening including phenolic contents and analysis of caffeine and metal elements are necessary. The objectives of this study were to (i) describe the changes in color, pH, and acidity that occur during roasting of date seeds, (ii) evaluate the sensory attributes of date seed extracts and compare them with traditional Arabic coffee, and (iii) evaluate the safety of the roasted date seed extracts with respect to phytochemical components including phenolic compounds, caffeine, and metal elements.

Materials and Methods

Raw date seeds characterization

Fifteen kilograms of seed samples from each of the three date varieties (Khalas, Khunaizy and Fard) were supplied by Al Foah Company (Al-Saad, UAE). The seeds were manually separated from date fruits, thoroughly washed with normal tap water and then with distilled water to remove the adhering dirt, and finally dried in an oven set at

*Corresonding author: Sami Ghnimi, Department of Food Science, College of Food and Agriculture, United Arab Emirates University, P.O. Box 15551, Al-Ain, UAE
Email: sami_ghnimi@uaeu.ac.ae

80°C for 8 hours. Date seeds from these three varieties were weighed, their lengths and diameters were measured by a Vernier caliber, volumes were determined by mixing with poppy seeds in a measuring cylinder, and their densities were calculated as weight/volume. The hardness of the seeds was determined by a Vickers surface micro-hardness device (HMV 2000; Shimadzu, Kyoto, Japan) using 50 g load for 15 seconds to determine the Vickers hardness number (HV/50). This test uses a microscope to measure the depression caused by a diamond indenter with a load of kilograms on a polished surface and determines HV as the force divided by the surface area of the indentation by the following equation:

$$HV = 1.854 * (F/d^2)$$

With the constant value of 1.854 being calculated from the specific geometry of the indenter; F is the indentation load in kilograms-force (kgf), and d is the diagonal of the indentation in millimeter [8]. In this case, kernel hardness was determined as HV/50 with 50 kgf being the load applied for 30 seconds.

Laboratory roasting of date seeds

Washed and dried date seeds were roasted at 220°C for 6 hours and samples were collected to study the changes induced by roasting with respect to color and acidity/pH. The roasted date seeds were grinded and preserved at 15°C until further analysis. The color of the date seed powders were evaluated every 30 min during roasting by using Hunter Tristimulus colorimeter (Colortec PCM, Clinton, USA). The instrument was standardized against a black and a white tile before sample measurement. Samples of the powder (4-5g) were wrapped in polyethylene plastic wrap folded to form a closed pack in which the samples were evenly spread out. The color parameters L*, a*, b* were evaluated in duplicates.

For the determination of pH and acidity, the ground roasted date seeds (1g) were extracted with 50 ml of boiling distilled water for 2 min using a magnetic stirrer. The extract was centrifuged at 4000 rpm for 5 min. using Sigma 2-16 centrifuge (Sigma labrzentrifugen D-37620 Osterode am Harz, Germany). The pH of the extract was determined using an electronic pH meter (Crison, model Micro pH 2002, SA, Barcelona, Spain). The titratable acidity was determined by titrating 10 ml of extract against 0.01 N sodium hydroxide using phenolphthalein as an indicator and the results were expressed as a percentage of citric acid.

Safety evaluation of commercial roasted date seed coffee alternatives

Three commercial roasted date seed powders (coffee alternatives) were purchased from the kingdom of Saudi Arabia (KSA Light and KSA Dark) and from United Arab Emirates (UAE). These powders were analyzed for their elemental composition and they were screened for phytochemicals, phenolic compounds, and caffeine. Elemental analysis determinations of major (Ca, K, Mg, Na, P) and minor (Al, Co, Cu, Fe, Mn, and Zn) trace elements were performed using the Thermo Scientific™ iCAP Q™ ICP-MS in a single operation mode. A Thermo Scientific™ Dionex™ ICS-5000 was coupled to the iCAP Q ICP-MS for these determinations. The roasted date seed powders were extracted by boiling in water at a ratio of 1:10 (w/v) for phytochemical analyses. The extract was centrifuged and screened for the presence of tannins, terpenoids, saponins, steroids, coumarins, alkaloids, anthraquinones, and anthocyanins according to Yadav et al. [9]. DPPH radical scavenging activity was determined according to the method of Aoshima et al. [10]. Total phenolic contents were determined according to the method used

by Singleton and Rossi [11] and ferric reducing antioxidant power (FRAP) was determined in the sample extracts according to method of Iris et al. [12]. Both total phenolic content and FRAP values were expressed as tannic acid equivalents.

For the determination of caffeine, date seed powder (2g) was weighed in triplicates and extracted with 100 ml of boiling distilled water for 5 minutes with stirring. The solution was cooled, filtered and 5 ml of the filtrates were diluted to 50 ml. Caffeine analysis was conducted as described by Wanyika et al. [13] by high performance liquid chromatography (RP-HPLC) equipped with a 1525 Binary HPLC pump, a 717 plus auto-sampler, and 2487 Dual-absorbance detector (Waters, Miliford, MA, U.S.A). Separation was performed on a reverse phase column Symmetry (C18, 250 × 4.6 mm i.d., 5 μm, Waters) at a flow of 1 ml/min. The injection volume was 20 μl and the mobile phase was water, methanol, acetic acid (79.9:20:0.1, v/v/v), and the detector was VU set at 278 nm.

Sensory evaluation of commercial roasted date seed coffee alternatives

A group of 10 trained students were recruited as consumer panelists at the Department of Food Science, United Arab Emirates University, UAE. All sensory work was carried out in the sensory laboratory at the UAE University, which fulfills requirements according to the International Standards (ISO, 1988). A panel of two experts defined the attributes to be evaluated in coffee-like beverages. Panelists were informed and agreed to taste the samples before the tests, and they were informed of the type of product being tested and asked about their coffee consumption habits. Tap water was provided between samples to cleanse the palate. The selected panelists were provided two training sessions to investigate their ability to identify odors and the five basic tastes before the final recruitment.

The sensory properties of the three commercial roasted date seed extracts samples were evaluated and compared to Arabic coffee. Beverage preparations were made from roasted date seed powders and Arabic coffee by boiling 45 g of the powders in 100 mL of water for 2 minutes. The sample presentation order was randomized for each panelist. A 15-point hedonic scale was used to quantitate each sensory attribute, where zero indicates the absence of intensity, and fifteen corresponds to an extreme intensity. The attributes evaluated include sourness, bitterness, sweetness, flavor, aroma, cloudiness, color, mouth feel and overall quality.

Statistical analysis

All experiments were run in triplicate. The experimental data were subjected to Analysis of Variance (ANOVA) and the differences between means were evaluated by Duncan's New Multiple Range Test. Data analysis was performed using a SPSS package (SPSS 14.0 for Windows, SPSS Inc, Chicago, IL, USA).

Results and Discussions

Date seeds

The weight, dimensions, density, and hardness data for date seeds from three date varieties are presented in Table 1. Hardness, defined as material resistance to permanent deformation under load, was determined by measuring the Vickers hardness number as HV/50. The values obtained for Khalas, Khunaizy and Fard were 46.5, 45, and 51. Similar to coffee beans, date seeds differ in their hardness with high-density beans having more cells per cubic millimeter causing a hard cell structure. Both dates and coffee kernels contain galactomannan, which is a hard type of fiber [14,15]. High density kernels are more resistant to heat during roasting requiring adjustment of the roasting temperature and time and they require stronger milling machines. This is an important aspect when considering seeds from different date varieties for coffee making.

Laboratory roasting of date seeds

Figure 1 presents the decrease in seed weight during roasting as a result of evaporating water and other volatiles. In all of the three cultivars tested (Khalas, Khunaizy and Fard), there was a reduction in weight of the seed during the first hour of roasting and then the weight became stable as the roasting process continued. The reduction observed here is comparable to the reduction of 13-17% observed for different types of coffee beans after roasting [16]. The reduction in weight is followed by remarkable changes in the color parameters of the roasted seeds as shown in Figure 2. Prior to roasting, the lightness (L* value) of the three varieties averaged around 47 and this value decreased to 15 after 6.5 hours of roasting (Figure 2a), representing 47% decrease in the original value which indicate progressive darkening of seeds. Roasting increased the red color (a* values) as the seeds changed from bright red to dark red (Figure 2b). On the other hand, a sharp decrease in the b* values was noticed with prolonged roasting time indicating dark yellow (Figure 2c), which is in agreement with what has been reported in coffee roasting [5]. In the first 30 minutes of roasting, the pH dropped markedly (Figure 3a) and the titratable acidity increased in all three varieties (Figure 3b) as a result of decreased moisture and possibly due to hydrolysis of some of the organic acids present in the seeds. This observation is similar to what has been reported for coffee beans and explained by the hydrolysis of some amines [16]. After this time, the roasting of the three varieties date seeds was accompanied by a decrease in pH and increase in titratable acidity possibly due to decomposition

Variety	weight (g)	length (cm)	diameter (cm)	volume (cm³)	density (g/cm³)	Hardness HV/50
Khalas	0.72	1.82	0.51	0.80	0.91	46.5
Khunaizy	0.57	1.88	0.45	0.66	0.90	45
Fard	0.53	1.67	0.53	0.61	0.88	51

*Values are means of two determinations. Hardness is reported as Vickers hardness (HV/50) with 50 being the load used in kilograms-force (kgf).

Table 1: Dimensions and hardness of date seeds from three different varieties*.

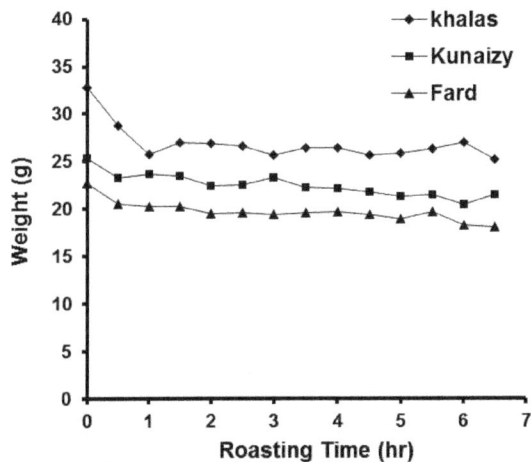

Figure 1: Reduction of weight during roasting of three date seed varieties.

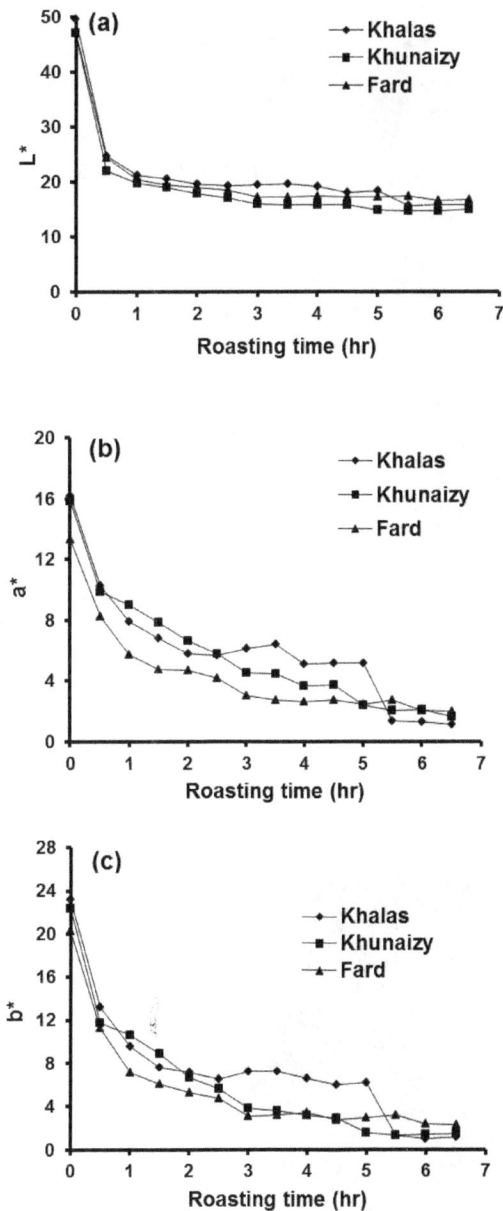

Figure 2: Changes in color parameters (L*, a* and b*) of date seed extracts.

of some of the organic acids in the seeds.

Safety evaluation of commercial roasted date seed coffee alternatives

Figure 4 presents photographs of Arabic coffee and three commercial roasted date seed coffee alternatives: KSA light, KSA dark, and UAE as purchased and used for the preparation of beverages tested in this study. In this part of the world, coffee beans are coarsely grinded and are boiled in water to prepare clear beverages; therefore finely milled coffee is not preferred. Table 2 presents the results of elemental analysis of the three commercial roasted date seed powders as compared to ranges found in Arabic coffee. These results show that with the exception of cadmium, the levels found for the selected elements in the date seed powders are within the ranges reported for Arabic

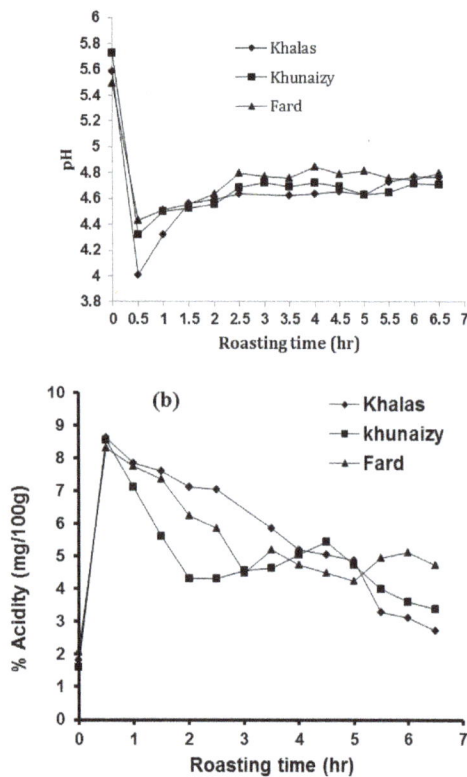

Figure 3: Changes in pH and acidity during roasting of date seeds.

Figure 4: Photographs of (1) Arabic coffee, and three commercial roasted date seed coffee alternatives (2) KSA light, (3) KSA dark, and (4) UAE as coarsely grinded and used for the preparation of beverages tested in this study.

Metal	KSA Light	KSA Dark	UAE	Arabian Coffee
Sodium	57	54	18.7	6.6–1467
Potassium	2147	2167	2396	11,400–29,100
Calcium	356	355	306	490-2200
Copper	6.7	6.4	5.5	0.4-30
Iron	25.4	79.4	18.1	12-617
Manganese	12.4	12.6	10.6	6.6–320
Zinc	15.3	15.7	14.6	1.2–803
Aluminum	21.2	18.5	5.0	3-200
Chromium	0.9	9.7	1.5	0.02–1.3
Cadmium	0.42	0.19	0.16	0.001- <0.1
Lead	<0.01	<0.01	<0.01	0.021–<2.6

*Values for roasted date seed powders (KSA light, KSA dark, and UAE) are means of two determinations. The ranges in Arabian coffee were taken from Pohl et al. [28]

Table 2: Concentrations of selected metal elements in three commercial roasted date seed powders compared with ranges in Arabian coffee (mg/Kg)*.

effect was attributed to estrogene negative feed-back mechanism on the pituitary and/or hypothalamus level.

The 2,2-diphenyl-1-picrylhydrazyl (DPPH) radical scavenging activity ranged from 65-73% and was comparable between Arabic coffee and roasted date seed extracts (Table 3). According to the results of the Folin-Ciocalteu assay, the level of total phenolic compounds in Arabic coffee (corresponding to about 2600 mg of tannic acid equivalents/100g of powder) is approximately three folds of what have been found in the roasted date seed extracts (Table 3). Similarly, the Arabic coffee recorded the highest value (8800 mg of tannic acid equivalents/100g) in the ferric ion reducing antioxidant power (FRAP) assay while the roasted date seed extracts have relatively much smaller values. High levels of phenolic compounds in date seeds were reported before ranging 3100-4400 mg gallic acid equivalents/ 100 g and 580-930 µM Trolox Equivalents Antioxidant Activity (TEAC) [21]. In this study, the difference in total phenolic responses between Arabic coffee and roasted date seed extracts may be due to the higher content or higher extractability from coffee beans compared to date seeds. Recently, it was shown that the phenolic compounds in date seeds belong mainly to the proanthocyanidins, condensed tannins with limited extractability [22]. High levels of phenolic compounds can have beneficial effects by acting as antioxidants or anti-nutritional effects by acting as metal scavengers reducing the bioavailability of iron and by acting as phytoestrogens. Therefore, these results should be interpreted with care especially in communities drinking high amounts of coffee/coffee alternatives.

Figure 5 presents an overlay of a chromatogram of standard caffeine solution and extracts from roasted date seed powder showing an absence of the caffeine peak in the roasted date seed extracts. The absence of caffeine in roasted date seed extract is an advantage for consumers who have considerable concern with caffeine. It has been reported that healthy adults may benefit from taking a small amount of caffeine by increasing alertness or ability to concentrate. However, some people depending on their physiological conditions are more sensitive to caffeine and for them; a small amount could cause reduced sleep, headaches, irritability and nervousness. Health Canada [23] recommended that women of child-bearing age and children should reduce their caffeine intake while healthy adults should consume no more than 400 mg per day. For the Food and Drug Administration of the United State of America, the 1981 recommendation limits maximum daily caffeine intake during pregnancy at 300 mg [24]. In the United Kingdom, the most recent recommendation of the British

coffee. The higher level of cadmium in date seed powders may be of concern because of the known toxicity of this trace element [17]. The phytochemical screening of the four roasted date seed samples revealed the presence of steroids, tannins, and coumarins, while the other phytochemicals (terpenoids, saponins, alkaloids, anthraquinones, and anthocyanins) were not detected. The presence of steroidal hormones such as estrone in some date seed varieties have been reported [18,19] and the consumption of date seeds was shown to have some effects on sex hormones in animals. For example, Aldhaheri et al. [20] studied the effect of date pits seeds on the testicular and uterus weights and reproductive hormone levels of rats. Although the date seed-containing diet had no significant effects on the weights of the sex organs, it was found that diets containing 12.5 and 25 % date seeds can significantly reduce the oestradiol concentrations in the serum of female rats. This

	KSA Light	KSA Dark	UAE	Arabic Coffee
TPC (g of Tannic Acid equiv./100g of sample)	0.65 ±0.06[a]	0.82 ±0.01[b]	0.50 ±0.06[c]	2.53 ±0.04[d]
FRAP (g of Tannic Acid equiv./100g of Sample)	2.91 ±0.01[a]	3.03 ±0.13[b]	1.99 ±0.10[b]	8.87 ±0.20[c]
DPPH (% reduction in absorbance)	70.4 ±1.9[a]	71.1 ±2.4[a]	64.9 ±3.0[a]	73.0 ±2.0[a]

*Values are Mean ± SD of three determinations. Means within the same raw having different superscripts are significantly different at p ≤ 0.05

Table 3: Total phenolic content (TPC), Ferric reducing antioxidant power (FRAP), and 2,2-diphenyl-1-picrylhydrazyl (DPPH) radical scavenging activity of roasted date seed extracts and Arabic coffee.*

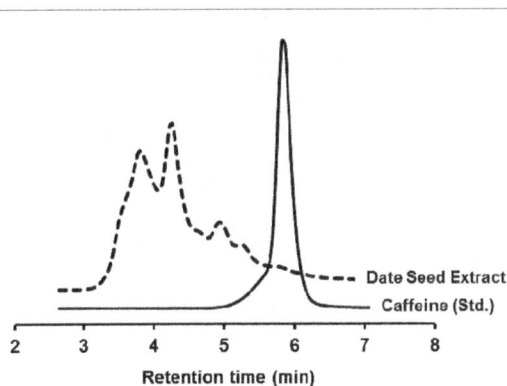

Figure 5: HPLC chromatogram showing the absence of caffeine in roasted date seed extracts with reference to caffeine standard (Detection: 278 nm).

Food Standard Agency limits intake to 200mg per day [25]. During pregnancy, caffeine is easily transmitted through the placenta to the fetus and by virtue of its immature liver the fetus is unable to metabolize caffeine effectively [26].

Sensory evaluation of commercial roasted date seed coffee alternatives

Consumers' affective test, such as the test run in this study, provides useful information on an existing product characteristics or the future commercial potential of a new developed food by quantifying its consumer preference or degree of liking/disliking [27]. The response of sensory panel for the three commercial roasted date seed extracts (KSA dark, KSA light and UAE) in comparison to Arabic coffee are presented in Table 4. There were significant differences (p<0.05) in color, cloudiness, bitterness, coffee flavor and overall quality between commercial roasted date seed extracts compared to Arabic coffee. Commercial date seed extracts are lighter in color, and have less cloudiness, bitterness, coffee flavor and overall quality compared to the Arabic coffee. With respect to the intensity of roasted aroma and sourness, the commercial roasted date seed extracts were comparable to Arabic coffee [13,27,28]. Roasted date seed extracts may be improved by altered production technique(s) and /or by addition of spices to induce better taste and higher nutritional and health benefits.

Conclusions

The physical, chemical, sensory and safety attributes of coffee-like product from roasted date seeds were evaluated in this study. The determined attributes allowed comparison between coffee-like beverages and traditional Arabic coffee. The phytochemical screening revealed the presence of steroids, tannins, coumarins, while caffeine, terponoids, saponins, alkaloids, anthraquinones, and anthocyanins

Attribute	KSA Light	KSA Dark	UAE	Arabic Coffee
Color	4.0 ± 1.4[a]	4.7 ± 1.5[ab]	5.5 ± 1.3[b]	7.7 ± 2.2[c]
Cloudiness	7.7 ± 2.3[a]	8.2 ± 1.7[ab]	9.4 ± 1.7[b]	10.7 ± 1.3[c]
Roasted aroma	6.5 ± 1.7[a]	5.6 ± 1.8[a]	6.3 ± 2.1[a]	6.8 ± 2.3[a]
Sourness	1.1 ± 0.9[a]	1.4 ± 0.6[a]	0.8± 0.8[a]	2.2 ± 2.1[a]
Bitterness	6.6 ± 2.7[a]	6.6 ± 3.0[a]	5.6 ± 2.5[a]	8.9 ± 2.8[b]
Coffee flavor	7.7 ± 2.5[a]	7.1 ± 2.6[a]	7.1 ± 2.3[a]	9.9 ± 2.0[b]
Overall quality	7.8 ± 2.1[a]	9.3 ± 1.6[ab]	8.7 ± 2.0[a]	10.9 ± 2.2[b]

*A 15-cm hedonic scale was used, where zero indicates the absence of intensity, and fifteen corresponds to an extreme intensity. Values are Mean ± SD of records by 10 panelists. Means within the same raw having different superscripts are significantly different at p ≤ 0.05

Table 4: Sensory score of the three commercial roasted date seed coffee alternatives compared to a sample of Arabic coffee.*

were not detected in the roasted date extracts. The presence of estrogenic compounds is a serious issue that needs to be investigated before date seed extracts can be recommended for human consumption. The absence of caffeine and the high levels of total phenolic compounds in the date seed extracts can serve as a strong motivating factor for those individuals who want to enjoy characteristic flavor of coffee without raising daily caffeine intake. The information obtained from elements analysis reveals that, with exception of cadmium, the levels in the date seed extracts are within the ranges reported for Arabic coffee. The sensory evaluation revealed that date seed extracts are acceptable and only slightly lower in quality compared to Arabic coffee. Therefore, future process of making coffee-like beverages from date seeds may be improved and other additives may be tested to improve the overall quality of the product.

Acknowledgement

The authors are grateful to Al Foah Company (Alsaad, AbuDhabi, UAE) for supplying the roasted date seeds and for Felix Guiabar Labata for performing the elemental analysis.

References

1. Food and Agriculture Organization FAO (2012) Crop production and trade data.

2. Hussein AS, Alhadrami GA, Khalil YH (1998) The use of dates and date pits in broiler starter and finisher diets. Bioresource Technology 66: 219-223.

3. Kamal-Eldin A, Hashim BI, Mohammed OI (2012) Processing and utilization of palm dates Fruits for edible application. Recent Patents on Food, Nutrition & Agriculture 4: 78-86.

4. Cohen S, Givataim H, Herzelia M, Shimshit G (2011) Date kernel preparation. Patent number US 2011/0143001A1.

5. Franca AS, Mendonca JCF, Oliveira SD (2005) Composition of green and roasted coffees of different cup qualities. Lebensmittel-Wissenschaft & Technologie 38: 709-715.

6. Sivetz M, Desrosier NW (1979) Coffee technology. Avi Publishing Company.

7. Sidel JL, Stone H (1993) The role of sensory evaluation in the food industry. Food Quality Preference 4: 65-73.

8. Craig JR, Vaughan DJ (1994) Quantitative Methods: Micro-indentation Hardness in Ore Microscopy and Ore Petrography. John Wiley & Sons Inc, New York, USA.

9. Yadav M, Chatterji S, Gupta SK, Watal G (2014) Preliminary phytochemical screening of six medicinal plants used in traditional medicine. International Journal of Pharmacy and Pharmaceutical Sciences 6: 539-542.

10. Aoshima H, Hideaki T, Hirofumi K, Yoshinobu K (2004) Aging of Wiskey increases 1, 1-Diphenyl-2-picryl hydrozyl Radical Scavenging Activity. Journal of Agricultural Food Chemistry 52: 5240-5244.

11. Singleton VL, Rossi JA (1965) Colorimetry of total phenolics with phosphomolybdic- phosphotungstic acid reagent. American Journal of Enology and Viticulture 16: 144-158.

12. Iris F, Benzie F, Strain JJ (1999) Ferric reducing/antioxidant power assay: Direct measure of total antioxidant activity of biological fluids and modified version for simultaneous measurement of total antioxidant power and ascorbic acid concentration. Methods in Enzymology 299: 15-27.

13. Wanyika HN, Gatebe EG, Gitu LM, Ngumba EK, Maritim CW, et al. (2010) Determination of caffeine content of tea and instant coffee brands found in the Kenyan market. African Journal of Food Science 4: 353-358.

14. Ishrud O, Zahid M, Zhou H, Pan Y (2001) A water-soluble galactomannan from the seeds of Phoenix dactylifera L. Carbohydrate Research 335: 297-301.

15. Nunes FM, Reis A, Domingues MR, Coimbra MA (2006) Characterization of galactomannan derivatives in roasted coffee beverages. Journal of Agricultural and Food Chemistry 54: 3428-3439.

16. Vasconcelos ALS, Franca AS, Glória MBA, Mendonça JCF (2007) A comparative study of chemical attributes and levels of amines in defective green and roasted coffee beans. Food chemistry 101: 26-32.

17. European Food Safety Authority EFSA (2009) Scientific opinion of the panel on contaminants in the food chain on a request from the European Commission on cadmium in food. European Food Safety Authority Journal 980: 1-139.

18. Elgasim EA, Alyousef YA, Humeida AM (1995) Possible hormonal activity of date pits and flesh fed to meat animals. Food Chemistry 52: 149-152.

19. Heftmann E, Bennett RD (1965) Identification of estrone in date seeds by thin layer chromatography. Naturwissenschaften 52: 431-438

20. Aldhaheri A, Alhadrami G, Aboalnaga N, Wasfi I, Elridi M, et al. (2004) Chemical composition of date pits and reproductive hormonal status of rats fed date pits. Food chemistry 86: 93-97.

21. Larrauri JA, Borroto B, Perdomo U, Tabares Y (1995) Manufacture of a powdered drink containing dietary fibre. Alimentaria 260: 23-25.

22. Habib HM, Platat C, Meudec E, Cheynier V, Ibrahim WH (2014) Polyphenolic compounds in date fruit seed (Phoenix dactylifera): characterization and quantification by using UPLC-DAD-ESI-MS. Journal of the Science of Food and Agriculture 94: 1084-1089.

23. Health Canada (2006) Caffeine

24. Higdon JV, Frei B (2006) Coffee and health: A review of recent human research. Critical Reviews in Food Science and Nutrition 46: 101-123.

25. Food Standards Agency (2008) New caffeine advice for pregnant women.

26. Maslowa E, Bhattacharya S, Lin SW, Michels KB (2011) Caffeine consumption during pregnancy and risk of preterm birth: A meta- analysis. The American Journal of Clinical Nutrition 92: 1120-1130.

27. Trigueros L, Sayas-Barberá E, Pérez-Álvarez JA, Sendra E (2012) Use of date (Phoenix dactylifera L.) blanching water for reconstituting milk powder: Yogurt manufacture. Food Bioproducts and Processing 90: 506-514.

28. Pohl P, Stelmach E, Welna M, Szymczycha-Madeja A (2013) Determination of the elemental composition of coffee using instrumental methods. Food Analytical Methods 6: 598-613.

26

Microstructure of a Third Generation Snack Manufactured by Extrusion from Potato Starch and Orange Vesicle Flour

Tovar-Jímenez X[1], Aguilar-Palazuelos E[1], Gómez-Aldapa CA[2] and Caro-Corrales JJ[1]*

[1]*Posgrado en Ciencia y Tecnología de Alimentos, Universidad Autónoma de Sinaloa, Culiacán, México*
[2]*Área Académica de Química, Instituto de Ciencias Básicas e Ingeniería, Ciudad del Conocimiento, Hidalgo, México*

Abstract

The objective of this work was to evaluate the effect of extrusion on microstructural properties of a third generation snack food expanded by microwaves manufactured from orange vesicle flour, commercial nixtamalized corn flour, and potato starch. A Brabender 20DN laboratory extruder was used to get the pellets (unexpanded extruded products). Viscosity profiles (RVA), scanning electron micrographs (SEM), X-ray diffraction patterns, and thermogram data (DSC) were obtained. Analyses were made on raw materials, unprocessed mixture, extruded pellets, and microwave expanded products to study the native starch structure changes. Analyzes suggest that the snacks obtained by the extrusion process were modified to a desirable microstructure for achieving physicochemical properties necessary for acceptance by the consumer.

Keywords: Microstructure; Extrusion; Snack; Biopolymers

Introduction

Extrusion is a continuous thermo-mechanical process where materials containing biopolymers are plasticized and cooked by the combined action of pressure, heat, and shear stress [1]. This thermo-mechanical process is very useful in producing low-fat snacks and has the advantage of increasing protein and starch digestibility, solubilizing fiber, inactivating toxins, anti-nutritional factors, and undesirable enzymes, such as lipo-oxygenases and peroxidases [2]. In addition, microwaves have been widely used for expanding this type of snacks and some researches on 3G snacks have focused on the effect of processing on different physical and physicochemical properties [3,4]. During extrusion, raw materials experience chemical and structural transformations such as starch gelatinization, protein denaturation, complex formation between amylose, lipids and/or proteins, and degradation of pigments and vitamins [5], but starch is the most important component due to the changes it undergoes by the thermal process as it affects expansion and final texture of the extrudate. Starch gelatinization can occur at levels from 12 to 22% moisture content; however, it has been indicated that at low moisture contents, gelatinization is accentuated because of the high shear stress, the heat generation, and the mechanical disruption of the granules [6]. Lee, et al. [4] mentioned that a partial gelatinization ($\approx$50%) is necessary for obtaining third generation snack foods (pellets), and a further degradation will reduce the size of sugar chains and thus the product stability after expansion will be lost. However, a lesser degradation will not be enough for opening the starch granules reducing the ability to adsorb water, which serves later as a means for expansion. For these reasons, it is important to know the microstructural changes after the extrusion process and microwave expansion of the pellets. Likewise, in Mexico, the citrus industry generates large amounts of residues that are normally discarded in landfills and left to decompose. Including these residues in technological processes can add value for industry and reduce contamination generated by residue decomposition; thus reducing their impact on ecosystems. Added to this, the use of by-products derived from the citric industry and mainly staple vitamins and fibers contained in the juice vesicle of oranges can be increased improving their functional properties. Dietary fiber acts as a bulking agent, normalizing intestinal motility and preventing diverticular disease. Some types may also be important in lowering serum cholesterol levels, in reducing colonic cancer and in preventing hyperglycemia in diabetic patients [7-9]. The objective of this work was to evaluate the effect of extrusion on microstructural properties of a third generation snack food expanded by microwaves manufactured from orange vesicle flour, commercial nixtamalized corn flour, and potato starch.

Materials and Methods

Raw material preparation

Orange vesicles were dried and milled in a hammer mill (Laboratory Mill 3100 Perten, Ireland). Commercial nixtamalized corn flour 10% (NCF), potato starch 79.9% (PS), and orange vesicle flour 10% (OVF) were homogenously and mixed with 0.1% monoglicerides. This formulation was obtained in preliminary studies.

Extrusion process

Extrusion was carried out in a laboratory extruder (Brabender 20DN, 8-235-00, Brabender OHG, Duisburg, Germany). A rectangular matrix with internal dimensions of 20 mm wide × 1.0 mm high × 100 mm long was used. Screw speed was 1.08 Hz (65 rpm), using a 2:1 L:D screw ratio and a feed rate of 33 g/min (dry matter), [0.42 Hz (25 rpm)]. Temperatures at the feed and out zones were 60 and 75°C. The transition zone temperature was 130°C and moisture content was 23%; these conditions were obtained from an optimization process

***Corresponding author:** Caro-Corrales JJ, Posgrado en Ciencia y Tecnología de Alimentos, Universidad Autónoma de Sinaloa, Blvd. De las Américas y Josefa Ortiz de Domínguez s/n, Ciudad Universitaria, Culiacán, Sin. C.P. 80000, México
E-mail: josecaro@uas.edu.mx

by response surface methodology considering expansion index, bulk density, penetration force, and total carotenoid content [10]. The extruded material was cut manually into approximately 1.5 cm long pellets, and dried at room temperature for approximately 24 h until reaching 9-13% moisture content.

Pellet expansion by microwave heating

Extruded products (pellets of 1.5 cm long) were expanded in a microwave oven (LG', R-501CW, 900 W, 2450 Hz) for 28 s. This condition was determined through expansion kinetics at different heating times [10].

Rapid visco-analyzer

Viscosity properties were evaluated using a Rapid Visco-Analyzer (RVA, 3C, Newport Scientific PTY Ltd., Sydney, Australia) following the instruction manual and suggestions from Aguilar-Palazuelos, et al. [11]. Two grams of sample were diluted with distilled water to get 28 g in the sample aluminum cup. Sample was continuously agitated and heated at 50°C for 1 min. Temperature was raised to 92°C at 5.5°C min^{-1}, held at 92°C for 5 min, lowered from 92 to 50°C at a rate of -5.6°C min^{-1}, and held at 50°C for 2 min (Newport-Scientific 1992). From RVA amylograph profiles, the next viscosities were measured: initial viscosity (V_{ini}), viscosity at 92°C (V_{92}) (maximum temperature), minimum viscosity (V_{min}) at 92°C, and final viscosity (V_{fin}) (higher viscosity for the cooling period). From these values, total retrogradation viscosity (V_r) (final viscosity minus minimum viscosity) was calculated [12].

Scanning electron microscopy

Analyses were made in raw materials, unprocessed mixture, extruded pellets, and microwave expanded products, according to methodology described by Zazueta-Morales, et al. [13] and Aguilar-Palazuelos, et al. [14]. A scanning electron microscope model JSM-6300 was used; a secondary electron detector and electron bombardment at 15 kV were used; samples were placed at high vacuum. The milled samples (60-mesh) were mounted on an aluminum 12 mm diameter holder, PIN type, previously prepared with carbon conductive double coated tapes and colloidal silver adhesive. Morphologies and particle size were observed.

X-ray diffraction

Analysis was performed on milled samples which pass through a 60-mesh sieve (250 μm), at moisture content between 9 and 13%. Samples were packed in a glass slide (0.5 mm deep) and mounted on a Rigaku X-ray diffractometer (Ultima D/Max-2100, Rigaku Denki Co. Ltd, Japan) according to procedures described by Zazueta-Morales, et al. [13]. Analyses were made in raw materials, unprocessed mixture, extruded pellets, and microwave expanded products, in order to determine the effect of processing on the crystallinity of starches.

Differential scanning calorimetry

Analyses were performed in raw materials, extruded pellets, and microwave expanded products. A calorimeter (Mettler-Toledo 821 e, Columbus, OH) was used following recommendations from Toro-Vazquez, et al. [15]. Samples of about 4 mg were added with 16 μl of distilled water and hermetically sealed in aluminum pans. Subsequently, they were vigorously shaken and allowed to stand for 18 h. Pans were heated from 40 to 100°C at a heating rate of 5°C/min; the heating chamber was vented with nitrogen at a flow rate of 20 ml/min. The enthalpy change, initial temperature (onset), (maximum)

peak temperature, and the final temperature were determined, using the software STARe Thermal Analizer.

Results and Discussion

Rapid visco-analyzer

Viscosity properties (Table 1) and viscosity profiles (Figure 1) are

Viscosity (mPa s)					
Material	V_{ini}	V_{92}	V_{min}	V_{fin}	V_r
NCF	47	3114	2191	4613	2422
PS	7	4809	2395	2404	9
OVF	37	43	43	49	6
Mixture	7	1059	981	1516	535
Pellet	120	133	97	127	30
Expanded	17	58	61	92.15	31.15

Table 1: Initial viscosity (V_{ini}), viscosity at 92°C (V_{92}), minimum viscosity (V_{min}) at 92°C, final viscosity (V_{fin}), and total retrogradation viscosity (V_r) for nixtamalized corn flour (NCF), potato starch (PS), orange vesicle flour (OVF), unprocessed mixture, extruded pellets, and microwave expanded product.

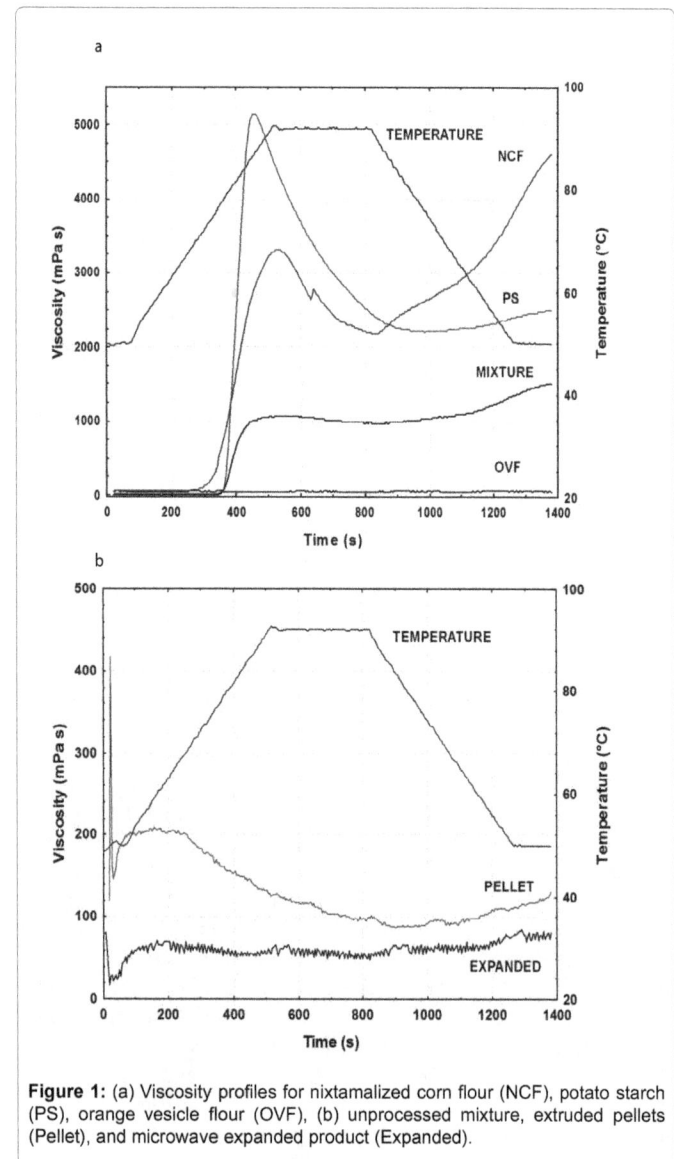

Figure 1: (a) Viscosity profiles for nixtamalized corn flour (NCF), potato starch (PS), orange vesicle flour (OVF), (b) unprocessed mixture, extruded pellets (Pellet), and microwave expanded product (Expanded).

shown for commercial nixtamalized corn flour (NCF), potato starch (PS), orange vesicle flour (OVF), unprocessed mixture, extruded pellets, and microwave expanded products. Viscosity of nixtamalized corn flour and potato starch increased from 47 (V_{Ini}) to 3114 (V_{92}) mPa s and from 7 to 4809 mPa s, respectively. The difference between NCF and PS viscosity may be due to chemical composition and the previous thermal treatment NCF has undergone; NCF contains gums and during the nixtamalization process, the more susceptible starch granules are partially gelatinized, which are generally the larger ones, remaining the most resistant granules practically unchanged [16]. This may explain why NCF showed a lower viscosity peak (V_{92}) of 3114 mPa s than PS of 4809 mPa s. Maximum viscosity (V_{92}) of PS is similar to that reported by Alvis, et al. [17]. Orange vesicle flour did not develop a maximum peak viscosity (V_{92}) during heating.

The unprocessed mixture showed a peak viscosity (V_{92}) of 1059 mPa s; this lower viscosity, compared to NCF and PS viscosity, may be due to the presence of fiber, affecting the development of viscosity and the type of link that hinder viscosity development and gel formation. Another reason may be water competition as pericarp can be chemically modified turning fiber from insoluble to soluble and competing for water; also gums in the mixture can be interacting with potato starch. The mixture showed minimum viscosity at 92°C (V_{min}) of 981 mPa s, which is lower than that for raw materials; this could be due to fiber affects starch gelatinization as starch molecules become less soluble and tend to aggregate [14].

Nixtamalized corn flour had a V_{min} of 2191 mPa s at min 14 and retrogradation started from this point. In like manner, potato starch had a V_{min} of 2395 mPa s at min 14 and retrogradation started at minute 18. This time difference for NCF and PS to start retrogradation could be due to amylose retrogrades faster, since as a result of its linear nature and polarity tends to form hydrogen bonds between hydroxyl groups of adjacent molecules (amilose and amilopectin); therefore, they start losing their hydration capability, which yields a partial shrinkage of starch [16,18].

In viscosity profiles for extruded pellets and microwave expanded products (Figure 1b), a slight increase in viscosity during retro gradation was observed, reaching their maximum viscosity at 50°C, after cooling. This result is similar to that reported by Zazueta-Morale, et al. [13] and Aguilar-Palazuelos, et al. [14]. These authors mentioned that the differences having an extruded starch, relative to a native starch, are the absence of a gelatinization peak, a high initial viscosity and a continuous fall in viscosity from 50 to 92°C. In addition, during cooling, a slight or no increase in viscosity is obtained, achieving its maximum value at 50°C. Similarly, the temperature used during extrusion (130°C) may have caused the dough to become less viscous, allowing the molecules become more susceptible to shearing action. In this way, a higher thermal and mechanical action is achieved, resulting in further degradation of starch and consequently lowering viscosity. This result is similar to that reported by Carvalho, et al. [19], who stated that viscosity is affected by the extrusion temperature (above 80°C) and moisture content. Lee, et al. [4] and Delgado-Nieblas, et al. [20] reported that the best physicochemical properties of third generation products, processed by extrusion, occur when there is a partial degradation of starch, which is confirmed by the above. The physicochemical (functional) properties of extruded starch products depend on the rheology of both the material during processing and the final product, since rheology is highly related to the microstructural changes the product underwent and hence to the ability the material

has to interact between the polymeric matrix and the other components of the extruded product.

Scanning electron microscopy

Figure 2 displays micrographs of raw materials (nixtamalized corn flour, potato starch, and orange vesicle flour) and unprocessed mixture used in the extrusion process. In the unprocessed mixture, different sizes of starch granules are observed, with circular and oval shapes due to the diversity of used materials. In addition, the presence of fiber attached to the starch granules can be seen. The corn flour starch granules are smaller than those of potato, with diameters from 15 to 25 µm and from 15 to 100 µm, respectively (Figure 2b and 2c). Similar data were reported by Medina and Salas [21], who reported that corn starch granules have a smaller size (2-30 µm) than those of PS (5-100 µm). Figure 2d shows the fiber fragments of OVF. Micrographs of extruded pellets obtained using the optimal processing conditions (130°C and 23% moisture content) are shown in Figure 3. It can be distinguished the starch granules were melted and plasticized; furthermore, certain channels and holes are observed in the product surface, which may be due to some leakage of water from the material, when leaving the extruder, as reported by Zazueta-Morales, et al. [13]. This type of structure is similar to that reported by Aguilar-Palazuelos, et al. [11] for third generation snacks. Also, it can be seen that pellet fragments are extended (Figure 3a and 3b); this may be due to an alignment effect of the short chain starches. The extrusion process is able to break the covalent bonds in biopolymers and both the intense structural disruption and mixing facilitate reactions that in other processes are limited [22]. Similarly, it appears that these products are covered by certain particles (Figure 3c), which can be very small size particles of the same material as suggested by Aguilar-Palazuelos, et al. [14]. Figure 4 shows the micrographs of the microwave expanded product. A porous (air cells with thin walls) and extended structure can be viewed (Figure 4a and 4b). This structure is very similar to that reported by Lee, et al. [4], with air cells of uniform size, for third generation products expanded by microwaves. This pattern can be caused by a sudden release of pressure generated by steam inside the pellet. Pellets were expanded by microwave, which caused the trapped water inside the extruded began to evaporate, creating an increase in pressure inside the product. When the internal pressure was high enough, the internal structure of the pellet could no longer endure, so that exploded and expanded, causing the air cells of the expanded product were elongated and with smooth surface walls with an approximate size of 1 mm (Figure 4a and 4c). The holes on the surface could be produced during the escape of steam; this is similar to that reported by Moraru and Kokini [23].

X-ray diffraction

X-ray patterns of nixtamalized corn flour, potato starch, orange vesicle flour, unprocessed mixture, extruded pellet, and microwave expanded product are shown in Figure 5. Nixtamalized corn flour showed an A-type diffractogram, which is typical for cereal starches, having two main peaks at approximately 2θ of 16 and 21 Å. These peaks have been reported by Zazueta-Morales, et al. [13] and Aguilar-Palazuelos, et al. [11]. For potato starch and the unprocessed mixture B-type patterns were obtained, which are typical for tubercle starches; the latter behavior was expected as the main component in the mixture was potato starch [20]. Extruded pellets and expanded products showed an amorphous structure. The presence of amorphous structure in extruded materials, at temperatures near 70°C and moisture contents above 20%, have been reported by Aguilar-Palazuelos, et al. [11]. These amorphous structures apparently occur because these

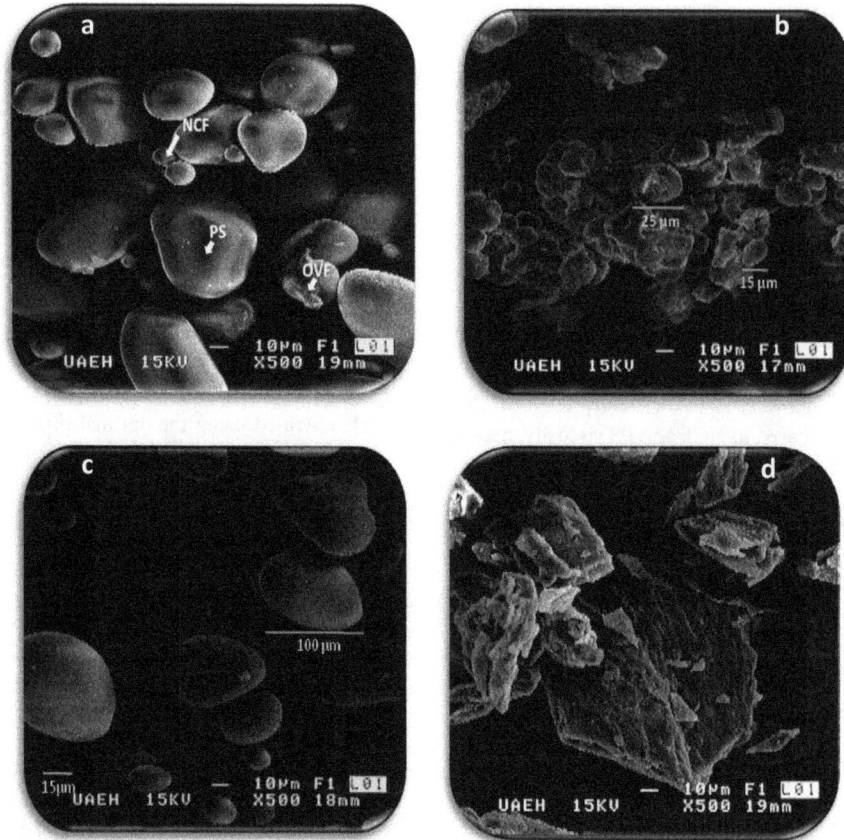

Figure 2: Scanning electron micrographs of (a) unprocessed mixture, (b) nixtamalized corn flour (NCF), (c) potato starch (PS), and (d) orange vesicle flour (OVF) used for the extrusion process.

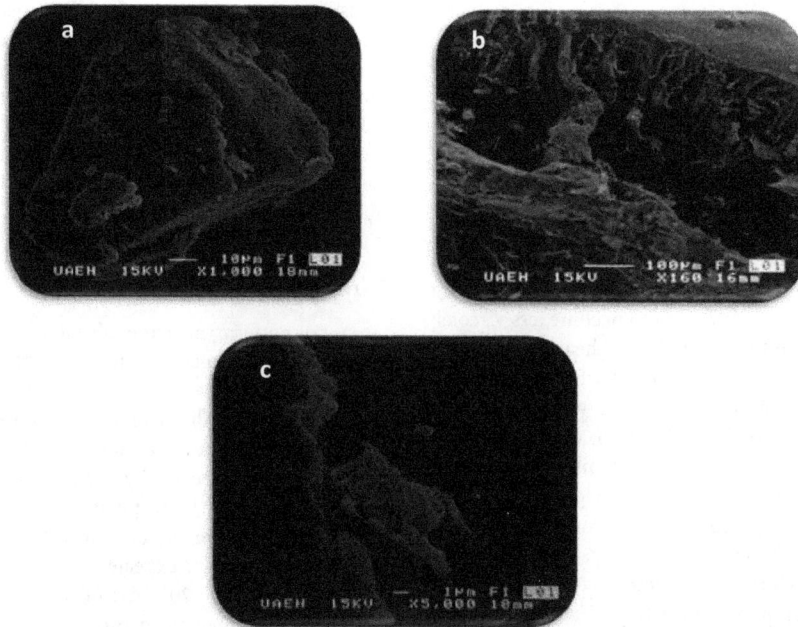

Figure 3: Scanning electron micrographs of milled extruded pellets obtained using the optimal processing conditions, 130 °C and 23% moisture content; a, b, and c correspond to different zones of the pellet.

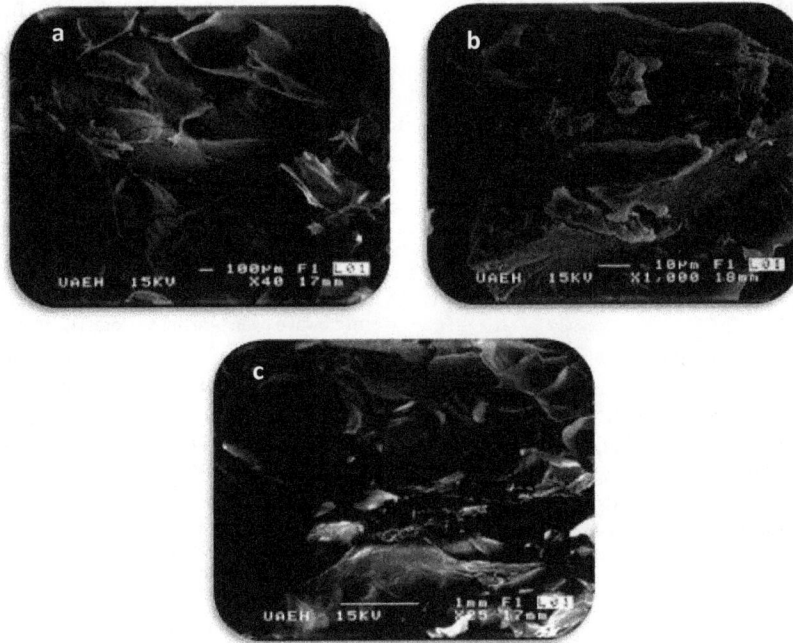

Figure 4: Scanning electron micrographs of the microwave expanded product; a, b, and c correspond to different zones of the expanded product.

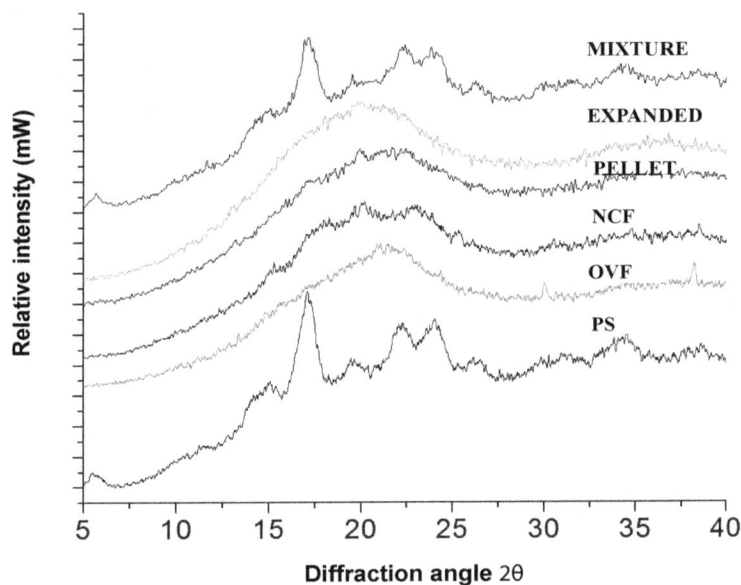

Figure 5: X-ray diffraction patterns of nixtamalized corn flour (NCF), potato starch (PS), orange vesicle flour (OVF), unprocessed mixture, extruded pellet, and microwave expanded product.

conditions are enough for modifying the native crystalline structure, but are not sufficient for the formation of new crystalline structures. Extrusion processing can partially or totally break the crystalline structure of starch, depending on the extrusion conditions such as moisture content, shear stress, and temperature. High temperatures can completely destroy starch structure, leading to the formation of an X-ray diffractogram typical of an amorphous state or can induce the formation of a new structure [24]. During extrusion, extruded materials show large amounts of plasticized and slightly damaged granules [25].

Differential scanning calorimetry

The thermogram data of raw materials [nixtamalized corn flour (NCF), potato starch (PS), orange vesicle flour (OVF)], unprocessed mixture, extruded pellet, and microwave expanded product are shown in Table 2. Nixtamalized corn flour had temperatures of 65.2°C for onset (T_i), 68.5°C for peak (T_g), and 69.7°C for final temperature (T_f). Potato starch and unprocessed mixture showed 59.5, 63.7, 69.5°C and 61.3, 64.9, 69.7°C for T_i, T_g and T_f respectively. Temperatures for

	T_i (°C)	T_g (°C)	T_f (°C)	ΔH (mJ)
NCF	65.2	68.5	69.7	-15.8×10^{-3}
PS	59.5	63.7	69.5	-107.08
OVF	--	--	--	--
Mixture	61.3	64.9	69.7	-55.48
Pellet	--	--	--	--
Expanded	--	--	--	--

(-- No gelatinization peak), T_i is initial temperature [onset], T_g is gelatinization temperature, T_f is final gelatinization temperature [endset], and ΔH is the transition enthalpy.

Table 2: Thermogram data of raw materials [nixtamalized corn flour (NCF), potato starch (PS), orange vesicle flour (OVF)], unprocessed mixture, extruded pellet, and microwave expanded product.

mixture are similar to those found for potato starch, which is attributed to potato starch is the main component in the mixture. The transition enthalpy was 26.8 and 13.9 J g^{-1} for potato starch and unprocessed mixture. The orange vesicle flour, extruded pellet, and microwave expanded product did not show a significant gelatinization peak, which can occur because starch is gelatinized. These results agree with those reported by Maninder, et al. [26] and Aguilar-Palazuelos, et al. [11]. Lee et al. [27] reported that the degree of gelatinization of starch, extruded under critical conditions, was higher as the extrusion temperature was lowered. They found that the extrudates at 80°C showed a small peak of gelatinization, meanwhile, at 90 and 100°C no endothermic peak was found. In our study, extrusion was accomplished at 130°C and 23% moisture content and no gelatinization peak was observed in the extrudates.

Conclusions

The analyzed optimum processing conditions were adequate for producing a third-generation snack from a mixture of corn starch and orange vesicle flour. These raw materials are a promising material for production of third-generation snacks and the analyses of the X-ray diffractograms, viscosity profiles, scanning electron microscopy and differential scanning calorimetry suggested that the extruded products presented changes mainly due to fragmentation, gelatinization and plasticization of the starch granule, causing the loss of the granular starch (modification in its native structure). The snacks obtained by the extrusion process and expanded by microwave were modified to a desirable microstructure for achieving adequate physicochemical properties for acceptance by the consumer.

References

1. Manrique-Quevedo N, Gonzales-Soto RA, Othman-Abn-Hardan M, Garcia-Suarez FJ, Bello-Perez LA (2007) Characterization of pregelatinized blends of mango and banana starches with different extrusion conditions. Rev Agrociencia 41: 637- 645.

2. Cheftel JC (1986) Nutritional effects of extrusion cooking. Food chem 20: 263-283.

3. Gimeno E, Moraru CI, Kokini JL (2004) Effect of xanthan gum and CMC on the structure and texture of corn flour pellets expanded by microwave heating. Cereal Chem 81: 100-107.

4. Lee EY, Lim KII, Lim JK, Lim ST(2000) Effects of gelatinization and moisture content of extruded starch pellets on morphology and physical properties of microwave-expanded products. Cereal Chem 77: 769-773.

5. Ding QB, Ainsworth P, Plunkett A, Tucker G, Marson H (2006) The effect of extrusion conditions on the functional and physical properties of wheat-based expanded snacks. J Food Eng 73: 142-148.

6. Vasanthan T, Yeung J, Hoover R (2001) Dextrinization of starch in barley flours with termostable alpha-amylase by extrusion cooking. Starch-Starke 53: 616- 622.

7. Larrea CMA, Chang YK, Martinez-Bustos F (2005) Effect of some operational extrusion parameters on the constituents of orange pulp. Food Chem 89: 301-308.

8. Hernández-Díaz JR, Quintero-Ramos A, Barnard J, Balandran-Quintana RR (2007) Functional properties of extrudates prepared with blends of wheat flour/pinto bean meal with added wheat bran. Food Sci Technol Int 13: 301-308.

9. Larrea CMA, Martínez-Bustos F, Yoon KC (2010) The effect of extruded orange pulp on enzymatic hydrolysis of starch and glucose retardation index. Food Bioprocess Technol 3: 684- 692.

10. Tovar-Jiménez X, Caro-Corrales J, Gómez-Aldapa CA, Zazueta-Morales J, Limón-Valenzuela V (2015) Third generation snacks manufactured from orange by-products: physicochemical and nutritional characterization. J Food Sci Technol 52: 6607- 6614.

11. Aguilar-Palazuelos E, Zazueta-Morales JJ, Martínez-Bustos F (2006) Preparation of high-quality protein-based extruded pellets expanded by microwave oven. Am Assoc Cereal Chem 83: 363-369.

12. Zeng M, Morris CF, Batey IL, Wrigley CW (1997) Sources of variation for starch gelatinization pasting, and gelation properties in wheat. Cereal Chem 74: 63-71.

13. Zazueta-Morales JJ, Martínez-Bustos F, Jacobo-Valenzuela N, Ordorica-Falomir C, Paredes-López O. (2002) Effects of calcium hydroxide and screw speed on physicochemical characteristics of extruded blue maize. J Cereal Sci 67: 3350-3358.

14. Aguilar-Palazuelos E, Zazueta-Morales JDJ, Jiménez-Arévalo OA, Martínez-Bustos F (2007) Mechanical and structural properties of expanded extrudates produced from blends of native starches and natural fibers of henequen and coconut. Starch-Stärke 59: 533-542.

15. Toro-Vázquez J, Herrera-Coronado V, Dibildox-Alvarado E, Charo-Alonso M, Gómez-Aldapa C. (2002) Induction time of crystallization in vegetable oils comparative measurements by differential scanning calorimetry and diffusive light scattering. J Food Sci 67: 1057-1064.

16. Salinas-Moreno Y, Herrera-Corredor JA, Castillo-Merino J, Pérez-Herrera P (2003) Cambios físico-químicos del almidón durante la nixtamalización del maíz en variedades con diferente dureza de grano. Arch Latinoam Nutr 53: 188-193.

17. Alvis A, Vélez CA, Villada HS, Rada-Mendoza M (2008) Análisis físico-químico y morfológico de almidones de ñame, yuca y papa y determinación de la viscosidad de las pastas. Inf Tecnol 19: 19-28.

18. Hoover R (2001) Composition molecular structure and physicochemical properties of tuber and root starches: A review. Carbohydrate Polymers 5: 253- 67.

19. Carvalho RV, Ascheri JLR, Cal-Vidal J (2002) Efeito dos parâmetros de extrusão nas propriedades físicas de pellets (3G) de misturas de farinhas de trigo arroz e banana Ciência Agrotecnica Lavras 26: 1006-1018.

20. Delgado-Nieblas C, Aguilar-Palazuelos E, Gallegos-Infante A, Rocha-Guzmán N, Zazueta-Morales JJ, et al. (2012) Characterization and optimization of extrusion cooking for the manufacture of third-generation snacks with winter squash (Cucurbita moschata d.) flour. Am Assoc Cereal Chem 89: 65-72.

21. Medina JA, Salas JC (2008) Morphological Characterization of Native Starch Granule: Appearance, Shape, Size and its Distribution. Rev Ing 27: 56-62.

22. Asp NG, Bjorck I (1989) Nutritional properties of extruded foods. En: Extrusion cooking. Mercier C Linko P y Harper JM (Eds) Am Assoc Cereal Chem Inc St Paul MN USA 14: 399-434.

23. Moraru CI, Kokini JL (2003) Nucleation and expansion during extrusion and microwave heating of cereal foods. Compr Rev Food Sci F 2: 120-138.

24. McPherson AE, Bailey TB, Jane J (2000) Extrusion of crosslinked hydroxypropylated corn starches I. Pasting properties. Cereal Chem 77: 320-325.

25. Della-Valle G, Vergnes B, Colonna P, Patria A (1997) Relations between rheological properties of molten starches and their expansion behaviour in extrusion. J Food Eng 31: 277-296.

26. Maninder K, Narpinder S, Kawaljit SS, Harmeet SG (2004) Physicochemical morphological thermal and rheological properties of starches separated from kernels of some Indian mango cultivars (Mangifera indica L.) Food Chem 85: 131-140.

27. Lee EY, Ryu GH, Lim ST (1999) Effects of processing parameters on physical properties of corn starch extrudates expanded using supercritical CO$_2$ injection. Cereal Chem 76: 63-69.

27

Estimated Iron and Zinc Bioavailability in Soybean-Maize-Sorghum Ready to Use Foods: Effect of Soy Protein Concentrate and Added Phytase

Akomo PO[1]*, Egli I[2], Okoth MW[3], Bahwere P[4], Cercamondi CI[2], Zeder C[2], Njage PMK[3] and Owino VO[5]

[1]Valid Nutrition, Nairobi, Kenya
[2]Laboratory of Human Nutrition, Institute of Food, Nutrition and Health, ETH Zurich, Zurich, Switzerland
[3]Department of Food Science, Nutrition and Technology, University of Nairobi, Kenya
[4]Valid International, Brussels, Belgium
[5]Technical University of Kenya, Nairobi, Kenya

Abstract

Efficacy and cost of nutritional supplements are critical in addressing malnutrition. Use of cheaper and locally available ingredients in manufacturing ready-to-use foods (RUF) can potentially reduce cost and increase access to supplements in resource-poor settings. Soy protein concentrate (SPC) is a cheaper source of protein and can potentially replace the more expensive milk powder in RUF. However, SPC contains phytic acid (PA) which inhibits mineral bioavailability. PA may be degraded by the enzyme phytase. This study aimed to determine the effect of replacing skim milk powder (MP) with SPC and of added phytase on bioavailability of iron and zinc in soybean-maize-sorghum RUF.

RUF samples were made using either SPC or MP. Phytase was added to food samples with either low (<5%) or high (>50%) moisture prior to estimation of bioavailability of iron and zinc by *in vitro* dialysability. Compared to samples with MP, SPC-based foods had significantly higher content of PA (0.84 g/100 g vs. 0.57 g/100 g; p<0.001); lower bioavailability of iron (2.79% vs. 4.85%; p<0.001) and lower zinc bioavailability (3.61% vs. 8.69% for zinc; p<0.001). After one hour of incubation at 35°C, 68% of PA in high-moisture foods and 10% of PA in low moisture foods were degraded. The data indicate that replacing MP with SPC in SMS RUF increases PA content with subsequent reduction of bioavailability of iron and zinc. Added phytase significantly reduces PA content in high moisture foods and may potentially remain active in the Stomach where moisture is high. Adding such a phytase could be a promising approach to increase iron and zinc bioavailability from SMS RUFs and provide cheaper locally produced formulations for addressing malnutrition in resource-poor settings.

Keywords: Iron and zinc bioavailability; Ready-to-use foods; Phytase

Introduction

Deficiencies of micronutrients are a major concern among children and adults globally with evidence that such deficiencies affect physical growth, cognitive development, reproduction, physical work capacity and risk of illness [1-3]. Iron and zinc deficiencies are most common in developing countries with the former being more widespread affecting up to 47% of children globally and up to 68% of pre-school age children and pregnant and lactating women in Africa [4-7]. Correlations have been identified between iron and zinc status of individuals. The highest deficiency prevalence of both iron and zinc has been reported in children from low-income families [4,8]. The prevalence of iron deficiency in resource-poor regions is exacerbated by a reliance on staple food crops which have low bioavailability of the minerals [5,9].

Ready to Use Foods (RUF) are fortified nutrient-dense formulations designed for prevention and treatment of malnutrition. In contrast to powdered foods such as Corn Soy Blend (CSB) and Wheat Soy Blend (WSB), RUF are high-energy, low-moisture lipid pastes that do not require further cooking. Their low-moisture characteristic makes them resistant to microbial contamination [10]. RUF formulations differ depending on target group and type of malnutrition to be addressed. Ready-to-use therapeutic food (RUTF) is designed to treat severe acute malnutrition (SAM) while ready-to-use supplementary foods (RUSF) are designed for prevention and treatment of moderate acute malnutrition (MAM). Ready-to-use complementary foods (RUCF) are designed to prevent chronic malnutrition when given to children 6-24 months of age. Peanut-based RUTF (P-RUTF) is the most widely used therapeutic food. P-RUTF is made from peanut paste, milk powder, vegetable oil, sugar, vitamins and minerals. It has been successfully used to address SAM in children and adults [11]. The success of P-RUTF has motivated development not only of other RUTF made from other ingredients but similar lipid-based foods to address other forms of malnutrition. Development of other RUF is aimed at reducing

cost and increasing variety. Soybean-maize-sorghum ready to use foods (SMS RUF) are similar lipid-based products meant to serve as cheaper alternatives for addressing different forms of malnutrition [12]. SMS, RUF are made from soybean, maize, sorghum, little or no milk powder, sugar, vegetable fat and micronutrient premix. Different formulations of SMS RUF have been developed. They include soybean-maize-sorghum ready-to-use therapeutic food (SMS RUTF), soybean-maize-sorghum ready-to-use supplementary foods (SMS RUSF) and soybean-maize-sorghum ready-to-use complementary food (SMS RUCF). Acceptability and efficacy of SMS-RUF have been tested for treatment and prevention of malnutrition [13-15] among children in resource poor settings. The impact of consumption of SMS-RUCF on breast milk intake has also been assessed [16]. A study among Congolese children in reduction of stunting and underweight found no significant difference between an SMS-based complementary food and a fortified blended food commonly referred to as UNIMIX [17]. Another study comparing effectiveness of a milk-free SMS-based therapeutic food with that of standard peanut-based food with milk in treating SAM among Zambian children did not confirm equivalence [14].

Bioavailability of minerals in cereal- and pulse-based foods can be significantly reduced by phytic acid (*myo*-inositol hexakisphosphate).

*Corresponding author: Akomo PO, Valid Nutrition, Nairobi, Kenya
E-mail: peterakomo@validnutrition.org

Phytic acid occurs naturally as the storage form of phosphorus in plants. It inhibits the bioavailability of the minerals by forming insoluble and indigestible complexes. Reduction in the inhibiting effect of phytic acid has been achieved through use of chelated forms of iron such as sodium iron ethelynediaminetetraacetate (EDTA) and by use of intrinsic or added phytase enzyme to degrade phytic acid [18,19]. The phytate/mineral molar ratio in foods has been used to estimate mineral bioavailability. A phytate/iron ratio of less than one is preferred in order to significantly improve iron absorption in cereal or legume-based foods without iron absorption enhancers [20]. Phytate/Zinc ratios of 15 and above are reported to greatly reduce zinc absorption resulting in negative zinc balance [21].

Cereals and pulses form about 50% by weight of the SMS-RUF. It is likely that the level of phytic acid in the foods is high enough to significantly reduce bioavailability of iron and zinc. Moreover, if SPC is to replace DSMP, mineral bioavailability is of critical interest. To our knowledge, no past studies have investigated bioavailability of the minerals in SMS RUF and how it is affected by added phytase and replacing disodium monophosphate (DSMP) with soy protein concentrate (SPC). Therefore, the aim of this study was to estimate the bioavailability of iron and zinc in SMS-RUF using an *in-vitro* method based on simulated gastric digestion followed by dialysis.

Materials and Methods

Food preparation

Processing of foods was done at Insta Products EPZ, Athi River Kenya. Insta Products EPZ is internationally certified by UNICEF to supply both corn soy blend (CSB) powder and ready to use therapeutic food (RUTF). Food preparation was done as described for SMS RUTF [13]. The process involved two main stages: extrusion cooking of a blend of whole soya, maize and sorghum grains and mixing of the blend with sugar, oil, fat, milk powder or soy protein concentrate, and premix containing vitamins and minerals. Figure 1 shows the process flow chart. Grains were cleaned and accurately weighed in the right proportions (Table 1) and blended manually. The full description of the procedure is as outlined by Owino et al. [13]. SPC was obtained from United Soy Board, USA while premixes of vitamins and minerals were obtained from Fortitech Strategic Nutrition, Denmark. According to the certificate of analysis from the supplier, 3 g of each premix contained among other micronutrients 487.7 µg retinol equivalents vitamin A as retinyl acetate, 16.5 mg tocopherol equivalents vitamin E as dl-alpha-tocopheryl acetate, 73 mg vitamin C as ascorbic acid, 7.3 mg iron as NaFeEDTA and 9.2 mg zinc as zinc sulfate. Foods were

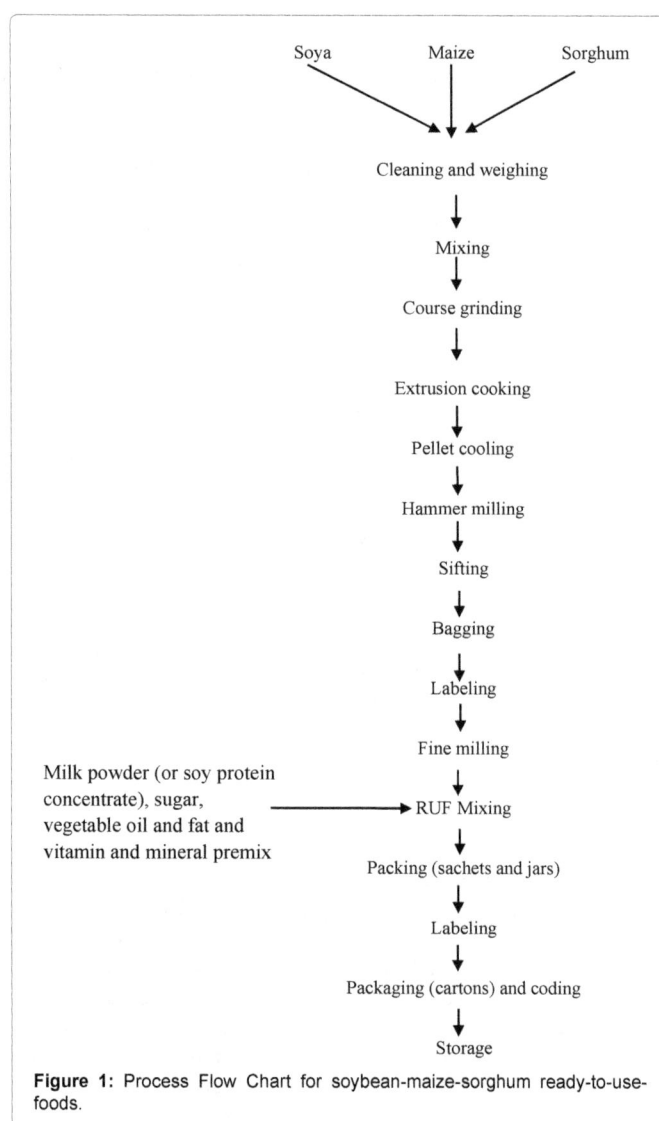

Figure 1: Process Flow Chart for soybean-maize-sorghum ready-to-use-foods.

packed in polyethylene terephthalate (PET) jars of 60 g and labeled appropriately.

Determination of iron and zinc dialysability

Analyses of iron and zinc dialysability were done by a modified method of Miller et al. [22] at the Human Nutrition Laboratory, ETH, Zurich, Switzerland. The method uses a simulated gastrointestinal digestion system in which test meals were enzymatically digested and dialyzable iron and zinc pass into a dialysis tube containing a $NaHCO_3$ solution. The quantity of dialyzed iron and zinc were then measured by atomic absorption spectrometry. The test meal was prepared as follows: 450 ml ultra-pure water (18 MΩ.cm) were added to 39.0 g semolina in a 1 L beaker. The mixture was then heated in a boiling water bath until the temperature of the semolina meal reached 90°C. After cooling down, the pH of the mixture was brought to 2.0 with 1 M HCl and the weight was adjusted to 600 g with ultra-pure water (18 MΩ.cm). Forty grams aliquots of this acidified meal were transferred into 100 ml Erlenmeyer flasks. About 5 g of RUF was weighed at accuracy of 0.1 g and added to the meal. A 1.3 ml volume of a porcine pepsin (Sigma-Aldrich, Buchs, Switzerland) solution (16 g/100 ml in HCl 0.1 M) was added to each

Ingredient	Grams per 100 g RUF	
	DSMP recipe	SPC recipe
Soybean	21.4	17.5
Maize	22.2	18.1
Sorghum	5.4	4.4
DSMP	10.0	-
SPC	-	19.0
Sugar	14.0	14.0
Palmolein	22.0	22.0
Palm stearin	2.0	2.0
Premix	3.0	3.0

SMP, Skim milk powder; SPC, Soy protein concentrate

Table 1: RUF recipes

aliquot, and the flasks were put in a shaking water bath at 37°C for 2 hours. For the second part of the digestion, a 28 cm cut of the dialysis tubing (SpectraPor 1, 6000-8000 Da, 20.4 cm $\varnothing$, Spectrum), filled with 25 ml of a NaHCO$_3$ solution was immersed in each flask. The amount of NaHCO$_3$ solution in the dialysis tubing was just enough to shift the pH of the digested meal to 7.5; this quantity was predetermined by titration of an aliquot of the acidified meal. After 30 min in the shaking water bath at 37°C, 5 ml of a pancreatin solution (containing 20 mg porcine pancreatin (Sigma-Aldrich) and 125 mg porcine bile (Sigma-Aldrich) extract in 0.1 M NaHCO$_3$) were added to the samples. This was followed by a 2-hour digestion at 37°C in the shaking water bath. Iron and zinc contents of the dialysates were then measured by graphite furnace and flame atomic absorption spectrometry (SpectrAA-240 Z and FS respectively, Varian Inc., Mulgrave, Australia) respectively. Each RUF was tested in duplicate and a reference sample containing about 2 mg Fe as ferrous sulfate was included in each batch of analyses.

Determination of phytic acid content

Analysis of phytic acid was done in triplicate using a modified method by Makower [23]. The method involved extraction, precipitation and mineralization of phytic acid followed by spectrophotometric measurement of the mineralized inorganic phosphate in a micro-titration plate according to Van Veldhoven and Mannaerts [24] and converted into phytate concentrations.

Determination of minerals content

Iron, zinc and calcium analysis was done in triplicate. An internal standard reference sample was used in each batch of analysis to check accuracy. Samples of food and ingredients were first mineralized by microwave digestion (MLS ETHOS plus, MLS GmbH; Leutkirch, Germany) using a nitric acid-hydrogen peroxide mixture (7:3 v/v).

Iron and zinc concentrations were measured by graphite furnace and flame atomic absorption spectrometry respectively, using external calibration curves.

Degradation of phytic acid

Phytase enzyme (20,000 FTU per gram; DSM Nutritional Products, Basel, Switzerland) was used to degrade phytic acid. The enzyme had two pH optima, one at pH 5 and the other just below pH 3 with activity of 100% and 60% respectively. FTU is the amount of enzyme that liberates 1 μmol of inorganic phosphorus per minute under defined conditions. The pH and moisture of three aliquots of RUF were adjusted using 1.2 M ascorbic acid (Table 2). pH 5 was specifically targeted because it is the optimum pH for 100% activity of the enzyme. Phytase was added at the rate of 5000 FTU per 100 g of food. Each aliquot was divided into two portions. One set of the portions was incubated at 25°C and the other at 35°C. Analysis of phytic acid was done to determine the initial and residual levels after 1, 3 and 14 days. To test for effect of food moisture content on the activity of the enzyme, food portion 3 (Table 2) was mixed 1:1 (w/w) with water. The mixture was incubated at 35°C in a water bath with intermittent mixing. Residual phytic acid was analyzed after 1 and 2 hours after stopping the degradation by addition of tetrachloroacetic acid.

Data entry, analysis and statistics

The data was entered in excel sheets (Excel 2010, Microsoft cooperation) from where variances were calculated. The data was subjected to analysis of variance using Genstat statistical package 13th edition (VSN International Limited, UK). Least significance difference (LSD) analysis was carried out at 5% level of significance to determine differences in treatment means. Zinc, iron and phytic acid levels were used to calculate phytate/mineral molar ratios in order to estimate their bioavailability according to Hallberg et al. [20] and Turnlund et al. [21].

Results and Discussion

Table 3 shows the compositional characteristics of ingredients of SMS RUF while Table 4 shows phytic acid content and iron and zinc bioavailability in SMS RUF. Soybean had significantly higher levels of iron, zinc and calcium (p<0.001) than maize and sorghum grains. PA content in soybean (1.61 g/100 g) was about three times higher than that in maize (0.59 g/100 g) and sorghum (0.56 g/100 g). DSMP was nearly five times higher in calcium content and twenty times lower in iron than SPC. The results of raw material composition are consistent

Food aliquot	pH	Added water	Added phytase (per 100g of food)
1	Native (6.8)	Nil	5000 FTU
2	5.1	Nil	5000 FTU
3	5.1	100% w/w	5000 FTU

FTU, phytase units defined as amount of enzyme that liberates 1 μmol of inorganic phosphorus per minute.

Table 2: Treatments for phytase study.

Parameter	Ingredient						
	Soybean	Maize	Sorghum	Raw SMS	Extruded SMS	SPC	DSMP
Fe, mg/100 g	6.8 ± 0.16[a]	1.8 ± 0.03[b,c]	3.9 ± 0.15[c]	4.3 ± 0.13[c]	15.0 ± 3.42[d]	10.1 ± 0.21[e]	0.6 ± 0.02[b]
Zn, mg/100 g	4.8 ± 0.02[a]	1.7 ± 0.03[b]	1.8 ± 0.04[c]	3.0 ± 0.02[d]	3.0 ± 0.03[d]	3.3 ± 0.00[e]	3.5 ± 0.06[f]
PA, g/100 g	1.61 ± 0.08[a]	0.59 ± 0.02[b]	0.56 ± 0.06[b]	1.09 ± 0.06[c]	0.897 ± 0.03[d]	2.09 ± 0.13[e]	Not detectable
PA/Fe molar ratio	20.0	27.9	12.3	21.6	1.60	17.5	-
PA/Zn molar ratio	32.9	35.1	30.4	36.3	29.2	63.0	-

SMS: Soybean-maize-sorghum blend; SPC: Soy protein concentrate; DSMP: Dry skim milk powder; PA: Phytic acid. Values with different superscripts in same row are significantly different (p ≤ 0.05).

Table 3: Compositional characteristics of ingredients of soybean-maize-sorghum ready-to-use foods (values are mean ± SD; n = 3).

RUF Recipe	Fe, mg/100 g	Zn, mg/100 g	PA, g/100 g	PA/Fe molar ratio	PA/Zn molar ratio	Fe dialysability, %	Zn dialysability, %
DSMP	11.90 ± 2.83[d]	5.84 ± 3.16[d]	0.57 ± 0.05[c]	4.05[a]	9.67[a]	4.85 ± 0.46[c]	8.69 ± 1.81[c]
SPC	15.11 ± 5.93[e]	5.60 ± 2.71[e]	0.84 ± 0.02[d]	4.70[a]	14.86[a]	2.79 ± 0.59[d]	3.61 ± 0.37[d]
Reference sample	2.00 ± 0.55	-	-	-	-	7.44 ± 2.86	-

SPC: Soy protein concentrate; DSMP: Dry skim milk powder; PA: Phytic acid. Values with different superscripts in same column are significantly different (p<0.05).

Table 4: Phytic acid content and iron and zinc dialysability in RUF.

with those in literature [25]. The PA/mineral molar ratios observed for soybean, maize, sorghum and the raw and extruded SMS were higher than those preferred for adequate mineral absorption [20,21] and thus indicate that all the grains are inhibitory for absorption of both iron and zinc. Moreover, bioavailability of iron in sorghum is also likely to be adversely affected by high content of polyphenols (2700 mg/100 g) [26]. Extruded SMS had significantly higher amounts of iron (15.03 mg/100 g; p<0.001) but lower amount of PA (0.897 g/100 g; p<0.001) than raw SMS (4.27 mg/100 g and 1.09 g/100 g for iron and PA respectively) (Table 3). Iron contamination from equipment during milling has been reported and is attributed to frictional wear and tear of the moving mill parts [27]. Because SMS processing was done in equipment traditionally used for fortified blended foods, the process steps that involve friction such as extrusion and milling are likely to have led to iron contamination, with consequent elevation of iron content in extruded SMS. The contaminant iron however was not studied for solubility and its bioavailability was unknown. It has been reported that heat treatment degrades some PA and is likely to explain the lower content of PA in extruded SMS compared to that in raw SMS.

Iron content was significantly higher in foods made from SPC recipe (15.11 mg/100 g) than in those from DSMP (11.90 mg/100 g; p<0.001) (Table 4). The significant difference in iron content between the recipes is attributed to higher iron content in SPC (10.07 mg/100 g) than in DSMP (0.55 mg/100 g; Table 4). PA was also significantly higher (p<0.001) in foods from SPC recipe (0.84 g/100 g) than those from DSMP recipe (0.57 g/100 g; Table 4). On the contrary, the PA/Fe molar ratios for the two recipes were statistically similar indicating that the high content of PA in SPC seems to be offset by the similarly higher content of iron. On the basis of PA/Fe ratios, it would be expected that iron bioavailability in both recipes would be similar. However, bioavailability of iron was significantly higher in foods made from DSMP (4.85%) than those made from SPC (2.79%; p<0.001; Table 4). The significantly lower iron bioavailability from SPC recipe can be explained in two ways. Firstly, because the iron in SPC and extruded SMS is non-heme of unknown form including contaminant iron, it appears that a large proportion did not enter the common iron pool for absorption. There is evidence that many forms of iron do not enter the iron common pool and remain unavailable for absorption. Secondly, soy protein [28] and milk proteins [29,30] have been associated with lower iron bioavailability even in the absence of phytic acid, but from these results, it appears that soy protein in SPC played a more significant role than milk proteins in DSMP to and further reduced iron bioavailability in SPC recipe. For adequate bioavailability of iron, the PA/Fe molar ratio should be <1 [30]. None of the recipes was of adequate iron bioavailability hence methods to improve iron bioavailability in the foods are needed.

Zinc bioavailability was significantly higher in foods from DSMP recipe (8.69; p<0.001) than in foods of SPC recipe (3.61%; Table 4). This was expected because DSMP foods had a lower PA/Zn molar ratio (9.67) than SPC foods (14.86; Table 4). SPC and DSMP had comparable zinc contents (3.29 and 3.46 mg/100 g respectively). However, the significantly higher PA content in SPC gives a much higher PA/Zn molar ratio in SPC foods than in DSMP foods, predicting higher inhibition in SPC foods. However, WHO [31] classifies bioavailability of zinc as low, moderate or high if absorption is 15%, 30-35% and 50-55%, respectively. Both DSMP and SPC foods therefore fall in the same category of low bioavailability and thus do not differ much nutritionally. However, due to their slightly higher zinc bioavailability, foods from DSMP recipe may offer better zinc complement than those from SPC recipe. The difference between 8.7% (for DSMP recipe) and 3.6% (for SPC recipe) is important in aiming to reach the Recommended Dietary Allowance.

Addition of phytase resulted in a significant reduction of phytic acid in high moisture foods (p<0.001). However, there was no effect of incubation time and temperature on PA content in foods without added water (Table 5). Phytic acid content of food with added water reduced from 0.68 to 0.22 g/100 g in the first one hour of incubation (Table 5). The reduction in PA content has significant potential to enhance iron and zinc bioavailability in the foods. As an enzyme, phytase requires moist environment to be active. Because one optimum of the enzyme is at pH 2, the enzyme can be utilized to degrade phytic acid in RUF in the gut. Such application of the enzyme in other inhibitory meals has been successfully studied [18]. However, retention of enzyme activity in the food matrix up to the end of shelf life needs to be established.

Conclusions and Recommendations

Replacing DSMP with SPC increased content of PA thereby decreasing the bioavailability of iron and zinc in SMS RUF. Adding phytase to the RUF resulted in degradation of PA and improved bioavailability of the minerals. However, phytase requires high moisture to be active hence its activity would predominantly take place in the gut at a second optimum pH of around 2. In settings where SPC is more affordable than MP, replacing MP with SPC in SMS RUF has potential of reducing cost and addition of phytase may bring the benefit of improving iron and zinc bioavailability in the foods. Research on retention of phytase activity in SMS RUF during storage is recommended and to verify improvement of bioavailability of the minerals in human absorption studies. Aspects of formulation would need to be appropriately considered to achieve target micronutrient and amino acid profiles since the two ingredients differ in nutritional composition.

Food aliquot	Food characteristics	Initial PA content (g/100 g food)	Residual PA content (g/100 g)								
			Incubated at 35°C		Incubated at 25°C			Incubated at 35°C		
			1 hr	2 hr	1 day	3 days	14 days	1 day	3 days	14 days
1	pH 5.9, no water	0.68 ± 0.03[c]	n.t.	n.t.	0.67 ± 0.02[b,c]	0.61 ± 0.01[a,b]	0.64 ± 0.00[b]	0.59 ± 0.04[a]	0.63 ± 0.01[a,b]	0.61 ± 0.00[a,b]
2	pH 5.1, no water	0.68 ± 0.03[d]	0.61 ± 0.01[b]	0.62 ± 2.81[b]	0.55 ± 0.01[a]	0.64 ± 0.05[b,c]	0.71 ± 0.02	0.67 ± 0.02[c,d]	0.60 ± 0.02[b]	0.58 ± 0.01[a]
3	pH 5.1, + water	0.68 ± 0.03[b]	0.22 ± 0.00[a]	0.19 ± 2.88[a]	n.t.	n.t.	n.t.	n.t.	n.t.	n.t.

PA: phytic acid; n.t.: not tested. Values with different superscripts in same row are significantly different (p<0.05).

Table 5: Initial and residual contents of phytic acid in foods incubated with phytase (values are mean ± SD; n = 3)

Acknowledgements

Contributions from Valid Nutrition Kenya staff Immaculate Wafula and Ben Otieno in manufacturing samples are appreciated. Technical support from Nicolai Petry and Jasmin Forman both of Human Nutrition Laboratory, ETH, Zurich in sample analysis are also appreciated.

Funding

This work was made possible through funding from Irish Aid and ETH Zurich.

References

1. Krebs NF (2007) Food choices to meet nutritional needs of breast-fed infants and toddlers on mixed diets. Journal of Nutrition 137: 511-517.

2. Bwibo NO, Neumann CG (2003) The need for animal source foods by Kenyan children. Journal of Nutrition 133: 3936-3940.

3. Singh M (2004) Role of micronutrients for physical growth and mental development. Indian Journal of Pediatrics 71: 59-62.

4. Diaz JR, De lasCagigas A, Rodriguez R (2003) Micronutrient deficiencies in developing and affluent countries. European Journal of Clinical Nutrition 57: 70-72.

5. Hunt JR (2003) Bioavailability of iron, zinc, and other trace minerals from vegetarian diets. American Journal of Clinical Nutrition 78: 633-639.

6. Benoist B, McLean E, Egll I, Cogswell M (2008) Worldwide prevalence of anaemia 1993-2005: WHO global database on anaemia. World Health Organization.

7. Stevens GA, Finucane MM, De-Regil LM, Paciorek CJ, Flaxman SR, et al. (2013) Global, regional, and national trends in haemoglobin concentration and prevalence of total and severe anaemia in children and pregnant and non-pregnant women for 1995-2011: a systematic analysis of population-representative data. The Lancet Global Health 1: 16-25.

8. Cole CR, Grant FK, Swaby-Ellis ED, Smith JL, Jacques A, et al. (2010) Zinc and iron deficiency and their interrelations in low-income African American and Hispanic children in Atlanta. American Journal of Clinical Nutrition 91: 1027-1034.

9. Brown K, Dewey K, Allen L (1998) Complementary feeding of young children in developing countries: a review of current scientific knowledge. World Health Organization, Geneva, Switzerland.

10. Owino V (2010) Why lipid-based ready to use foods (RUF) must be key components of strategies to manage acute malnutrition in resource poor settings. African Journal of Food, Agriculture, Nutrition and Development 10: 1-6.

11. Manary MJ (2005) Local production and provision of ready-to-use therapeutic food for the treatment of severe childhood malnutrition. Informal Consultation, Geneva, Switzerland.

12. World Food Programme (2010) WFP Nutrition Improvement Approach. Informal Consultation, Rome, Italy.

13. Owino VO, Irena AH, Dibari F, Collins S (2012) Development and acceptability of a novel milk-free soybean-maize-sorghum ready-to-use therapeutic food (SMS-RUTF) based on industrial extrusion cooking process. Maternal and Child Nutrition10: 126-134.

14. Irena AH, Bahwere P, Owino VO, Diop EI, Bachmann MO, et al. (2013) Comparison of the effectiveness of a milk-free soy-maize-sorghum-based ready-to-use therapeutic food to standard ready-to-use therapeutic food with 25% milk in nutrition management of severely acutely malnourished Zambian children: an equivalence non-blinded cluster randomised controlled trial. Maternal and Child Nutrition.

15. Dibari F, Bahwere P, Huerga H, Irena AH, Owino V, et al. (2012) Development of a cross-over randomized trial method to determine the acceptability and safety of novel ready-to-use therapeutic foods. Nutr 29: 107-112.

16. Owino VO, Bahwere P, Bisimwa G, Mwangi CM, Collins S, et al. (2011) Breast milk intake of 9-. Am J 10 month old rural infants given a ready to use complementary food (RUCF) in South Kivu, Democratic Republic of Congo Clin Nutr 93(6): 1300-1304.

17. Bisimwa G, Owino VO, Bahwere P, Dramaix M, Donnen P, et al. (2012) Randomized controlled trial of the effectiveness of a soybean-maize-sorghum-based ready-to-use complementary food paste on infant growth in South Kivu, Democratic Republic of Congo. American Journal of Clinical Nutrition 95: 1157-1164.

18. Troesch B, Egli I, Zeder C, Hurrell RF, Pee S, et al. (2009) Optimization of a phytase-containing micronutrient powder with low amounts of highly bioavailable iron for in-home fortification of complementary foods. American Journal of Clinical Nutrition 89: 539-544.

19. Hurrell RF, Reddy MB, Juillerat MA, Cook JD (2003) Degradation of phytic acid in cereal porridges improves iron absorption by human subjects. American Journal of Clinical Nutrition 77: 1213-1219.

20. Hallberg L, Brune M, Rossander L (1989) Iron absorption in man: Ascorbic acid and dose-dependent inhibition by phytate. American Journal of Clinical Nutrition 49: 140-144.

21. Turnlund JR, King JC, Keyes WR, Gong B, Michel MC, et al. (1984) A stable isotope study of zinc absorption in young men: Effects on phytate and R-cellulose. American Journal of Clinical Nutrition 40: 1071-1077.

22. Miller DD, Schricker BR, Rasmussen RR, Campen DV (1981) An in vitro method for estimation of iron availability from meals. American Journal of Clinical Nutrition 34: 2248-2256.

23. Makower RU (1970) Extraction and determination of phytic acid in beans (Phaseolus vulgaris). Cereal Chemistry 47: 288-295.

24. Vanveldhoven PP, Mannaerts GP (1987) Inorganic and Organic Phosphate Measurements in the Nanomolar Range. Analytical Biochemistry 161: 45-48.

25. U.S. Department of Agriculture, Agricultural Research Service (2012). USDA National Nutrient Database for Standard Reference, Release 25. Nutrient Data Laboratory Hom.

26. Cercamondi CI, Egli IM, Zeder C, Hurrell RF (2013) Sodium iron EDTA and ascorbic acid, but not polyphenol oxidase treatment, counteract the strong inhibitory effect of polyphenols from brown sorghum on the absorption of fortification iron in young women. Br J Nutr 111: 1-9.

27. Greffeuille V, Kayodé AP, Icard-Vernière C, Gnimadi M, Rochette I, et al. (2011) Changes in iron, zinc and chelating agents during traditional African processing of maize: Effect of iron contamination on bioaccessibility. Food Chemistry 126: 1800-1807.

28. Hurrell RF, Juillerat MA, Reddy MB, Lynch SR, Dassenko SA, et al. (1992) Soy protein, phytate, and iron absorption in humans. American Journal of Clinical Nutrition 56: 573-578.

29. Cook JD, Monsen ER (1976) Food iron absorption in human subjects. III. Comparison of the effect of animal proteins on nonheme iron absorption. American Journal of Clinical Nutrition 29: 859-867.

30. Hurrel RF, Lynch SR, Trinidad TP, Dassenko SA, Cook JD, et al. (1989) Iron absorption in humans as influenced by bovine milk proteins. American Journal of Clinical Nutrition 49: 546-552.

31. World Health Organization (1996) Trace elements in Human Nutrition and Health. Geneva, Switzerland.

28

Temperature Monitoring in the Transportation of Meat Products

Tomasz Jakubowski*

Associate Professor, University of Agriculture in Krakow, Poland

Abstract

This paper analyses air temperature changes inside a cooling chamber of a vehicle carrying meat products, depending on the monitoring system in terms of maintaining the cold chain. The research was carried out in the years 2014-2015 in Małopolska region. The object of the research was a food business whose main specialty is the purchase, slaughter of pigs and butchering pig carcasses, and then transport of the meat products. The research was focused on the product transportation as the critical process determining health safety of the foodstuffs being carried. Performance of the air temperature monitoring system was analysed during three rides inside cooling chambers of three vehicles. Air temperature monitoring inside the cargo hold was performed using certified, wireless and autonomous meters with data loggers at measuring steps of 60 s. Inside the cooling chambers of vehicles systems for monitoring temperature changes, based on thermocouple sensors (K and J type) and the Pt-1000 thermistors, were installed. One of the monitoring systems was provided with 4 temperature sensors located in the cooling chamber (middle of chamber, air inlet and outlet from the evaporator and product temperature), and 4 bistable signals (opening the side and rear doors, defreezing and operation of the cooling unit). According to the results of the tests, in none of the experimental combinations analysed (taking into account the vehicle, route and monitoring system) the recorded temperatures were found to have caused interruption to the cold chain. A significant difference was found in the values of measured temperatures recorded by the monitoring system with the Pt-1000 thermistor, in relation to the monitoring systems based on K and J type thermocouples.

Keywords: Temperature; Meat; Transport; Cold chain HACCP

Nomenclature

T_z: Temperature outside the insulated body (°C)

t_p : Time of ride (min.)

R_L: Number of unloading operations

S_d: Standard deviation

W_z: Coefficient of variation (%)

min., max., av.: Respectively, minimum, maximum and average value

Introduction

Meat products belong to the group of perishable foodstuffs. Meat is prone to natural, continuous and irreversible bio-physiochemical changes (high water content, the presence of protein, carbohydrates and fat promote the processes of oxidation and rancidity) [1,2]. To ensure the health safety of meat products, it is required to use suitable raw material, appropriately selected cooling methods (freezing), the storage and distribution conditions and a continuous monitoring and control, in accordance with the applicable quality assurance systems [3,4]. The regulations (EC) No 853/2004 of 29 April 2004 laying down hygiene rules for food of animal origin (Journal of Laws L139 / 55 dated 30 April 2004) requires that the transport of this product be carried out according to the following criteria: fresh meat (red)-below +7°C, poultry (white)-below +4°C, meat offal-below +3°C, minced meat products-below +2°C. The above legislation allows for the so-called "Limited periods", in which the temperature does not have to be controlled, and which are associated with the necessary logistic processes (shipment preparation, loading and unloading of goods) [5,6]. The above-mentioned products are transported in various types of refrigerated vehicles. Transportation in deep-freezer vehicles (of frozen products) is based on similar rules to the requirements for the transport of chilled products, however, the temperature should not exceed -18°C (with acceptable short-term upward fluctuations of not more than +3°C). In both forms of transportation discussed, it is required to document the temperature conditions during the shipping process [7]. The process of transport, especially of meat products, is subject to a special surveillance system in accordance with the applicable system for hazard analysis and determination of critical control points (HACCP) (Sperber). Chilled or deep-frozen foodstuffs are subject to the procedure of maintaining unchanged conditions in which they have to be kept (cold chain). This means that from production, transport and distribution until the consumption by consumers, both the chilled and frozen products should be stored at an appropriate temperature [8]. Exposing the product to conditions outside the range of recommended temperatures, in either of the links of the cold chain, can result in lowering the quality, as well as changes which may impact the health security. In the cold chain, there are many critical points at which its continuity can be interrupted [9]. For this reason, mechanical or electronic temperature monitoring systems are in use. To meet the sanitary requirements for food safety, entrepreneurs carrying animal products, are obliged to provide this service using appropriate means of transport. The carriers, for carriage of perishable foodstuffs, use such means of transport as insulated truck bodies (with limited exchange of heat between the environment and the interior), cold storage (using e.g. ice with the ability to keep the temperature low), as well as refrigerator and freezer trucks (equipped with special refrigerating units). In road transport, the most common cooling methods include mechanical cooling, cooling with liquid nitrogen and with eutectic plates [10]. As the practice shows [11] the most common reasons for exceeding the temperature (and humidity) limits of the carried foodstuffs include

---

*Corresonding author: Tomasz Jakubowski, Associate Professor, University of Agriculture in Krakow, Poland, E-mail: Tomasz.Jakubowski@ur.krakow.pl

the human factor, technical defects of means of transport, damage of components in the systems for monitoring critical parameters. The manufacturers of temperature monitoring systems dedicated to refrigerated transport offer ready solutions based on various types of sensors to measure this parameter. Also the monitoring organization can be different. The systems offered usually have the certificates and approvals required by law. The use of various types of sensors (with different sensitivity, linearity, tolerance, resistance to external conditions etc.) may affect the measurement values obtained, which can have an effect on maintaining the cold chain (e.g. switching on a chiller early enough) [12]. This paper analyses air temperature changes inside a cooling chamber of a vehicle carrying meat products, depending on the monitoring system in terms of maintaining the cold chain.

Material and Research Methods

The research was carried out in the years 2014-2015 in Małopolska region. The object of the research was a food business whose main specialty is the purchase, slaughter of pigs and butchering pig carcasses, and then transport of the meat products. The research was focused on the product transportation as the critical process determining health safety of the foodstuffs being carried. The company is equipped with motor vehicles for the transport of meat products (in accordance with the requirements of the ATP Agreement on the International Carriage of Perishable Foodstuffs and on the special equipment used for such carriage). The transport vehicles have common features in line with current standards and regulations. They are equipped with specialized cargo holds adapted to transport of food and designed to allow the maintenance of proper hygienic conditions along with the refrigeration system, and the systems for monitoring thermal conditions [12-14]. Vehicle operators were trained on the applicable HACCP standards (also for monitoring thermal conditions of cargo holds during transport and keeping appropriate documentation in this regard). Performance of the air temperature monitoring system was analyzed during three rides inside cooling chambers of three vehicles. A single ride lasted from 240 to 270 minutes, including the time from loading and closing the cooling chamber door until unloading the last batch of product. The objects of research (refrigerated trucks) differed as to the payload, cooling method of the cargo hold, the number of breaks for unloading, ride time and distance covered (Table 1) The heat transfer coefficient of the insulated body walls was 0.4-0.6 $W \cdot (m^2 \cdot °K)^{-1}$ (according to the ATP classification of the heat transfer coefficient in the range 0.4-0-7 $W \cdot (m^2 \cdot °K)^{-1}$ is the so called normal insulation) [15]. Such differentiation was deliberate and purported to reflect the actual conditions in transit of foodstuffs. Air temperature monitoring inside the cargo hold was performed using certified, wireless and autonomous (powered with lithium battery) meters with data loggers. Measurement step of 60 s. was assumed. Inside the cooling chambers of the vehicles, 3 systems for temperature monitoring were installed: system I-K type thermocouple, system II-J type thermocouple (systems I and II based on a thermocouple sensor) and system III-based on Pt-1000 resistance sensor. System III comprised a thermo hygrometer based on a Pt-1000 termistor, with 4 temperature sensors located in the cooling chamber (in the middle-sensor 1, evaporator air inlet and outlet-sensors 2 and 3 and product temperature-sensor 4). The thermo hygrometer was additionally provided with 4 bistable signals (opening the rear and side doors, defreezing and operation of the refrigeration unit). Processing characteristics and tolerance limits of thermocouples are described in the standard PN-EN 60584-1, and those of resistance sensors in PN-EN 60751. Temperature sensors of the system I and II were located in the middle part of the body (near the cz1 sensor of the system III). Thermocouple is a component of electrical circuit that uses See

beck effect (generating of electromotive force in a circuit containing two metals when their interfaces are in different temperatures) at the interface of two different materials. Thermocouple type K (sheet 06, NiCr-NiAl) in which the relationship between the emf and temperature is nearly linear and its sensitivity is 41 $\mu V \cdot °C^{-1}$, is used in the temperature range from -200 to +1,200°C. Thermocouple type J (sheet 04, Fe-CuNi) is used in the temperature range from -40 to +750°C, and its sensitivity is 55 $\mu V \cdot °C^{-1}$.

Pt-1000 is a resistance sensor reacting to temperature variations by changing resistance of an integrated resistor (with temperature, increases the vibration amplitude of atomic nuclei and the probability of collision of free electrons and ions which, due the inhibition of electron motion, causes the resistance increase). The microprocessor built in the thermo hygrometer calculated the current measurement results (knowledge of non-linear characteristics of sensors and digital calibration data saved during calibration of the instrument was used). The meter can create a histogram containing statistical information on the occurrence of measurement values recorded in the ranges specified. As a standard, the measuring range for temperature was divided to 63 parts, each with a span of 2°C. It was possible to set limit values of the parameter measured and to indicate its exceeding during the recording session. The monitoring system III was characterized by the following technical specifications: measuring range from -30 to +80°C, measurement uncertainty +/- 0.2°C, resolution 0.1°C. Such system is dedicated to refrigerated transport vehicles with individual refrigeration unit which, at an average ambient temperature +30°C, makes it possible to lower and hold temperature inside the body (cooling chamber) from +12 to 0°C inclusive-class A [10,15]. Recorder programming, reading, presentation and printing of the data saved was performed by means of a PC with installed dedicated software and a reader connected to the USB port. Data transmission between the recorder and USB reader connected to the PC was wireless in both systems with the use of an optical infra-red interface. The analysis of the temperatures measured included differences between the data sequences recorded. The differences were determined by the value of basic statistical measures and variance analysis. During each ride of a vehicle, the ambient temperature was also monitored immediately outside the insulated body. The test results were analyzed using STATISTICA 10 software at the significance level assumed α=0.05. In carrying out a preliminary analysis of the measuring series obtained during the tests (air temperature inside the refrigerated chamber), the λ^2 test was used to check the equality of groups. The distribution normality in tests was determined with Kolmogorov-Smirnov test, and homogeneity of variance with Brown-Forsythe test. Significance of differences was studied using the variance analysis with F-Snedecor test. For statistically significant quality predictors, multiple comparisons were performed to distinguish homogeneous variable groups (Tukey test).

Vehicle designation and description		Route description			Route designation
		T_z (°C)	t_p (min)	(R_L)	
P1	Refrigerated van, volume 8.9 m³, compressor refrigeration unit, heat transfer coefficient 0.6 $W \cdot (m^2 \cdot °K)^{-1}$	24.1	241	4	T1
		2.4	255	5	T2
		11.0	253	4	T3
P2	Refrigerated container truck, volume 13.8 m³, absorption refrigeration unit, heat transfer coefficient 0.6 $W \cdot (m^2 \cdot °K)^{-1}$	23.1	244	5	T1
		3.5	270	5	T2
		9.6	236	4	T3
P3	Hook refrigerator truck, volume 53.5 m³, compressor refrigeration unit, heat transfer coefficient 0.4 $W \cdot (m^2 \cdot °K)^{-1}$	19.0	240	1	T1
		3.3	246	1	T2
		12.2	241	2	T3

Table 1: Description of the transport vehicles and routes covered.

Results and Discussion

The calculated values λ^2 of Kolmogorov-Smirnov and Brown-Forsythe tests were insignificant. Due to the fact that the test probability value p=0.057 for the statistics λ^2 was close to the limit value of the significance level assumed, for multiple comparisons Spjotvoll-Stoline procedure (generalisation of the HSD Tukey tests for samples of different sizes) was applied. Analysis of basic statistics (tab) characterising the relationships between the vehicle used and monitoring system and the temperature value measured inside the refrigerated body clearly points that in neither of the experimental combination was the average temperature exceeded during transport of meat. 10 to 15 minute interruptions (Figure 1) of the critical limit 7°C occurred (indicated in Table 2 as the maximum value). Such short-term temperature exceeding is acceptable, which is caused mainly by opening the cooling chamber doors and unloading [5]. The values of variability factor (W_z, the ratio of standard deviation and average value expressed in %) indicate that the variability within the measured temperature range was: 15.2% for system I, 13.7% for system II and 17.4% for system III. Such a great variability for system III was due to the fact that the system was based on measuring sensors placed in different points of the cooling chamber, thus recording extreme temperature values. The value of the variability factor within the measured temperature range, based exclusively on the transport vehicle, was: 15.9%-vehicle 1, 15.2%-vehicle 2 and 14.8% for vehicle 3 (maximum difference in the variability exceeded slightly 1%). For vehicle 3, the result obtained is justified as this was a hook refrigerator truck with large volume of the cooling chamber, which is one the reasons for reduced heat exchange (inertia) during transit and single unloading.

The variance analysis result presented in Table 3, relating to the differences in air temperature measurements inside the cooling chamber indicates clearly that only the monitoring systems used in the experiment influenced significantly the parameter being measured. Therefore, it has to be concluded that the means of transport (equipped with various cooling systems) assumed in the research and the following rides (with different distances, duration, ambient temperature and number of unloading) had no effect on the temperature changes inside the refrigerated body. The result of Spjotvoll-Stoline procedure (Table 4) proved existence of 2 homogeneous groups whose distribution indicates that temperature measurement values obtained by system 3 were different than those of systems 1 and 2. In this case, it can be concluded that the monitoring system based on the Pt-1000 thermistor provides significantly different temperature readings (about 8% greater) in relation to the systems based on the K and J thermocouples. To verify the result obtained, at the following stages, only the temperature data acquired by the monitoring system based on the Pt-1000 thermistor was analyzed. As already mentioned, the monitoring system using the Pt-1000 thermistor was based on 4 temperature sensors located in the cooling chamber space. Variance analysis performed in this respect, taking also into account the spacing of sensors, proved a significant effect of both, the route of the ride and location of sensors on the temperature values measured (Table 5). Multiple comparisons of differences in air temperatures obtained with the Pt-1000 thermo hygrometer inside the cooling chamber, depending on the route and sensor location (Tables 6 and 7) indicate that the differences achieved are determined mainly by the location of measuring sensors and also by the route covered by the vehicle (frequency of unloading, ride time and ambient temperature). Such a result of the analysis of variance and arrangement of homogeneous groups would indicate that not without significance is the type of temperature meter installed in the refrigerated vehicle. To verify this conclusion, the relationships

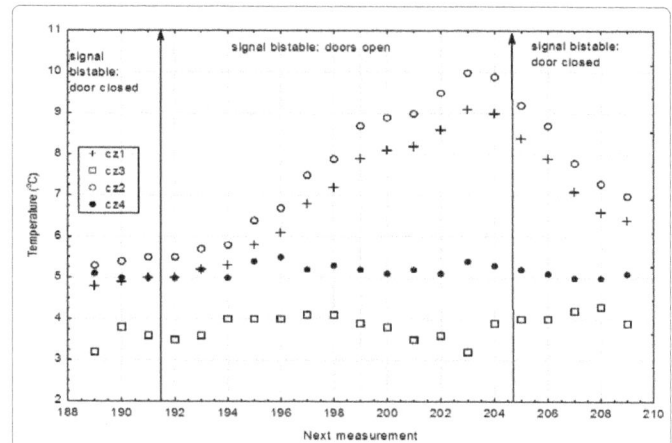

Figure 1: Temperature curve inside the refrigerated vehicel measured by sensores (cz1, cz2, cz3, cz4 - sesonrs 1-4 respectively) of the monitoring system equipped with Pt-1000 thermistor (selected example).

Vehicle	Route	System	T_w (°C) min	T_w (°C) max	T_w (°C) av	Sd (°C)	Wz (%)
1	T1	I	4,1	8,1	6,1	0,9	14,8
		II	3,9	7,9	5,9	0,9	15,3
		III	3,4	8,4	5,9	1,1	18,8
	T2	I	4,2	8,8	6,5	1,0	15,4
		II	4,2	8,1	6,2	0,9	14,6
		III	3,5	7,9	5,7	1,0	17,5
	T3	I	3,8	7,7	5,8	0,9	15,1
		II	4,1	7,6	5,9	0,8	13,3
		III	3,4	8,4	5,9	1,1	18,6
2	T1	I	4,4	8,6	6,5	0,9	13,8
		II	4,1	7,9	6,0	0,8	13,3
		III	3,4	8,8	6,1	1,2	19,7
	T2	I	3,7	7,1	5,4	0,8	14,8
		II	3,9	8,1	6,0	0,9	15,0
		III	3,5	7,9	5,7	1,0	17,5
	T3	I	4,2	7,7	6,0	0,8	13,4
		II	4,2	7,4	5,8	0,7	12,1
		III	3,5	8,2	5,9	1,0	17,1
3	T1	I	4,1	7,9	6,0	0,8	14,1
		II	4,4	7,8	6,1	0,8	12,4
		III	4,0	8,5	6,3	1,0	16,0
	T2	I	3,5	8,2	5,9	1,0	17,1
		II	3,6	7,8	5,7	0,9	15,8
		III	3,8	8,2	6,0	1,0	16,7
	T3	I	4,2	8,1	6,2	0,9	14,6
		II	4,4	7,6	6,0	0,7	11,7
		III	4,0	7,9	6,0	0,9	15,1

Table 2: Basic statistics characterizing the relationships between the vehicle used, monitoring system and the measured temperature value inside the refrigerated body.

Independent variables	Number degrees of freedom	Value of F-Snedecor test	Value of test probabilisty
Free word	3	26568,63	0,0000
vehicle	6	0,02	0,9997
Route	6	0,13	0,9931
System	6	21,91	0,0000

Table 3: Variance analysis result - differences in air temperature measurements inside the cooling chamber depending on the vehicle, covered distance and monitoring system.

System	Temperature(°C)	Homogenous groups	
		1	2
System II	5,66	****	
System I	5,75	****	
System III	6,24		****

Table 4: Multiple comparisons for the differences in air temperature measurements inside the cooling chamber depending on the vehicle, covered distance and monitoring system.

Independent variables	Number degrees of freedom	Value of F-Snedecor test	Value of test Probabilisty
Free word	1	14668,76	0,000000
Vehicle	2	3,83	0,4215
Route	2	33,46	0,000000
Sensor	2	31,31	0,000000

Table 5: Variance analysis result - differences in air temperature measurements made with the thermo hygrometer Pt-1000 inside the cooling chamber depending on the vehicle and covered distance.

Route	Temperature (°C)	Homogenous groups		
		1	2	3
route 1	4,88	****		
route 2	5,33		****	
route 3	5,94			****

Table 6: Multiple comparisons for differences in air temperature measurements made with the thermo hygrometer Pt-1000 inside the cooling chamber depending on the ride route.

Sensor	Temperature (°C)	Homogenous groups			
		1	2	3	4
sensor 3	4,85	****			
sensor 4	4,93		****		
sensor 1	5,3			****	
sensor 2	5,91				****

Table 7: Multiple comparisons for differences in air temperature measurements made with the thermo hygrometer Pt-1000 inside the cooling chamber depending on the sensor location.

Independent variables	Number degrees of freedom	Value of F-Snedecor test	Value of test probabilisty
Free word	3	22381,02	0,0000
vehicle	6	1,12	0,3674
Route	6	1,29	0,3584
System (I, II, III cz1)	6	1,71	0,2954

Table 8: Variance analysis result - differences in air temperature measurements in the middle of the cooling chamber depending on the vehicle, covered distance and monitoring system (I, II, III cz1).

between readings of the sensors of three monitoring systems located in close vicinity would have to be studied either. Bearing in mind the above, the significances of differences in temperature readings from the monitoring systems I and II and the sensor 1 of the system III (middle part of the refrigerated body) were analyzed. The analysis of variance performed (Table 8) showed no significant differences between the grouping variables assumed in the experiment (insignificant values of F-Snedecor test). Therefore, there are no grounds to conclude that the monitoring systems used in the experiment could deliver different temperature readings and thereby disturb the function of the cold chain. The tests performed allow a conclusion that for an agricultural and food business transporting the foodstuffs, it is a less important issue to choose between the sensor with a thermocouple and thermistor, but more important is the number of sensors, their location

and provision of the monitoring system with bistable indication system giving the information on the cold storage functioning (monitoring organization). The readings of the sensor no. 4 (system III), responsible for temperature measurement of the cargo being carried, are important for the safety of food and control of critical points. Temperature of the meat transported was around 5°C, even during unloading [16-18].

Conclusions

i. In none of the experimental combinations analyzed (taking into account the vehicle, route and monitoring system) the recorded temperatures were found to have caused interruption to the cold chain.

ii. In neither of the combinations studied did the temperature of cargo (meat) transported exceed the value 7°C.

iii. The means of transport used and distances covered (unloading frequency, ride time and ambient temperature) were found not to have negative impact on the temperature values measured inside the refrigerated bodies.

iv. A significant difference was found in the values of measured temperatures recorded by the monitoring system with the Pt-1000 thermistor, in relation to the monitoring systems based on K and J type thermocouples.

References

1. Ziembińska A, Krasnowska G (2007) Ensuring safety of health in trade inks wild game. Technologia. Ziembinska 1: 16-25.

2. Zhou GH, Xu XL, Liu Y (2010) Preservation technologies for fresh meat (a review). Meat Science 86: 119-128.

3. Domaradzki P, Skałecki P, Florek M, Litwińczuk A (2011) Impact on storage-freezing Physical and chemical properties BEEF Vacuum-packed. Nauka. Technologia. Jakość 4: 117-126.

4. Cegielska-Radziejewska R, Kijowski J, Nowak E, Zabielski J (2007) Effect of tempeture on the dynamics of changes number bacteria in selected sausages stored under wholesale and retail trade. Nauka. Technologia. Jakość 4: 76-88.

5. Choroszy K, Tereszkiewicz K (2013) Zarządzanie higieną i jakością mięsa oraz jego przetworów. Modern Management Review 20: 9-25.

6. Danyluk B, Pyrcz J (2012) Health safety of meat and meat products. Gospodarka Mięsna 1: 12-14.

7. Śliwczyński B (2008) The system of traceability in the supply chain - a guarantee of safety, quality and fast response. Przem. Spoż 7: 2-8.

8. Czarniecka-Skubina E, Nowak D (2012) System tracking and tracing food as tool ensure consumer safety. Żywność. Nauka. Technologia. Jakość 5: 20-36.

9. Bauman HE (1995) The origin and concept of HACCP. Advances in Meat Research 10: 1-7.

10. Bieńczak K (2011) Providing security for food consumer in transport link of refrigeration chain. Maintenance and Reliability 1: 16-26.

11. Hsu C, Liu K (2011) A model for facilities planning for multi-temperature joint distribution system. Food Control 22: 1873-1882.

12. Piekarska J, Kondratowicz J (2011) Wykorzystanie technologii chłodniczej w transporcie żywności. Chłodnictwo 4: 44-47.

13. Piekarska J (2012) Transportation of food - a key link in the cold chain. Chłodnictwo 5: 18-22.

14. Schnotale J, Steindel M (2005) Trends in development of testing and certification of refrigerated trucks to transport food szybkopsującej up in the light of the findings of 60 sessions of the Working Group on Land Transport Committee of the United Nations. Refrigeration 3: 34-37.

15. Góral D, Kluza F, Kozłowicz K (2013) Balance of heat loss refrigeration trailers as the basis for proper selection chille. Technica Agraria 12: 21-30.

16. Gajana CS, Nkukwana TT, Marume U, Muchenje V (2013) Effects

of transportation time, distance, stocking density, temperature and lairage time on incidences of pale soft exudative (PSE) and the physico-chemical characteristics of pork. Meat Science 95: 520-525.

17. Konstantinos PK, Gougouli M (2015) Use of time temperature integrators in food safety management. Trends in Food Science and Technology 43: 236-244.

18. Jol S, Kassianenko A, Wszol K, Oggel J (2007) The cold chain, one link in Canadas food safety initiatives. Food Control 18: 713-715.

29

Nutritional Evaluation and Utilization of Pea Pod Powder for Preparation of Jaggery Biscuits

Meenakshi Garg*

Assistant Professor, Food Technology Department, University of Delhi, India

Abstract

Food industry generates massive waste, which is a concern not only to environment but also loss of valuable biomass. The foremost benefit of industrial waste is that it is available at zero cost and in immense quantities. Pea pods which otherwise are discarded in bins or at finest used for animal feed is exploited for its nutritional benefits in present study. Pea powder formulated was found rich in crude protein, fibre and ash with exceptional good amounts of iron. Composition of powder was 5% ash, 0.43% fat, 14.88% protein, 77.86% crude fibre, 61.43% total carbohydrates and 309.11Kcal energy content The concept of healthy eating is addressed in current analysis whereby, value added biscuits are devised by substituting pea pod powder in place of refined wheat flour at 10%, 20% and 30% level. Sensory evaluation conducted using a 9-point hedonic scale revealed 20% level as optimum level of incorporation. Water activity decreased slightly on storage. The biscuits are advantageous for people suffering from lifestyle diseases as these contain high amount of fibre and minerals. In this way peels can also be utilized for human consumption otherwise this important source of nutrients goes wasted.

Research Highlights

a. Food industry waste, pea pods are transformed to powder.

b. Pea pod fibre thus, obtained has high protein, fibre and minerals especially iron

c. Value added biscuits are prepared by substituting wheat flour with pea pod powder at various concentrations

d. Sensory and physical evaluation of biscuits revealed high acceptability

Keywords: Pea pod; Nutritional evaluation; Sensory evaluation; Biscuits; Jaggery

Introduction

Outsized amount of waste either solid or liquid in nature is being produced by food industry nowadays on regular basis. The unwanted artifacts pose not merely disposable concerns but furthermore massive loss of nutrients. Many studies indicate food-processing waste might have potential for reutilizing into raw useful products or by-products of higher value [1]. Fruit and vegetable waste has been established to stand good source of numerous polyphenols [2-7]. Vegetable waste yet remains unexploited source of phytonutrients in comparison to fruit waste. Carrot peels [8], outer leaves of cabbage have been used to produce antioxidant rich dietary fibre powder [9]. Arora and Camire [10] reported potato peels as dietary fibre supplement in muffins and cookies. Cauliflower by-products incorporation increased not only dietary fibre but also protein content in ready-to-eat snacks [11]. A lot of work has been done on antioxidant properties of tomato peel [12-14]. There are many other vegetable waste sources, which are yet to be discovered and utilized like pea pod. India is the second largest producer of green peas next to China [15]. Established upon supposition, 30% of the total pea weight is owing to pea pods (fresh weight basis). Thus based on India's yearly production of pea, more than 1 million ton of pea pod waste is generated annually alone in India, of which sizeable extent is discarded as waste [16]. Use of pea pod waste for cellulolytic enzyme production), feed for goat bucks [17], ruminants [18-21] has been reported. Literature reveals pea pods have high protein and dietary fibre [22].

Amid gigantic changes in working environments and economics, there is an equal comparable change in lifestyle. The demand for processed food has sky rocketed. Baked products have made their space in Indian kitchen since time immemorial. Baked foods especially biscuits are low in cost; can be stored for long time at room temperature. Consumer today wants to consume food, which can fit into their concept of holistic and healthy eating. Manufacturers and researchers around the globe are working to replace unhealthy ingredients and supplement them with therapeutic value [23-28]. Aim of the present work was to develop pea pod powder and showcase it as a potential source of nutrient. Pea pod powder produced thus is used as a prospective dietary fiber supplement in biscuits with sugar replaced by jaggery to further enhance the nutritive value.

Materials and Methods

Commercial refined wheat flour (brand Rajdhani Flour Mill Ltd) was purchased from local supermarket (Sanjay Store, Possangipur, New Delhi, India). Peas were purchased from local vegetable market (Azadpur mandi, New Delhi, India). Seeds were removed from pea pods. Pea pods were subjected to standard washing using Veg Fru Wash. Cleaned pods were dried in tray drier (Assembled at Agro Life Science Solutions Pvt Ltd., New Delhi, India) at 65°c for 5 hrs [20].

***Corresonding author:** Meenakshi Garg, Ph.D. Assistant Professor, Food Technology Department, University of Delhi, India
Email: meenagargbcas@gmail.com

Preparation of pea pod powder

Pods after drying were subjected to grinding (Model MXAC555, Panasonic India) to obtain fine pea pod powder.

Composition of biscuit flour

Blends of pea pod powder with refined wheat flours were prepared by combinations of 90:10, 90:20 and 90:30 of refined wheat flour with pea pod powder.

Preparation of biscuits

Biscuits were prepared from composite flours of wheat flour, pea pod powder and other ingredients such as shortenings, jaggery, salt and sodium bicarbonate (Table 1). Ingredients were weighed using an electric balance (Model ATX series, Shimadzu Corporation, Tokyo). Pea pod powder was incorporated at 10, 20 and 30% level in the standardized formula of biscuits by substituting refined flour. Dough was prepared and rolled. Square pieces of 0.6 cm thickness were cut and baked at 180°C for 15 min.

Chemical analysis of pea pod powder

Pea pod powder was analyzed for moisture, ash, protein and fat, according to the standard AACC methods [19]. Water activity was calculated using water activity meter (Model 4TE, Aqua Lab, Pullman, WA, USA).

Evaluation of biscuits

Physical properties of biscuits such as weight and thickness were determined by AACC, 1955. The biscuits were evaluated for sensory attributes by panel of 10 semi- trained judges using 9-point hedonic scale. The biscuits were evaluated for color, appearance, aroma, texture and taste by the judges and mean of the scores for all the sensory characteristic was expressed as overall acceptability.

Results and Discussion

All formulations showed good dough handling characteristics, sheeting and cutting. Average weight and thickness of standard biscuit was 10 gram and 0.6cm respectively. Figures 1-4 shows biscuits prepared from composite flours.

Chemical analysis of pea pod powder

Nutritional composition of pea pod powder is given in Table 2. Data presented shows that main component was carbohydrate. Crude protein content of pea pod powder was estimated to be 14.88%, which was found to be higher than reported by Aparicio et al. [22]. The results also showed that pea pod powder is rich in ash content which is 5% ash and higher than other vegetable waste like tomato peel fibre [12]. Our studies indicate low amount of crude fat making them healthier. Literature reports high amount of calcium in pea pods in comparison to by-products broad bean pod and okara [22]. The results indicate that pea pod powder is a good source of crude fibre with 7.86 % content. The calcium, iron and zinc content computed is 0.83 %, 0.83 % and

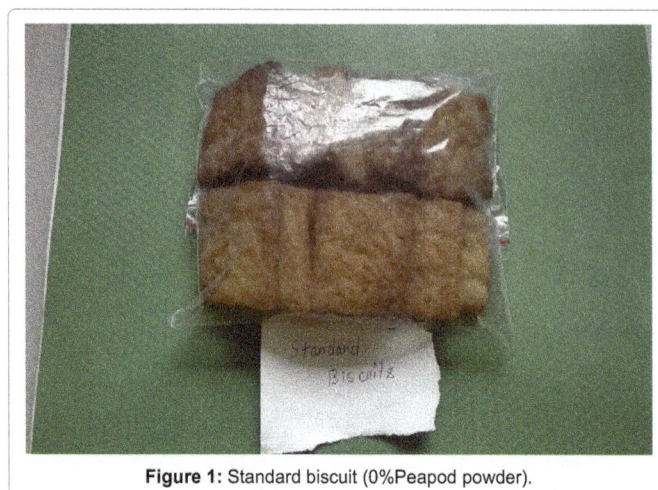

Figure 1: Standard biscuit (0%Peapod powder).

Figure 2: 10% Peapod powder Biscuits.

Figure 3: 20% Peapod powder Biscuits.

0.56%, respectively which is in agreement with previous findings by Aparicio et al. [22]. Analysis confirms that pea pod powder has very good amount of minerals particularly iron.

Physical evaluation of biscuits

Biscuits prepared using 0%, 10%, 20% and 30% of pea peapod powder were evaluated for various physical and sensory characteristics. The thickness was slightly affected with increase in level of pea pod powder (Table 3). The thickness of biscuits incorporated with pea pod

Ingredients	Quantity (%)
Composite flour	100
Jaggery	50
Fat	50
Baking powder	0.01
Milk	0.2

Table 1: Ingredients used in biscuit preparation.

powder decreased from standard but highest thickness was observed with 20% pea pod powder. Further decease observed at 30% level might be due to dilution of gluten, also reported by Ajila et al. [29].

Sensory evaluation of biscuits

Sensory evaluation of biscuits prepared by substituting wheat flour with pea pod powder is presented in Table 4. The overall acceptability declined at 30% level. No significant difference was observed in color up till 20% level of addition. The taste and flavor of biscuits improved on incorporation of pea pod powder at 20%. Considering the parameters: color, appearance, aroma, texture and taste, it could be deduced that 20% level of incorporation of peapod powder was ideal.

Influence of pea pod powder on water activity of biscuits

At 30°C, water activity of standard biscuits was 0.35, 10% pea pod biscuits had 0.38, 20% pea pod biscuits contained 0.42 and 30% biscuits had 0.46 water activity (Figure 5). This showed increased value of water activity on increasing the percentage of peapod powder. This also showed that product can be stored for long time at room temperature as it has low water activity.

Conclusions

This study illustrates the feasibility of producing powder rich in protein, minerals and dietary fibre from industry waste, pea pod. Pea

Figure 4: 30% Peapod powder Biscuits.

Nutrient	Content (g/100g dry matter)
Protein	14.88 ± 0.44
Fat	0.43 ± 0.03
Ash	5 ± 0.50
Total carbohydrates	61.43 ± 0.03
Crude fibre	7.86 ± 1.0
Calcium	0.83 ± 0.03
Iron	0.83 ± 0.08
Zinc	0.56 ± 0.06
Energy	309.1kcal

Table 2: Nutritional composition (g/100g) of pea pod po wder.

Pea pod powder level (%)	Weight (gm)	Thickness (cm)
0 (standard)	10 ± 1.00	0.6 ± 0.2
10	9.66 ± 0.67	0.55 ± 0.09
20	11.06 ± 0.45	0.62 ± 0.05
30	10.73 ± 0.55	0.61 ± 0.04

Table 3: Physical characteristics of biscuits.

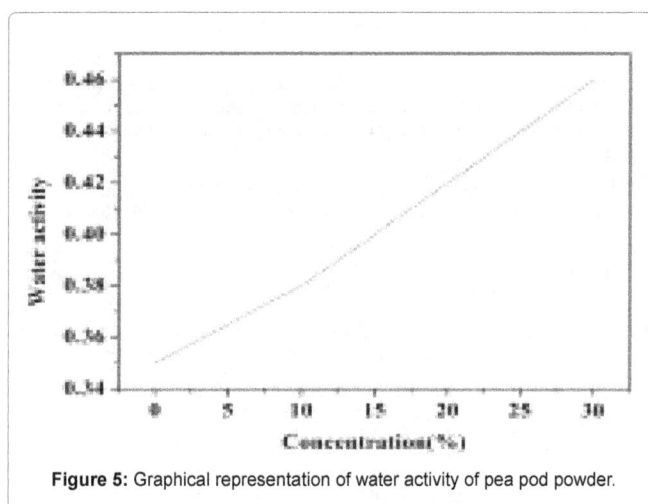

Figure 5: Graphical representation of water activity of pea pod powder.

Pea pod powder level (%)	Color	Appearance	Aroma	Texture	Taste	Overall acceptability
0	7.1 ± 0.8	7 ± 0.66	6.2 ± 0.63	8 ± 0.66	7.2 ± 0.63	7 ± 0.66
10	7.5 ± 1.08	7.6 ± 0.84	6.1 ± 0.73	7.2 ± 0.78	7.0 ± 0.66	7 ± 0.66
20	6.9 ± 0.73	8.1 ± 0.73	7.2 ± 0.63	7.0 ± 0.66	7.9 ± 0.56	7.9 ± 0.56
30	6.3 ± 0.48	6 ± 0.66	6.8 ± 0.63	6.1 ± 0.56	7.1 ± 0.56	6.1 ± 0.73

Table 4: Sensory characteristics of pea pod powder biscuits.

pod powder can be exploited to enhance the nutritive value of biscuits. Biscuits containing 20% pea pod powder were highly acceptable. Thus, it can be concluded that pea pods which otherwise are discarded in bins or used in animal feed can be used for value addition in various food.

References

1. Laufenberg G, Kunz B, Nystroem M (2003) Transformation of vegetable waste into value added products: (A) the upgrading concept; (B) practical implementations. Bioresource Technol 87: 167-198.

2. Peschel W, Rabaneda FS, Diekmann W, Plescher A, Gartzía I, et al. (2006) An industrial approach in the search of natural antioxidants from vegetable and fruit wastes. Food Chem 97: 137-150.

3. Shea NO, Arendt EK, Gallagher E (2012) Dietary fibre and phytochemical characteristics of fruit and vegetable by-products and their recent applications as novel ingredients in food products. Innov Food Sci Emerg Technol 16: 1 -10.

4. Stajčić S, Ćetković G, Brunet JC, Djilas S, Mandić A, et al. (2015) Tomato waste: Carotenoids content, antioxidant and cell growth activities. Food Chem 172: 225 -232.

5. Babbar N, Oberoi HS , Uppal DS, Patil RT (2011) Total phenolic content and antioxidant capacity of extracts obtained from six important fruit residues. Food Res In 44: 391 -396.

6. Massias A, Boisard S, Baccaunaud M, Calderon FL, Paternault PS, et al. (2015) Recovery of phenolics from apple peels using CO_2 + ethanol extraction: Kinetics and antioxidant activity of extracts. J Supercrit Fluids 98: 172 -182.

7. Amado IR, Franco D, Sánchez M, Zapata C, Vázquez JA, et al. (2014) Optimisation of antioxidant extraction from Solanum tuberosum potato peel waste by surface response methodology. Food Chem 165: 290 -299

8. Chantaro P, Devahastin S, Chiewchan N (2008) Production of antioxidant high dietary fiber powder from carrot peels. Food Sci Technol 41: 1987-1994.

9. Nilnakara S, Chiewchan N, Devahastin S (2009) Production of antioxidant dietary fibre powder from cabbage outer leaves. Food Bioprod Processc 87: 301 -307.

10. Arora A, Camire ME (1994) Performance of potato peels in muffins and cookies. Food Res Int 27: 15-22.

11. Stojceska V, Ainsworth P, Plunkett A, İbanoğlu E, İbanoğlu S, et al. (2008) Cauliflower by-products as a new source of dietary fibre, antioxidants and proteins in cereal based ready-to-eat expanded snacks. J Food Eng 87: 554-563.

12. González IN, Valverde VG, Alonso JG, Periago MJ (2011) Profile, functional and antioxidant properties of tomato peel fiber. Food Res Int 44: 1528-1535.

13. Rizk EM , El-Kady AT, El-Bialy AR (2014) Characterization of carotenoids (lyco-red) extracted from tomato peels and its uses as natural colorants and antioxidants of ice cream. Ann Agri Sci 59: 53-61.

14. Elbadrawy E, Sello A (2011) Evaluation of nutritional value and antioxidant activity of tomato peel extracts. Arab J Chem.

15. FAO Statistical yearbook (2013) Food and Agriculture Organization of the United Nations, Rome, Italy.

16. Sharma R, Rawat R, Bhogal RS, Oberoi HS (2015) Multi-component thermostable cellulolytic enzyme production by *Aspergillus niger* HN-1 using pea pod waste: Appraisal of hydrolytic potential with lignocellulosic biomass. Process Biochem 50: 696-704.

17. Wadhwa M, Kaushal S, Bakshi MPS (2006) Nutritive evaluation of vegetable wastes as complete feed for goat bucks. Small Ruminant Res 64: 279-284.

18. Wadhwa M, Bakshi MPS (2005) Vegetable wastes - a potential source of nutrients for ruminants. Indian J Anim Nutr 22: 70-76.

19. AACC (1995) Approved methods of the American Association of Cereal Chemists (9th eds). AACC International.

20. Garg, M, SharmaS, Varmani GS, Sadhu DS (2014) Drying kinetics of thin layer pea pods using tray drying. Int J Food Nutr 3: 61-66.

21. Gupta R, Chauhaun TR, Lall D (1993) Nutritional potential of vegetable waste products for ruminants. Bioresource Technol 44: 263-265.

22. Aparicio IM, Cuenca AR, Suárez MJV, Revilla MAZ, Sanz MDT, et al. (2010) Pea pod, broad bean pod and okara, potential sources of functional compounds. Food Sci Technol 43: 1467-1470.

23. Umesha SS, Manohar RS, Indiramma AR, Akshitha S, Naidu KA, et al. (2015) Enrichment of biscuits with microencapsulated omega-3 fatty acid (Alpha-linolenic acid) rich Garden cress (Lepidium sativum) seed oil: Physical, sensory and storage quality characteristics of biscuits. Food Sci Technol 62: 654-661.

24. Tyagi SK, Manikantan MR, Oberoi HS, Kaur G (2007) Effect of mustard flour incorporation on nutritional, textural and organoleptic characteristics of biscuits. J Food Eng 80: 1043-1050.

25. Kaur M, Sandhu KS, Arora AP, Sharma A (2015) Gluten free biscuits prepared from buckwheat flour by incorporation of various gums: Physicochemical and sensory properties. Food Sci Technol 62: 628-632.

26. Vujić L, Čepo DV, Dragojević IV (2015) Impact of dietetic tea biscuit formulation on starch digestibility and selected nutritional and sensory characteristics. Food Sci Technol 62: 647-653.

27. Hooda S, Jood S (2005) Organoleptic and nutritional evaluation of wheat biscuits supplemented with untreated and treated fenugreek flour. Food Chem 90: 427-435.

28. Reddy V, Urooj A, Kumar A (2005) Evaluation of antioxidant activity of some plant extracts and their application in biscuits. Food Chem 90: 317-321.

29. Ajila CM, Leelavathi K, Prasada Rao UJS (2008) Improvement of dietary fiber content and antioxidant properties in soft dough biscuits with the incorporation of mango peel powder.

30

Extruded Pet Food Development from Meat Byproducts using Extrusion Processing and its Quality Evaluation

Javeed Akhtar* and Mohd Ali Khan

Department of Post-Harvest Engineering and Technology, Faculty of Agriculture Science, AMU, Aligarh, India

Abstract

The studies were carried out to development and quality evaluation of extruded pet food by utilization of buffalo meat byproducts like livers, trims and agro wastes namely sorghum, oat and corn flour. The quality of the extruded pet food was evaluated on the basis of qualities characteristics namely pH, fat content, protein content, ash content, and TBA number. Protein content and fat content of fresh pet food were found in the following ranges 15.84% and 10.64% respectively. The pH of extruded pet food was significantly decreased. The ash content, TBA number and pH content were 2.43%, 0.605 mg/kg and 6.29% in fresh condition. During ambient storage pH values was found to decrease consistently. Protein content was found to decrease significantly. The Sorghum, oat and corn flour incorporation, significantly affected on quality of extruded pet food samples.

Keywords: Extrusion technology; Pet food; Chemical treatment; Quality attributes and storage

Introduction

Extrusion cooking technology is used for the manufacture of commercial dry food for canine and feline diets: about 95% of dry pet foods are extruded [1]. In this processing technology a mixture of ingredients is steam conditioned, compressed and forced through the die of the extruder [2]. The reason for the widespread use of extrusion cooking to produce pet diets is the versatility of this technology to mix diets and functionally improve, detoxify, sterilize and texturize a large variety of food commodities and food ingredients. A combination of moisture, pressure, residence time, temperature and mechanical shear is used for these reactions and transformations [3]. Pet food by extrusion processing focuses on the production of feeds for animals with a high nutritional value but low economic usefulness. Nonetheless, balancing the components and gentle processing are at the center of the related quality considerations. Extrusion is therefore increasingly proving to be the right tool for modern pet food.

Extrusion has become a very important process in the manufacture of pet food. Extrusion machines use single or twin screws to transport, mix, knead, shear, shape, and/or cook multiple ingredients into a uniform food product by forcing the ingredient mix through shaped dies to produce specific shapes and lengths. Extrusion provides the foundation for continuous production. Food extruders are used to produce pasta and other cold formed products, cereals, snacks, pet food, feed, confectionery products (including chewing gum, licorice, and marsh mallows), modified starches for soup, baby food, and instant foods, beverage bases, and textures vegetable proteins. Single-screw extruders have been in use for continuous cooking and in the forming of ready-to-eat (RTE) cereals as a one-step process since the 1960s. Twin-screw extruders were in common use in food production by the 1980s. A gradual shift towards prepared dog food was observed over the review period and now dog owners have been more willing to spend on food for their pet than ever before. This trend has spawned an entire industry, with a growing emphasis on dog food.

Now dog food is a multibillion dollar industry. The growth difference in India is 30-40%. India has an estimated population of 1.5 million pedigreed dogs [4]. The export market is a big revenue earner with around Rs.15 crore turnovers compared to Rs.2.5-3 crore from the domestic market (Newspaper report, Times of India, 18 June 2005). According to American President George Bush's (May 5, 2008, HT), Indians spent 20 million dollars in feeding their pets in 2004, a figure that up to million dollars in 2007. This gradual increase in commercial pet food market was because of two factors this is the role of veterinarians in educating consumers and selling pet food with vets currently accounting for as much as 44% sales of commercially prepared foods (HT, May 5, 2008). Second, thing is that most of the mid-priced and premium pet food brands in India are imported.

The government reduction of the import duties on pet food in 2007 provides a boost to both manufacturer and importers. This reduction in import duties also signals the first indication of government's softening towards the pet food and pet care products industry. The food industry offers strong growth opportunities in the forecast period. Our country is diversified and developing country here those ingredients which makes the pet food more nutrition, safe, healthy and cheap are easily available as a by products and waste of food industries and slaughterhouse wastes. In developing countries of Asia where meat from ruminants constitute only about 21.0% of the total meat production, buffalo meat is about 11.52% of the total ruminant meat, and about 2.7% of all meat produced in the region. The average annual growth rate in production was about 1.3%. Undoubtedly, majority of world's buffalo meat is Asian, representing 91.89% and with volume of 3.08M tons in 2008 (FAO, 2010). World Bank survey India is the largest producer of food grains and milk and second largest in fruits and vegetables in the world. Yet nearly 40% of its food is wasted it means country losses more than Rs.58,000 crore worth of agricultural food items, due to lack of post harvesting infrastructure such as cold chains, transportation and storage facilities [5].

The present study was undertaken for the development and quality evaluation of agro-processing byproducts incorporated extruded dog

*Corresponding author: Javeed Akhtar, Department of Post-Harvest Engineering and Technology, Faculty of agriculture Science, AMU, Aligarh, India
E-mail: er.jakhtar@gmail.com

food. Agro-processing byproducts used were meat byproducts along with grains, fruit and vegetable. Quality evaluation has been done in terms of physicochemical analysis. The quality of fresh extruded dog food was evaluated on the basis of physico-chemical characteristics namely pH, fat content, protein content and TBA number. The samples of dog food were packed using heat sealing machine in combination film and kept at ambient temperature for 120 days. It was found that the shape of extrudates was stick form.

Materials and Methods

Materials collection

Sample collection of agro processing byproducts: Freshly slaughtered edible or inedible byproducts namely, liver, kidney, blood and bones of Indian buffalos of 5 years of age were collected. The samples were immediately transferred to the laboratory in ice for further processing. The samples were stored in a deep freezer at-20°C until used.

Sample collection of agro wastes: All the collected agro processing byproducts including sorghum, wheat, corn and oat were also collected from the market of Aligarh. All the collected material firstly cleaned, sieved washed and dried in tray drier oven at 60°C temperature.

Manufacturing process of extruded pet food

The buffalo meat byproducts were fed into mincer for making liquid material and all cereals ingredients sorghum, oat, corn flour were grinded in grinder to obtained fine powder. The liquid meat byproducts materials and cereals fine powder were mixed in the mixer to obtain good mixed materials after complete mixing of both materials added other ingredient are added and mixed then fed into twin screw extruder for preparation of rod type extruded pet food. Extruder was operated at 170°C and at 300 rpm screw speed product was complete cooked drawn the process of extrusion and types extruded pet food were collected and packed by heat sealing method in combination film/pouch and stored at ambient temperature for studying their quality characteristics as a function of time.

pH measurement

10 g of developed dog food samples were taken along with 50ml-distilled water homogenized in a mixer grinder. The ground sample was filtered and the pH was determined by dipping the combined glass electrode of a digital pH meter (Khera model, Indian make) into the filtrate.

Determination of ash content

According to Association of Official Analytical Chemist (AOAC) 1990 method, dried sample was weighted into crucible and ignited at 550°C for 12-18 hrs in the muffle furnace (make Tango). It was then taken out and allowed to cool for a moment and then placed in desiccators until cooled and finally weighed to constant weight. Following formula was used to calculate ash content (Figure 1).

$$Ash\ content\ (\%) = \frac{final\ weight\ of\ ash}{initial\ weight\ of\ sample} \times 100$$

Determination of fat content

The Soxhlet method has been suggested by Association of Official Analytical Chemist (AOAC) of year 1971 was used for dog food product fat extraction.

The following formula used to express fat content of the sample.

$$Fat\% = \frac{Gram\ of\ fat\ in\ sample}{Gram\ of\ sample} \times 100$$

Determination of protein content

The protein content of dog food sample was evaluated according to Association of Official Analytical Chemist (AOAC, 1990) method. In this method Kjeldhal apparatus was used. The dog food sample was digested with sulphuric acid in the presence of catalyst. The total organic nitrogen was converted to ammonium sulphate. The digestion was neutralized with alkali and distilled into a boric acid solution. The borate anions formed were titrated with standardized acid, which was then converted to nitrogen in the sample. The result of the analysis represented crude protein of the food since nitrogen also comes from non-protein component.

% Protein = %N × 6.25

A reagent blank should be run to subtract nitrogen from the sample nitrogen.

$$\%N = \frac{NHCl \times Corrected\ acid\ value}{g\ of\ sample} \times \frac{N}{mole} \times 100$$

% N × 6.25 = % Protein

Measurement of thiobarbutyric acid (TBA) number

For Thiobarbituric acid number (TBA), evaluation TBA reagent was prepared according to Pearson by dissolving 0.2883 g of Thiobarbityric acid in sufficient quantity of 90% and slight warming, the volume being made up to 100 ml with 90% acetic acid. TBA number was measured by the method described by Strange et al. For this 5 ml of TCA extract was mixed with 5ml of TBA reagent in a test tube. The test tube was kept in a water bath at 100°C for 30 minutes along with another test tube containing a blank of 5 ml of 10% TCA and 5 ml TBA reagent. After cooling the tube in running water for about 10 min, the absorbance was measured at 530 nm in spectrophotometer and reported as TBA number.

Statistical Analysis

Standard deviation

The best and most commonly used statistical evaluation of the precision of analytical data is the standard deviation. The standard deviation measures the spread of the experimental values and gives a good indication of how close the values are to each other. Samples were prepared in three replication and data obtained for selected quality parameters were analyzed for mean and standard deviations using following formula:

$$SD = \pm \sqrt{\Sigma (X_i - X)^2 / n}$$

Where,

X_i = individual sample values

X = mean of individual samples

n = total population of sample

Results and Discussion

Effect on pH of extruded pet food

The pH values of the agro processing waste incorporated extruded dog food were in the neutral range (6.28-6.18). The extruded dog food

Figure 1: Developed extruded pet foods.

control sample (SCL) which was developed by incorporation of buffalo meat byproducts along with the grains and vegetable shows the pH value 6.31 after preparation. The effects of treatment were noticed on pH values of agro processing waste incorporated extruded dog food and it was observed that the pH of samples STA (with 0.5% α-tocopherol acetate), SPS (0.5% potassium sorbet), STAPS (combination of 0.5% α-tocopherol + 1% potassium sorbet) and SPSTA (combination of 0.5% potassium sorbet + 1% α-tocopherol acetate) were 6.28, 6.27, 6.28 and 6.26 respectively on day 1 of storage (Figure 2).

As a function of time the pH values were significantly changed during storage with respect of treatment. The pH of buffalo meat byproducts incorporated extruded dog food control sample (SCL) reached to 6.27 whereas the pH of STA (with 0.5% α-tocopherol acetate treated), SPS (0.5% potassium sorbet treated), STAPS (combination of 0.5% α-tocopherol +1% potassium sorbet treated) and SPSTA (combination of 0.5% potassium sorbet + 1% α-tocopherol acetate treated) after 120 days of storage reached to 6.16, 6.20, 6.23 and 6.21. The least fall in the pH value was noticed in STAPS (combination of 0.5% α-tocopherol+1% potassium sorbet) in comparison to the other treated extruded dog food samples. Mild denaturation of proteins during extrusion cooking may be responsible for the declined of pH during storage of 120 days. During storage the pH was declined significantly.

Effect on Protein content of extruded pet food

The protein component in pet foods can constitute between 25 and 70% of the dry matter (DM) [2]. Dogs required relatively high proportion of protein because of carnivorous nature. Effects of extrusion on the protein component of buffalo meat byproducts incorporated extruded dog foods was evaluated (Figure 3).

The protein content of buffalo meat byproducts incorporated control sample (SCL) was initially found 15.84 which on further storage

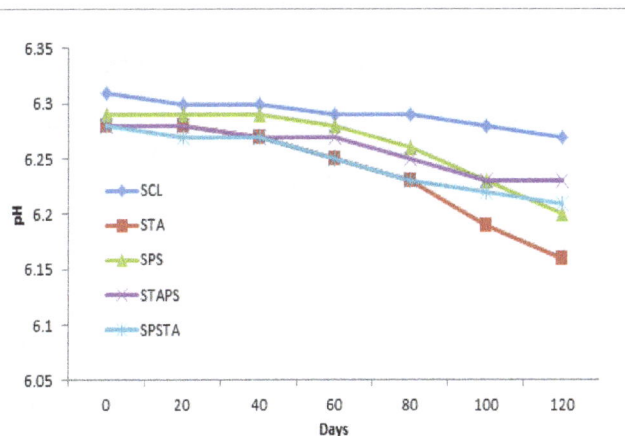

Figure 2: Effect on pH of extruded pet food during storage of 120 days at ambient temperature.

continuously decreased with increased storage period. Where the 0.5% α-tocopherol acetate treated sample (STA) shows no significant change in protein content up to 60 days of storage but later the reduction were with the increasing storage time the similar condition were also noticed in 0.5% potassium sorbet treated extruded dog food. The preservative system STAPS (combination of 0.5% α-tocopherol +1% potassium sorbet treated) and SPSTA (combination of 0.5% potassium sorbet + 1% α-tocopherol acetate treated) both shows the marginally declined protein content in the composition of extruded dog food.

An undesirable effect of heat treatment during extrusion causes protein denaturation because of Millard reactions which can make them more susceptible to digestive enzymes and therefore improve the digestibility of protein of extruded dog food. It is observed that

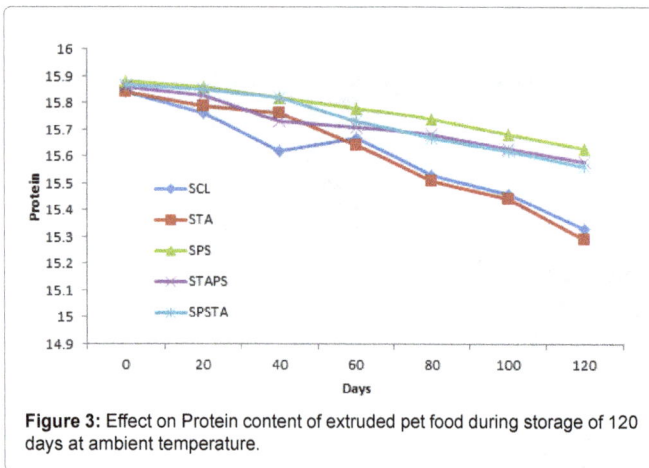

Figure 3: Effect on Protein content of extruded pet food during storage of 120 days at ambient temperature.

in all four treated samples more or less reduction in protein content during storage may be because of weak antioxidant systems namely α-Tocopherol in combination or individual level as it was found unable to inhibit the protein oxidation in atmospheric packaging Marianne et al [6] also noticed that protein oxidation during storage beef patties treated with antioxidants namely ascorbic acid and α-Tocopherol and packed with atmospheric packaging.

Effects on Fat Content of extruded pet food

The effect of preservative namely α-Tocopherol and potassium sorbet individual or in combination treated buffalo meat byproducts incorporated in combination with grain, vegetable and fruits extruded dog was evaluated and discussed below. It may be noted from the figure that the fat content of treated or non-treated extrudates significantly degraded during storage study of 120 days. The buffalo meat byproducts incorporated control sample (SCL) on day one shows 10.64 fat content which on further storage rapidly reduced with increasing storage period.

The fat content values of 0.5% α-tocopherol acetate treated (STA), 0.5% potassium sorbet treated (SPS), combination of 0.5% α-tocopherol + 1% potassium sorbet treated (STAPS) and combination of 0.5% potassium sorbet + 1% α-tocopherol acetate treated (SPSTA) were 10.65, 10.66, 10.68 and 10.67 respectively fresh samples and finally reached to 10.46, 10.49, 10.43and 10.39 respectively after 120 days of storage. It is observed that the preservative system in combination works quite better than individual because initially the reductions in fat content were negligible. It is also concluded that from the data shown in Figure 3 at initial level the preservative system works well but as increasing storage period these system were unable to provide protection and that samples slowly get degraded in presence of known antioxidant α-tocopherol acetate which is not able to work in atmospheric system of packaging. Marianne et al., 2006 also noticed that protein and fat oxidation was occurred in presence of antioxidants namely ascorbic acid and α-Tocopherol treated beef patties in atmosphere packaging system during storage. Oxidation rate is affected by many factors such as fat type, fat content, moisture content and expansion degree where the unsaturation in fats increases the preservation challenge [7,8] (Figure 4).

Effect on ash content of extruded pet food

Ash has been shown to be an effective content for dog foods. Due to the high dietary levels of animal by-products, the levels of ash content in

our dog foods were high which shows the higher mineral content. The ash content of five types of dog foods is shown in figure the ash content of SCL control sample without (α-tocopherol acetate and potassium sorbet) treatment was 2.44% on day one. Where the ash content of STA (with 0.5% α-tocopherol acetate treated), SPS (0.5% potassium sorbet treated), STAPS (combination of 0.5% α-tocopherol, 1% potassium sorbet treated) and SPSTA (combination of 0.5% potassium sorbet, 1% α-tocopherol acetate treated) were 2.43%, 2.42%, 2.42%, and 2.42% respectively on day 1 of storage. As a function of time the reduction in ash content were noticed during storage study of 120 days. The samples treated with combination of 0.5% α-tocopherol, 1% potassium sorbet treated (S4TAPS) significantly shows the minimum degradation of mineral content with reflecting marginal decrease in ash content values in throughout storage. There were significant differences ($p<0.05$) in ash content [9,10]. The ash contents were higher. Ash constitutes the organic matter which is also difficult for animal digestion but required in trace amounts. As ash does not contain any energy, so it would naturally lower the overall energy of the feed (Figure 5).

Effect on TBA values of extruded pet food

Thiobarbituric acid number (TBA), of treated and untreated extrudates to measures the oxidative rancidity, was analyzed and shown in figure. TBA reactive materials were measured as milligrams of malondialdehyde per kg of sample (Pearson 1971). Malondialdehyde is released from the endoperoxides that are formed from lipid peroxidation of polyunsaturated fatty acids.The fresh extruded dog food (control sample, SCL) had a TBA value of 0.605 mg/kg significantly increased with increasing storage period and finally reached to 0.822 mg (Figure 6).

The fresh extruded dog food of STA (with 0.5% α-tocopherol acetate treated), SPS (0.5% potassium sorbet treated), STAPS (combination of 0.5% α-tocopherol, 1% potassium sorbet treated) and SPSTA (combination of 0.5% potassium sorbet, 1% α-tocopherol acetate treated) were 0.602 mg/kg, 0.604 mg/kg, 0.601 mg/kg and 0.603 mg/kg respectively but were not significantly different among preservative system in throughout the 120 days storage study [11,12]. As a function of time the TBA values is increased with increasing storage study of 120 days. The samples treated with combination of 0.5% α-tocopherol, 1% potassium sorbet treated (STAPS) significantly shows the marginally inclined TBA values in throughout storage.

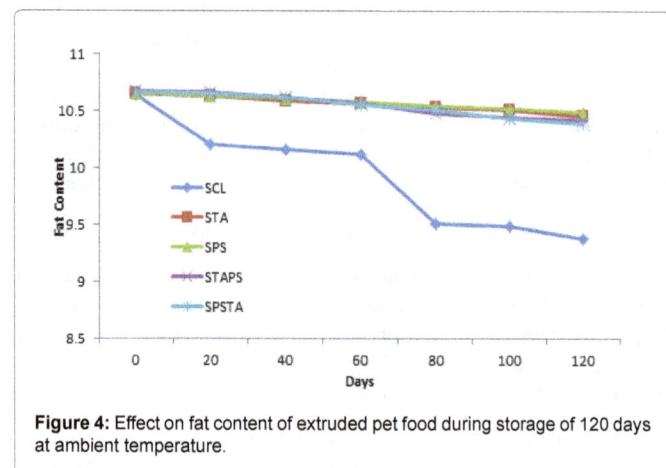

Figure 4: Effect on fat content of extruded pet food during storage of 120 days at ambient temperature.

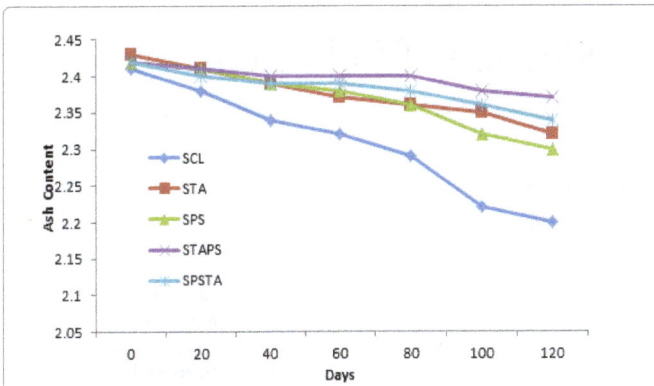

Figure 5: Effect on ash content of extruded pet food during storage of 120 days at ambient temperature.

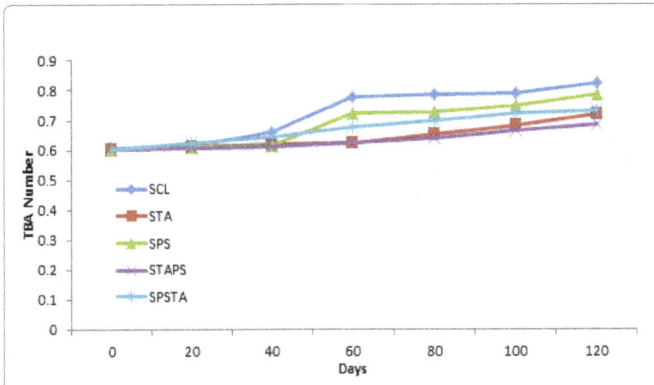

Figure 6: Effect TBA numbers of extruded pet food during storage of 120 days at ambient temperature.

with increasing storage period.

The increasing trend in TBA values indicated that natural antioxidant α-Tocopherol individual or in combination of potassium sorbet was less stable in atmosphere packaging. Finally, after 120 days of storage there was no quality deterioration found in any sample but the rancidity value of meat byproducts incorporated extrudates were increasing with increasing storage period which later will not be suitable for the dog health.

Acknowledgments

The authors thank Staffs from Department of Post-Harvest Engineering and Technology, Faculty of agriculture Science, A.M.U, Aligarh, India for its help in the conduction of experimental procedures. Also, the authors are grateful for the financial support of Aligarh Muslim University, Aligarh India.

References

1. Spears JK, Fahey GC (2004) Resistant starch as related to companion animal nutrition. JAOAC Int 87: 787-791.

2. Rokey G, Plattner B (1995) Process description: pet food production. Wenger Mfg Inc, Sabetha, KS USA.

3. Cheftel JC (1986) Nutritional effects of extrusion cooking. Food Chem 20: 263–283.

4. Atrey Arun (2001) Petcare Industry: Challenges and the road ahead.

5. Mishra PK (2008) 72 percent of India's fruit, vegetable produce goes waste, New Delhi.

6. Marianne NL, Marchen SH, Leif H (2006) The combined effect of antioxidants and modified atmosphere packaging on protein and lipid oxidation in beef patties during chill storage. International Journal of Meat science 76: 226-233.

7. Lin S, Hsieh F, Huff HE (1997) Effects of lipids and processing conditions on lipid oxidation of extruded dry pet food during storage. Anim Feed Sci Technol 71: 283-194.

8. Deffenbaugh (2007) Optimizing pet food, aquatic and livestock feed quality. Agrimedia GmbH, Clenze, Germany: 327-342.

9. AFFCO (2007) Compendium of Association of American Feed Control Officials.

10. Parsons CM (1997) Protein and amino acid quality of meat and bone meal. poult sc76: 361-368.

11. Ranganna S (2000) In Hand Book of Analysis and Quality Control for Fruit and Vegitable Products. Tata Mc Grawhill Pubco, New Delhi.

12. Shrivastva RP, Sanjeev K (2003) The Book of Fruit and vegetable preservation principles and practices by (I.B.D. CO) Lucknow.

Conclusions

Cooking at high temperature, dehydration and packaging etc. of buffalo meat byproducts extruded dog food in combination with preservatives namely potassium sorbet and α-Tocopherol were proved to be effective method for controlling microflora and make them shelf stable. The deterioration of protein and fat content among these characteristics reflect that antioxidant system was not able to work in atmosphere packaging this also shows that α-Tocopherol, slowly lose its working efficacy in air packaging individual or in combination. Where potassium sorbet was successfully inhibit the mold growth but not work as a powerful barrier for against rancidity.

The decreasing in pH was not related to spoilage of the product. The fall in pH values in both treated as well as untreated (control) samples is due to denaturation or oxidation of lipid and protein during extrusion cooking at high temperature and atmospheric pressure. All these factors may be responsible to alter the acidity as a result pH decreased

31

Persistence and Effect of Processing on Reduction of Chlorpyriphos in Chilli

Kumari B*and Chauhan R

Department of Entomology, Chaudhary Charan Singh Haryana Agricultural University, Hisar-125004, Haryana, India

Abstract

Persistence behavior of chlorpyriphos in chilli was studied following application of formulation of 20 EC @ 160 (single dose) and 320 (double dose) g a.i.ha^{-1} at fruiting stage. Samples of green chilli and soil under crop were drawn at different time intervals and quantified by gas liquid chromatography equipped with electron capture detector. The initial deposits of residue on the chilli fruit were 0.397 and 1.021 mg kg^{-1} at single and double dose, respectively. Residues of chlorpyriphos dissipated to more than 75% after 10 days at both the dosages. The half-life period of chlorpyriphos in chilli was recorded to be 6.02 days at single dose and 5.67 days at double dose. Washing process was found effective in reducing the residues by 67.75 and 71.60 percent at, respective doses.

Keywords: Chlorpyriphos; Chilli; Residues; Persistence; Half-life; Washing; Soil

Introduction

Pesticides help the farmers to increase their income by improving the quality of crop, saving crop losses, increase crop productivity and reduce cost of production. In recent years, the role and contribution of pesticides have increased much more, especially in the country like India because of fast growth of population; the demand for food supply will continue to grow steadily. Due to regular use of pesticides, the residues in food commodities seldom exceed the maximum residue limits (MRLs) set by the food authorities. As a result, consumption of pesticide treated food commodities become risky. So it becomes necessary that, pesticide should be effective against pest along with its toxicologically acceptable residues in food commodity [1]. For this reason, proper assessing of pesticide residues is very important for reducing health hazard to consumers. Chilli (*Capsicum annum*) is an essential pillar of the cuisines of India. It is attacked by various insect pests resulting in yield losses. The crop is attacked by various insect pests resulting 51 species of insects and 2 species of mites. Thrips, mites and pod borer are serious among its pests [2].

Chlorpyriphos [O,O-diethyl-O-(3,5,6-trichloro-2-pyridinyl) phosphorothionate] is an organo phosphorus broad spectrum insecticidal active ingredient registered for application to more than 40 different food commodities. It is a stable compound in neutral and acidic conditions [3]. It kills insects by disrupting their nervous system and is effective against both sucking and chewing insects and has been widely used to control pests of various vegetables [4]. It is non-systemic, fairly persistent, and highly soluble in organic solvents like acetone, xylene, and methylene chloride [5]. Cholinesterase inhibition is the mode of action of chlorpyriphos and is the cause of potential toxicity in human [6,7].

There is currently an increasing concern and awareness about the hazards of pesticides to consumers. Even with the adoption of integrated pest management, farmers believe in the control of pests using pesticides because of their quick effect. Therefore, the present study was designed to determine the residual persistence and effect of processing on reduction of chlorpyriphos residues in chilli.

Materials and Methods

Chemicals and reagents

Chlorpyriphos formulation (20 EC) used for field application was procured from local market. Solvents and reagents like dichloromethane, acetone, sodium chloride, and anhydrous sodium sulfate all were procured from Merck (Darmstadt, Germany). Before use all the common solvents were redistilled in glass apparatus. By running reagent blanks the suitability of all the solvents was ensured before actual analysis.

Field experiment

The crop was raised at the Research Farm of Chaudhary Charan Singh Haryana Agricultural University (CCSHAU), Hisar following recommended agronomic practices. Plot size was 25 m^2 with spacing 60 × 30 cm (rows × plants) in triplicate with randomized block design (RBD). Before spraying, chilli fruits (*Variety*: HPH-2024) in all plots/replicates were tagged and sprayed with chlorpyriphos 20 EC using knap sap sprayer at single dose(160 g a.i ha^{-1}) and double dose (320 g a.i ha^{-1}) along with a control plot where no insecticide was applied. The volume of water used at the time of spray was @ 500 Lha^{-1}.

The soil under crop was of light texture with low content of organic matter; other relevant properties of the soil were EC 2 dSm^{-1}; K 10.08, P$_2$O$_5$ 15 kg ha^{-1} with pH 7.6 and organic carbon 0.67 percent.

Sampling

The composite sample about 250 g of tagged chilli fruits were collected randomly separated from the control and treated plots of each treatment at 0 (1 hr after spray), 1, 3, 5, 7, 10, 15 and 30 days after the application of insecticide, packed in paper bags, and brought to the laboratory for processing. In the laboratory the samples were divided in two lots, one was processed as such and other was kept to study the effect of washing on reduction of residues. After thorough mixing and quartering of each part, 25 g representative sample of chilli was used for processing and analysis.

---

*Corresonding author: Kumari B, Department of Entomology, Chaudhary Charan Singh Haryana Agricultural University, Hisar-125004, Haryana, India, E-mail: beenakumari.958@rediffmail.com

Soil samples (500 g) under crop were also collected separately from 5-10 different sites with the help of a steel auger from the depth of 0-15 cm. Soil collected from different sites was pooled and sieved to remove extraneous matter, including stones/pebbles. After thorough mixing, a sub sample of about 250 g was taken from pooled sample of each treatment and transported to the laboratory. The samples were processed and analyzed at the Pesticide Residue Laboratory, Department of Entomology, CCS HAU, Hisar.

Extraction

Residues of chlorpyriphos from chilli were extracted by adopting the method of Bhardwaj et al. [7]. Representative 25 g sample of chilli was chopped into small pieces and macerated in mixer grinder and shaken with acetone (100 ml) on mechanical shaker for 1 hr and kept overnight in an Erlenmeyer flask. Extract was filtered through 2–3 cm layer of anhydrous sodium sulphate and the filter was washed with acetone (2×10 ml). The filtrate was subjected to liquid-liquid partitioning in 1-L separator funnel and diluted with 4–5 times brine solution (10% sodium chloride solution), and the contents partitioned first with dichloromethane (75,75 ml) and then with hexane (75,75 ml). Combined both the fractions and treated with 300 mg activated charcoal powder for about 2-3 hrs at room temperature. When the solution became clear, it was filtered through Whatman filter paper no.1. The clear extract was concentrated using a rotary vacuum evaporator. The extract was finally made up to 2 ml and added to the liquid – solid chromatography column.

Clean-up

Clean-up of samples was done by using column chromatography. Glass columns (60 cm × 22 mm i.d) were packed compactly with silica gel sand-witched in between two layers of anhydrous sodium sulphate. Prepared column were pre-washed using 40 ml hexane. The column was loaded with the concentrated extract and eluted with a solution of hexane: acetone (9:1 v/v). Combined the organic phases and concentrated to about 5 ml on a rotary vacuum evaporator. Finally, the extract was concentrated to dryness on gas manifold evaporator and final volume was made to 2 ml in n-hexane and analyzed by GC.

Washing

The whole fruits of chilli were washed under running tap water for 1 min by gentle rubbing with hands and the water was discarded. These washed samples were kept on blotting paper just to remove the excess of water following method of Walter et al. [8]. Then the samples were extracted, cleaned up in a similar manner as the raw samples were processed and analyzed.

Soil

Ground, sieved and dry representative (15 g) sample was mixed with 1-2 drops of ammonia solution and left for half an hour. Mixed the soil thoroughly with 0.3 g activated charcoal, 0.3 g Florisil and 10 g of anhydrous sodium sulphate and packed compactly in a glass column (60 cm × 22 mm i.d) in between two layers of anhydrous sodium sulphate as per method of Kumari et al. [9]. Residues were eluted with 125 ml solution of hexane: acetone (9:1 v/v) at flow rate of mL min⁻¹. The elute was concentrated on rotary vacuum evaporator and made the final volume to 2 ml in n-hexane for GC analysis.

Estimation by gas liquid chromatography

The cleaned extracts analyzed on gas chromatograph (Shimadzu-2010) equipped with ^{63}Ni electron capture detector

(ECD) and HP-1 capillary column provide good results. Operating conditions were as per details: Temperature (ºC): Oven: 150 (5 min⁻¹) →8 min⁻¹→190 (2 min) →15 min⁻¹→280 (10 min), Injection port: 280ºC and detector: 300ºC. Flow rate of nitrogen (carrier gas) was maintained at 60 ml min⁻¹ and through column 2 ml min⁻¹ with split ratio 1:10. Under these operating conditions, the retention time of chlorpyriphos was found to be 13.991 min. The residues of chlorpyriphos in samples were identified and quantified by comparing retention time and area of sample chromatograms with that of standards run under identical conditions.

Chilli and soil samples were spiked with chlorpyriphos at two concentration levels (0.010 and 0.025 mg kg⁻¹) processed and analyzed as per the methodology described above to check the validity of the method. Percent recoveries in chilli were 90.50 and 92.10 while in soil were 92.70 and 93.95 at two fortification levels, respectively. As the percent recovery obtained were more than 90%, therefore, the results have been presented as such without applying any correction factor. Limit of determination/quantification (LODe /LOQ) was 0.010 mg kg⁻¹.

Results and Discussion

The results of chlorpyriphos residues after application @ 160 and 320 g a.i.ha⁻¹ detected in chilli samples collected from the Research Farm of CCS HAU, Hisar are presented in Table 1. The results indicated that at raw stage the highest chlorpyriphos residues were found in chilli. The initial deposits of chlorpyriphos at single and double dose were observed to be 0.397 and 1.021 mg kg⁻¹, respectively. In the start, residues dissipated slowly i.e. 0.385, 0.278 and 0.180 in case of single dose while 0.997, 0.695 and 0.485 in case of double dose on 1, 3 and 5th day after application. The residues reached below detectable levels (0.010 mg kg⁻¹) on 15th and 30th days of treatment in single dose and double dose, respectively, showing per cent dissipation of 97.48 and 98.14 per cent. Residue data were subjected to statistical analysis for computation of regression equations, half-life ($t_{1/2}$) values and percent degradation. The residues dissipated with half-life period of 6.02 days at single dose and 5.67 days at double dose. Residues of chlorpyriphos did not follow the first order kinetics. But the residues followed the pseudo first order kinetics (R^2) with 0.900 and 0.856 for single and double dose, respectively. Subhash et al. [10] studied the persistence behavior of chlorpyriphos, cypermethrin and monocrotophos in okra. The residues of all the three insecticides reached below detection limit (BDL) showing complete dissipation with in 15, 17 and 19 days respectively, when it was applied 100, 200 and 300 g a.i. h⁻¹. Jyot et al. [2] reported that the average initial deposits of chlorpyriphos on chilli were 0.59 and 2.02 mg kg⁻¹ following the application of chlorpyriphos at 500 and 1,000 ga.i.ha⁻¹. These residues reached below the determination limit of 0.01 mgkg⁻¹ after the 10th and 15th with half-life period of 4.43 and 2.01 days at single and double doses, respectively. Samriti et al. [11] observed that when chlorpyriphos applied on okra @ 200 and 400 g a.i ha⁻¹, residues on 7th and15th day of application reached below detection limit (BDL) of 0.010 mg kg⁻¹ in single and double dose, respectively and reported half-life period of 3.15 days at single dose and 3.46 days at double dose following first order kinetics. Similar type of observation with chlorpyriphos in chilli was reported by Waghulde et al. [12]. Present results seem to be in conformation with earlier reports.

Effect of processing

Chilli fruits were subjected to processing like washing in order to investigate the reduction of residues up to 5th day of sampling. The results pertaining to the effect of washing on the removal of chlorpyriphos residues applied at 160 and 320 g a.i. ha⁻¹ on chilli fruits are presented in Table 2. It has been found that washing was found effective in reducing the residues on 0 (1 h) days after application. The average initial deposits of chlorpyriphos was 0.397 and 1.021 mg kg^{-1} at single and double dose on chilli fruits were reduced to 0.128 and 0.290 mg kg^{-1} as a result of simple washing with tap water showing the loss of 67.75 and 71.60 percent at, respective doses. In case of third day samples, the residues were reduced to 0.180 and 0.425 mg kg^{-1} thereby, accounted 35.25 and 38.84 percent residues loss at single and double doses, respectively. Sunayana et al. [13] studied the effect of washing alone with tap water to check the percent reduction in residues of chilli, when fipronil was applied on chilli crop @ 50 and 100 g a.i.ha^{-1}. As a result of simple washing percent reduction was observed 42.05 and 45.42 at single and double dose respectively. Rani et al. [14] reported that processing was very effective in reducing the residues of chlorpyriphos in tomato fruits when applied @ 400 and 800 g a.i. ha^{-1}. Samriti et al. [11] reported that due to washing 13–35% reduction of chlorpyriphos was observed in okra when applied @ 200 and 400 g a.i. ha^{-1}.

Chlorpyriphos residues in soil samples

In soil, initial deposit of chlorpyriphos on 0 (1 h after treatment) days at single dose was 0.439 mg kg^{-1} and 0.903 mg kg^{-1} at double dose (Table 3). Under study period of 30 days, residues reached to the levels of 0.037 and 0.088 mg kg^{-1} at single and double dose, respectively. The dissipation after 30 days was observed to be 91.57 per cent for single dose and 90.25 per cent for double dose.

Jyot et al. [2] reported that residues of chlorpyriphos and cypermethrin were found to be <0.01 mg kg^{-1} for both these insecticides at the single (500 g a.i ha^{-1}) and double dosages (1000 g a.i ha^{-1}) collected 15 days after the last spray. Persistence of chlorpyriphos in soil under tomato crop was studied by Rani et al. [14] with active application of chlorpyriphos at 400 and 800 g a.i. ha^{-1}. In soil samples, residues of chlorpyriphos reached below detectable level of 0.010 mg kg^{-1} after 5 and 10 days after spray at single and double dose, respectively. Samriti and Kumari [3] reported that the residues of chlorpyriphos in soil under okra crop dissipated below determination level of 0.005 mg kg^{-1} on 5 days at single dose (200 g a.i ha^{-1}) and 7 (400 g a.i ha^{-1}) days at double dose reported.

Conclusion

From the above outcome, it is obvious that the application of pesticides in agriculture is necessary for better crop production against the possible health hazards arises due to the presence of pesticide residues in food. The half-life values for chlorpyriphos following three applications at the single and double dose on chilli fruits were observed to be 6.02 and 5.67 days, respectively. The residues of chlorpyriphos studied in chilli at two doses, reached to below detectable value of 0.01 mg kg^{-1} on 15th and 30th in single and double dose, respectively. Meticulous processing with simple household practices like washing, dislodged the residues by 68 -72 per cent and thereby ensures the safety of the chilli fruits to the consumers.

Days after treatment	Residue (mg kg^{-1})				
	Single Dose (160 g a.i. ha^{-1})		Double Dose (320 g a.i. ha^{-1})		
	Average	% Dissipation	Average	% Dissipation	
0(1 h)	0.397 ± 0.101	-	1.021 ± 0.226	-	
1	0.385 ± 0.017	3.02	0.997 ± 0.098	2.35	
3	0.278 ± 0.033	29.97	0.695 ± 0.111	31.92	
5	0.180 ± 0.016	54.65	0.485 ± 0.015	52.49	
7	0.134 ± 0.010	66.24	0.360 ± 0.085	64.74	
10	0.085 ± 0.012	78.58	0.258 ± 0.032	74.73	
15	0.010 ± 0.007	97.48	0.019 ± 0.007	98.14	
30	BDL	-	BDL	-	
	t$_{1/2}$ = 6.02 days		t$_{1/2}$ = 5.67 days		

BDL: 0.01 mg kg^{-1}

Table 1: Persistence and dissipation of chlorpyriphos residues in chilli.

Days after treatment	Residue (mg kg^{-1})					
	Single Dose (160 g a.i. ha^{-1})			Double Dose (320 g a.i. ha^{-1})		
	Initial residues ± SD	Washing ± SD	% Reduction	Initial residues ± SD	Washing ± SD	% Reduction
0(1 h)	0.397 ± 0.101	0.128 ± 0.110	67.75	1.021 ± 0.226	0.290 ± 0.031	71.60
1	0.385 ± 0.017	0.157 ± 0.012	59.22	0.997 ± 0.098	0.354 ± 0.101	64.49
3	0.278 ± 0.033	0.180 ± 0.016	35.25	0.695 ± 0.111	0.425 ± 0.111	38.84
5	0.180 ± 0.016	BDL	-	0.485 ± 0.015	0.405 ± 0.075	16.49

Table 2: Effect of processing on chlorpyriphos residues (mg kg^{-1})* in chilli.

Days after treatment	Residue (mg kg^{-1})				
	Single Dose (160 g a.i. ha^{-1})		Double Dose (320 g a.i. ha^{-1})		
	Average	% Dissipation	Average	% Dissipation	
0(1 h)	0.439 ± 0.011	-	0.903 ± 0.012	-	
3	0.304 ± 0.013	30.75	0.635 ± 0.006	29.67	
7	0.149 ± 0.008	66.05	0.315 ± 0.023	65.11	
15	0.078 ± 0.006	82.23	0.165 ± 0.010	81.72	
30	0.037 ± 0.006	91.57	0.088 ± 0.007	90.25	

Table 3: Persistence and dissipation of chlorpyriphos residues in soil.

Acknowledgments

The authors wish to express their gratitude to the Head, Department of Entomology for providing research facilities.

References

1. Singh G, Singh B, Battu RS, Jyot G, Singh B, et al. (2007) Persistence of ethion residues on cucumber, *Cucumis sativus* (Linn.) using gas chromatography with nitrogen phosphorus detector. Bull Environ Contam Toxicol 79: 437-439.

2. Jyot G, Kousik M, Battu RS, Singh B (2013) Estimation of chlorpyrifos and cypermethrin residues in chilli (*Capsicum annuum* L.) by gas-liquid chromatography. Environ Monit Assess 185: 5703-5714.

3. Chauhan SR, Beena K (2012) Persistence of chlorpyrifos in Okra (*Abelmoschus esculentus*) fruits and soil.Toxicol Environ Chem 94: 1726-1734.

4. Atif MR, Anjum MF, Ahmed A, Saqib RM (2007) Field incurred chlorpyrifos and 3,5,6-trichloro-2-pyridinol residues in fresh and processed vegetables. Food Chem 103: 1016-1023.

5. (2000) Office of prevention, pesticides and toxic substance chlorpyrifos; Revised product and residue chemistry chapters. US Environmental Protection Agency, Washington DC.

6. Oliver GR, Bolles HG, Shurdut BA (2000) Chlorpyriphos: Probabilistic assessment of exposure and risk. Neurotoxicology 21: 203-208.

7. Bhardwaj U, Kumar R, Kaur S, Sahoo SK, Mandal K, et al. (2012) Persistence of fipronil and its risk assessment on cabbage, *Brassica oleracea* var. *capitata* L. Ecotoxico Environ Safety 79: 310-308.

8. Walter JK, Arsenault TL, Pylypiw HM, Mattina MJI (2000) Reduction of pesticide residue on produce by rinsing. J Agric Food Chem 48: 4666-4670.

9. Kumari B, Madan VK, Kathpal TS (2008) Status of insecticide contamination of soil and water in Haryana, India. Environ Monit Assess 136: 239-244.

10. Subhash C, Mukesh K, Anil MN, Shinde LP (2014) Persistence pattern of chlorpyriphos, cypermethirn and monocrotophos in Okra. Internat J Advanc Res 2:738-743.

11. Reena SC, Beena K (2011) Persistence and effect of processing on reduction of chlorpyriphos residues in Okra fruits. Bull Environ Contam Toxicol 87:198-201.

12. Waghulde PN, Khatik MK, Patil VT, Patil PR (2011) Persistence and dissipation of pesticides in chilly and Okra at North Maharashtra Region. Pestic Res J 23: 23-26.

13. Sunayana S, Beena SK (2015) Persistence and effect of processing on fipronil 1 and its metabolites residues in Chilli (*Capsicum Annum* Linn). Pestic Res J 101409.

14. Mamta R, Sunayana S, Beena K (2013) Persistence and effect of processing on chlorpyriphos residues in tomato (*Lycopersicon esculantum Mill*). Ecotoxicol Environ Safety 95: 247-252.

32

Food Protein Powder from *Eisenia foetida*: Dearomatization Using Food Grade Solvents and Controlled Storage Conditions

Elias Bou-Maroun and Nathalie Cayot*

Food Process and Microbiological Unit, UMR A 02.102, AgroSup Dijon/Burgundy University, France

Abstract

Delipidation was used to dearomatize protein powder of *Eisenia foetida*. The remaining volatile fraction after delipidation was studied over a period of three months. Volatile fraction dramatically increased between the first and the second month of storage. Four volatile compounds, chosen as tracers, were studied, namely: benzaldehyde, 2-pentyl furan, o-xylene, and limonene. Controlled conditions of storage are very efficient to limit volatile compound increase: 10% or less for three among the four volatile compounds chosen as tracers. The main parameter to control is temperature. To obtain a food grade protein powder, delipidation was done using ethyl acetate/ethanol instead of chloroform/methanol mixture. The remaining volatile fraction after delipidation ranged from 6% to 18%. Dearomatization was improved using an additional drying after delipidation. Solvent residues amounted to a few mg/g and conformed to EU regulations concerning solvent residues in food stuffs and food ingredients.

Keywords: Protein powder; Dearomatization; Delipidation; Off-flavors; Solvent residues

Introduction

Delipidation was studied as a way to dearomatize a non-conventional protein powder obtained from *Eisenia foetida* earthworms by Romero et al. [1]. These authors showed that an ultrasound extraction method using a chloroform/methanol mixture could be used to extract lipids efficiently and, at the same time, to extract a wide range of volatile compounds. In fact, the volatile compounds detected in the headspace of the delipidated powder represented less than ¼ of the volatile compounds detected in the headspace of the non-delipidated powder (named regular powder).

Nevertheless, some lipids still remain in the powder, and can evolve during storage to produce off-flavors. If the final goal is to use the delipidated protein powder for food supplementation, the efficiency of dearomatization must be checked during storage in order to guarantee protein powder with a flavor as neutral as possible for future uses.

Another point is the use of organic solvent in the delipidation step in comparison with solvent-free techniques such as the supercritical CO_2 extraction. This latter method is, however, costly threatening the economic viability of the powder. Thus the chosen process must use simple equipment with reasonable overall cost and delipidation must be improved by choosing a more environment-friendly solvent. From that point of view, ultrasound-assisted leaching is interesting as it allows the extraction of a wide variety of compounds, whatever their polarity, because it can be used with any solvent [2].

Finally, to be able to conform to regulations and consumer demand for safer products, the residual amount of solvent must be checked.

Thus, the present study was done to answer to these three questions:

a. What are the best storage conditions to maintain dearomatization? For that purpose an experimental design was done to check the impact of oxygen, light, and temperature on the evolution of the volatile fraction during a three month storage period.

b. Could the chloroform/methanol be replaced by safer solvents for delipidation/dearomatization using ultrasound assisted extraction? For that purpose, ethyl acetate/ethanol mixtures were tested. The polarity of the mixture is similar to the one of chloroform/methanol: polarity of ethyl acetate, ethanol,

chloroform, and methanol are respectively 4.4, 5.2, 4.1, and 5.1 [3]. Moreover, EU regulations mention that ethyl acetate and ethanol can be used as 'extraction solvents in compliance with good manufacturing practice for all uses' i.e. the presence of residues in technically unavoidable quantities is acceptable [4,5].

c. What is the amount of solvent residues in the delipidated powder? For that purpose, the standard addition method was used to quantify solvent residues in delipidated powders.

Materials and Methods

Samples and reagents

Earthworm protein powder obtained from *E. foetida* was kindly donated by Professor Ana Luisa Medina of Los Andes University, Merida (Venezuela). The protein powder was obtained from earthworms fed a diet of organic waste compost. The earthworms were washed with abundant water before they were killed with boiling water. They were then dried at 60°C for 4 hours and ground to homogeneous powder designated as raw protein powder, RPP. The detailed procedure is described in a previous paper [1].

Chloroform (>99%) and ethyl acetate (≥99.9%) were purchased from Sigma-Aldrich (Saint-Quentin Fallavier, France). Dichloromethane (≥99.8%), methanol (>99.9%) and ethanol (≥99.8%) were bought from Carlo Erba Reagents (Val De Reuil, France).

Delipidation of the protein powder

The delipidation of the protein powder was done by ultrasound extraction using a US bath Bransonic Mod 3210 (Branson Europe

***Corresonding author:** Nathalie Cayot, Food Process and Microbiological Unit, UMRA 02.102, AgroSup Dijon/Burgundy University, 1 esplanade Erasme, Dijon, France, E-mail: nathalie.cayot@agrosupdijon.fr

B.V.) with an ultrasound fixed-frequency of 47 kHz ± 6%. In a typical experiment, 30 g of RPP was extracted with 150 ml of solvent for 3 minutes in an ultrasound bath. After the first extraction step, the delipidated protein powder (DPP) was filtered off. This process was repeated twice more using the DPP obtained from the previous step of extraction. The three solvent fractions were gathered, the solvent was evaporated under reduced pressure at 45°C using a rotary evaporator and the extracted lipids were weighed.

Different solvents were used for delipidation:

I.   Chloroform/methanol mixture as published previously [1]. In this case, the residual solvent of DPP was evaporated overnight at room temperature. The obtained powder was designated DPPa1-8 and stored in different conditions before analysis of the volatile fraction (Table 1).

II. Ethyl acetate/ethanol mixture was used in order to replace chloroform/methanol: solvent proportions were 67/33 or 87/13; powder/solvent ratio was 2/5 or 1/5. Extracted lipids and volatile fraction were determined in the samples obtained with ethyl acetate/ethanol, designated as DPPb1-4. The best results were obtained with DPPb3 (Table 2). Two ways of solvent elimination were tested on this sample: powder dried overnight under a hood, or powder dried overnight in an oven at 60°C. Additionally, DPPb3 was also washed using ethanol (1 hour with a powder/ethanol ratio 1/5 (w/v)) before being oven dried. Solvent residues were determined in these samples.

Study of the volatile fractions using headspace solid phase micro extraction, gas chromatography-mass spectrometry (HS-SPME-GC-MS)

For the HS-SPME analysis, 1 g of each sample was transferred into a 10 ml headspace vial which was immediately sealed with a Teflon-lined septum and screw cap. After an equilibration time of 24 h at 4°C in obscurity incubation was operated at 30°C for 30 minutes in a thermostated water bath. The headspace volatiles were then extracted using an SPME fiber (2 cm–50/30 μm DVB/Carboxen/PDMS/StableFlex, Supelco, USA) for 60 minutes at 30°C. The fiber was desorbed in an injector of a 5973 Hewlett–Packard, Palo Alto, CA, USA chromatograph in splitless mode. It was equipped with a fused-silica capillary column (30 m × 0.32 mm ID, 0.5 μm film thickness) coated with a DB-Wax stationary phase (J and W Scientific, USA). Helium

was used as the carrier gas with a constant flow of 1.5 ml/min. The initial oven temperature was maintained for 3 minutes at 40°C. Then, it was increased from 40 to 180°C at a rate of 4°C/min and from 180 to 240°C at a rate of 10°C/min. The final temperature was maintained for 10 minutes. Mass spectrometry was taken in the electron ionization mode at 70 eV and the scan range was between 29 and 350 amu. The ion source was set at 230°C and the transfer line at 240°C. Compounds were identified by comparison with mass spectra libraries (WILEY138, NIST, and INRA database) and by the calculation and comparison of the GC retention index of a series of alkanes (C8–C30) with the retention index from published data calculated under the same conditions. The quantitative data were obtained by electronic integration of the TIC peak areas with the ChemStation program.

Impact of storage conditions on the evolution of the volatile fraction of DPPa samples during storage

For this experiment, delipidation was done using the chloroform/methanol mixture, i.e. DPPa samples. After delipidation, 1 g of DPPa was transferred to 25 ml hermetic Hungate tubes. The tubes were immediately closed with open top screw caps having a rubber septum. Then, samples were stored under various conditions of temperature, light, and oxygen. A 2^3 factorial experimental design (Table 1) was carried out to map the effects of three qualitative factors on the percentage of the volatile fraction of the DPPa sample as compared to the total volatile fraction of RPP and to determine the optimal conditions for the storage of the delipidated powder. The selected factors were considered at two levels: light (obscurity or day light), gas storage (nitrogen or air) and temperature (4°C or ambient). Storage was at room temperature or at 4°C in a refrigerator. The tubes were put in an opaque box for the experiments done in the dark. The refrigerator used had a glass door enabling light to enter. In order to have a low-oxygen atmosphere, the tubes were placed for 10 minutes in a nitrogen glove-box before their storage.

The quantitative observed variable (percentage of volatile fraction) was calculated from results obtained after HS-SPME-GC-MS analyses. The main effects and interactions were calculated on variables as described by Box et al. [6].

Extraction and quantification of the residual solvent by the standard addition method

For this experiment, delipidation was done using the ethylacetate/

Sample labels	Factors			Observed variables: Remaining volatile fraction, RVF (% of the cumulative peak areas obtained for DPPa to the one obtained for RPP by HS-SPME-GC-MS)			
	Light	Gas	Temperature	Storage period (days)			
				13	33	58	93
DPPa1	- (obscurity)	- (N2)	- (4°C)	3.5	3.7	14.8	7.5
DPPa2	+ (day light)	- (N2)	- (4°C)	3.5	2.7	16.8	13.6
DPPa3	- (obscurity)	+ (air)	- (4°C)	3.1	2.1	5.4	5.9
DPPa4	+ (day light)	+ (air)	- (4°C)	3.0	2.6	5.2	6.3
DPPa5	- (obscurity)	- (N2)	+ (ambient)	5.6	9.7	17.1	19.7
DPPa6	+ (day light)	- (N2)	+ (ambient)	6.9	2.4	18.9	18.1
DPPa7	- (obscurity)	+ (air)	+ (ambient)	5.8	10.0	20.9	15.5
DPPa8	+ (day light)	+ (air)	+ (ambient)	6.7	10.0	20.6	19.5
Calculated effects (in bold if significant)							
Light				0.3 ± 0.3	-1.0 ± 0.9	0.4 ± 0.8	1.1 ± 0.9
Gas				-0.1 ± 0.3	0.8 ± 0.9	**-1.9 ± 0.8**	**-1.5 ± 0.9**
Temperature				**1.5 ± 0.3**	**2.6 ± 0.9**	**4.4 ± 0.8**	**4.9 ± 0.9**

Table 1: Experimental conditions, response values and main effects of the 2^3 experimental design used to study the impact of storage on the volatile fraction.

ethanol mixture, i.e. DPPb samples. Using a micro syringe, 0,1 or 2 μl of ethylacetate or ethanol were added to a 10 ml headspace vial containing 0.4 g of a delipidated protein powder. The vials were closed using an aluminum capsule with a rubber septum coated with PTFE and then subjected to ultrasonic vibrations for five minutes. After an equilibration period of one night at room temperature, the headspace was exposed to a SPME fiber (DVB/carboxen/PDMS) for 30 minutes at room temperature. The fiber was then desorbed in the GC-MS apparatus equipped with a fused-silica capillary column (30 m × 0.32 mm ID, 0.5 μm film thickness) coated with a DB-Wax stationary phase. Helium was used as the carrier gas at a constant flow of 1.5 ml/min. The initial oven temperature was maintained for 2 minutes at 35°C. Then, it was increased to 60°C at a rate of 2°C/min and from 60 to 240°C at a rate of 10°C/min. The final temperature was maintained for 10 minutes. All analyses were duplicated.

Results and Discussion

Impact of storage conditions on the evolution of the volatile fraction of DPPa samples during storage

Factors of the 2^3 factorial experiment and corresponding percentages of volatile fraction are reported in Table 1. Tested factors were chosen for their known effect on the oxidation phenomenon and other degradation reactions: light, O_2, and temperature. The observed variable is the volatile fraction of the samples. The remaining volatile fraction recovery (RVF) was calculated as the percentage of the cumulative peak areas obtained for a given sample of delipidated protein powder (DPPa) compared to the cumulative peak areas obtained for the raw protein powder (RPP). RVF was measured during a three-month storage period with measurements at 13, 33, 58, and 93 days of storage. RVF ranged from 2.1% to 20.9% with mean values of 4.7, 5.4, 15.0, and 13.3, for 13, 33, 58, and 93 days of storage respectively. As hypothesized, the volatile fraction continues to evolve after delipidation. Whatever the conditions of storage, RVF dramatically increased between the first and second month of storage. After 2 months, it seemed quite stable.

The main effects of the different factors of the experimental design on RVF were calculated and are given in Table 1. Main effects showing contrasts higher than the experimental error were thus significant and reported in bold letters in the table. At 13 days of storage, only temperature had had an effect on RVF. As time went by, more and more factors became significant. Finally, at the end of three months storage, all the factors had a significant impact. Throughout storage, temperature had a positive effect: an increase of storage temperature induced an increase of RVF.

Surprisingly, in the presence of air, RVF decreased significantly at storage of two months and beyond. Despite this, the effect of air on RVF during storage remains low: -1.9 ± 0.8 after 58 days storage and -1.5 ± 0.9 after 93 days storage.

In the present experiments, light had little effect on RVF. After 93 days storage, a slight effect of light appears (1.1 0.9) indicating that this factor is not the most important one to be controlled in the storage of the delipidated protein powders. However, it should be noted that a part of UV light, especially UVB, was stopped by the refrigerator glass door. UVA (315<λ<400 nm) are able to induce lipid peroxidation [7].

Temperature increase can accelerate all types of reactions. For example, Maillard reactions, which occur slowly under normal conditions, can be accelerated by a temperature increase. A relative humidity (RH%) around 60 to 70% is optimum for these reactions

which produce furfurals and aldehydes. As the water content of samples is quite low (under 7% w/w [8], RH% is low too and does not correspond to the optimum for Maillard reactions. On the contrary, this low RH% is known to favor oxidation reactions. Some volatile compounds coming from lipid oxidation (aldehydes and ketones) are able to interact with amines and give a similar effect to reducing sugars in Maillard reactions [9]. Oxidation reactions and Maillard reactions are thus closely linked and detected volatile compounds can come from both types of reactions. Some volatile compounds chosen as tracers were studied more closely. Their relative quantities determined in DPPa1 (sample stored in the dark, at low temperature, and under nitrogen: the storage conditions supposedly the most favorable to low volatile fraction) and DPPa8 (sample stored in the light, at ambient temperature, and in presence of oxygen: the storage conditions supposedly the less favorable to low volatile fraction) are reported in Figures 1a and 1b respectively. At first sight, it seemed that amounts of volatile compounds found in the DPPa1 sample were far lower than the ones found in DPPa8, which was consistent with hypotheses. For each volatile compound, amounts increased as storage went on. This means that degradation reactions were not totally stopped using controlled storage conditions and that, as delipidation was not total, some substrate was still present in the powder for degradation reactions.

Benzaldehyde comes from Strecker degradation and is known to be a Maillard reaction product [10,11]. Amounts found in DPP were lower than 20% of what was found in RPP. In DPPa8, amount of benzaldehyde

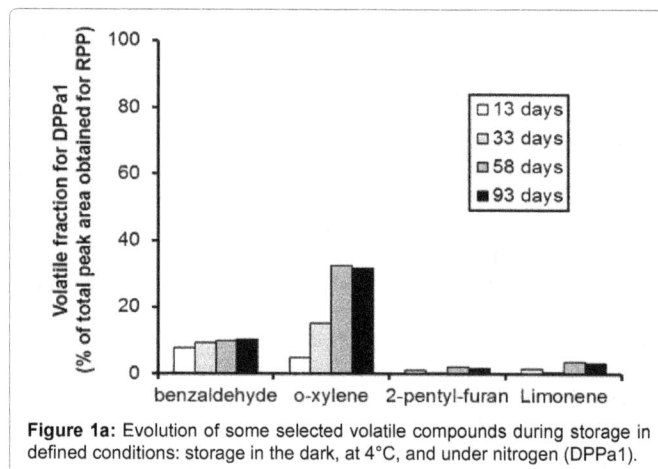

Figure 1a: Evolution of some selected volatile compounds during storage in defined conditions: storage in the dark, at 4°C, and under nitrogen (DPPa1).

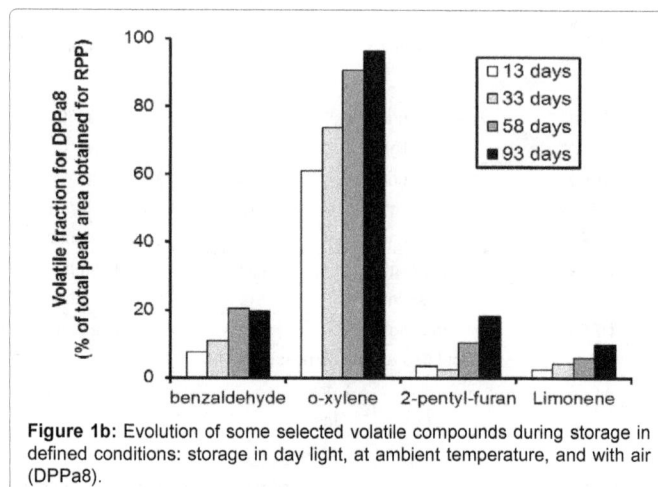

Figure 1b: Evolution of some selected volatile compounds during storage in defined conditions: storage in day light, at ambient temperature, and with air (DPPa8).

increased as storage went on, but seemed to be stable after 63 days of storage. The increase was feeble for DPPa1. This Maillard product was clearly favored by temperature increase and this is consistent with literature. Hexanal has been reported to be the most abundant compound resulting from lipid oxidation [12]. Unfortunately, in the present experiments, it was not detected as it was eluted at the same time as the residual solvent in the protein powder. Thus, it could not be selected as tracer of lipid oxidation, but other compounds linked with lipids degradation could be observed.

2-pentyl furan was studied originating from the thermal degradation of linoleic acid [13]. In the DPPa1 sample, 2-pentyl furan was found in very small amounts: 1 or 2% of what was found in RPP. In DPPa8, amounts of 2-pentyl furan increased exponentially during storage and were at 18% for 93 days storage. As expected, delipidation was not total and residual amounts of fatty acids can produce degradation products. As foreseen, thermal degradation of linoleic acid was favored by storage at ambient temperature. Nevertheless, controlled storage conditions can limit very efficiently the production of 2-pentyl furan.

Terpenes, such as limonene, are known as compounds that originate from animal feed; they can be directly transferred from grass to animal tissue and can be considered a green forage indicator [14]. Limonene was also reported to be a product generated from enzymatic hydrolysis-mild thermal oxidation of tallow [15]. In the DPPa1 sample, limonene was found in very small amounts: 1 to 4% of what was found in RPP. In DPPa8, amounts of limonene increased regularly during storage and were at 10% for 93 days storage. Limonene production was favored by storage at ambient temperature. So, it can be concluded that part of the limonene detected in Eisenia foetida powder comes from fat oxidation. Nevertheless, controlled storage conditions can limit and stabilize very efficiently the production of limonene.

Aromatic hydrocarbons, such as o-xylene, are reported as one of the largest classes of volatile compounds found in cooked meat, and are derived from the thermal degradation of lipid by thermal homolysis or autoxidation of long-chain fatty acid [16]. In the present experiments, o-xylene was found in huge amounts in DPP samples. In DPP8, amounts found after 93 days of storage were almost 100% of what was found in RPP. Nevertheless, controlled storage conditions succeeded in reducing the production of this volatile because it seemed to be stable after 58 days and represented 33% of the amount found in RPP.

Extraction by ethyl acetate/ethanol and quantification of solvent residues by the standard addition method

Experiments described here above showed the efficiency of delipidation using chloroform/methanol to obtain a dearomatized powder. In order to obtain a food grade ingredient, the extraction solvent must be changed for a safer one. For this purpose, an ethyl acetate/ethanol mixture was chosen on the basis of its similar polarity compared to the chloroform/methanol mixture. The amount of extracted lipids and the RVF determined for DPPb samples are reported in Table 2. The amount of extracted lipids depends on the powder/solvent ratio and on the solvent proportion of the ethyl acetate/ethanol mixture. When the proportion of solvent towards the powder mass increased, the amount of extracted lipids increased. The 67/33 ethyl acetate/ethanol solvent proportion, which is similar to the Folch solvent proportion (a reference in lipid extraction [17]; 2/1, chloroform/methanol), extracted better the protein powder lipids in comparison with the 87/13 ethyl acetate/ethanol solvent mixture.

The amount of extracted lipids significantly varied with experimental conditions and ranged from 8.3% to 9.5%. These values were twice as low as what was obtained with the chloroform/methanol mixture [1]. Associated RVF aligned with the variations of the amount of extracted lipids, and ranged from 6% to 18% which is quite similar to what is obtained with the chloroform/methanol mixture. The sample showing the best results i.e. the highest amount of extracted lipids and the lowest RVF was DPPb3. It was kept for the following experiments.

In order to eliminate solvent residues, samples were dried overnight, either under a hood, or in an oven at 60°C. RVF was found at 4.8% ± 0.2 for DPPb3 under hood and 0.20% ± 0.01 for DPPb3 oven dried. It was then possible to reduce drastically the volatile fraction of the delipidated samples using a safe solvent and using an overnight drying. To ensure the safety of the delipidated powder, solvent residues were determined in DPPb3-oven dried. This was done using a standard addition method. Known amounts of ethyl acetate and ethanol were added to the samples and total peak areas were determined as previously done by HS-SPME-GC-MS.

Resulting curves were presented in Figure 2a for ethyl acetate and Figure 2b for ethanol.

Simple linear regressions were calculated for ethyl acetate and for ethanol from the following equation: y=ax+b; where x is the amount of pure compound added, y is the corresponding peak area. Using the regression models, the concentration of both solvents in the delipidated samples (x-intercept) was calculated. The concentration of ethyl acetate was found to be 3.072 mg/g of the sample, and of ethanol 1.164 mg/g.

Acute toxicity is lower for ethyl acetate than for ethanol: LD50 (orally tested on rat) is 7060 mg/kg for ethanol [18] and 11300 mg/kg for ethyl acetate [19], but sub-chronic toxicity of ethyl acetate is higher than that of ethanol. NOAEL values are respectively 900 mg/kg/day and 2400 mg/kg [20,21]. An additional experiment was then done to remove as much ethyl acetate as possible. DPPb3 was washed with ethanol before being oven dried. In this case, the concentration of ethyl acetate measured in the sample was far lower: 0.008 mg/g. As a consequence of ethanol washing, ethanol concentration increased and was found at 4.791 mg/g.

Conclusion

It is possible to obtain a dearomatized protein powder using delipidation and storage under controlled conditions. Storage experiments conducted up to 93 days showed limited volatile compounds evolution: 10% or less for three among the four volatile

Sample labels	Solvent proportion Ethyl acetate/ethanol (v/v)	Powder/solvent ratio(w/v)	Extracted lipids (% w/w of RPP, mean value ± standard deviation)	Remaining volatile fraction, RVF (% of the cumulative peak areas obtained for DPPb to the one obtained for RPP by HS-SPME-GC-MS, mean value ± standard deviation)
DPPb1	67/33	2/5	8.3 ± 0.3	12 ± 2
DPPb2	87/13	2/5	7.8 ± 0.0	18 ± 7
DPPb3	**67/33**	**1/5**	**9.5 ± 0.0**	**6 ± 2**
DPPb4	87/13	1/5	8.6 ± 0.5	9 ± 3

Table 2: Experimental conditions and response values for lipid extraction and volatile compounds recovery using ethyl acetate/ethanol mixture.

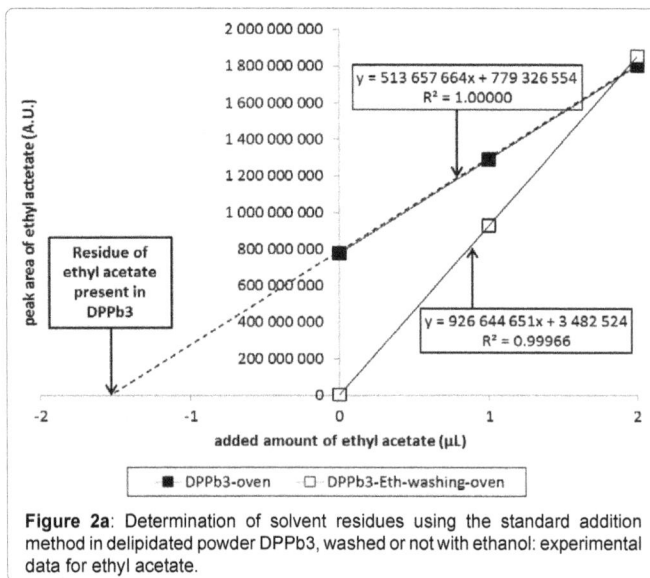

Figure 2a: Determination of solvent residues using the standard addition method in delipidated powder DPPb3, washed or not with ethanol: experimental data for ethyl acetate.

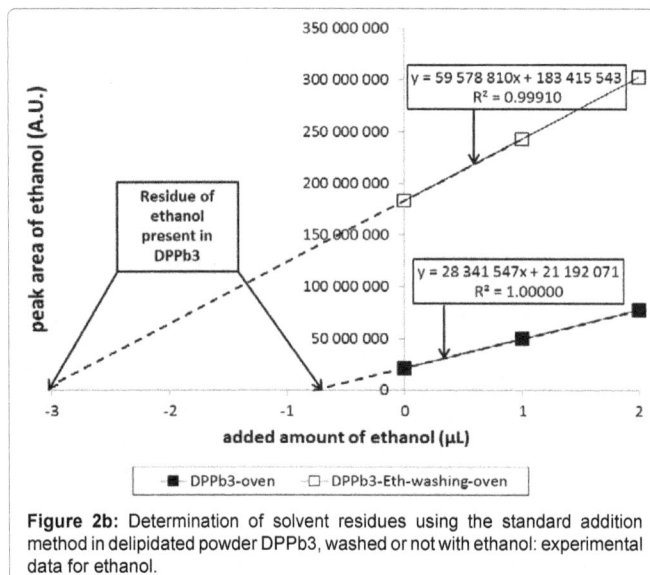

Figure 2b: Determination of solvent residues using the standard addition method in delipidated powder DPPb3, washed or not with ethanol: experimental data for ethanol.

compounds chosen as tracers. The main parameter to be controlled is temperature. There is no need to use storage under modified atmosphere. Additionally, it is possible to use a solvent mixture compatible with food grade products and conforming to EU regulations concerning solvent residues in food stuffs and food ingredients. De-aromatization was completed using an additional drying after delipidation.

References

1. Romero BA (2010) Impact of lipid extraction on the dearomatisation of an Eisenia foetida protein powder. Food Chemistry 119: 459-466.

2. Luque-García JL, Luque de Castrov MD (2003) Ultrasound: a powerful tool for leaching. TrAC Trends in Analytical Chemistry 22: 41-47.

3. Sadek PC (2002) The HPLC solvent guide (2ndedn.) Wiley Interscience p: 664.

4. The European Parliament and Council Directive (2009) Approximation of the laws of the member states on extraction solvents used in the production of food stuffs and food ingredients (2009/32/EC). Official Journal of the European Union.

5. The European Parliament and Council Directive (2010) Amending Directive on the approximation of the laws of the member states on extraction solvents used in the production of food stuffs and food ingredients (2010/59/EU). Official Journal of the European Union.

6. Box GEP, Hunter WG, Hunter JS (1978) Statistics for experimenters. Design, data analysis and model building. Wiley and Sons Inc., New York.

7. Bose B, Agarwal S, Chatterjee SN (1989) UV-A induced lipid peroxidation in liposomal membrane. Radiation and Environmental Biophysics 28: 59-65.

8. Cayot N (2009) Physico-chemical characterisation of a non-conventional food protein source from earthworms and sensory impact in arepas. International Journal of Food Science & Technology 44: 2303-2313.

9. Lorient D (1998) Modifications biochimiques des constituants alimentaires. Techniques de l'ingénieur Procédés biochimiques et chimiques en agroalimentaire: F 3400.

10. Adamiec J (2001) Minor Strecker degradation products of phenylalanine and phenylglycine. European Food Research and Technology 212: 135-140.

11. Mancilla-Margalli NA, Lópezc MG (2002) Generation of Maillard Compounds from Inulin during the Thermal Processing of Agave tequilana Weber Var. azul. Journal of Agricultural and Food Chemistry 50: 806-812.

12. Bou-Maroun E, Cayot N (2011) Odour-active compounds of an Eisenia foetida protein powder. Identification and effect of delipidation on the odour profile. Food Chemistry 124: 889-894.

13. Mandin O, Duckham SC, Ames JM (1999) Volatile Compounds from Potato-like Model Systems. Journal of Agricultural and Food Chemistry 47: 2355-2359.

14. Vasta V, Priolo A (2006) Ruminant fat volatiles as affected by diet. A review. Meat Science 73: 218-228.

15. Shi X (2013) Identification of characteristic flavour precursors from enzymatic hydrolysis-mild thermal oxidation tallow by descriptive sensory analysis and gas chromatography–olfactometry and partial least squares regression. Journal of Chromatography B 913-914: 69-76.

16. Domínguez R (2014) Effect of different cooking methods on lipid oxidation and formation of volatile compounds in foal meat. Meat Science 97: 223-230.

17. Folch J, Lees M, Stanley GHS (1957) A simple method for the isolation and purification of total lipides from animal tissues. Journal of Biological Chemistry 226: 497-509.

18. Wiberg GS, Trenholm HL, Coldwell BB (1970) Increased ethanol toxicity in old rats: Changes in LD50, in vivo and in vitro metabolism, and liver alcohol dehydrogenase activity. Toxicology and Applied Pharmacology 16: 718-727.

19. Smyth HF (1962) Range-Finding Toxicity Data: List VI. American Industrial Hygiene Association Journal 23: 95-107.

20. US EPA (1986) Rat oral subchronic study with ethyl acetate. United States Environmental Protection Agency, Office of Solid Waste, Washington DC.

21. OECD-SIDS (2004) SIDS Initial assessment report Ethanol in SIAM, UNEP Publications, Berlin.

Permissions

All chapters in this book were first published in JFPT, by OMICS International; hereby published with permission under the Creative Commons Attribution License or equivalent. Every chapter published in this book has been scrutinized by our experts. Their significance has been extensively debated. The topics covered herein carry significant findings which will fuel the growth of the discipline. They may even be implemented as practical applications or may be referred to as a beginning point for another development.

The contributors of this book come from diverse backgrounds, making this book a truly international effort. This book will bring forth new frontiers with its revolutionizing research information and detailed analysis of the nascent developments around the world.

We would like to thank all the contributing authors for lending their expertise to make the book truly unique. They have played a crucial role in the development of this book. Without their invaluable contributions this book wouldn't have been possible. They have made vital efforts to compile up to date information on the varied aspects of this subject to make this book a valuable addition to the collection of many professionals and students.

This book was conceptualized with the vision of imparting up-to-date information and advanced data in this field. To ensure the same, a matchless editorial board was set up. Every individual on the board went through rigorous rounds of assessment to prove their worth. After which they invested a large part of their time researching and compiling the most relevant data for our readers.

The editorial board has been involved in producing this book since its inception. They have spent rigorous hours researching and exploring the diverse topics which have resulted in the successful publishing of this book. They have passed on their knowledge of decades through this book. To expedite this challenging task, the publisher supported the team at every step. A small team of assistant editors was also appointed to further simplify the editing procedure and attain best results for the readers.

Apart from the editorial board, the designing team has also invested a significant amount of their time in understanding the subject and creating the most relevant covers. They scrutinized every image to scout for the most suitable representation of the subject and create an appropriate cover for the book.

The publishing team has been an ardent support to the editorial, designing and production team. Their endless efforts to recruit the best for this project, has resulted in the accomplishment of this book. They are a veteran in the field of academics and their pool of knowledge is as vast as their experience in printing. Their expertise and guidance has proved useful at every step. Their uncompromising quality standards have made this book an exceptional effort. Their encouragement from time to time has been an inspiration for everyone.

The publisher and the editorial board hope that this book will prove to be a valuable piece of knowledge for researchers, students, practitioners and scholars across the globe.

List of Contributors

Sivasakthi M and Sangeetha N
Department of Food Science and Technology, Pondicherry University, Puducherry-605014, Tamil Nadu, India

Heshe GG and Woldegiorgis AZ
Center for Food Science and Nutrition, College of Natural Sciences, Addis Ababa University-1176, Ethiopia

Haki GD
Department of Food Science and Technology, Botswana College of Agriculture, Private Bag 0027, Gaborone, Botswana

Indu Sharma and Kashmiri Das
Microbiology Department, Assam University, Silchar-788011, India

Nadia Bayar, Imen Ghazala, Nadhem Sayari and Ellouz-Ghorbel R
Unité Enzymes et Bioconversion, Ecole nationale d'Ingénieurs de Sfax, Université de Sfax, Tunisie

Assaâd Sila
Unité Enzymes et Bioconversion, Ecole nationale d'Ingénieurs de Sfax, Université de Sfax, Tunisie
Institut Régional de Recherche en Agroalimentaire et Biotechnologie: Charles Viollette, Equipe ProBioGEM, Université Lille 1, France

Ellouz-Chaabouni S
Unité Enzymes et Bioconversion, Ecole nationale d'Ingénieurs de Sfax, Université de Sfax, Tunisie
Unité de service commun bioréacteur couplé à un ultrafiltre, Ecole nationale d'Ingénieurs de Sfax, Université de Sfax, Tunisie

Ali Bougatef
Unité Enzymes et Bioconversion, Ecole nationale d'Ingénieurs de Sfax, Université de Sfax, Tunisie
Institut Supérieur de Biotechnologies de Sfax, Département de Technologies Alimentaires, Tunisie

El-Zeini Hoda M and Moneir El-Abd M
Dairy Science and Technology Department, Faculty of Agriculture, Cairo University, Cairo, Egypt

Mostafa AZ and Yasser El-Ghany FH
Food Technology Research Institute, Agriculture Research Center, Giza, Egypt

Santos Vaz AB, Aline G Ganecco, Juliana Lolli MM, Mariana P Berton, Cássia RD, Hirasilva Borba and Pedro A de Souza
Faculty of Agrarian and Veterinary Sciences, University Estadual Paulista, Jaboticabal, São Paulo, Brazil

Greicy Mitzi BM
Federal University of Alagoas, Arapiraca, Maceió, Brazil

Marcel M Boiago
State University of Santa Catarina, Chapecó, Santa Catarina, Brazil

Luciana Miyagusku
Federal University of Mato Grosso do Sul, Campo Grande, Mato Grosso do Sul, Brazil

Amaraegbu A, Adewale P and Ngadi M
Department of Bioresource Engineering, McGill University, Canada

Sobowale SS and Bamgbose A
Department of Food Technology, Moshood Abiola Polytechnic, Abeokuta, Ogun State, Nigeria

Adeboye AS
Department of Food Science, University of Pretoria, Hatfield, Pretoria, South Africa

Nwanekezi EC, Ekwe CC and Agbugba RU
Department of Food science and Technology, Faculty of Agriculture and Veterinary Medicine, Imo State University, Owerri, Sudan

Virendra Singh and Lakshmi BK
Department of Food Process Engineering, Vaugh School of Agricultral Engineering and Technology, Sam Higginbottom Institute of Agriculture, Technology and Sciences, Allahabad, Uttar Pradesh, India

Abdel Hafeez HH
Department of Anatomy, Embryology and Histology, Faculty of Veterinary Medicine, Assuit University, Assuit, Egypt

Zaki RS
Department of Food Hygiene, Faculty of Veterinary Medicine, Branch of New Valley, Assiut University, Assuit, Egypt

Abd El-Magiud DS
Department of Forensic Medicine and Toxicology, Faculty of Veterinary Medicine, Branch of New Valley, Assiut University, Assuit, Egypt

Mawada Mahfoudh, Sadok Boukhchina and Hajer Trabelsi
Biology Department, Lipids Biochemistry Unit, Science Faculty, Tunisia

Khaled Sebei
Biology Department, Higher Institute of Applied Biological Sciences, Tunisia

Sunil L, Prakruthi A, Prasanth Kumar PK and Gopala Krishna AG
Department of Traditional Food and Sensory Science, CSIR-Central Food Technological Research Institute (CSIR-CFTRI), Mysore, India

Alabi KP
Department of Food, Agriculture and Bio-Engineering, College of Engineering and Technology, Kwara State University, Kwara State, Nigeria

Olaniyan AM
Department of Agricultural and Bioresources Engineering, Faculty of Engineering, Federal University Oye Ekiti, Ekiti State, Nigeria

Odewole MM
Department of Food and Bioprocess Engineering, Faculty of Engineering and Technology, University of Ilorin, KwaraState, Nigeria

Nazni P and Shobana Devi R
Department of Food Science and Nutrition, Periyar University, Salem, India

Lavieri NA, Sebranek JG, Cordray JC, Dickson JS and Horsch AM
Department of Animal Science, Iowa State University, 215 Meat Laboratory, Ames, IA, USA

Jung S
Food Science and Human Nutrition Department, Iowa State University, 1436 Food Science Building, Ames, IA, USA

Manu DK, Mendonça AF and Brehm Stecher B
Food Science and Human Nutrition Department, Iowa State University, 3399 Food Science Building, Ames, IA, USA

Mekuria B and Emire SA
Food, Beverage and Pharmaceutical Industry Development Institute, Ministry of Industry, Private mailbag 33381, Addis Ababa, Ethiopia

Tess M, Bhaduri S, Ghatak R and Navder KP
CUNY School of Public Health at Hunter College, 2180 Third Avenue, 10035, New York, USA

Muresan C, Socaci S, Suharoschi R, Tofana M, Muste S and Pop A
Faculty of Food Science and Technology, University of Agricultural Sciences and Veterinary Medicine, Cluj-Napoca, Romania

Covaci A
Toxicological Center, University of Antwerp, Universitetsplein, Wilrijk B-2610, Belgium

Mohammed AA
Department of Quality Control, Seen Milles, Khartoum North, Sudan

Babiker EM and Khadir EK
Department of Food Science and Technology, Faculty of Agriculture, University of Khartoum, Khartoum North, Shambat, Sudan

Khalid AG
Department of Nutrition and Food Technology, Faculty of Science and Technology, Omdurman Islamic University, Omdurman, Sudan

Mohammed NA
Faculty of Applied Medical Science, Department of Community Health, Albaha University, Albah City, Kingdom of Saudi Arabia

Eldirani
Department of Food Science and Technology, Faculty of Agriculture, Omdurman Islamic University, Omdurman, Sudan

Sanaullah Noonari and Irfana Noor Memon MS
Assistant Professor, Department of Agricultural Economics, Faculty of Agricultural Social Sciences, Sindh Agriculture University, Tandojam, Pakistan

Abdul Sami Kourejo
Research assistant, Department of Agricultural Economics, Faculty of Agricultural Social Sciences, Sindh Agriculture University, Tandojam, Pakistan

Jahanzeb M, Atif RM, Shehzad A, Sidrah and Nadeem M
National Institute of Food Science and Technology, University of Agriculture, Faisalabad

Ahmed A
Wheat Research Institute, Ayyub Agricultural Research Institute, Faisalabad

Tariq Kamal
Department of Agricultural Extension Education and Communication, the University of Agriculture, Peshawar, Pakistan

Matilda Gill
Department of National Institute of Food Science and Technology, University of Agriculture, Faisalabad, Pakistan

Ismail Jan
Department of Statistics and Mathematics, the University of Agriculture Peshawar, Pakistan

Taimur Naseem
Department of Soil Sciences, the University of Agriculture, Peshawar, Pakistan

Vivek K
Department of Food Process Engineering, National Institute of Technology, Rourkela, Odisha, India

Pratibha Singh
Indian Institute of Crop Processing Technology, Thanjavur, Tamil Nadu, India

Sasikumar R
Department of Agri-Business Management and Food Technology, North Eastern Hill University, Tura Campus, Tura, Meghalaya, India

Sami Ghnimi, Raisa Almansoori, Baboucarr Jobe, Hassan MH and Kamal-Eldin A
Department of Food Science, College of Food and Agriculture, United Arab Emirates University

Tovar-Jímenez X, Aguilar-Palazuelos E and Caro-Corrales JJ
Posgrado en Ciencia y Tecnología de Alimentos, Universidad Autónoma de Sinaloa, Culiacán, México

Gómez-Aldapa CA
Área Académica de Química, Instituto de Ciencias Básicas e Ingeniería, Ciudad del Conocimiento, Hidalgo, México

Akomo PO
Valid Nutrition, Nairobi, Kenya

Egli I, Cercamondi CI and Zeder C
Laboratory of Human Nutrition, Institute of Food, Nutrition and Health, ETH Zurich, Zurich, Switzerland

Okoth MW and Njage PMK
Department of Food Science, Nutrition and Technology, University of Nairobi, Kenya

Bahwere P
Valid International, Brussels, Belgium

Owino VO
Technical University of Kenya, Nairobi, Kenya

Tomasz Jakubowski
Associate Professor, University of Agriculture in Krakow, Poland

Meenakshi Garg
Assistant Professor, Food Technology Department, University of Delhi, India

Javeed Akhtar and Mohd Ali Khan
Department of Post-Harvest Engineering and Technology, Faculty of Agriculture Science, AMU, Aligarh, India

Kumari B and Chauhan R
Department of Entomology, Chaudhary Charan Singh Haryana Agricultural University, Hisar-125004, Haryana, India

Elias Bou-Maroun and Nathalie Cayot
Food Process and Microbiological Unit, UMR A 02.102, AgroSup Dijon/Burgundy University, France

Index